Introduction to Financial Mathematics

Textbooks in Mathematics

Series editors:
Al Boggess, Kenneth H. Rosen

Introduction to Financial Mathematics
With Computer Applications
Donald R. Chambers, Qin Lu

Linear Algebra
An Inquiry-based Approach
Jeff Suzuki

Mathematical Modeling in the Age of the Pandemic
William P. Fox

Games, Gambling, and Probability
An Introduction to Mathematics
David G. Taylor

Financial Mathematics
A Comprehensive Treatment in Discrete Time
Giuseppe Campolieti, Roman N. Makarov

Linear Algebra and Its Applications with R
Ruriko Yoshida

Maple™ Projects of Differential Equations
Robert P. Gilbert, George C. Hsiao, Robert J. Ronkese

Practical Linear Algebra
A Geometry Toolbox, Fourth Edition
Gerald Farin, Dianne Hansford

An Introduction to Analysis, Third Edition
James R. Kirkwood

Student Solutions Manual for Gallian's Contemporary Abstract Algebra, Tenth Edition
Joseph A. Gallian

Elementary Number Theory
Gove Effinger, Gary L. Mullen

https://www.routledge.com/Textbooks-in-Mathematics/book-series/CANDHTEXBOOMTH

Introduction to Financial Mathematics

With Computer Applications

Donald R. Chambers
Emeritus Walter E Hanson KPMG Chair in Finance at Lafayette College
Qin Lu
Associate Professor in Mathematics at Lafayette College

CRC Press
Taylor & Francis Group
Boca Raton London New York

CRC Press is an imprint of the
Taylor & Francis Group, an **informa** business
A CHAPMAN & HALL BOOK

First edition published 2021
by CRC Press
6000 Broken Sound Parkway NW, Suite 300, Boca Raton, FL 33487-2742

and by CRC Press
2 Park Square, Milton Park, Abingdon, Oxon, OX14 4RN

Library of Congress Cataloging-in-Publication Data

Names: Chambers, Donald R., author. | Lu, Qin, 1967- author.
Title: Introduction to financial mathematics : with computer applications / Donald R. Chambers, Emeritus Walter E. Hanson KPMG Chair in Finance at Lafayette College, Qin Lu, CFA, Associate Professor in Mathematics at Lafayette College.
Description: First edition. | Boca Raton : Chapman & Hall/CRC Press, 2021. | Series: Textbooks in mathematics | Includes bibliographical references and index. Identifiers: LCCN 2020050771 (print) | LCCN 2020050772 (ebook) | ISBN 9780367410391 (hardback) | ISBN 9780367814427 (ebook) Subjects: LCSH: Business mathematics. | Economics, Mathematical. Classification: LCC HB135 .C43 2021 (print) | LCC HB135 (ebook) | DDC 332.64/57015192--dc23 LC record available at https://lccn.loc.gov/2020050771LC ebook record available at https://lccn.loc.gov/2020050772

ISBN: 978-0-367-41039-1 (hbk)
ISBN: 978-0-367-75278-1 (pbk)
ISBN: 978-0-367-81442-7 (ebk)

Typeset in Palatino
by MPS Limited, Dehradun

Dedicated to my wife, Suzanne, and in memory of Robert, Marjorie, and Katie Chambers (Don)

Dedicated to my husband, Ji Jonathan Li, kids Samuel Li, Yiwei Lyu, Daniel Li and Lu (Qin)

Contents

Preface to the Instructor

This textbook is primarily designed for a one-semester course for students who have learned calculus-based probability. Many books on financial derivatives require advanced training in stochastic calculus, which most undergraduates do not have and is hard to acquire through self-study. This book presents the fundamental theories of financial mathematics based on undergraduate calculus-based probability courses. Unlike other undergraduate financial mathematics books, a key feature of this book is its focus on applying models in three programming languages, which is discussed next.

We focus on a variety of computer methods to meet the spectrum of challenges in understanding and applying financial derivative models. Many of the concepts regarding financial derivatives are best understood by a "hands-on" approach using several types of computer applications. This book offers students the opportunity to develop skills in applying three important computer applications: R, Mathematica, and EXCEL. The reason for using multiple computer applications is that each of the three approaches offers unique advantages. In addition, students and other readers develop basic coding skills.

The "hands-on" aspect of the book guides the reader into putting the concepts into their own computer programs. For example, the reader is invited to create their own spreadsheet of a binomial tree by entering formulas, learning when to copy the formulas from one cell to another and learning why the formula might need to be changed. In doing so, the reader learns the mechanics of the trees and the models.

This book includes numerous computer-based activities. Each of this book's computer applications is carefully introduced and requires little prior programming background. This book demonstrates how each programming approach offers advantages. However, the main learning objectives of this book do not require every student to practice all three computer approaches.

Most of the students in our financial math course pursue a career in the financial industry. The content of this book provides these students with excellent preparation for their professional aspirations. Our book is designed to meet the needs of professionals who apply models and must understand the intuition and workings of the models. It carefully teaches the understanding and intuition of each model by using numerous exercises involving the three programming approaches. The objective is to guide the reader to learn the mathematics and financial economics of the model – not just how to plug the numbers into formulas or software.

The book's exercises in R offer excellent experience in computer programming and basic coding. Mathematica offers experience in utilizing well-developed programs that are especially designed for the types of applications found in mathematics. EXCEL, like other spreadsheets, offers highly intuitive visualizations of potentially confusing problems such as the construction of binomial trees. This book demonstrates the application of these three different programming approaches to those financial problems for which they are most appropriate. This book is about how to select and apply an application to better solve each type of challenge.

For example, this book shows students an incredibly efficient method to form a large binomial tree without requiring VBA or macros. All that the students need to do is type a few crucial cells into EXCEL. Then, the many additional cells of a large tree can be formed quickly using simple copying and pasting from the original cells. The approach introduced in this book can be easily implemented from scratch to form an EXCEL spreadsheet containing hundreds of time periods (steps)! This book's tree-building shortcuts can be applied it to many different financial products and many different financial models.

To be specific, students only need to input five cells – all the other cells of a large tree are formed by simple copying and pasting of those five cells! Further, by changing one of the five cells, the spreadsheet can go from solving only European options to being able to solve American options. Also, by changing another cell, the tree can solve for either calls or puts or even other payoffs. Finally, by changing just three parameters in the cells (u, d, and Q) a CRR tree can be changed to an alternative tree.

We use the above exercises as a foundation for more sophisticated applications. For example, the spreadsheet can be built to apply the finite difference method. The explicit finite difference method is shown to require that only seven initial cells be created. The approach can be learned and manipulated further in just a couple of minutes.

We also provide students with the necessary foundational instructions to get started using R. Section 1.10 details how to access and install R. Exercises involving R are designed to be relatively short and approachable. R is useful for developing better understanding of material such as simulations.

The text also provides students with the necessary foundational instructions regarding how to get started with Wolfram Mathematica. Section 1.10 details how to access and install Mathematica. Exercises involving Mathematica are designed to be relatively short and approachable. For instructors who have never included programming in their mathematics courses, this is a good starting point.

Mathematica code 2.1 uses Mathematica to derive an efficient frontier using ten lines of code (provided). Exercise 18 at the end of Chapter 2 calls on the students to modify the program for a new set of data. Chapter 3 uses Mathematica for finding the first four moments of a normal distribution and Chapter 6 uses matrix programming features in Mathematica for plotting efficient frontiers. All Mathematica codes and R codes in this book can be copied directly from Lafayette College Mathematics department's website in the section of Professor Lu. Mathematica is also good for generalizing the Black-Scholes model to other derivative pricing models where numerical integration technique is essential.

R is a useful programming tool that facilitates statistics, but R's built-in integration packages are not as good as Mathematica's packages. Mathematica contains very efficient algorithms of integration. EXCEL can offer valuable insights through visualizing the models and as an easy and effective tool in communicating concepts to others. Programming tools such as R save data in arrays, which can only be visualized by the programmer.

This book's coverage of various financial derivatives and their models has been carefully developed to explore the most important products. We cover the major, fundamental derivative models. Given that this book has relatively deeper mathematics and computational exercises, the coverage of models is focused on deep understanding of the foundational models.

The book uses Chapters 1 and 2 to introduce financial terms, which is especially important to STEM majors who have not been exposed to this area. Chapters 3 and 4 discuss option pricing. Chapter 5 focuses on topics in risk management, and Chapter 6 is on portfolio management. Chapter 7 summarizes the main ideas of the HJM model of interest rates into a single chapter that is accessible to undergraduate students. The HJM model, used throughout the financial industry for fixed income valuation, is presented using binomial trees in EXCEL spreadsheets. Chapter 8 discusses modeling of credit risk. Some instructors may decide to include Chapters 7 or 8 but not both because it may be difficult to include both chapters in a single semester.

Sample Course Schedule:

- Week 1: Sections 1.3, 1.5, 1.1, 1.2
- Week 2: Sections 2.1, 2.2, 2.3, 5.6
- Week 3: Sections 1.4, 1.6, 1.7, **Test 1**
- Week 4: Sections 3.2, 3.3.6
- Week 5: Section 3.3
- Weeks 6–7: Section 3.4, **Test 2**
- Weeks 8–9: Chapter 4
- Weeks 9–10: Section 5.3, 5.4, 5.5
- Weeks 11–12: Chapter 6
- Weeks 13–14: Chapters 7 or 8, **Test 3, and/or Final Exam**

(Note: some or all of the sections not included above can be assigned as optional readings).

The financial derivative models that are included in this book are virtually identical to those covered in the top financial professional certificate programs in finance. Like this book, those programs focus on the most essential and foundational knowledge and skills that the financial industry demands including models on options, forwards, and swaps on equities, bonds, and currencies.

The coverage of other models (including portfolio management, risk management, and credit derivatives) substantially overlaps with those models found in the CFA program (Qin Lu is a CFA charterholder). This book's coverage also matches much of the coverage of financial models found in the CAIA program (Don Chambers is a CAIA charterholder). Finally, the coverage also closely matches the content of the curriculum for actuarial exams provided by the Society of Actuaries (SOA). The overlap of financial models between the three programs and our book is broad and deep. We should note, however, what distinguishes our book from these programs (and most popular books on financial derivatives) is the mathematical depth of our book and our extensive exercises using diverse computer programming applications.

The authors have found that this book's content easily fills up a full semester course due to the depth of mathematical content and its many diverse computational exercises. Chapters 3–8 of this book end with sections that describe models and other material that may be of interest for further study. The purpose of these sections is to point readers toward extensions and other advances that build on the foundational models detailed in this book.

The authors believe that *Financial Mathematics* is important and increasingly vital to a modern financial system, which in turn is vital to a productive and efficient economy. Financial derivatives are powerful tools that can increase economic efficiency, but have also been shown to exacerbate financial crises. In order to maximize the opportunities and minimize the threats, investment professionals need to better understand the underlying financial mathematics. This book bridges the knowledge gap with its innovative use of computer applications. The book's goal is to help its readers contribute to deeper mathematical knowledge and better application of that knowledge throughout the financial industry.

Preface to the Student

Few things can be more satisfying than having a career that is fascinating, challenges one's mind, and generates value to society. For many, financial mathematics and financial engineering using financial derivatives provide an ideal area of interest. This textbook is primarily designed to help its readers develop the knowledge and skills to understand and apply financial mathematics.

This financial mathematics book has two especially important features: (1) it is designed to be accessible by only requiring foundational mathematical training in calculus-based probability and (2) it emphasizes hands-on training in programming and applying models in three different programming languages. The book focuses on this variety of computer methods to meet the spectrum of challenges in understanding and applying financial derivative models. Specifically this book offers students the opportunity to develop and apply skills in R, Mathematica, and EXCEL – each of which offers unique advantages.

R offers excellent experience in computer programming and basic coding. Mathematica offers experience in utilizing well-developed programs that are especially designed for the types of applications found in mathematics. EXCEL, like other spreadsheets, offers highly intuitive visualizations of potentially confusing problems such as the construction of binomial trees. This book demonstrates the application of these three different programming approaches to those financial problems for which they are most appropriate. This book is about how to select and apply an application to better solve each type of challenge. Its numerous activities use easily accessible computational tools that are carefully introduced and require little prior programming background. However, the main learning objectives of this book do not require every student to practice all three computer approaches.

The "hands-on" aspect of the book guides the reader into putting the concepts into their own computer programs. For example, the reader is invited to create their own spreadsheet of a binomial tree by entering formulas, learning when to copy the formulas from one cell to another and learning why the formula might need to be changed. In doing so the reader learns the mechanics of the trees and the models. For example, the book's approach to tree modeling in spreadsheets only needs to input five cells – all the other cells of a large tree are formed by simple copying and pasting of those five cells! Further, by changing one of the five cells, the spreadsheet can go from solving only European options to being able to solve American options. By changing another cell, the tree can solve for either calls or puts or even other payoffs.

Chapters 1 and 2 introduce financial terms and foundational concepts. Chapters 3 and 4 discuss option pricing. Chapter 5 focuses on topics in risk management, and Chapter 6 discusses portfolio management. Chapter 7 summarizes the main ideas of the HJM model of interest rates, used throughout the financial industry for fixed income valuation, into a single chapter that is accessible to undergraduate students. Chapter 8 discusses modeling of credit risk.

Many of the readers of this book will pursue a career in the financial industry and our book is designed to meet their needs as they learn and apply modeling. This book carefully teaches the understanding and intuition of each model by using numerous exercises involving the three programming approaches. The objective is to guide the reader to learn both the mathematics and financial economics of the model – not just how to plug the numbers into formulas or software.

This book's coverage of various financial derivatives and their models has been carefully developed to explore the most important products and models. Many students deepen and broaden their knowledge of finance and enhance their career trajectory by pursuing well-regarded certifications in finance. The financial derivative models that are included in this book are virtually identical to those covered in the top financial professional certificate programs in finance, including options, forwards, and swaps on equities, bonds, and currencies. The coverage of this book substantially overlaps with those models found in the CFA program (Qin Lu is a CFA charterholder), the CAIA program (Don Chambers is a CAIA charterholder), and the curriculum for actuarial exams provided by the Society of Actuaries (SOA). While the overlap of financial models between the three programs and this book is broad and deep, what distinguishes this book from these programs (and most popular books on financial derivatives) is the mathematical depth of the book and its extensive exercises using diverse computer programming applications.

Developing skills in an area that one finds interesting and that is essential to a vibrant economy increasingly requires mastery of underlying theory as well as the ability to apply that knowledge through information technology. This book is unique. It combines underlying mathematics, financial economic intuition, and computer applications – serving as a gateway to rewarding careers that serve the vital role of facilitating well-functioning capital markets.

Acknowledgments

The authors would like to first thank our editor, Bob Ross, who provided us with encouragement and outstanding advice. Without his continuous help, this book would not be possible. The authors also thank John Hull for his inspiring book *Options, Futures and Other Derivatives,* which Qin uses for her financial mathematics class and the book's framework which the authors used to help develop many classroom projects. The authors also thank Robert A. Jarrow for his pioneering work that serves as a foundation for Chapter 7 of this book. The authors appreciate the editing work of Daniel Crowley, a former student at Lafayette College, on the initial drafts of our book. The authors thank all reviewers for their wonderful suggestions and Lafayette College Provost John Meier who gave us guidance regarding potential publishers and advice throughout the submission process. The authors are also grateful to Professors Jeffrey Liebner and Trent Gaugler who improved the quality of many of the graphs in the book and the Office of the Provost of Lafayette College for funding.

Qin in particular thanks her colleagues in the Lafayette College Mathematics Department who make Qin's academic life enjoyable. Qin would also like to thank her friend, Joey Thompson, for help in writing the Preface and students in her financial math classes including Ryan Nadire, Mathew Curry, Rindra Randriamanantena, and Ran Cao for pointing out typos. Last, but not least, Qin thanks her Lord for his many blessings.

About the Authors

Donald R. Chambers served as the Walter E. Hanson KPMG Chair in Finance at Lafayette College in Easton, Pennsylvania, for 25 years. During that time, he worked closely with economics and math-economics undergraduate majors, providing him with an understanding of the needs and abilities of students interested in the intersection of math and finance. Professor Chambers has authored or coauthored approximately 50 research papers in scholarly journals and six books. Professor Chambers served previously as Associate Director of Programs at the CAIA Association, a risk management consultant to the Bank of New York in Manhattan and as a senior portfolio strategist with Karpus Investment Management. He currently serves as an investment strategiest at Biltmore Capital Advisors. These experiences have provided him with a deep and broad knowledge of the practical applications of mathematical finance.

Qin Lu has taught mathematics at Lafayette College in Easton, Pennsylvania, for the last 21 years. Trained as an algebraic topologist, Professor Lu began her journey in mathematical finance in 2003 by taking CFA (Charted Financial Analyst) exams. By passing three rigid tests during a three-year period, Professor Lu became CFA charterholder in 2006. There are very few CFA charterholders who are working at colleges/universities, most of them are working in investment industry. During these years at Lafayette, Professor Lu has taught financial mathematics course many times. In addition, she has been NSF REU (Research Experiences for Undergraduates) PI and mentor for multiple years and has guided a lot of undergraduate research through honors thesis and REU program. Professor Chambers and Professor Lu have coauthored eight papers, one of which was published in a top-three finance journal and had an undergraduate student coauthor.

1

Introduction to Financial Derivatives and Valuation

Applied mathematics is vital to a deep understanding of financial derivatives. The use of mathematics to understand our world has exploded in recent decades and is sure to continue to change everything from medicine to mechanical engineering and even to our understanding of the mind. From Bernoulli's eighteenth century use of utility functions to explain risk aversion, through Markowitz's groundbreaking use of probability and statistics in portfolio selection, to Black and Scholes development of derivative pricing based on risk-neutral modeling, applied mathematics has changed economics and finance, which in turn has changed our world.

After reviewing foundations in economics and finance, this chapter introduces the basic idea of valuing financial contracts, as well as two important types of financial derivatives known as forward contracts and options. The use of forward and option contracts and other financial derivatives has soared throughout the world over the last few decades because of their usefulness in performing tasks such as risk management. Both Chapters 1 and 2 focus more on finance than mathematics to familiarize readers with financial terms. The end-of-chapter exercises are also straightforward and do not require sophisticated mathematical tools. Later chapters introduce a wide spectrum of financial derivatives and demonstrate the use of mathematical applications to address the challenges of valuation and risk management. Students and practitioners who are familiar with investments may skip the first two chapters. Instructors could cover Sections 1.3–1.7 in class while assigning Sections 1.1–1.2 in this chapter for reading. Section 1.8 is completely optional.

1.1 Foundations in Economics and Finance

Financial securities, financial markets, and financial institutions are at the center of the rapid acceleration of economic growth throughout the world – economic growth that plays the key role in reducing starvation, increasing life expectancy, expanding educational opportunities, facilitating travel and communication, and generally increasing the choices available to people throughout the world. Even the richest people on earth two centuries ago could not have even imagined the healthcare, transportation, communications, entertainment, and conveniences that are widely available today. Markets, and in particular financial markets, are essential building blocks that have addressed the problems of the past and that will address the challenges of the future.

This first section of Chapter 1 directly addresses the key question as to why it is important that our societies employ substantial numbers of talented employees to develop and operate financial systems. In other words, when a student embarks on a career in finance does he or she become a parasite on society or a vital contributor to the mosaic of talents required to maintain and innovate a modern economy?

The foundation of a modern economic system is capital – resources that have been accumulated in order to facilitate the production of additional resources in the future. The efficient creation, maintenance, and utilization of capital rely on understanding two essential concepts: the time value of money and the management of risk. Financial mathematics centers on facilitating our understanding of the economics of time and risk.

1.1.1 The Role of Exchanging Real Assets

Real assets are resources that directly enhance our ability to produce and consume goods. Real assets are often viewed as being tangible assets – assets that have physical form such as buildings, land, and equipment. But intangible real assets – assets such as technologies, patents, copyrights, and trademarks – are playing an increasingly important role in modern economies.

Without trade, people must make every good that they consume; an incredibly inefficient and ultimately unsustainable system. The ability of people to trade assets is a foundation for an economy. Trade allows people to focus their skills on producing a few products based on their confidence that they can exchange their production for the goods produced by others. When people exchange real assets, they can specialize in producing those goods that best utilize their skills and preferences. More importantly, when people specialize they can discover ways to improve the efficiency of their production – skills which others may adopt. In doing so, the technologies underlying an economy rapidly evolve toward the marvels of today. For example, in the last 150 years, the percentage of the American workforce toiling in agriculture has declined from over 50% to about 2%, allowing the workforce to provide new and expanded services in areas such as information technology, healthcare, and higher education.

1.1.2 The Role of Financial Assets

Financial assets are the complement to real assets. ***Financial assets*** are contractual or *indirect* claims. Where a real asset directly provides consumption, a financial asset is typically a claim to cash flows and is therefore an indirect claim on consumption. Chapter 2 introduces two major types of financial assets: bonds (as well as other fixed income securities) and stocks.

Bonds and stocks are financial assets. There is one other major and important type of financial asset: financial derivatives. ***Financial derivatives*** are financial contracts involving two parties: a buyer or long position, and a seller or short position. Each contract represents a zero-sum game wherein the buyer's gain is the seller's loss and vice versa. Finally, the derivative's cash flows (payoffs) depend on the uncertain price of the contract's underlying asset at a specified future date.

Accordingly, financial derivatives differ from traditional stocks and bonds in key ways. Financial derivatives are contracts between two (and in a few cases more) parties. The payoff(s) between the parties is derived from (hence the name derivative) or determined by the value of a specified asset that underlies the derivative. For example, an investor who holds bonds of an airline company may enter into a financial derivative with an investment bank that requires the investor to make a series of fixed payments to the bank in return for a promise by the bank that if the airline company defaults on the bonds the bank will make a large cash payment to the company to offset the losses due to the default. In this case, the investor is using a financial derivative to buy financial protection against default from an investment bank.

Financial securities and other financial assets are part of the financial system of a modern economy. Figure 1.1 illustrates the role of the financial system as a conduit between people and the real assets that meet their needs and desires for consumption.

Financial securities, financial markets, financial institutions, corporations, and even governments are simply concepts in our minds that help organize and structure a society by determining who has the rights to the benefits generated by real assets. Corporations, governments, and other institutions do not produce or consume goods – people do. People are more efficient at producing goods and services, and benefiting from those good and services, when they organize themselves using concepts such as corporations, unions, governments, financial institutions, and so forth. The hallmark of societies with highly successful economies is that they have well-developed financial systems and institutions.

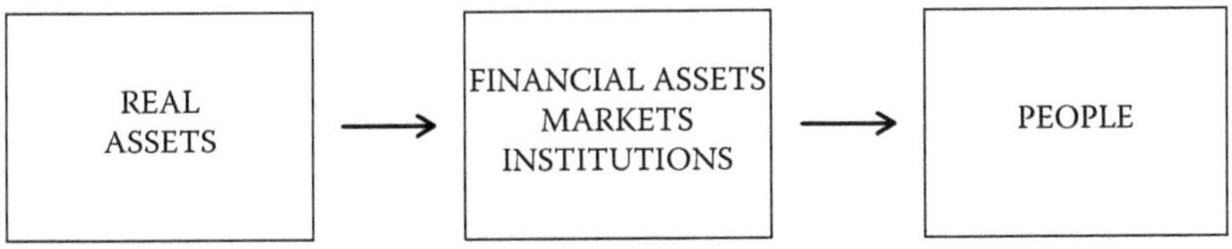

FIGURE 1.1
Relationship Between Real Assets, the Financial World, and People.

While our financial systems do not directly produce the resources that people utilize, our financial systems (along with other systems such as educational and governmental institutions) play key roles in improving our ability to be productive and to thrive. Financial markets allow people to accumulate savings efficiently for future needs such as college education or retirement. Financial markets allow people to borrow to buy houses or automobiles. And most importantly, although often unappreciated, financial markets allow people to manage and control financial risks and provide pricing information that serves as a communication network enabling people to cooperate efficiently.

1.1.3 The Roles of Financial Mathematics and Financial Derivatives

Financial mathematics has provided powerful tools for the development, understanding, and use of the wide spectrum of innovative financial products that have exploded in availability since the 1970s. These financial products have tremendously expanded our abilities to exchange, manage, control, and understand economic risk. Economic risk is intangible and tends to be difficult to understand. Yet it is clear that those economies that develop the greatest skills and tools for dealing with economic risk are best able to harness the incredible power of economic trade and growth to meet the needs and wants of a large society. Financial mathematics lies at the heart of that past success and our ability to create future opportunities.

For example, a large operating firm such as an airline company faces a number of huge economic uncertainties regarding their revenues and expenses. What effects will fuel costs, labor costs, and financing costs have on their profitability and, ultimately, the firm's ability to continue to provide transportation services? What factors will determine the revenues for forthcoming quarters and years? Will exchange rates and airplane prices change in directions that will prevent the airline company from purchasing new equipment to replace aging aircraft or to open new routes?

Financial derivatives can help the airline company control for risks external to the firm such as changing energy costs, interest rates, and exchange rates. By offsetting or ***hedging*** the effects of these otherwise uncontrollable external variables, the company can focus their attention on dealing with those matters over which they have direct control: operating their firm with efficiency, safety, and high-quality service.

Financial derivatives have also played roles in creating or exacerbating financial crises at the international, individual investor, and firm levels. Clearly, derivatives are powerful tools that when used improperly can be as damaging as they are beneficial when used properly.

The key to effective management of financial risk is effective valuation. In finance, valuation of assets in general and financial derivatives in particular involves valuing prospective cash flows based on the timing of those cash flows and their risk. A good definition of finance is that it is *the economics of time and risk*. This chapter begins with highly simplified examples that ignore the effects of the timing of cash flows and the risk of cash flows on current values. Then the analyses will be expanded to include the time value of money (using forward contracts as an example) and the potential effects of risk on asset values (using options as an example).

1.1.4 The Roles of Arbitragers, Hedgers, and Speculators

Hull (2015) discusses three types of traders:[1] arbitragers, hedgers, and speculators. Arbitragers, hedgers, and speculators, along with investors, borrowers, and entrepreneurs, all commonly use markets for financial securities and financial derivatives. Understanding the roles of arbitragers, hedgers, and speculators is helpful in understanding why derivatives have emerged and evolved into diverse forms.

An important concept before discussing these roles is short selling. The idea that a trader can take a short position in a financial derivative was introduced. It makes sense that, as a contract between two people regarding the price of an asset one of the parties to the contract will take the long side and one will take the short side. Having the long side of the contract means that one will have the risk exposure more similar to owning the asset with the other side having an exposure similar to *owing* the asset. ***Short selling*** is the process of obtaining the opposite exposure of owing an asset by borrowing the asset and selling it into the market at the current market price. The short seller of an asset therefore is inversely exposed to the price changes of the underlying asset: the short seller gains if the underlying stock or other asset's market price falls and loses market value if the asset rises in value. Eventually, the short seller purchases the asset and returns it to the entity from which it was borrowed, terminating the short sale. Ignoring dividends and the time value of money, the short seller of a stock gains $X if the stock declines $X over the life of the short sale and loses $X if the stock rises by $X.

Arbitragers are traders who establish a *long* (or buy) position in one or more assets that they believe are relatively underpriced and one or more *short* (or sell) positions in assets that they believe are relatively overpriced. Arbitragers hold these positions based on their belief that the prices will revert to more proper price levels that allow the arbitrager to exit both positions at a net profit. When markets are poorly or *inefficiently* priced, arbitragers can earn riskless or virtually riskless profits. Successful arbitragers serve the market by correcting pricing errors thereby (and unintentionally) ensuring that less informed market participants are transacting at prices that are nearer to their proper or *equilibrium* values. The distinguishing feature of arbitragers is that their trades are designed to take very little risk and tend to be short term.

Hedgers are market participants that use financial derivatives and other instruments to offset risks that they do not wish to bear. For example, an agricultural firm may use derivatives to lock in the revenues from their output such as food products, lock down their costs of production such as energy costs, and minimize their exposure to exchange rates from their contracts to export their products. Financial derivatives allow hedgers to hedge or "lay off" those risks that they cannot control and allow them to focus on those factors over which they may have control.

Speculators are similar to arbitragers except that they typically take substantially more risk by establishing one or more positions with unhedged risks. Speculators take positions in anticipation of profits if their forecasts of future price movements are accurate. The key to successful speculators is that they gain money and are able to continue speculating only when they are able to consistently buy assets that are underpriced and sell assets that are consistently overpriced. In doing do, they can consistently earn profits while acting to balance supply and demand, drive prices toward their true values, and stabilize markets and prices.

1.2 Introduction to the Valuation of Financial Contracts

In this section, we will discuss three topics in the valuation or pricing of financial contracts: (1) how to deal with randomness and to calculate expected value, (2) how to deal with the time value of money by introducing a time value discount factor as well as an interest rate, and (3) how to deal with risk by using a risk premium in valuation. At the end of this section, we also introduce one of the most important and fascinating breakthroughs in financial economics: in financial derivative valuation, the risk premium usually does not matter.

1.2.1 Market Prices, Risk, and Randomness

Risk is a fascinating topic that is central to financial economics. What causes risk, and what causes security prices to change? The answers to these questions and the study of financial risk in general have progressed primarily using financial math.

The terms *value* and *price* (as well as the terms *valuation* and *pricing*) are often used interchangeably in finance. This book tends to use the term *value* to describe how much a particular person believes an item is worth and uses the term *price* to describe the amount of money that people receive or pay when they exchange the item. However, the term *asset pricing* is used frequently in financial economics to describe very important models even when the context is more clearly described as involving asset valuation.

Since the general meaning of the terms are quite similar, this book often uses the terms *price* and *pricing* to describe values and valuation in cases where price and pricing are conventionally used.

The value of an asset tends to change through time because the preferences of people change through time and the abilities of a modern economy to meet those preferences changes through time. Agricultural prices change due to factors such as weather, energy prices change due to factors such as economic activity, and stock prices change due to factors such as predictions of future revenues and expenses. Prices respond to changes as soon as the information about those changes is revealed. The effects of a frost on orange harvests begin changing market prices of orange juice when the weather is forecasted, not just when the damage is done.

The ***spot price*** is the price today for delivery today. The spot price of an asset such as a stock or bond reflects a consensus in the marketplace with regard to the future benefits that the asset offers. As time passes, the asset price changes because the market's predictions of future conditions change. More generally, asset prices change because of the *arrival of new information.* Good news for a stock such as higher forecasts of earnings causes a positive price change, while bad news such as a disappointment regarding the sales of a new product causes a negative price change. To the extent that market participants rationally and efficiently process *available* information to form the current price of an asset, it follows that future price changes will be based on the arrival of *new* information.

Therefore, security prices can be viewed as random variables that change through time based on the arrival of new information. Simply put, the future price of an asset can be modeled as being equal to its future expected value plus or minus price changes due to new information that becomes available to market participants.

1.2.2 Expected Value as a Foundation of Asset Valuation

The starting point for valuing a financial asset is that the price of the asset should reflect the expected cash inflows that the asset offers. We know that future cash flows are uncertain. In order to find expected cash flows, we have to use probability models to model the likelihood of cash flows occurring. Let's review some mathematics with a focus on expected value.

Expected value's intuitive meaning is that it is a long-run average. If a random variable is sampled many times, the long-run average of these sampled values is the ***expected value*** of the variable. We begin by reviewing how to calculate expected value from probability model perspective.

Mathematics Review 1.1

Warm-up 1.1

Let X be a discrete random variable with a set of possible values x and probability $p(x)$. The expected value is $E(X) = \mu_X = \sum_x x * p(x)$. This expected value exists if $\sum_x |x| * p(x) < \infty$. The expected value is a probability weighted average of all possible realizations.

Warm-up 1.2

Let X be a continuous random variable with p.d.f. (probability density function) $f(x)$. The expected value is $E(X) = \mu_X = \int_{-\infty}^{\infty} x * f(x)dx$. This expected value exists if $\int_{-\infty}^{\infty} |x| * f(x)dx < \infty$.

In a competitive market, a financial asset's price is determined by supply and demand. To simplify an example, let's consider an asset that will distribute its final cash flow to the owner of the asset immediately (e.g., before the end of the day). Financial analysts compete to identify and *purchase* assets with current market prices below the asset's expected distribution (i.e., final payoff) and to identify and *sell* assets with current market prices above their expected payoff. Thus the expected value of the random payoff of a financial asset drives the current market price of the asset. Throughout this book, we assume that financial analysts have estimates of the probability distribution of these random variables (i.e., asset or security prices). Without going into detail here about how the values of expected cash flows are adjusted for time (i.e., for the delay between buying an asset and receiving its payoffs) and risk (i.e., the uncertainty of the size and timing of future payoffs), the value of a financial asset or contract is based on expected cash flows as indicated in Basic Principle 1.1.

Basic Price Principle 1.1

The value of a financial asset or contract will equal the sum of the expected cash flows (discounted for time and risk) that the asset or contract offers.

Let's look at some simple examples of assets or contracts with random future payoffs and calculate their expected values. To simplify the first example, future cash flows are not discounted for time or risk. Later examples in this chapter include discounting for time and risk.

Example 1.1: Abbi plays a game with Betti. Abbi tells Betti that she is going to roll a normal die with equally likely outcomes of 1 to 6. If the value 1 appears, Abbi will pay Betti \$1; if 2 appears Abbi will pay Betti \$2; and so forth up to paying \$6 if the value 6 appears. However, Betti must pay Abbi by buying a ticket to play the game. How much is the fair ticket price ignoring the valuation effects of time and risk aversion?

The starting point is to calculate the expected value of the gamble.

Because we know that the die is balanced fairly, we can predict that if Betti rolls the die 60,000 times, approximately 10,000 times she will get \$1; 10,000 times she will get \$2; and so forth. So, the long-run average payoff to the gamble is

$$\frac{10{,}000 \times 1 + 10{,}000 \times 2 + 10{,}000 \times 3 + 10{,}000 \times 4 + 10{,}000 \times 5 + 10{,}000 \times 6}{60{,}000}$$

After simplification, we get a formula matching the above probability model, namely, the expected value is the probability weighted average.

$$= 1/6 \times 1 + 1/6 \times 2 + 1/6 \times 3 + 1/6 \times 4 + 1/6 \times 5 + 1/6 \times 6 = 3.5$$

Abbi will not offer to gamble for a price of less than \$3.50 because she knows that she would be expected, on average, to lose money. Similarly, Betti will not gamble at a price above \$3.50 because she does not want to lose money on average. Therefore, we might expect that the supply and demand for the gamble would force the equilibrium market price to be equal to its expected value, $E(X)$ where X is the payoff, which is a random variable. So,

$$\textbf{Value} = \textbf{Market Price} = \textbf{E(Payoff)} \tag{1.1}$$

Note that what Betti gives Abbi (the ticket price) should be equal to the expected value of what Abbi gives to Betti (the payoffs). Note that we did not consider time value of money and risk aversion for Betti.

Example 1.2: Abbi plays a new type of game with Betti. Abbi tells Betti that she will give Betti X^2 dollars, where X is the result of a single observation of a variable distributed $N(2,1)$, which is a normal distribution with a mean of 2 ($EX = 2$) and a variance of 1($\text{Var}X = 1$). Betti must make a payment (i.e., buy a ticket) to play. How much is the fair price for the ticket, ignoring valuation of the effects of time and risk aversion?

$$X \sim N(2, 1)$$

Betti's average payoff is EX^2, which is the fair price of the ticket. One way to find EX^2 is with the integral with respect to the p.d.f. of normal distribution:

$$\int_{-\infty}^{\infty} x^2 \times f(x)dx = \int_{-\infty}^{\infty} x^2 \times \frac{1}{\sqrt{2\pi}} e^{-\frac{(x-2)^2}{2}} dx$$

More simply, we can use the definition of variance to solve the problem.

$$\text{Var}X = EX^2 - (EX)^2$$

We know $EX = 2$, $\text{Var}X = 1$ by $N(2,1)$. Inserting these numbers into the above equation and solving for $EX^2 = 1 + 4 = 5$. The ticket value is equal to the expected value ($5). This exercise demonstrates that even complex payoffs such as a nonlinear payoff can be easily valued. Note that this exercise did not consider the time value of money or the effects of risk aversion for Betti.

Mathematically, a ***probability measure*** for a random experiment is a real-valued function with values between 0 and 1, defined on the collection of events, and satisfying *countable additivity*. Note that in the above two examples, the probability distributions are the real-world probability distributions, which we call the ***P-measure probability distributions***. Later chapters, in the context of pricing financial derivatives, use ***Q-measure probability distributions***. Q measures are probability measures determined such that an asset's market price is equal to the discounted expectation of cash flows. Another way to understand Q measures is that they assume that all investors live in a ***risk-neutral world,*** where investors have no aversion to risk or they ignore their aversion to risk. Therefore, each asset price is exactly equal to the discounted expectation of the share price without consideration of investors' aversion to risk. In other words, people don't receive or require a compensation for taking risk. P-measure expected values (based on P-measure probability models) discounted for risk and time are equal to the derivative values and Q-measure expected values (based on Q-measure probability models) discounted only for time are also equal to the derivative values. Discounting for risk (using P-measure) is equivalent to including a ***risk premium,*** a compensation of investors for risk which we will define formally later. Risk premiums are hard or even impossible to observe, so risk-neutral valuation using Q measures (and discounting only for time) has provided simplicity without loss of accuracy that has created a revolution in derivative valuation.

1.2.3 Discounted Expected Value and the Time Value of Money

The previous simple example deals with randomness; however, it ignores two potentially important factors: time and risk. If the payoff to a security takes place one or more days into the future, or if the payoff contains a non-trivial degree of risk, then the equilibrium price of an asset will tend to be less than the expected payoff of the asset. The factors of time and risk can be very important. The issue of risk is an especially important issue that will be discussed extensively throughout this book. In this section, we turn out attention to the effect of time value while ignoring the risk discounting, noting that proper asset valuation requires the consideration of both the timing and risks of the asset's cash flows, which we will discuss in Section 1.2.4.

Basic Price Principle 1.2

If we ignore the aversion of investors to risk, the value of a financial asset or contract will equal the sum of the expected cash flows that the asset or contract offers *discounted for the time value of money.*

The ***time value of money*** is the relationship between money at different points in time in the absence of risk. In finance, ignoring risk aversion, the time value of money is measured using the market prices of very low risk sovereign debt such as U.S. Treasury securities. Even though no assets are absolutely riskless, finance professionals describe short-term U.S. Treasuries as being riskless and describe their promised returns as "riskless" and their yields as ***risk-free*** rates.

Mathematical finance uses observations of the market prices of riskless debt securities to estimate the current price of $1 to be received at various points in the future. For example, the price of a three-month U.S. Treasury bill that offers $10,000 in exactly three months might be observed to have a price of $9,950. (An investor pays $9,950 now and receives $10,000 in three months at maturity.) We therefore estimate the market value of $1 in three months to be $0.995 today. Financial practitioners can use this price to infer the prices of cash flows to be received in three months. The value "0.995" is a ***time value discount factor***, which is a multiplying factor used to discount any future cash flow to its present value at today. We might also observe that the price of a six-month U.S. Treasury bill that offers $10,000 in exactly six months might be $9,875. We therefore estimate the value of $1 in six months to be $0.9875 today. The time value discount factor for six months is 0.9875. The value of an asset with two final cash flows, one in three months and one in six months can be found by multiplying each cash flow by its corresponding time value discount factor to form the present value of each cash flow and then summing the two present values.

Financial practitioners use interest rates as an intuitive metric of the time value of money. ***Interest rates*** can be viewed as specifying the speed with which investments grow through time and is often represented as an annual percentage. The exact specification of interest rates requires an assumption regarding the time interval at which investments begin to earn interest on interest (i.e., how often is interest compounded). Chapter 2 discusses interest rates and compounding in greater detail. Financial math is often performed under the assumption of continuous compounding of interest. The formula for the present value of a \$1 cash flow using continuous compounding at the interest rate r over the time interval T, which is called the value or price of \$1 due in T years, is

$$\textbf{Present value} = \textbf{\$1} \times e^{-rT} \tag{1.2}$$

The expression e^{-rT} is the time value discount factor for T years and it should be viewed as the present value or price of \$1 due in T years when the current market interest rate is r.

For example, if the current continuously compounded market interest rate on two-year riskless debt is 5%, then the value of \$1 due in two years with certainty is \$0.9048, found as $e^{-2 \times 0.05}$.

Example 1.3: Abbi offers a new game with Betti in which Abbi does not make any payoffs to Betti until one year after Betti pays for a gamble and rolls the die. As before, Abbi rolls a normal die but in this example the payoff will be paid and received with a delay of one year. The ultimate size of the payoff is the same as before with \$1 if 1 appears, \$2 one year later if 2 appears, and so forth. The only difference between this gamble and the previous gamble is that although Betti needs to buy the ticket to play today, she will not receive the payoffs until one year later. How much is the fair ticket price given the one-year delay before the payoff is made, ignoring valuation of the effects of risk aversion?

We need to consider time value of money. Chapter 2 discusses the concepts of the time value of money and interest rates in much more detail. At this stage, let's simply use the following formula:

$$\textbf{Value} = e^{-rT} \times \textbf{E(Payoff)} \tag{1.3}$$

$$\text{Value} = e^{-r} \times 3.5$$

where r is some interest rate and $T = 1$.

Note that if the interest rate, r, is zero, the answer is \$3.50. But if the interest rate (i.e., the time value of money) is positive, the equilibrium price of the gamble will be less than \$3.50. In effect, Abbi is borrowing money from Betti, and Betti (and other potential gamblers) will demand compensation in the form of an expected gain for waiting for their payoffs.

Note that if we ignore the aversion of investors to risk, we can set r equal to the risk-free rate. However, the cash flow of \$3.5 is risky, compared with someone getting \$3.5 for sure one year later, so we may want to use a rate higher than the risk-free rate to include a risk premium, as discussed in the next section.

After considering an example of discounting for the time value of money, the next example makes the payoff more complicated.

Example 1.4: Let's return to the original game where Abbi plays with Betti and makes an immediate payoff but let's add an optional second roll. As before, Abbi will roll a normal die and will pay Betti \$1, if 1 appears, \$2 if 2 appears, and so forth up to \$6 if 6 appears. However, Abbi offers the bet with an added feature this time: If Betti is unhappy with the first roll of the die, Betti can choose to reject the first roll and to have the die rolled a second and final time. If Betti accepts the first die roll, she receives whatever number appears on that first roll. After seeing the first roll, if Betti opts for a second roll she does not receive any payoff based on the first roll and must accept the payoff of the second roll (the payoff of a second roll is still \$1 if a 1 appears, \$2 if a 2 appears, and so forth). If Betti must pay for the ticket today and will receive the payoff today, how much is the fair ticket price, ignoring the valuation effects of risk aversion? What if Betti will not receive the payoffs until one year later?

Note that this example begins like the first example but adds a right (i.e., an option) for Betti. This right or option allows Betti to make a new decision after information has been revealed (i.e., after the first roll takes place). Under what circumstances should Betti opt for a second roll?

Ignoring risk aversion, Betti's rational decision will be to turn down the first roll when 1, 2, or 3 happens, and to opt to roll again (and opt not to roll again with a 4, 5, or 6 on the first roll). How do we know this? It is actually simple. As previously shown, the expected value of any single roll is \$3.50. If the first roll produces 1, 2, or 3, Betti can improve her expected payoff to \$3.50 by opting for a second roll. If the first roll is a 4, 5, or 6 Betti will decide not to exercise her option for a second roll since she already earned a payoff greater than \$3.50.

Let's look at the expected value of the game with the option for a second roll. Since the two trials are independent, the chance to get a \$1 in the second roll as $1/6 \times 1/2 = 1/12$ (note that the probability of having a second roll is ½ and the probability of a 1 given that there is a second roll is 1/6). Note that the probability of a 4, 5, or 6 is the sum of the probability of that value occurring on the first roll and the probability of that value occurring on the second roll. So the probability table will be Table 1.1.

TABLE 1.1

Probability Distribution Table

X	1	2	3	4	5	6
$P(x)$	1/12	1/12	1/12	1/12 + 1/6	1/12 + 1/6	1/12 + 1/6

$$\text{Value} = E(X) = 1/12 \times 1 + 1/12 \times 2 + 1/12 \times 3 + 1/4 \times 4 + 1/4 \times 5 + 1/4 \times 6$$

Note that the \$4.25 expected value of this example (the game with an option for a second roll), exceeds the \$3.50 expected value of the first example (the game with an immediate payoff and no option for a second roll), by \$0.75. Thus the *option* has a value of \$0.75. Options are an important type of financial derivative. Returning to the previous example of an airline company attempting to manage its risks, it is a common for an airline company to negotiate an option to buy additional jets in the future when they commit to a contract to buy jets for immediate delivery. The option protects the airline from the risk of increases in the prices of additional jets once they have committed to adding that type of jet to their fleet.

If Betti will not receive the payoffs until one year later, then the price should be adjusted with a time discount factor:

$$\text{Value} = e^{-r} \times 4.25$$

where r is the interest rate for one year. Note that this example did not consider risk aversion for Betti.

1.2.4 Adjusting for Risk through Discounting

Most financial securities such as stocks and bonds are claims to cash flows emanating from underlying real assets such as the assets of a corporation. For example, a shareholder or bondholder in an airline company invests in order to receive cash flows generated by the successful operations of the firm. The value of one of these securities is equal to the expected cash flows offered by the security discounted for the time value of money and adjusted for the riskiness of the cash flows.

Modern economies rely heavily on assets such as buildings, equipment, and technologies that are acquired using capital provided by investors. The providers of that capital demand compensation both for the time value of money and also a premium for bearing risk. Accordingly, the present or current value to an asset is based on the expected values of the asset's future cash flows discounted for both time and risk. Put differently, the expected return or expected cash flows received by an investor from a risky asset can be divided into two components: compensation for the use of the capital through time (the time value of money) and an expected risk premium to compensate the investor for bearing risk.

Asset valuation for time and risk may be viewed somewhat like the process of renting a moving truck. Often a truck rental company will charge both a daily rate for the renter to possess the truck and a rate per mile for the renter to drive the truck. Similarly, providers of capital demand to be compensated for the time value of money with compensation made at the riskless interest rate and for bearing the risks that the actual cash flows received will differ from their expected values. Investors pay less money for risky assets than safe assets when both assets have the same expected value of future cash flows. The higher average returns demanded by investors in risky assets relative to the returns demanded by investors in riskless assets is termed a *risk premium*. A ***risk premium*** may therefore be viewed as an expected rate of return above and beyond the riskless interest rate that is a component of the return demanded by investors for holding risky assets. Here is the concept in equation form:

$$\textbf{Risk premium of an asset} = \textbf{Expected rate of return of an asset} - \textbf{Riskless rate} \quad (1.4)$$

Let's look at two payoffs: one payoff is $10,000 with a probability of ½ and $0 with a probability of ½; the other payoff is $5,000. The expected values are the same. However, a risk-averse person may prefer the second payoff because it has a lower risk. So, we need to adjust for the risk of the first payoff. In other words, the ticket for the first payoff should be less than that of the second one for a risk-averse person. Therefore, there are three factors in valuing financial products: expected values (based on P-measure probabilities), the time value of money (based on risk-free rates), and risk premiums (based on risk aversion).

Basic Price Principle 1.3

The value (fair price) of a financial asset or contract will equal the sum of the expected cash flows that the asset or contract offers discounted for the *time value of money* and discounted by a corresponding *risk premium*.

In general, risk premiums are unobservable, so it is very hard to determine a value (fair price) for a financial asset or contract. The good news is that in a later chapter shows that the valuation of financial derivatives can be simplified. In financial derivative valuation, P-measure probabilities and risk aversion can be replaced with Q-measure probabilities and the time value of money (based on risk-free rates). Even though Q-measure probabilities are not the actual probability of various cash flows occurring in the real world, they combine and carry the effects and information of P-measure probabilities and risk-aversion together. Q measures are a very convenient way to value financial derivatives.

In summary, to value risky securities other than financial derivatives it is necessary to incorporate the riskiness of the asset into the determination of the asset's value by discounting future risky cash flows by a rate that includes a risk premium. However, one of the most astounding and important breakthroughs in modern finance was the discovery that in the case of many types of financial derivatives, the value of the financial derivative can be determined using the riskless interest rate and the price of the derivative's underlying asset rather than risk premiums.

Basic Price Principle 1.4

The value of a financial derivative is based simply on the value of the derivative's underlying asset or, alternatively, on the asset's expected cash flows under a risk-neutral world and the observed riskless interest rates without regard for risk premiums. This is extremely important since risk premiums are unobservable!

1.2.5 Valuing Financial Derivatives

Most financial derivatives are a zero-sum game wherein any cash flows received by one side of the contract must be delivered by the other side of the contract. In the previous examples of Abbi and Betti betting on the roll of a die, the contract was a zero-sum game. In that example, the roll of the die may be viewed as symbolizing the economic outcome of a risky business venture. If instead of betting on the roll of a die, the contract between Abbi and Betti could depend on the future market price of a stock, a bond, a commodity, or a market index, or could depend on a future rate such as a currency exchange rate or an interest rate. In those cases, their contract would clearly be a financial derivative.

The remainder of the book models the random behavior of the payoff function in a risk-neutral world. If the payoff distribution is known, it can be used to calculate the expected value of the payoffs. If the payoff is a discrete variable, we can use the probability weighted average of those payments as the expected value. In Chapter 3, a method known as a binomial tree is used to provide a simple but powerful method with which to model discrete payoffs. If the payoff is a continuous random variable, we might evaluate an integral to calculate the expected payoff. Chapter 3 demonstrates an example of that method known as the Black-Scholes formula using Q-measure probabilities.

To summarize, in order to find the price of a financial derivative, we need to do three things:

1. Model the distribution of the underlying asset in a future date in a risk-neutral world
2. Calculate the Q measure expected value of the future payoff
3. Discount the expected value back to a present value by risk-free rate(s)

There are a lot of technical issues. How do we model the distribution of the underlying asset's price and cash flows at future dates? How do we estimate the risk-free interest rates to be used to discount the cash flows?

Chapter 3 models the distribution of the underlying asset two ways: as a binomial distribution and as a lognormal distribution. These two methods correspond respectively to the binomial tree approach for discrete payoffs and Geometric Brownian motion for continuous payoffs. The next few sections introduce several financial derivatives.

1.3 Details Regarding Forward Contracts

This section introduces one of the major classifications of financial derivatives: forward contracts. We use forward contracts to demonstrate how the time value of money is incorporated into the valuation of financial derivatives. Later in this chapter we will discuss options contracts and use options contracts to demonstrate how risk is modeled in the valuation of financial derivatives. This section covers a variety of issues related to forward contracts such as spot prices, futures contracts, and settlement.

1.3.1 Spot Prices and the Mechanics of Forward Contracts

A ***spot price*** of an asset such as a financial asset or commodity is the current price for immediate delivery of that financial asset or commodity. For example, if an airline company purchases jet fuel for immediate delivery and with immediate payment it is known as a spot transaction and is executed in a competitive market (the spot market) at the spot price of jet fuel. In this section, we discuss guaranteed contracts for purchase and delivery known as forward contracts in which the agreements require delivery and payment on a substantially delayed basis such as weeks, months, or even years.

A ***forward contract*** is an agreement where the ***long position*** or buyer in the contract agrees to buy a specified asset on a specified date at a specified price and the ***short position*** or seller of the contract agrees to deliver that asset to the buyer under the specified terms. The agreed date T at which delivery is scheduled to take place is known as the ***delivery date*** (or ***settlement date*** or ***expiration date***). The pre-specified price F at which the exchange will take place is the ***forward price*** or ***delivery price***. The forward contract specifies terms of the delivery such as the quantity and quality of the asset to be delivered as well as the time and location at which delivery is to take place.

For example, on January 1, 2022, an airline company may enter into a long position in a forward contract to purchase 100,000 gallons of jet fuel on July 1, 2022, at $2.50 per gallon. An energy refiner may enter into the short side of the contract by promising to deliver the fuel for $2.50 per gallon. Both sides of the forward contract are reducing or *hedging* their business risk by locking in prices on January 1 rather than taking the risk that the price may move against them by July 1. The airline can sell airplane tickets with confidence regarding its future fuel costs while the energy refiner may arrange to purchase crude oil with confidence that the refiner will be able to sell its refined fuel at a profitable price.

1.3.2 Cash Settlement as an Alternative to Actual Physical Settlement

Consider a forward contract requiring the delivery of a share of stock T years from now (time 0) at the forward price F. The price of the stock at any time t is denoted as S_t, so the price of the stock when delivery of the forward contract is scheduled to take place is S_T.

Let's examine what happens if $S_T > F$ on the delivery date. In many such derivatives the parties to the contract can demand to take or make delivery. In this case, the short position in the contract delivers the stock to the long position in the contract in exchange for the payment F.

The ***payoff of the derivative*** is the net gain or loss for each party at the expiration of the derivative contract. Since the actual price of the stock at time T is likely different from F, the economic result of the contract at time T is that the long side gains (if $S_T - F$ is positive) or loses (if $S_T - F$ is negative). Specifically, the long position has a payoff of $S_T - F$ and the short position has a payoff of $F - S_T$. Note that the payoff of the long and short sides of the contract sum to 0.

$$\textbf{Forward payoff for a long position} = S_T - F \tag{1.5}$$

$$\textbf{Forward payoff for a short position} = F - S_T \tag{1.6}$$

The two parties to the forward contract might be using the contract to control risk, not with the desire to ultimately exchange the stock. In that case the parties may do a cash settlement rather than a physical settlement. In a cash settlement the entire transaction is terminated or settled at time T by exchanging the cash $S_T - F$. When $S_T > F$, the cash payment is from the short side of the contract to the long side, and vice versa.

Let's return to the previous example of a forward contract between an airline company and a refinery regarding jet fuel. Rather than actually buying the jet fuel for the pre-specified price directly from the short side of the contract at the time of settlement, the two parties may agree that the long side of the contract (the airline company) may receive S_T–F from the short side when $S_T - F > 0$ and then buy the fuel in the spot market at the price S_T. Putting together the value $S_T - F$ from the forward contract and the expenditure of S_T to purchase the jet fuel in the spot market generates a net expenditure of F for the jet fuel, which is exactly what the airline company expected. If $F > S_T$ on the delivery date, we may view the long side as paying $F - S_T$ to the short side and buying the fuel in the spot market at the price S_T which nets again to an expenditure of F for the jet fuel, as the airline company expected. When $F = S_T$, the contract terminates with no value to either party. The analysis for the short side of the contract reveals that participation in the forward contract may be viewed as having the payoff $F - S_T$ which when combined with selling the jet fuel that they produced at S_T nets sales proceeds of F which is the price that was fixed through the forward contract.

The above discussion is more than just an academic point. In practice, actual physical delivery of the underlying asset for forward contracts (and futures contracts to be discussed later) is often replaced with cash settlement wherein the long side receives $S_T - F$ from the short side when $S_T > F$, and the short side receives $F - S_T$ from the long side when $F > S_T$. The reason is simply that the parties often prefer to take and make delivery of assets through their preferred channels rather than through the method specified in the contracts. So in the case discussed, the airline company likely would prefer to purchase the jet fuel from local suppliers and the energy refiner would prefer to distribute its jet fuel through its distribution network. The forward contract's primary purpose was likely to transfer price risk, not to negotiate new channels for receipt or distribution of goods.

1.3.3 Forward Contracts and Futures Contracts

A ***futures*** contract is similar to forward contract. Generally, the contracts differ in the following major ways: (1) futures contracts tend to be traded on organized exchanges while forward contracts are often privately negotiated as ***over-the-counter*** (***OTC***) transactions, (2) futures contracts tend to be uniform using standardized specifications determined by exchanges while forward contracts are often negotiated by the parties and may be tailored to the needs of one or both of the parties, (3) the integrity of futures contracts is generally guaranteed by the clearinghouse of the exchange while forward contracts are often more exposed to the risk that the counterparty (i.e., the other side) to the contract will default, and (4) futures contracts are marked-to-market on a daily basis where forward contracts typically involve no cash flows prior to settlement. When a contract is *marked-to-market* in this context, it means that changes in the value of the contract are settled daily between the parties as discussed later. This book focuses on the mathematics of forward contracts.

The differences between futures and forward contracts have blurred in recent years, especially from regulatory pressures subsequent to the 2007–2008 global financial crisis. The relationship between the pricing of futures and forward contracts can be complex. The privately negotiated nature of forward contracts combined with lack of being marked-to-market introduces greater counterparty risk than most futures contracts which are generally well backed by brokerage firms and the clearinghouses of futures exchanges. ***Counterparty risk*** to a derivative is the risk that one side of the contract will default on promises to the other side. The private, non-standardized nature of many forward contracts tends to cause greater illiquidity than is experienced with market-traded futures contracts. As an introductory book, the counterparty risk of derivatives is ignored. Even without considering counterparty risk, the pricing of futures and forward is different as is discussed in Chapters 4 and 7. ***Illiquidity*** in this context refers to the potentially higher cost and longer time that is often required to initiate or close a position in an illiquid asset. The more actively and continuously securities are traded in a market the greater the liquidity. Illiquidity and counterparty risk can add substantial complexity to derivative valuation. As this is an introductory book, the modeling of illiquidity is ignored.

Finally, the ***marking-to-market*** of futures contract has implications for the pricing of otherwise identical futures and forward contracts. Marking-to-market is the process of *changing the delivery price* of existing futures contracts, usually on a daily basis, to equal the current market price of new futures contracts. At the end of any day when futures prices for *new* futures contracts have risen, each *existing* contract's delivery price is increased to match the current futures price and the short position is required to make an equally-sized cash transfer payment to the long side of the contract (and vice versa for price decreases). So, the payoff to the long position is $f_2 - f_1$ where f_i is the future price at ith day. The payoff to the short position is $f_1 - f_2$. The process mitigates the counterparty risk of a crisis at maturity. A crisis at maturity occurs when a relatively large obligation comes due and it is discovered that the debtor may not be able to make the payment. The accelerated realization of profits and losses caused by the process of marking the contracts to market can have an effect on the relationship between futures and forward prices when interest rates of different maturities differ.

$$\textbf{Futures payoff of a long position} = f_2 - f_1 \tag{1.7}$$

$$\textbf{Futures payoff of a short position} = f_1 - f_2 \tag{1.8}$$

where f_i is the future price at ith day.

For most of the remainder of this book, the focus will be on forward contracts because they lack some of the complexities of futures contracts (primarily marking-to-market) and are quite easily modeled when illiquidity and counterparty risk are ignored.

1.3.4 Applications of Forward Contracts

Let's look at a variety of examples to illustrate the diverse uses of forward contracts to manage risks.

Example 1.5: A U.S. importer of British beer establishes a long position in a three-month forward on 100,000 British pounds at $1.35 per pound. Assume today is August 26, 2022, and the spot exchange rate of the British pound price is $1.31 per pound.

The importer establishes the long position to lock in the cost of importing the beer which is priced in the brewer's local currency. The importer's long position in the forward contract is fixing the costs of imports in three months in terms of U.S. dollars. Perhaps the short side of the contract is a speculator who has predicted that the exchange rate will fall substantially and will be settled with a positive value to the speculator.

Example 1.6: A speculator predicts that the U.S. stock market will soon decline and so she enters a short position in five 1-month S&P 500 futures contracts at $2,967.50 (each contract is on $250 times the index).

Assume today is August 26, 2022, and the spot price of the S&P 500 contracts is $3,068.50. The speculator hopes to make a profit when the stock market declines. The long side of the contract is an arbitrager who notes that the S&P 500 futures contract appears to be trading at a low price relative to the cash S&P 500 index. The arbitrager simultaneously shorts the stocks underlying the S&P 500 and takes a long position in the forward contract hoping to profit when the forward contract and underlying stock prices converge to relative values that the arbitrageur believes are consistent with an equilibrium. (The S&P 500 is a market index in United States, which is weighted average of the prices of 500 U.S. stocks in different industry sectors.)

Example 1.7: A major agricultural firm establishes a short position in 500,000 bushels of grain using forward contracts with delivery scheduled to coincide with the firm's anticipated harvest date. The agricultural firm is hedging its risk that grain prices might decline prior to their harvest. The long position in the contracts is taken by a hedge fund that seeks to make profits by serving as a counterparty to a variety of participants in the grain market. The hedge fund manager believes that through superior analysis of weather patterns and potential demand for grain the fund will be able to consistently predict when grain prices are going to move in a particular direction.

1.4 The Arbitrage-Free Pricing of Forward Contracts

One of the most important objectives of this book is to demonstrate the use of financial mathematics in modeling the prices of financial derivatives using the simplifying assumptions that prices will tend toward an equilibrium in which market participants will be unable to earn arbitrage profits and that many financial derivatives can be accurately valued with risk-neutral valuation models. ***Arbitrage*** is the simultaneous buying and selling in different markets or of slightly different products in order to take advantage of differing prices for the same or similar assets. In mathematical idealization, arbitrage is the process of earning riskless or virtually riskless profits. In reality, arbitrage profits are only possible when arbitragers can establish positions in multiple assets that are relatively mispriced and when hedges can be formed such that their portfolios that have little or no risk.

1.4.1 Efficient and Frictionless Markets as a Foundation for Forward Price Models

How can the airline company and the energy refiner previously discussed determine an appropriate forward price F (i.e., delivery price) for the jet fuel six months prior to delivery? The answer at first may appear to require forecasting of the likely spot price of fuel on July 1, 2022. However, this first chapter focuses on forward contracts with underlying assets that trade in competitive markets. We will show that when a derivative's underlying asset is a financial asset that trades at a market price that reflects all available information, the forward price of that financial asset is very easy to determine.

Let's begin with the concept of a ***frictionless market*** – sometimes called a perfect market. In a frictionless market, there are no transactions costs, taxes, or other impediments to trade. In a frictionless or perfect financial market, all market participants are able to buy, sell, and borrow cash and securities at market prices.

Let's move to the concept of an ***informationally efficient market***. In an informationally efficient market with costless access to information, arbitragers and other market participants will compete to utilize available information to earn superior risk-adjusted profits. The result of that competition drives market prices to reflect all available information.

While no market is frictionless or perfectly efficient, the assumption that markets are frictionless or even informationally efficient can be a useful tool for deriving valuation models. Models are well described as being abstractions from reality. Models are formed using assumptions that are unrealistic. The power of assumptions is when by assuming away realities the model can reveal a clear picture of the focus of a researcher's attention. The simplifying view generated by models often allows us to better explain the past and predict the future. If the models do not perform well in explaining the past or predicting the future then the assumptions should be examined to determine if they should be relaxed or modified.

Arbitrage-free pricing models describe pricing relationships that will exist in a frictionless market when the actions and potential actions of arbitragers ensure that all securities are trading at prices that prevent riskless profits in excess of the riskless interest rate.

In a single period framework, the absence of arbitrage opportunities can be described as follows. Assume that $\vec{w}$ is a vector of portfolio weights. $\overrightarrow{P_0}$ is a vector of the known current prices and $\overrightarrow{P_1}$ is a vector of the random future payoffs for each security at the end of the period. The vector of payoffs differs across various economic states that may occur. An arbitrage free market is when no portfolio can be constructed that requires zero investment (i.e., $\vec{w}^T\overrightarrow{P_0} = 0$) and that generates end-of-period payoffs for the portfolio that are non-negative in each potential state and that has at least one state with $\vec{w}^T\overrightarrow{P_1} > 0$. The previous sentence describes a ***no arbitrage condition.***

1.4.2 Simplified Arbitrage-Free Model

This section uses an example of spot market foreign exchange rates to illustrate the idea of arbitrage. Exchange rates, the ratios at which monetary currencies trade, are increasingly important in the world's economy as larger and larger sums of currencies are exchanged in the process of international trade and international payments. Exchange rates express the relative market values between two currencies and can be expressed in terms of either currency. If the exchange rate between U.S. dollars and Euros is such that it takes \$1.25 to purchase one Euro, then the exchange rate is 1.25 dollars per Euro or 0.80 Euros per dollar.

To illustrate the power of arbitrage free modeling, consider three exchange rates: (1) U.S. dollars to Euros, (2) Euros to Japanese yen, and (3) yen to U.S. dollars. Arbitragers in foreign exchange markets are continuously scanning market exchange rates to detect if and when the no arbitrage condition at the end of Section 1.4.1 fails to hold. For example, consider the following three rates:

1. U.S. dollars to Euros : 1.5 U.S. dollars per Euro
2. Euros to Japanese yen : 0.01 Euros per yen
3. Yen to U.S. dollars : 80 yen per U.S. dollar

An arbitrager is aware that: (1) exchanging \$150 from U.S. dollars to 100 Euros, (2) then exchanging the 100 Euros to 10,000 yen, and (3) finally, exchanging the 10,000 yen back to \$125 U.S. dollars will cause a loss of \$25. Going in the other direction must make money. Starting with \$150 and: (1) exchanging the dollars to 12,000 yen, (2) exchanging the 12,000 yen to 120 Euros, and (3) finally, exchanging the 120 Euros to \$180 will make a profit of \$30. The only arbitrage-free pricing relationship is when the product of all three rates is 1.0 (e.g., $1.5 \times 0.01 \times 80 \neq 1$).

As financial mathematics is used to develop more and more sophisticated arbitrage-free models, financial market prices are driven closer and closer to prices that optimally organize production and consumption. A key insight is that market prices drive everyone's analysis of the benefits and costs of all decisions in a market-based economy. More accurate pricing means better decision making, and better decision making leads to greater efficiency and greater economic growth.

1.4.3 An Arbitrage-Free Model of Forward Prices

This section discusses a slightly more complex arbitrage-free pricing model than the model of spot market foreign exchange rates in the previous section: a model of forward prices on financial securities. The model is more complex because unlike the arbitrage using the three exchange rates discussed in the previous section which all take place instantaneously, this model of forward prices involves the passage of time and models the time value of money using riskless or risk-free interest rates.

In order to determine the forward price F of a financial asset, we begin by recognizing the similarities and differences between owning a financial asset and having a long position in a forward contract on that financial asset. Basically, the difference is the time value of money since buying the financial asset involves a cash outlay while entering a forward contract on that asset does not. This difference in financing can be easily modeled using observable market rates (i.e., riskless interest rates).

Let's look at a simple illustration involving tickets to an event. One way to arrange to attend an event such as a play or a football game is to buy a ticket ahead of time – paying cash and receiving the ticket perhaps a few weeks before the event. Another way might be to reserve a ticket by guaranteeing payment for the ticket on the day of the event. In both cases, payment is guaranteed and receipt of the ticket is guaranteed. In the first case the transaction is a spot transaction with immediate cash payment. In the second case, the transaction is a forward contract since both the payment and the issuance of the ticker were deferred. In theory, the ticket with deferred payment should sell at a higher price since deferred payment allows the ticket buyer to earn interest on the money up until payment is made on the day of the event. Specifically, the ticket with deferred delivery should sell for more by a factor of e^{rT} where r is the riskless market interest rate and T is the length of time in years for which payment is deferred by using a forward contract.

Consider a share of stock with a current market price of S_0. An arbitrager borrows an amount of money equal to S_0 and uses the loan proceeds to buy one share of the stock at S_0 per share. At the same time, the arbitrager enters a short position requiring delivery of one share of the same stock at time T at the delivery price of F. Assume that the forward contract is for 0.5 years, S_0 is \$60, and the arbitrager is able to borrow at the riskless interest rate, r, equal to 4% per year (continuously compounded). What value of forward price F prevents an arbitrage (i.e., sets the arbitragers profit to zero)? Note that when the forward contract settles, the arbitrager will repay the accrued loan amount of S_0e^{rT} and will receive F for delivering the share of stock to satisfy the forward contract. In a frictionless and competitive market, the profit to riskless arbitrage (with no net investment) must be zero.

$$\text{Arbitragers profit} = F - S_0\, e^{rT} = 0$$

Therefore, $F = S_0e^{rT}$ is the no-arbitrage price of a forward contract on a financial security that generates no interim cash flows such as dividends or coupons. Using the assumed values, the price of the forward contract should be \$61.21 (rounded). In an imperfect market with a forward price that is too high, the arbitrager takes a short position in the forward contract, borrows money, buys the stock, waits T years, delivers the stock to satisfy the forward contract, receives the delivery price, repays the loan, and pockets a riskless profit. If the forward price is too low, the transaction is reversed using a transaction known as a short sale of the stock, which is discussed further in Chapter 2. A ***short sale*** involves borrowing an asset and selling it with the goal of later purchasing the asset at a lower price and returning it the lender.

Note that the delivery price of the forward contract is fully identified by r, T, and S_0. It is not necessary to know the distribution or expectation of S_T. If investors are risk neutral (i.e., they do not demand a risk premium for taking risk), then $E_Q(S_T)$ must equal S_0e^{rT} (because all assets would be expected to grow at the risk-free rate since no one would demand or offer a premium return for risk)[2]. Investors are not risk neutral in real life, but for forward valuation purposes, risk neutrality can be assumed. The forward contract has a payoff that includes receiving S_T. The forward price is equal to the Q-measure expected value of S_T, $F = E_Q(S_T) = S_0e^{rT}$. So, the derivative price is the Q-measure expected value (there is no need to consider a risk premium). Note that there is no reason to discount F for time because F occurs at time T, not today. The forward price F is

$$\mathbf{F} = E_Q(S_T) = S_0 e^{rT} \tag{1.9}$$

The formula derived above for the delivery price of a forward contract applies to the extent that markets are competitive, markets have little or no frictions and forward contracts have an underlying asset that is a financial security with no intervening cash flows such as dividends. The model can easily be adjusted to handle intervening cash flows, but complexities arise when the underlying asset is a real asset such as a commodity that involves storage costs, transportation costs, and potentially offers convenience to the owner (such as the desirability of having physical ownership of a commodity such as gold rather than a piece of paper acknowledging that the gold is in a vault somewhere being held in the name of the owner).

1.4.4 Analytics of Forward Contracts

Recall from the previous section that derivative contracts are often settled on a cash basis that generates a more convenient but roughly economically equivalent result for both parties. The cash settlement view provides a more intuitive basis on which to value a forward contract. Mathematically, the long position in the contract may be viewed as having the payoff: $S_T - F$, where S_T is the asset's spot price at the future delivery date T, and F is the delivery price which is specified in the contract today. As with almost all derivatives, the contract is a zero-sum game such that the payoff to the short position (seller) is $F - S_T$.

Forward contracts are often entered with no initial cash payment between the long and short sides. However, sometimes there is an initial payment such as when an investment bank negotiates to receive an upfront payment when offering to take the opposite side of a forward contract tailored to the needs of an operating firm. For example, an initial payment from the buyer to the seller of $e^{-rT}E(S_T - F)$ may be negotiated. However, a pure forward contract has no initial payment (although collateral may be required). So, $F = S_0e^{rT}$.

Let us look at an example:

$$\text{If } S_0 = 60,\ r = 5\%,\ T = 0.35 \text{ then } F = \$61.06$$

For another example, suppose that a portfolio manager enters a six-month forward contract ($T = 0.5$) on 1,000 of Microsoft stock (the underlying asset) at $160 (the deliver price) when the continuously compounded interest rate on riskless six-month bonds is 2.5%. What must be the current price of Microsoft (MSFT) stock? The solution is found as Fe^{-rT}, which is approximately $158.01. If MFST's price goes to $161 in six months, the long position in the forward contract receives $1 at settlement and the short side loses $1. If MFST is $157 in six months, the short position in the forward contract gains $3 and the long side loses $3. Note that the stock price when the forward contract is established is not directly related to the computation of the payment at settlement. However, today's underlying asset price enters into the determination of the delivery price of the contract which in turn is related to the probability distribution of S_T on the delivery date of the forward contract.

So, the long position of the forward contract hopes that underlying asset rises (assuming it is unhedged), so that long side can gain $S_T - F$. Note that an unhedged long side will generally gain through time when the price of the asset underlying the contract rises, while the short side will generally gain when the underlying asset price declines. The payoff to the long side (solid line) and the payoff to the short side (dashed line) are mirror images (relative to the horizontal axis), as illustrated in Figure 1.2.

Sometimes the entity taking a long position in a forward contract will be described as "buying" the forward contract and the holder of the short side will be described as "selling" the contract. While buyer and seller is a good description in cash markets (i.e., spot transactions), the preferred terms in forward markets is establishing a long side or long position and establishing a short side or short position. Other terms often used in trading forward contracts include *opening, entering, holding, closing, offsetting,* and/or *liquidating.*

Note that forward contracts and futures contracts are not *opportunities* to transact; they are *obligations*. A long position in a forward or futures contract is a commitment to buy the underlying asset. In an economic sense the holder of the long position in a forward contract *effectively owns* the underlying asset even though delivery has yet to take place. Thus, in a fundamental economic sense there is little difference between being long a forward contract on 50,000 ounces of silver or actually having 50,000 ounces of silver in storage. There are somewhat minor differences in that the latter requires an immediate cash outlay, allows immediate access to the silver, and requires storage of the silver. But whether an individual physically possesses 50,000 ounces of silver or holds a long position in forward contracts on 50,000 ounces of silver the most important result is the same: the individual ultimately gains $50,000 for each dollar by which the price of

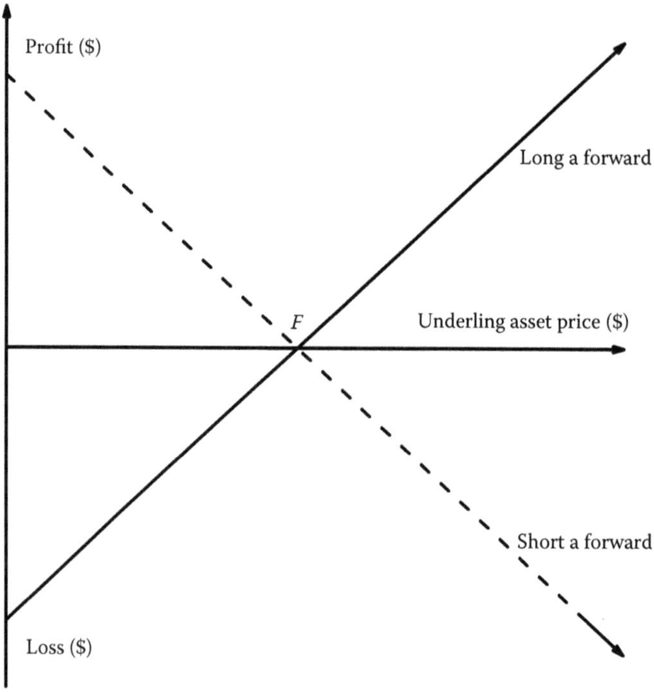

FIGURE 1.2
Payoff of a Long Forward Position and a Short Forward Position.

silver rises and loses $50,000 for each dollar by which the price of silver declines. Finally, note that the differences between physical ownership and *synthetic* ownership through forward contracts (e.g., storage costs, financing costs and convenience yield) are reflected in the relationship between the forward price (i.e., the delivery price) of the commodity and the spot price of the commodity in more advanced models than the model of forward prices demonstrated in this section.

1.5 Introduction to Option Contracts

This section introduces another major classification of financial derivatives: option contracts. Unlike forward contracts, which are commitments to buy or sell, ***option contracts*** are opportunities or rights to buy or sell. A call option confers on its owner the right to buy an underlying asset while a put option is the right to sell an underlying asset.

1.5.1 The Mechanics and Terms of Options

A ***call option*** gives the holder of that contract the right (but not the obligation) to buy an underlying asset by a certain future date (***expiration*** or ***maturity date***, T) for a certain price (***strike price*** or ***exercise price***, K). Note that a call option contract has two parties: a buyer (the holder of the long position, also called the call holder or owner) and a seller or ***writer*** (the holder of the short position). The entity selling or writing the call option is obligated to deliver the underlying asset at a price of K to the option holder at the option holder's demand. The buyer of the call option also has the right to do nothing at the expiration date and to let the option expire.

In competitive and frictionless markets, the holder of a call option on a share of stock may be viewed as having the following potential payoff at the expiration of the option:

$$\textbf{Call option payoff to a long position at expiration} = \textbf{Max}(S_T - K, 0) \tag{1.10}$$

where S_T, is the value of the stock at the expiration of the option. When the stock price at the expiration of the call, S_T, is less than the strike or exercise price, K, the call expires worthlessly. If the stock price exceeds the strike price, the call holder may exercise the call and receive the economic value of the difference $(S_T - K)$.

The holder of the short side of the call option has the payoff: $-\text{Max}(S_T - K, 0)$. Entities holding long or short positions in options often can choose to exit their positions by closing their positions into the market, rather than taking or making delivery of the underlying asset. If an entity closes their position on the expiration date by exiting the option position at the market price, then in a perfect market the cash received or paid will equal the payoffs described above (that would occur by exercising the option, if valuable). The reason that delivery of the underlying asset often does not take place is that participants in option markets often do not view the options contracts as vehicles to buy or sell the underlying assets, they view options as vehicles to transfer and manage risk exposures.

A ***put option*** gives the buyer the right to sell an asset by a certain future date (the expiration or maturity date, T) for a certain price (the strike price or exercise price, K). The long position in a put option has the right to deliver one unit of an asset to the seller and in that event the seller has the obligation to pay the strike price K for the underlying asset. The put holder also has the right to do nothing at the expiration date and let the option expire worthlessly. Mathematically, the long position in the put option receives the economic benefit:

$$\textbf{Put option payoff to a long position at expiration} = \textbf{Max}(K - S_T, 0) \tag{1.11}$$

The put option seller has the payoff: $-\text{Max}(K - S_T, 0)$. As with call options, the longs and shorts in put options can choose, effectively, to cash settle by closing their option positions into the market which in a perfect market generates the same economic result without involving delivery of the underlying assets.

The price paid for buying an option, or received for selling an option, is often termed an ***option premium*** or ***option price***. Note that unlike forward contracts that are appropriately described as being "entered into," options are usually described as being bought, written, or sold. A person selling an option to establish a short position in that option is often described as having ***written*** the option because the transaction literally creates an option that did not previously exist. A trader selling an option that she previously purchased would be described as having sold that option.

There are two major kinds of call and put options: European and American (although both types of options trade on both continents). A ***European-style option*** is where the long position cannot exercise the option prior to the option's expiration or maturity date. An ***American-style option*** allows its holder to exercise the option at *or before* the expiration date.

Options on shares of stock usually trade in contracts, each of which represents an option on 100 shares of stock. Thus, a trader wishing to have the option to buy 1,000 shares of stock would enter a trade to purchase 10 call option contracts. Option exchanges do not usually offer options on a single share of stock. However, mathematical option models are usually based on a single share of stock.

Finally, when $S_T - K > 0$ for a call option or $K > S_T$ for a put option the option is described as being ***in the money*** since it would have value if exercised immediately. The value of an in-the-money option if exercised early is known as its ***intrinsic value***. The excess of the actual option price over its intrinsic value is known as the ***option's time value***. When $S_T - K < 0$ for a call option or $K < S_T$ for a put option the option is ***out of the money***. Finally, when $S_T = K$ options are described as ***at the money*** (or "near the money" if the difference is small).

1.5.2 Applications of Financial Options

One of the most obvious applications of options is their ability to facilitate the management of positions in the assets underlying the options.

Example 1.8: Based on an optimistic forecast for software sales, a portfolio manager buys a six-month (European) call option on Microsoft with a strike price of \$160 for a purchase price of \$3 per share. If the price of Microsoft's common stock price at the end of the six months is \$165, the buyer can exercise the call option to receive a benefit of \$5 per share by being able to purchase a \$165 stock for \$160. The option buyer nets a gain of \$2 (the \$5 benefit at exercise less the \$3 cost at purchase) while the option seller loses $2. If Microsoft price fell to $157 in six months, the buyer does not exercise his call, losing her $3 cost of purchase while the option writer retains the initial $3 option price or premium as a profit.

Note that an option buyer's worst case is losing the premium paid. A call option writer receives money when the option is sold (written) but then must fulfill the obligation to deliver the underlying asset if so demanded regardless of the price of the asset. Here is an example of a common use of puts.

Example 1.9: A former employee of UPS Corporation has a large position in UPS stock currently trading at $105 per share and is concerned that the stock may drop if the economy softens. She purchases three-month put options to sell all of her stock for $100 per share. Each put option costs her $3. Her worst loss in the portfolio from a huge decline in UPS stock limited to $8 per share (the cost of the option and the exercise price that is $5 below the current market price). This is known as a protective put strategy.

1.5.3 The Payoffs of Calls and Puts

Consider a wiring company projecting use of 100,000 pounds of copper in the next quarter. The company wishes to lock in its copper acquisition costs. The firm can buy a three-month call option on 100,000 pounds of copper with an exercise or strike rate of $2.31 per pound. Let's look at the payoffs of the firm's call option position while ignoring the firm's other cash flows.

The payoff of the call option to the call buyer at expiration (ignoring the option's initial cost or premium) is

$$\text{Call payoff to buyer} = \text{Max}(S_T - K, 0)$$

where K is the strike price and S_T is the value of the underlying commodity at the option's expiration date and where $\text{Max}(S_T - K, 0)$ takes on the greater of the value $S_T - K$ or 0.

For a copper producer, the decision may be to buy a put option to sell its copper production next quarter.

The payoff of the put to the put buyer at expiration (ignoring the option's initial cost or premium) is

$$\text{Put payoff to buyer} = \text{Max}(K - S_T, 0)$$

Note that the payoff of the put option is not the mirror image of the payoff of the call. The payoff of the put is 0 when $S_T > K$, while the payoff of the call is 0 when $S_T < K$. Let's look at the payoffs of the call and put to the party that writes or sells the options.

$$\textbf{Call payoff to seller} = -\,\textbf{Max}(S_T - K, 0) \tag{1.12}$$

$$\textbf{Put payoff to seller} = -\,\textbf{Max}(K - S_T, 0) \tag{1.13}$$

The call buyer hopes that the price of the asset underlying the option will rise above K, so that she can gain $S_T - K$. The put buyer hopes that the underlying asset will fall below K, so that she can gain $K - S_T$ at the expiration of the option.

Note that the option seller's payoff is the opposite of the buyer's payoff because the option is a zero-sum game. An option's payoff is not a deterministic number; It is a function of S_T and therefore depends on the future asset price S_T. It is this uncertainty that makes options interesting to study and useful to manage risk.

1.5.4 Profit and Loss Diagrams for Analyzing Option Exposures

Option profit and loss diagrams express the potential profits and losses to an option or a portfolio that contains one or more options measured at the expiration date of the option. The profits and losses in these diagrams typically ignore the time value of money and frictions such as transactions, costs, or taxes. Option premiums are not included in option payoff diagrams but they are included in profit-loss diagrams.

Figure 1.3 illustrates the potential profits and losses to a simple long position in a call option as of the date that the option expires. Note that the call buyer suffers a loss when S_T is less than or equal to K. That loss will be equal to the price or premium paid to purchase the option. For each dollar by which the price of the underlying asset exceeds the strike price, the call option buyer receives \$1. The breakeven point (ignoring the time value of money) is when the underlying asset's price exceeds the strike price by the cost of the option to the buyer (i.e., the premium paid).

The price of the option just prior to the moment that it expires will be driven by arbitragers in a frictionless market to be equal to the payoff described in the previous section (e.g., $\text{Max}(S_T - K, 0)$) for an option buyer. If the market price of the option differs from $\text{Max}(S_T - K, 0)$ at expiration, then an arbitrager can buy an underpriced in-the-money option, exercise that option and collect a profit. If the option is overpriced, the arbitrager can write the option, buy the underlying asset for S_T, and deliver the asset and receive K. In both cases, the arbitrager would profit without risk. Therefore, options tend to be priced at expiration such that market participants can

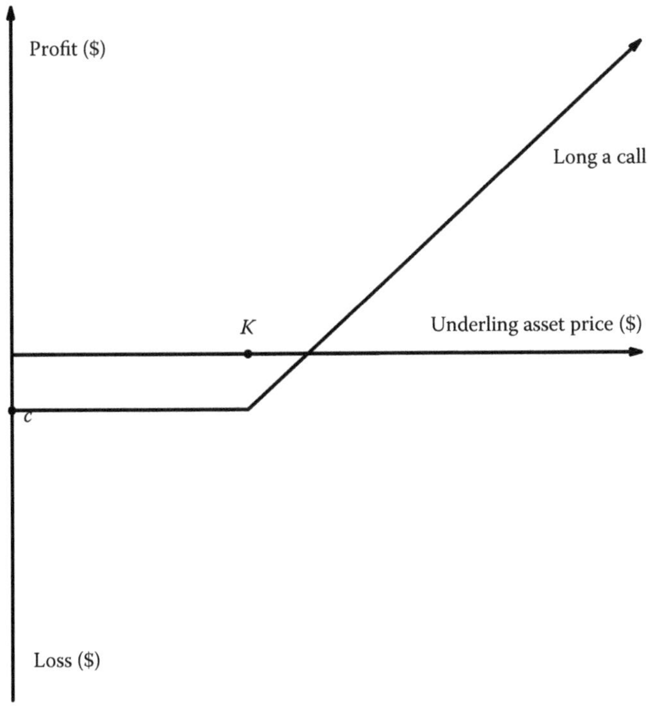

FIGURE 1.3
Profit of a Long Position in a Call.

receive appropriate value by closing their positions rather than taking or making delivery, and can often lower their transactions costs at the same time. As is often the case with derivatives, many or most market participants close their derivative positions without taking or making delivery of the derivative's underlying asset because their goal in establishing a derivative position is usually to obtain or hedge exposure, not to obtain or deliver an asset.

Note that the kink in the profit-loss diagram occurs directly below the strike price or exercise price of the option. All profit and loss diagrams for options will experience kinks directly above or below strike prices – and only there. A long position in a call option exhibits the potentially attractive profile of having losses limited to the cost of the option while offering unlimited profits. Of course, in a competitive market, the price of the option reflects the higher probabilities of modest losses and the relatively low probabilities of extraordinary profits. The buyer of a call option on a stock is a bullish position on the stock in that it tends to rise in value when the stock rises in value, which tends to occur when the overall stock market is a bull market.

Figure 1.4 is the diagram of a short position in a call option, which is a mirror image of the long position in a call. As the other side of a zero-sum

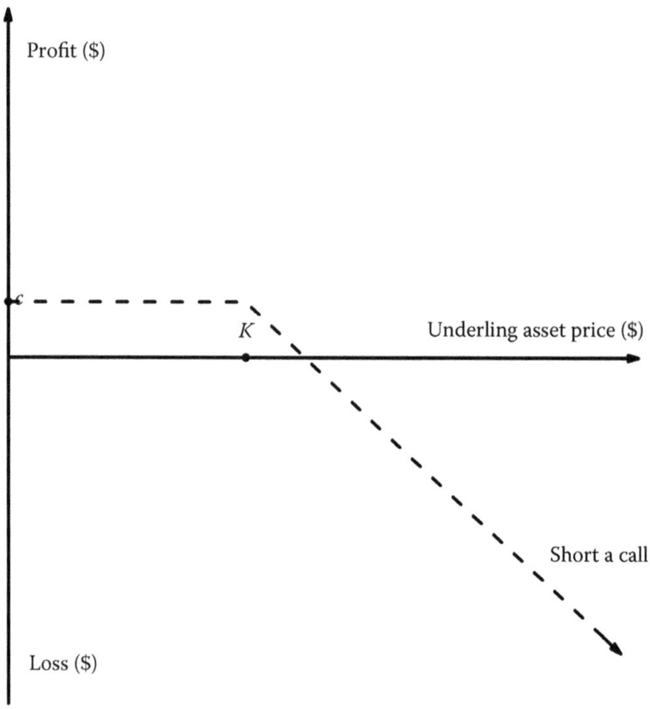

FIGURE 1.4
Profit of a Short Position in a Call.

game, the option writer faces the prospects of limited profits and unlimited losses. Most option writers hedge this risk by having a long position in the stock underlying the option (a position known as a covered call).

Figure 1.5 simultaneously diagrams both the long and short sides of a put option with the long position expressed as a solid line and the short position as a dashed line. Note that the long position in the put option is bearish and that the long and short positions are mirror images of each other. Note also that while a long position in a call option offers theoretically unlimited profit potential, the potential payoff to a put buyer is limited to K and the buyer's profit is limited to K minus the price or premium paid for the put.

The option diagrams in this section describe the four foundational option positions: long or short a call or put. These relationships can be used to diagram the profit and losses of portfolios that contain multiple options and/or that include a long or short position in the underlying asset.

Finally, calls and puts can both be viewed as *options to exchange*, with calls being exercised by the long side by exchanging cash for the underlying asset and puts are exercised by the long side exchanging an asset for cash. While calls and puts initially appear to be opposites (the right to buy vs. the right to sell), consider the case of an option to exchange one currency for another

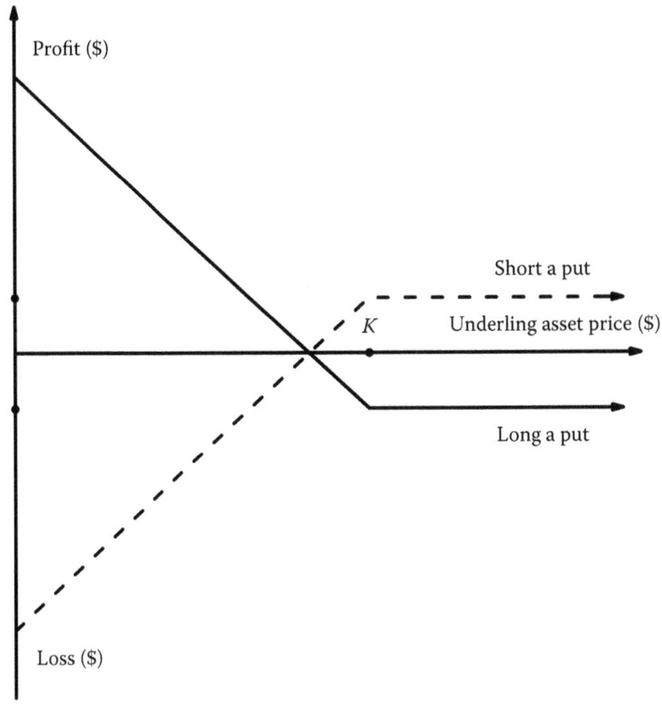

FIGURE 1.5
Profits and Losses of a Long Position and a Short Position in a Put.

(foreign currency or FX options). Since both sides of the exchange are cash, the distinction between calls and puts as buying vs. selling is ambiguous. While from the perspective of a particular party the delivery of domestic cash to purchase a foreign currency acts like a call option, the description of the option as a call or put tends to be driven by convention.

1.5.5 Applications of Options

Options are initially and correctly viewed as offering directional exposures to the value of the underlying asset. A long call position tends to rise and fall with the market price of the underlying asset (although the call position also responds to other effects such as the passage of time and changes in the anticipated volatility of the underlying asset). An important role of call options is to allow a market participant to benefit from large upward movements in the underlying stock while having losses limited to the price paid for the option (often described as having "leverage" even though no borrowing is used). Of course the price of the option reflects this asymmetric exposure so that there is no "free lunch" to the call option buyer.

A common purpose of a long position in a put tends to be to protect the owner of the put option's underlying asset from the downside risk of that underlying asset. Asset owners tend to buy puts when they want to hedge downside risk perhaps due to a belief that the asset has an unusually high chance of falling during the life of the option. The put option serves as a hedge because the put option price tends to gain when the underlying asset's price falls substantially. If the put option is used to hedge the risk of the underlying asset, we can think of the put option as being similar to buying insurance such as car insurance. A person buys car insurance so that if a bad thing happens to her car the insurance company covers the loss. In fact, car insurance may be viewed as a put option on a car such that the car owner can sell or "put" the car to the insurance company at the car's previous value after the car has been damaged and has declined in value.

If there were only call buyers and put buyers we could not have a market in options since someone needs to take the other sides of the zero-sum game by serving as option writers or sellers. The benefit of writing a call is to gain cash (the call price) today. Call option writers may write calls because they believe that the price of the underlying asset is unlikely to rise. The purpose of writing a put is also to gain cash (the put price) today and is often done when the put writer thinks there is a relatively small risk of the asset falling dramatically.

The motivations discussed above for writing options emphasized the motivation of speculation: an option writer may speculate in options in the hopes that their evaluation of the likelihood of various price changes in the underlying asset is superior to the evaluations reflected in the market prices. Option markets therefore facilitate efforts of market participants to speculate that the market prices of various assets are either too high or too low. To the extent that these speculators are correct, their actions bring asset prices closer to their true values. As asset prices are driven closer to their true values, their new prices signal information throughout the entire economy that provides better guidance as to the benefits and costs of decisions regarding those asset prices.

Financial market prices signal the benefits and costs of society's use of capital, commodities, and other goods and services, silently coordinating the economic decisions of the entire economy.

The efficiency of a market-based system depends on the extent to which the prices properly incorporate relevant information, and the actions of speculators tend to enhance that process. Successful speculators buy underpriced assets, sell overpriced assets, stabilize market prices, and increase their wealth. In doing so their "voice" becomes larger in the process of determining market prices. Unsuccessful speculators buy overpriced assets, sell underpriced assets, destabilize market prices, and lose their wealth. In doing so, their "voice" becomes smaller in the process of determining market prices. This tends to lead to a meritocracy where the people who are best at valuing assets exert more control over market prices than those who are poor at valuing assets. Since market prices guide the economic decisions of a market economy, it is important that the market prices be accurate and efficient.

Option contracts provide unique and valuable ways for speculators and others to express and utilize their information and beliefs with regard to valuation. Option contracts also provide convenient vehicles for market participants to manage their risks. Those are the two primary purposes served by option markets.

Finally, note that the profit and loss diagrams in this section focus on the relationship between the price of the underlying asset and the value of the option at the time of the option's expiration. Prior to expiration other variables affect the profitability of an option strategy. Option prices for a call and a put depend on at least five variables:

S_0 *(underlying asset price today)*
K (the strike price)
T (the time to expiration)
r (the riskless interest rate)
σ *(the volatility of the underlying asset's returns which is also the standard deviation of the underlying asset's returns)*

The volatility of the returns of the option's underlying asset is an important variable that will be discussed in detail later in this book. In a subsequent chapter, the issue of dividends being paid on an option's underlying asset is discussed. In that case, a sixth variable is needed to value an option: the anticipated dividend.

1.5.6 Option Hedging Strategies

The previous section mostly discussed speculators in option contracts. Options can also be valuable opportunities for hedgers whose primary purpose is to reduce risk. The following examples illustrate the use of options for hedging purposes while illustrating the variety of options available to hedgers.

A producer of rings for college graduations may by concerned that the price of gold is going to rise prior to the spring when most graduation ceremonies are held. However, the jeweler needs to announce prices for the rings earlier in the year before the orders for rings are placed. If the manufacturer knew how many rings would be ordered in the months prior to graduation, the manufacturer could buy the gold now or could enter a forward contract to purchase the gold. But the manufacturer does not want to be heavily exposed to the risk that their forecasts may be quite erroneous and the firm would end up with a large surplus or shortage of gold. Options can come to the rescue. The firm might enter forward contracts for a low estimate of their gold usage and purchase a call option on the additional amount of gold to reach the high end of their estimates. The purchase price of the call option on gold may be reasonable when compared to the potential outlay involved with ordering too much or too little gold in the spot, forward, or futures markets.

Next, consider the uncertainty faced by farmers, food processors, and food exporters with regard to the prices of grains near an important harvest season. The farmer fears that the price of her crop will fall before it can be harvested and sold. But the farmer is concerned about entering a short position in a forward contract since if her crop fails she will still be responsible for settling the promise made to deliver the crop to satisfy the terms of the forward contract. A potentially attractive solution to the farmer's dilemma is to purchase a put option on some or all of the anticipated harvest. The put option will lock in a minimum price while limiting the farmer's downside risk to the purchase price of the option.

Consider a cereal producer that is concerned about maintaining a stable price of its product in order to retain loyal customers and therefore wishes to enter contracts with food wholesalers that include long-term prices for its products. However, the cereal producer is concerned only about a very large increase in the price of grains since the firm has the resources to cover small price increases in the cost of the grain that it purchases. The cereal producer can purchase out-of-the-money call options on grains that lock in an upper ceiling to its grain acquisition expenses.

Finally, consider a grain exporter that has arranged the sale of a large quantity of grain at a guaranteed price for export a few months after the anticipated harvest. The grain exporter is concerned about several things: the importer of the grain might default on the agreement to purchase the grain, the price of the grain at harvest might be very different from current grain prices, and the foreign currency offered by the importer may change in value relative to the domestic currency that the exporter will use to purchase the grain. Options can address all these problems. The exporter can buy put options to sell the grain and foreign exchange options to convert the foreign exchange to the domestic currency of the exporter. These options allow the exporter to hedge or lay off some or all of the price risks regarding grains and foreign exchange while potentially reducing downside risk to the costs of the options. There is even a sophisticated financial derivative known as a credit default swap, discussed much later in this book, which can help the exporter hedge the risk that the importer will default.

Better hedging against risks over which an economic agent has little or no control allows that agent to focus on those risks that the agent can control, to focus her energies into creating wealth and to enjoy lower levels of total risk.

1.6 Put-Call Parity and Arbitrage

An important relationship exists between call option prices, C, and put option prices, P. The price of a European put option and European call option with the same underlying asset, same strike price K, and same expiration date T must satisfy the following relationship known as ***put-call parity:***

$$C + Ke^{-rT} = P + S_0 \tag{1.14}$$

From left to right, Equation 1.14 can be expressed as "owning a call plus a bond is equivalent to owning a put plus the stock." This well-known equation is often arranged differently or expressed with different notation. This particular arrangement is convenient for demonstrating why the relationship must be true in a well-functioning market with no arbitrage. The left and right sides of Equation 1.14 can be viewed as payoff to two portfolios, A and B.

Portfolio A (left side): Buy a call on a stock + buy a zero-coupon bond that pays K at time T.

Portfolio B (right side): Buy a put on the same stock + buy the stock.

Note that the two payoff columns on the right side of Table 1.2 demonstrate that the total payoff to Portfolio A (S_T or K) equals the total payoff to Portfolio B (S_T or K) whether the *call* option is exercised at expiration (i.e., $S_T > K$), or whether the *put* option is exercised ($K > S_T$). Obviously, when $S_T = K$ neither option has a value at expiration and portfolios A and B have equal values (K). Since under all scenarios both portfolios have the same total

TABLE 1.2

Values of Portfolios at Purchase and Payoffs at Option Expiration

		Purchase Cost Today (Negative for Money Paid)	**Payoff If $S_T > K$**	**Payoff If $S_T < K$**
Portfolio A	Long a Call option	$-C$	$S_T - K$	0
	Long a bond	$-Ke^{-rT}$	K	K
	Total	$-(C + Ke^{-rT})$	S_T	K
Portfolio B	Long a Put Option	$-P$	0	$K - S_T$
	Long the stock	$-S_0$	S_T	S_T
	Total	$-(P + S_0)$	S_T	K

payoffs, they must therefore be worth the same today. This means that Equation 1.14 holds:

$$C + Ke^{-rT} = P + S_0.$$

Another way to explain put-call parity is that the payout at time T for call (C) minus put (P) is the same as $S_T - K$. (This can be seen by combining the payout diagrams). The cost today to replicate $S_T - K$ by buying the stock today and borrowing Ke^{-rT} dollars from the bank is $S_0 - Ke^{-rT}$.

If the stock underlying the options distributes a dividend, D, between now and the option's expiration with present value $I = PV(D)$, then the put-call parity relationship with a cash dividend is

$$\boldsymbol{C + Ke^{-rT} = P + S_0 - I} \tag{1.15}$$

The logic of Equation 1.15 is that the current value of the stock, S_0, represents the sum of the present value of the next dividend, I, and all remaining dividends (the principle that the value of a stock is the sum of the present values of all of its anticipated dividends is discussed in Chapter 2). In the case of a dividend anticipated being paid on an option's underlying stock prior to an option's expiration, the option's underlying asset should be viewed as being a hypothetical asset having rights to all the dividends except I, with value $S_0 - I$.

Equation 1.16 expresses put-call parity in the case of a dividend-paying stock with dividends approximated as occurring in a continuous stream at the dividend yield rate, q:

$$\boldsymbol{C + Ke^{-rT} = P + S_0 e^{-qT}} \tag{1.16}$$

If put-call parity fails to hold in a well-functioning market then there would be an arbitrage opportunity. For example, suppose that $C = 3$, $S_0 = 31$, $T = 0.25$, $r = 10\%$, $K = 30$, and $q = 0$. What is the arbitrage possibility when $P = 2.25$?

The key to identifying the arbitrage trade based on put-call parity is always to buy a relatively underpriced asset and sell a relatively overpriced asset. Solving the put-call parity equation for P:

$$P = C + Ke^{-rT} - S_0$$

$$P = 3 + (30 \times e^{-0.25 \times 0.10}) - 31 = 1.25$$

If $P = 2.25$ in the market, then it is overpriced by \$1. The overpriced side should be sold and the underpriced side should be bought. Therefore, a speculator could sell the put for 2.25 and establish an offsetting position $C + Ke^{-rT} - S_0$ (buying a call and a bond while shorting the stock) for a net cost of 1.25. The resulting positions are fully hedged and generate a \$1 profit (+2.25 – 1.25) at the expiration date of the options.

Example 1.10: Suppose that $C = 3$, $S_0 = 31$, $T = 0.25$, $r = 10\%$, $K = 30$, and $q = 0$ What is the arbitrage possibility when $P = 0.25$? As before, the price of P based on put-call parity should be 1.25. Since it is 0.25 in the market, the put is underpriced by \$1. The underpriced put should be bought and the overpriced equivalent positions should be sold. Buying the put for 0.25 and establishing an offsetting position $S_0 - C - Ke^{-rT}$ (buying the stock while shorting the call and bond) for proceeds of 1.25. The resulting positions are fully hedged and generate a \$1 profit (+1.25 – 0.25).

While put-call parity holds for European options in a perfect market, in the case of American options (prices are written as c and p, respectively, to denote American options) on a non-dividend-paying underlying common stock the formula is based on inequalities:

$$S_0 - K \leq c - p \leq S_0 - Ke^{-rT} \tag{1.17}$$

Inequality 1.17 indicates that the spread between the call and put must lie between the present value of the immediate exercise payoff (the left side) and the present value of the payoff for exercise deferred until the option expiration date (the right side).

Finally, the primary usefulness of put-call parity is that given the price of a call (or put), the price of the put (or call) can be easily calculated using Equation 1.14 and the easily observable values of the stock and bond. Also, analytics such as risk measures for one type of option (e.g., a put) are calculated with ease using the risk measure for the other type of option (e.g., a call) and the put-call parity formula. Note that rearranging Equation 1.14 demonstrates that the *spread* between a call price and a put price (assuming no intervening dividends and European-style options), $C - P$, depends only on the stock and bond price, not directly on the anticipated volatility of the stock:

$$C - P = S_0 - Ke^{-rT}$$

When $S_0 = Ke^{-rT}$ the option prices will be the same, $C = P$, assuming of course that the options are European with the same strike price K, expiration date T, and underlying asset S. The condition $S_0 = Ke^{-rT}$ describes an option as being at the money only if $r = 0$. When $r > 0$, analysts refer to the condition $S_0 = Ke^{-rT}$ as describing an option that is *at the discounted money*.

A ***synthetic long position*** in an asset using options is the simultaneous purchase of a call option and writing of a put option on the same underlying asset with the same strike price and same expiration date. A ***synthetic short position*** using options reverses the signs of the options: the simultaneous writing of a call option and purchase of a put option on the same underlying asset with the same strike price and same expiration date. Note that the idea that a portfolio that consists of a long position in a call option and a short position in a put option (both European and with the same underlying asset, expiration date and strike price) has the same risk exposure as owning the underlying asset is consistent with put-call parity. The only difference between the synthetic long position and owning the actual underlying stock is that the synthetic position requires less cash. That is why the put-call parity relationship includes the present value of a zero coupon bond with a face value equal to the exercise price of the two options.

Note also that the synthetic long stock position using options is equivalent to a long position in a forward contract on the same asset or a financed position in the spot market (cash market).

Conversely, the synthetic short position using options is equivalent to a short position in a forward contract on the same asset or a short sale of the underlying asset into the spot or cash market with the proceeds of the sale being invested in a riskless bond.

Example 1.11: Suppose that p_1, p_2, and p_3 are the prices of three European put options with strike prices K_1, K_2, and K_3, respectively, and where $K_3 > K_2 > K_1$ and $K_3 - K_2 = K_2 - K_1$. All options have the same expiration date. Show that

$$p_2 \leq 0.5(p_1 + p_3)$$

We need to show that if $p_2 > 0.5(p_1 + p_3)$, then there is an arbitrage opportunity which is assumed to be impossible in a perfect market.

Assuming that $p_2 > 0.5(p_1 + p_3)$, a arbitrager can sell two puts from the expensive side, p_2 (the left side of the inequality), and buy the cheap right side (buy one put with a strike K_1 and buy one put with strike K_3). The arbitrager will receive the following payoff which can be shown to be non-negative:

$$2p_2 - p_1 - p_3$$

Table 1.3 demonstrates the payoffs of the put options at maturity over four ranges for S.

Note that the possible payoffs include two cases of 0 and two cases of positive values. There is a contradiction: the portfolio has a potential positive *payoff*, no potential payouts, but can be created today at a negative cost. This would mean that arbitrage would be possible; therefore, p_2 cannot be greater than $0.5(p_1 + p_3)$.

TABLE 1.3

Payoffs at Maturity Over Four Ranges for S_T

	Today (Negative for Money Out)	Payoff If $S_T < K_1$	Payoff If $K_1 < S_T < K_2$	Payoff If $K_2 < S_T < K_3$	Payoff If $K_3 < S_T$
Buy a put at K_1	$-p_1$	$K_1 - S_T$	0	0	0
Buy a put at K_3	$-p_3$	$K_3 - S_T$	$K_3 - S_T$	$K_3 - S_T$	0
Sell two puts at K_2	$2p_2$	$-2(K_2 - S_T)$	$-2(K_2 - S_T)$	0	0
Total		$K_1 + K_3 - 2K_2 = 0$	$K_3 + S_T - 2K_2 > K_3 + K_1 - 2K_2 = 0$	$K_3 - S_T$	0

1.7 Arbitrage-Free Binomial Option Valuation

One of the most fascinating models of arbitrage-free derivative valuation is a binomial option pricing model. A binomial model allows only two possible outcomes to the value of a share of stock over a single time period. Each possible outcome is termed a "state," which is often described as being an "up state" or a "down state."

For example, consider a share of stock currently trading at $10 per share that is viewed as having two possible outcomes at the end of one year: $16 if things go well (the up state) and $7 is things go poorly (the down state). We denote the current price of the stock as S_0, the value of the stock price in the up state as S_u, and the value of the stock in the down state as S_d. This simplest of all binomial models is illustrated in Figure 1.6 for a single time period.

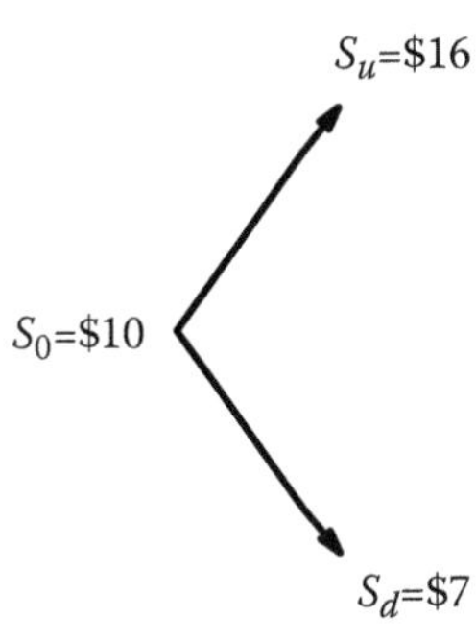

FIGURE 1.6
Stock Tree.

Now consider a call option on that stock that expires in one period and has a strike price of \$13. While we do not yet know the market value of that option, C_0, we do know that the value of the option at the end of the year (C_u in the up state and C_d in the down state) from the payoffs of \$3 in the up state (found as \$16 – \$13) and \$0 in the down state (because the option is not exercised) as depicted in Figure 1.7.

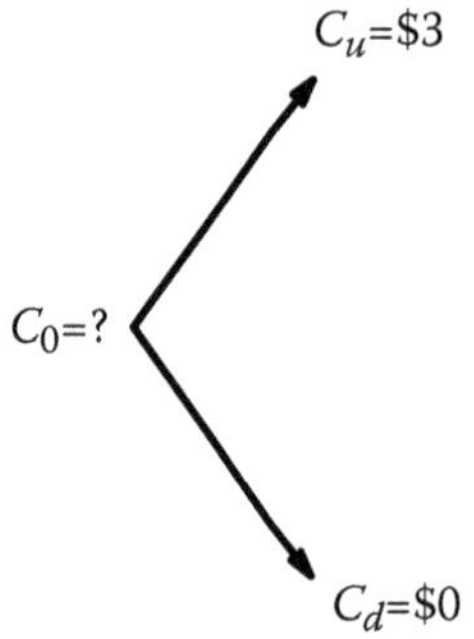

FIGURE 1.7
Option Tree.

Note that the stock price varies by \$9 between the two states (i.e., \$16 – \$7) at the same time that the option price varies by only \$3. The key to solving for the current option price (C_0) begins by realizing that the payoffs to the option are perfectly correlated with the payoffs to the stock. The only differences are that the stock price varies three times as much and the stock price has a worst case value of \$7 while the option has a worst-case value of \$0. This means that an arbitrager can construct two portfolios with identical payoffs, which is called *replicating* one portfolio using the other. Here, two portfolios are constructed: (1) one or more call options plus one riskless bond and (2) one share of the stock. The two portfolios are constructed to have the same payoffs at the option's expiration. If their payoffs are identical, then their current values must be identical.

To replicate the stock using a call option and a bond, the payoffs must be the same. The first step is to note that if the options expire worthlessly the riskless bond must have the same payoff as the stock in the down state (in this case \$7). Therefore, the riskless bond must have a face value of \$7. The second step is to determine the number of call options which, combined with the bond, will offer the same payout at the option's expiration as the stock in the upstate (\$16). The number of call options, h, must be such that:

$$\text{Payoff in upstate} = \$16 = \text{bond}+(h \times \text{call}) = \$7+(h \times \$3)$$

$$h = (\$16 - \$7)/\$3 = 3.$$

Specifically, a riskless bond with a payoff in one year of \$7 combined with three calls (with possible payoffs of \$3 each in one year) will offer the same payoffs as the stock:

Payoff of option and bond portfolio in up state = (3 × \$3) + \$7 = \$16.
Payoff of option and bond portfolio in down state = (3 × \$0) + \$7 = \$7.

Finally, let's assume that each period represents one year and that continuously compounded interest rates are 7.41% so that the price of a riskless bond that pays \$7 in one year is \$6.50 (found as \$7 $e^{-0.0741}$). Since the payoffs to the option and bond portfolio are identical to the payoffs of a single share of stock, the price to establish the portfolio with three options and one bond must equal the price of the stock:

$$\text{Price of option and bond portfolio} = S_0$$
$$(3 \times C_0) + \$6.50 = \$10$$
$$C_0 = (\$10.00 - \$6.50)/3 = \$1.17$$

Three call options costing a total of \$3.50 plus one bond costing \$6.50 sums to the same price as the share of stock (\$10) and offers the same payoffs (\$7 or \$16) depending on the state of the world).

Note that the computation of the current call price was driven by the current price of the stock, but that the option pricing problem did not indicate the probability that the stock price was going to rise or fall. Thus, the model does not require or describe the expected value or return of the stock in the real world. This simple example demonstrates a point so important that it is discussed in more detail below for emphasis.

The Binomial Model and Risk-Neutrality Modeling

Here is the amazing breakthrough that the example using the binomial model illustrates. The solution to the option price did not depend on the probabilities that the stock price will rise or fall in the real world. The probabilities of each state, if known, would reveal the stock's expected price and expected return. But neither the probabilities nor expected spot prices were needed to find the option price. In other words, the option price is *independent* of the expected rate of return in the underlying stock in the real world. Since in general the prices of financial assets depend on risk premiums, which in turn depend on the risk of the financial asset, this means that, under some conditions, the price of an option is independent of the *risk premiums*! In other words, option prices do not depend on the degree to which markets require higher returns for bearing risk. In fact, the price of this option under any assumption regarding risk aversion must equal the price of the option when there is no risk aversion (i.e., risk neutrality). This breakthrough in financial economics is entitled risk-neutral modeling and it has changed the world by simplifying derivative pricing, which in turn has tremendously facilitated risk management and risk control by simplifying the pricing of the products (i.e., derivatives) that risk managers use. In many cases, the equilibrium values of derivatives in the real world can be determined using riskless interest rates and the assumption that market participants are risk-neutral since risk aversion does not alter the price of the financial derivative.

One final note on the binomial option pricing example: the above option pricing model determines one price (i.e., the current call price) based on the value of another price (i.e., the current stock price) and is therefore a ***partial equilibrium model*** because it only explains some of the prices. A model that uses exogenous information such as the preferences of all participants and the productive capabilities of all participants to generate the prices of all goods is termed a ***general equilibrium model***. Because derivative payoffs are derived from the potential prices of a market-traded asset, the partial equilibrium models to value the derivatives can be quite simple and robust. Simply put, the necessary information regarding risk and expected returns to value the option is contained in the price of the underlying asset.

1.8 Compound Options and Option Strategies

Option traders often establish single long or short positions in call options or put options on various individual assets. But other traders establish multiple option positions on the same asset. This section discusses many of the most common such strategies. This section only discusses strategies in the context of a single underlying asset. Obviously there are many strategies involving options that can span two or more underlying assets.

1.8.1 Popular Option Strategies with Multiple Positions

Two of the most popular strategies involving options are covered calls and protective puts. Both of these strategies were illustrated in the examples in Section 1.5.1. A ***covered call*** is the simultaneous writing of a call option while having a long position in the underlying asset. A ***protective put*** is the simultaneous purchase of a put option while having a long position in the underlying asset. These two portfolios are depicted in Figure 1.8 and Figure 1.9.

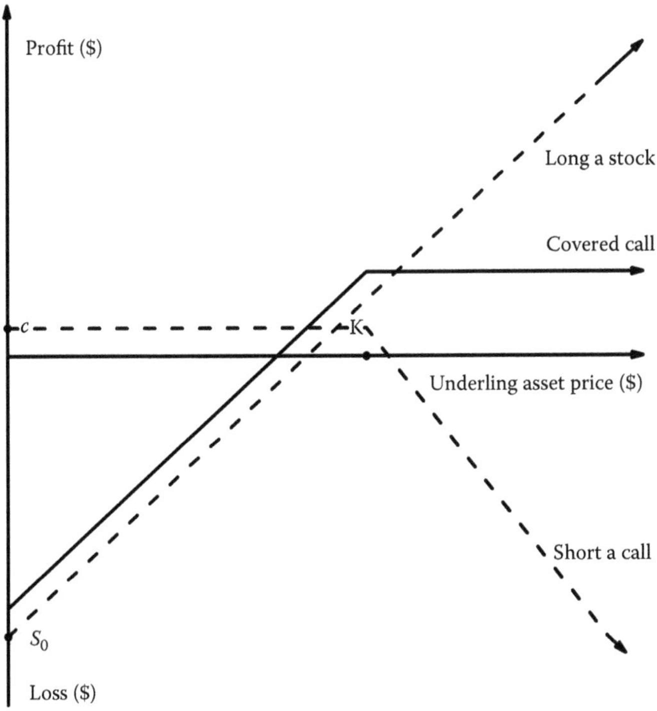

FIGURE 1.8
Covered Call.

Note that the portfolio profits and losses in Figures 1.8 and 1.9 can be formed by summing the profits and losses of the portfolio's component positions. In other words, at each point on the horizontal axis (i.e., each value of S_T), the profit or loss of the covered call or protective put is found by summing the profits and/or losses of the two positions that form them.

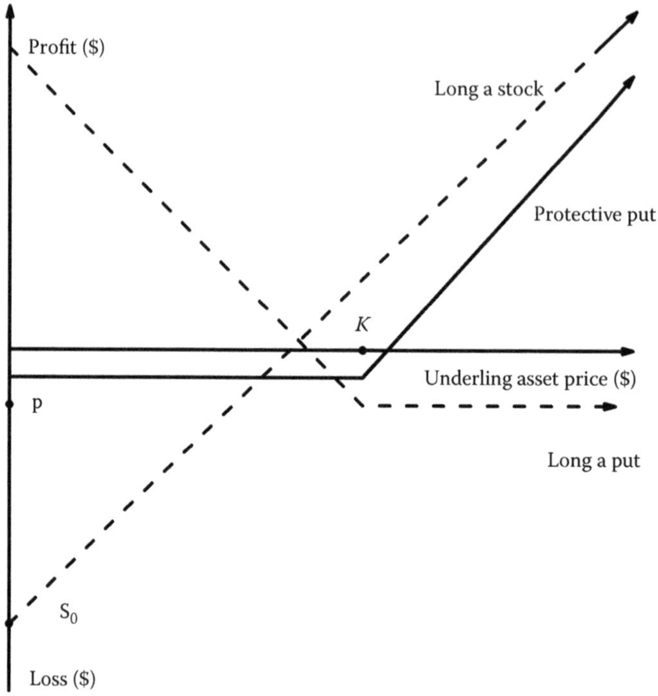

FIGURE 1.9
Protective Put.

1.8.2 Option Spread Strategies

Option spreads are simultaneous long and short positions in either different call options or different put options (but not both calls and puts). Three general types of option spread strategies are vertical spreads, horizontal spreads, and diagonal spreads. The terms describing the three spreads relate to the visualization of a matrix of option prices with strike price forming the vertical axis and expiration date forming the horizontal axis. Thus, in a ***vertical option spread,*** the options differ by strike price; in a ***horizontal option spread,*** the options differ by expiration date; and in a ***diagonal spread,*** they differ by both strike price and expiration date. This section focuses on the strategies with the same expiration date – vertical spreads.

Figure 1.10 illustrates a vertical call spread known as a ***bull spread***. The payout of a bull spread at expiration is positively related to the underlying asset, hence it is "bullish" with respect to the price of the underlying asset. The bullish nature of a bull spread occurs when the long call option position is in the option with the lower strike price and the short option position is in the asset with the higher strike price. Interestingly, the same bullish diagram can be generated using put options with the same

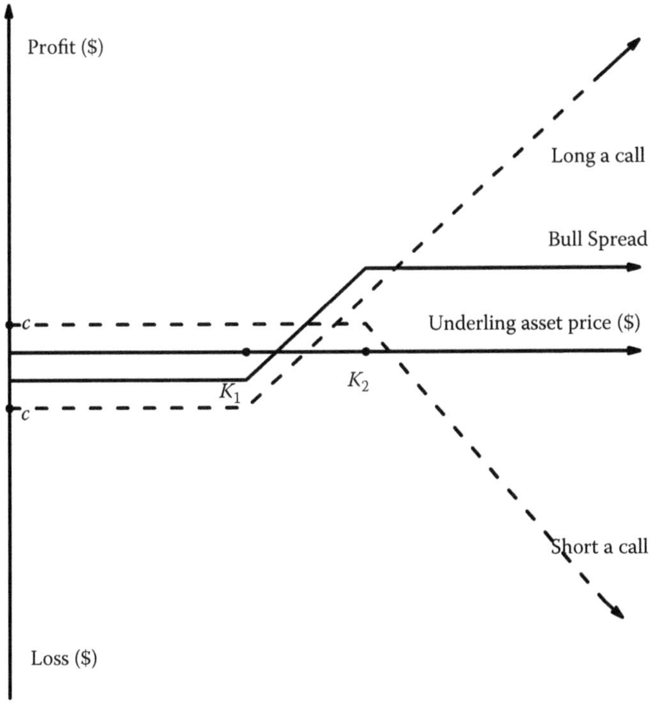

FIGURE 1.10
Bull Spread.

structure: the long put option position is in the option with the lower strike price and the short option position is in the asset with the higher strike price.

Figure 1.11 illustrates a vertical call spread known as a ***bear spread***. Bear spreads reverse the direction of the options by establishing the long call option position in the option with the higher strike price and the short option position is in the asset with the lower strike price. Like in the case of bull spreads, the same bearish diagram can be generated using put options with the same structure: the long put option position is in the option with the higher strike price and the short option position is in the asset with the lower strike price.

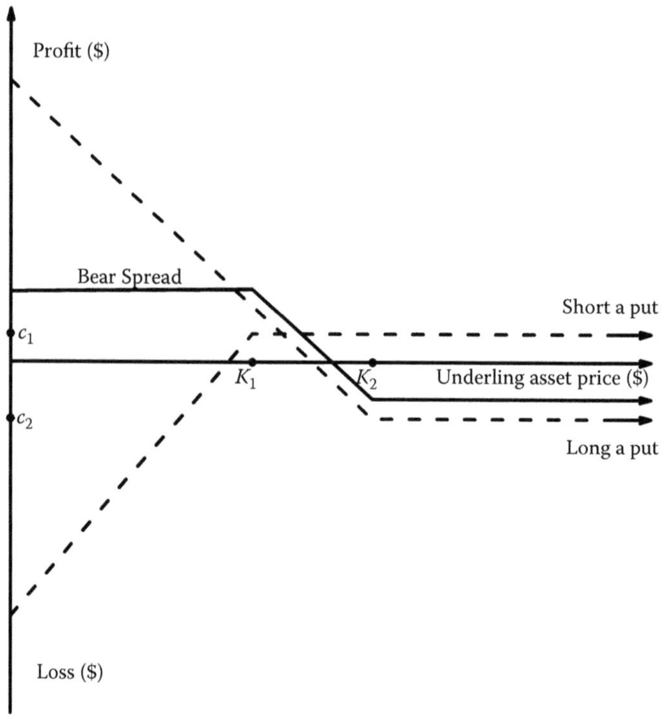

FIGURE 1.11
Bear Spread

1.8.3 Option Combination Strategies

Option combinations involve simultaneous positions in at least one call option and at least one put option. This section focuses on combinations of calls and puts with the same expiration date and underlying asset. The synthetic long positions in the underlying asset using options (long a call and short a put) and short positions (long a put and short a call) discussed in Section 1.8.1 are option combinations. This section discusses others.

Option straddles have equally sized simultaneous long positions or simultaneous short positions in call and put options with the same strike price (and same expiration date). Figure 1.12 illustrates the profitability of a long straddle at expiration with a solid line and shows the underlying components of long a call and long a put with dashed lines. The value of a long straddle is positively exposed to the volatility of the underlying asset. A short straddle is the mirror image (i.e., the maximum profit is above K with losses on the far right and left).

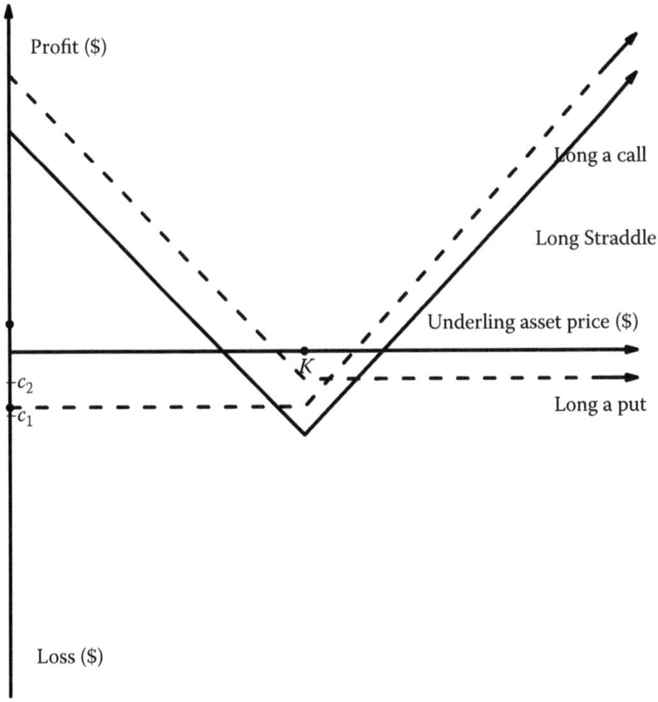

FIGURE 1.12
Long Straddle.

Option strangles are similar to option straddles, except the call and put options have different strike prices. Figure 1.13 illustrates a long strangle. Other options strategies can involve option portfolios such as ratio spreads with more calls than puts or vice versa such that the payoffs differ in different directions.

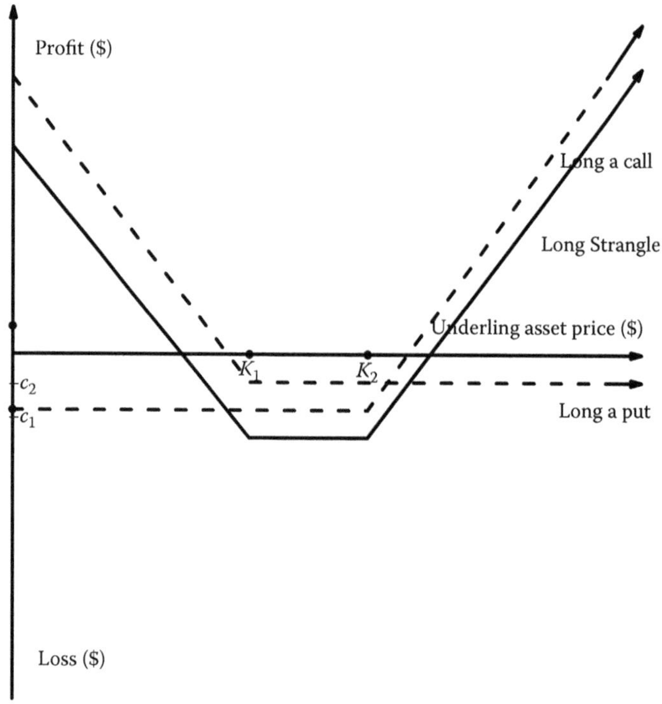

FIGURE 1.13
Long a Strangle.

1.8.4 Other Options

There are many types of stand-alone options on financial assets that differ from simple calls and puts. Some options allow exercise at several specific points in time (a Bermuda option), some are based on functions of prices such as averages or extremes (e.g., ***Asian options*** have payoffs based on averaged prices of the underlying asset). Some options cease to exist if the underlying asset reaches a particular level or become exercisable if the underlying asset reaches a particular level such as ***barrier options***.

Financial markets and economic activity in general are full of options. There are options on real assets such as options to: buy land, rent space, return products, extend warrantees, take early retirement, terminate contracts, and on and on.

There are non-traded financial options such as options to pre-pay mortgages and other loans, options to break some bank certificates of deposits (i.e., receive early termination), options to rollover some bank deposits, options to cash out insurance policies, options to increase insurance coverage, and so forth.

The common stock of a leveraged corporation can be viewed as a call option on the corporation's stock. There are even options on options known as ***compound options***.

Two things to keep in mind are this: (1) there are countless implicit and explicit options in a modern economy because those options serve important purposes of allowing participants to manage and control their risk exposures and (2) the astounding variety and complexity of these important contracts creates demand for people with mathematical modeling skills, especially those who can create innovations or at least understand and model the innovations. This book touches on the most common options in existence today. No one can yet imagine the options that will emerge in the future to meet the changing needs of our rapidly changing world.

1.9 The Limits to Arbitrage and Complete Markets

The models and intuition of earlier sections often relied on or described prices in frictionless markets wherein numerous investors such as speculators and arbitragers compete to earn higher returns. Financial economists often describe this condition of informational market efficiency as being when all available information becomes reflected in market prices such that it is not possible to utilize that information to consistently earn a risk-adjusted abnormal profit (i.e., market participants cannot consistently identify mispriced securities).

A well-recognized problem with the theory of informationally efficient markets is that if available information is instantaneously incorporated into market prices then there will be *no incentive* for market participants to gather information and integrate that information into their investment decisions. If no one searches for mispriced securities then prices will not be efficient. In a perfectly efficient market everyone would adopt passive investment strategies which are buy-and-hold strategies with no attempt to trade in an effort to gain from mispriced securities.

Clearly no market can be perfectly efficient. The only meaningful issue is the extent to which markets approach informational efficiency.

The concept of *inefficiently efficient markets* is that securities are mispriced just enough and just often enough to attract a moderate number of active investment managers (managers that execute trades for the purpose of trying to improve risk-adjusted return) and active individual investors. The benefits and costs of active investing reach an equilibrium that results in a level of market inefficiency that sustains this equilibrium level of information analysis.

The primary purpose of derivatives is to facilitate risk management. Financial derivatives help to *complete* a market. Perfect completion of a market means that there are enough distinct investment opportunities available that investors can establish long and short positions in existing securities in a way that allows them to position their portfolio exactly as they desire. As an example, if a grocery store mixes apples, bananas, and cherries into three different types of fruit baskets, a customer may be inconvenienced by being unable to purchase one basket with exactly the amount of each type of fruit that she desires unless by chance one of the baskets exactly meets her preferences. However, if a customer is allowed to trade with the store and can buy and sell the different types of fruit baskets without trading costs, she will be able to obtain exactly the numbers of each type of fruit that she desires so long as the number of distinct fruit baskets being traded equals the number of different types of fruit (i.e., the market is complete). In a similar way, derivatives are created to move the market closer to completion so that market participants are better able to establish positions that move the participants closer to their desired risk exposures.

Financial derivatives can also be used to provide arbitragers, speculators, and investors with powerful tools with which to attempt to enhance their risk-adjusted returns through superior processing of available information. When those market participants with the greatest abilities to identify mispriced securities are enabled with superior tools such as derivatives to best utilize their abilities to buy underpriced assets and sell overpriced assets, the market prices of assets will tend to be better driven toward their intrinsic values. Because market prices provide the signals that guide production and consumption decisions throughout an economy, these arbitragers, speculators, and investors are unwittingly driving the decision making throughout the entire economy into being more and more efficient. Therefore, derivatives can play a role in increasing the efficiency in production and consumption decisions which in turn means improved economic growth and economic utility.

1.10 Computing Systems: Mathematica and R

Mathematica and R are two programming languages used in this book. The logical structure of the programming languages is quite similar. The sample codes guide the user to understand the purpose of the codes. This book does not spend a lot of time on basic programming language skills such as how to use an "*if* statement" and how to use a "*for* loop." However, here is a brief introduction to Mathematica and R.

Mathematica has a help button through which users can find the syntax of codes and a lot of helpful information. In R, type in "? help" to obtain information on how to find the syntax of codes and a lot of other helpful information.

- Mathematica
  Wolfram Mathematica is a modern technical computing system that is used in many mathematical, computing, scientific, and engineering fields. Mathematica is widely used in college/university calculus and engineering courses. However, a Mathematica license is required. Many colleges and universities have an on-site license available for students. If not, Mathematica desktop can be purchased at $160 per year.
- How to run Mathematica

1. Double-click the left mouse button (i.e., press the button two times, relatively quickly, in succession) when the arrow is on the *Mathematica* icon.
2. It is a good idea to begin each of your sessions by giving Mathematica a simple command to execute. (This will load the kernel, the part of the program that does all of the work.) Type:

   1 + 1

   and then press the regular ENTER key while you are holding down the SHIFT key. You can use CTRL-C (simultaneous pressing of the Ctrl and C keys) to copy any Mathematica codes in electronic copies of this book and use CTL-V to paste the codes in Mathematica. Then, press the regular ENTER key while you are holding down the SHIFT key to run the code.

- R
  R is a programming language and free software environment for statistical computing and graphics supported by the R Foundation for Statistical Computing. R is widely used in college/university statistics and data science courses. Developed primarily for use by members of academia, its use has grown over the years. Employees of leading technology firms often use R. Stock traders often use R to process transactions. A recent survey of job advertisements that sought specialty in a form of analytical software showed that R was the fifth most sought-after specialty (ranking behind only Java, SAS, Python, and C).
- How to install R

Download R to any personal computer. R is supported for Windows, Mac, and Unix platforms. Simply go to the website http://www.r-project.org and follow the directions to download the program:

1. You need to select a CRAN mirror. Click on CRAN on the left panel. CRAN is a collection of sites that carry identical materials and were created as mirror sites to lessen the load on any one server.

2. Click on one of the USA links, such as http://lib.stat.cmu.edu/R/CRAN/. The link takes you to Carnegie Mellon University's Statlib mirror site. In theory, all the USA CRAN mirrors should work.
3. In the "Download and Install R" box, click on the Download R for Windows link (if using Windows).
4. Click on the "base" link.
5. Right-click on the Download R 3.6.1 for Windows link and choose "Save Link As".
6. Save the .exe file to your preferred location, or simply on your Desktop.
7. Double-click on the .exe icon and follow the instructions. Only run this for a Windows-style installer. It contains all the R components, and you can select what you want installed. When asked to "Select the components you want to install", choose the (default) "User" installation. Don't worry about "customizing the startup options". In general, you should install R perfectly by just clicking on the "Okay", "Next", or "Finish" buttons at each step and letting the R setup use the default choices.

For Mac users, in Step 3 choose "Download R for MacOS X", and follow a similar process.

- How to run R?

Once R is opened, there is a screen called "R Console" with some quick descriptions of R. There is also a ">" on the screen. Use Ctrl-C to copy any R codes in this book (if electronic) and use Ctrl-V to paste the codes after ">". There is no need to run the code as it will automatically run.

One important thing is that if there are some packages you wish to install, click on "Packages" and "Install packages" (usually pick USA(PA 1)). All the packages are listed. For example, "quadprog" is a commonly used package for Chapter 6 application. Click on quadprog, then go to "Packages", pick "load package", pick "quadprog", and the package should be loaded. You can copy and paste any code in the book; for example, run "R Code 6.2: Find Markowitz Frontier with no risk-free asset, short sales allowed". It should generate a hyperbola like the one shown as Figure 6.8 in the book.

Chapter Summary

Chapter 1 introduces foundational concepts in financial markets and provides an introduction to derivative analysis using forward contracts and option contracts. It starts with the role of real assets, financial assets, financial mathematics, and financial derivatives, as well as arbitragers, hedgers, and speculators. It then discusses principles of asset valuations. Although expected value is a foundation of asset pricing, we also need to include the time value of money (by discounting future cash flows) and adjust for risk by incorporating a risk premium to reflect investor risk aversion to the extent that the future cash flows are risky. Chapter 1 introduces several financial derivatives: forward contracts, call options, and put options. The intuition and application of put-call parity is introduced, along with basic derivative trading strategies and the concept of arbitrage opportunities. Finally, a simplified model of equity prices is used to demonstrate that the equilibrium market price of a call option depends on the price of the underlying stock but does not otherwise depend on the market's aversion to the risk of the stock. Therefore, the value of an option relative to the price of the stock is the same in a world with risk-neutral investors as it is in the real world in which investors are risk-averse. This concept of the risk-neutral principle is stressed in Chapter 1 and in various derivative valuation models throughout the book. The ease and precision of risk-neutral financial derivative valuation is central to the crucial role that financial derivatives serve as effective risk management tools in an increasingly complex (and productive) economy.

Chapter Demonstrating Exercises

A U.S.-based export firm will receive 10 million British pounds in three months. The firm can tolerate an exchange rate between $1.25 to $1.34 U.S. dollars per pound to convert the pounds to its domestic currency (U.S. dollars), but is unwilling to bear the risk of converting the foreign exchange to U.S. dollars at an exchange rate of $1.25 or lower. On the other hand, the firm will be very content with an exchange rate of $1.34 per British pound. How can a financial derivative strategy be designed to meet the needs of this U.S. export firm?

To hedge the downside risk, the firm can enter an option contract to exchange the 10 million British pounds for U.S. dollars at $1.25 per pound. The company can raise some or all of the money to finance this option purchase by writing a call option allowing the option buyer to convert 10 million British pounds to U.S. dollars at a rate of $1.34. From the perspective of U.S. dollars the option with a strike rate of $1.25 might be more clearly viewed as a long put (allowing the U.S. company to put foreign exchange for U.S. dollars) and the option with a strike price of $1.34 is short a call (allowing the counterparty to buy the British pounds from the U.S. company for $1.34 per pound). But in practice the conventions for designating FX options as calls or puts vary and are driven by local conventions.

Strategy: Long a 1.25$/pound put, short a 1.34$/pound call

Table 1.4 illustrates financial outcomes over the range of possible exchange rates.

TABLE 1.4

Payoff per Pound Unit (Units in Tens of Millions)

	$S_T < 1.25$ $/pound	$1.25 < S_T < 1.34$ $/pound	$S_T > 1.34$ $/pound
Long a 1.25 $/pound put	$1.25 - S_T$	0	0
Short a 1.34 $/pound call	0	0	$-(S_T - 1.34)$
In exchange market, sell pound	S_T	S_T	S_T
Net	1.25	S_T	1.34

End-of-Chapter Problems (Fundamental Problems)

1. Abbi plays a new type of game with Betti. Abbi tells Betti that she will give Betti X^2 dollars where X is the result of a single observation of a variable Exp(2), which is an exponential distribution with parameter of 2. Betti must make a payment (i.e., buy a ticket) to play.
   a. What is the fair price for the ticket ignoring the valuation of time and risk?
   b. If the payment is delayed for one year, what is the fair price for the ticket ignoring the valuation of risk while assuming that the risk-free rate is 5% a year?
   c. If Cindy offers a deterministic payment of the value $E(X^2)$, compare Abbi's offer with Cindy's offer. If Betti is a risk-averse investor, which offer will Betti take? What ticket price does Abbi need to offer in order to attract Betti to the game?

2. Abbi allows Betti to roll a die up to a maximum of three times. After each roll, Betti needs to decide if this is her final roll. Once she decides, the final roll outcome will determine the payoff and the game is over. Abbi will pay Betti $1, if 1 appears; $2 if 2 appears; and so forth up to $6 if 6 appears. Betti must pay for the ticket today and will receive the payoff today. What is the fair ticket price? What is the fair ticket price if Betti will not receive the payoff until one year later? Find those fair ticket prices for each of the following strategies:
   a. If the first roll is less than or equal to 3, Betti continues. If the second roll is less than or equal to 3 Betti continues.
   b. If the first roll is less than or equal to 3, Betti continues. If the second roll is less than or equal to 4 Betti continues.

c. If the first roll is less than or equal to 4, Betti continues. If the second roll is less than or equal to 3 Betti continues.

d. If the first roll is less than or equal to 4, Betti continues. If the second roll is less than or equal to 4 Betti continues.

3. What is a P measure and what is a Q measure and what is their difference? Which is more commonly used in derivative valuation?

4. The spot price of an asset (i.e., an investment) that provides no dividend or interest income is $30 and the risk-free rate for all maturities (with continuous compounding) is 10%.

   a. What, to the nearest cent, is the asset's three-year forward price?

   b. Next, assume that the asset provides an income distribution of $2 at the end of the first year. What, to the nearest cent, is the three-year forward price?

5. The price of a stock is $35 and the price of a three-month call option on the stock with a strike price of $36 is $1.80. Suppose a trader has $3,600 to invest and is trying to choose between buying call options on 2,000 shares or buying 100 shares of stock.

   a. Do a scenario analysis on four different stock prices at maturity (at a three-month expiration date): $51, $41, $31, and $26.

   b. How high does the stock price have to rise for the investment in options to be equally profitable as the investment in the stock in terms of total dollar profit?

6. Assume that an investor has a little over $5,000 and is confident that Coca-Cola stock is going up in three months. However, the investor worries about the downside risk. Assume that KO stock price is $50 today, a three-month-strike 50 call price is $1.68 per share, and a three-month-strike 50 put price is $1.31 per share. Assume there is no time value of money (i.e., $r = 0$).

   a. Investment manager A suggests buying 100 KO stock shares and buying the KO puts on 100 shares. How much is today's cash outflow? How much is the value (i.e., payoff) of the puts three months later if KO goes up to $60? What is the net gain/loss for this strategy? How much is the payoff of the puts three months later if KO goes down to $45? What is the net gain/loss for the strategy? How much is the payoff of the puts three months later if KO stays at $50? What is the net gain/loss for the strategy?

   b. Investment manager B suggests buying the KO calls on 1,000 shares and buying the KO puts on 1,000 shares. How much is today's cash outflow out? How much is the payoff of calls and puts three months later if KO goes up to $60? What is the net gain/loss for this strategy? How much is the payoff of the calls and puts three months later if KO goes down to $45? What is the net gain/loss for this strategy? How much is the payoff of calls and puts three months later if KO stays at $50? What is the net gain/loss for this strategy?

c. What recommendation would you give to the investor based on your analysis?

7. A one-year European call option on a stock with a strike of $30 costs $3, and a one-year European put option on the stock with a strike price of $30 costs $4. The risk-free interest rate is 10% (continuously compounded). The current stock price is $28.

   a. Identify the arbitrage opportunity. Use a table to prove the arbitrage.

   b. Suppose that a trader buys two call options and one put option. What is the trader's payment today? What is the break-even stock price at which the trader makes no profit or loss? (Don't forget to include the trader's payment today but don't consider the time value of money.)

8. The price of a European call option on a non-dividend-paying stock with a strike price of $100 is $9. The price of a European put option on the same stock with a strike price of $100 is $1. The stock price is $102, the continuously compounded risk-free rate (for all maturities) is 6% and the time to expiration of the options is one year. Identify the arbitrage opportunity. Use a table to demonstrate the arbitrage.

9. Assume that it is April and a farmer plans on harvesting and selling her crop of corn in November. She does not like the uncertainty regarding the future price of corn.

   a. If the farmer knew how many pounds of corn she will harvest and sell, what is the appropriate hedge strategy?

   b. If the farmer does not know how many pounds of corn she can sell, what is the hedge strategy? Assume that she has an estimated range of the harvest.

   c. If the farmer knows how many pounds of corn she can harvest and sell and she is willing to pay some immediate premium so that she doesn't lose her upside potential in case the price of corn goes up, what can she do with derivatives to modify the "perfect" hedge strategy discussed in (a)?

10. A U.S.-based exporter will receive 50,000 Euros in three months. The firm does not like the U.S.-dollar uncertainty of the value of the Euros. Compare the advantages and disadvantages of the following strategies (assuming that derivative contracts in various jurisdictions

are based on their home currencies). Assume that the current spot and forward prices of Euros in terms of U.S. dollars is $1.20.

a. Sell (i.e., establish a short position in) a three-month forward on 50,000 Euros

b. Buy a three-month put on 50,000 Euros in the United States.

c. Buy (i.e., establish a long position in) a three-month forward on 60,000 US dollars in Europe.

11. An investor bets that MSFT is going up in three months by buying a large number of MSFT shares. If the investor doesn't want to take enormous down side risk, what strategy can the investor use:

    a. if the investor is willing to pay some upfront premium?

    b. if the investor does not want to pay any upfront premium?

    c. and what are the advantages and disadvantages to the above two strategies?

12. An investor owns a diversified portfolio benchmarked to S&P 500 index. The investor believes that the market is likely to go up for the next three months, but does not want to take downside risk over those three months. The investor is considering either buying some three-month puts on the S&P 500 index or establishing a short position in forward contracts on S&P 500 index. Which strategy is more appropriate and why?

13. An investor shorts a stock (based on a negative opinion about the prospects of this particular company), but worries about the overall market rising. What derivative-based strategy should he/she take?

14. An investor has a position designed to benefit if the stock market goes up during the next month. But the investor does not want to take large risks that the stock market may move to the downside. What derivative-based strategy can he/she take?

15. An investor thinks the stock market is either going to move down or remain unchanged during the next few months. The investor seeks a derivative-based strategy that best benefits if the investor's predictions are correct and that generates some cash now. What strategy can the investor take? What is the risk of taking this strategy?

16. A stock price is currently $50. Assume that over the next three-month period it is certain that the stock will either go up by 10% or down by 10%. The risk-free interest rate is 10% per annum with continuous compounding.

    a. What is the value of a three-month European put option with a strike price of $52?

    b. How many shares of stock could hedge a position in this put option?

17. The current U.S. dollar to British pound exchange rate is 1.31\$/£. An investor wants to gain if the future exchange rate is above 1.35\$/£ or below 1.27\$/£. The investor is willing to take some relatively minor losses if the exchange rate is between \$1.27 and \$1.35. What strategy should the investor take? Use a table to show that your strategy generates the desired exposures.
18. The current U.S. dollar to British pound exchange rate is 1.31\$/£. An investor wants to benefit if the exchange rate rises above 1.35\$/£ and is willing to be exposed to losses if the rate falls below \$1.27. What strategy can the investor implement? Use a table to demonstrate that the strategy meets the investor's preferences.
19. If an investor wants to receive the following payoff with ($K_1 < K_2$): $\text{Max}(K_1 - S_T, 0) - \text{Max}(S_T - K_2, 0)$.
    a. What strategy should the investor take?
    b. Draw a table to analyze the payoff while, $S_T < K_1$, $K_1 < S_T < K_2$, and $S_T > K_2$.
    c. Is this strategy very risky? Why or why not?
20. If an investor wants to receive the following payoff with ($K_1 < K_2$): $\text{Max}(S_T - K_1, 0) - \text{Max}(S_T - K_2, 0)$
    a. What strategy should the investor take?
    b. Draw a table to analyze the payoff while, $S_T < K_1$, $K_1 < S_T < K_2$, and $S_T > K_2$.
    c. Is this strategy very risky? Why or why not?

Notes

1 Hull, John C. Options, Futures, and other Derivatives, Ninth Edition. 2015, Pearson Education.

2 Where $E_Q(S_T)$ is the expected value based on Q measures (i.e., based on risk-neutral modeling).

2

Introduction to Interest Rates, Bonds, and Equities

This chapter reviews the foundations of investment analysis with a focus on the two primary types of traditional investments that underlie many financial derivatives: (1) ***fixed income securities,*** which provide investors a return in the form of fixed periodic payments and eventual return of principal at maturity (e.g., bonds), and (2) ***equities,*** which represent ownership interest of a company (i.e., stock). Accordingly, this chapter focuses more on finance than mathematics. An important emphasis is on financial terms. The end-of-chapter exercises are straightforward and do not require sophisticated mathematical tools. Later chapters introduce a wide spectrum of financial products and demonstrate the use of mathematical applications to address the challenges of valuation, risk management, and optimization. Students and practitioners who are familiar with fixed income securities and equity securities may skip this chapter. Instructors might cover Sections 2.1–2.3 in class while assigning Sections 2.4–2.5 as a reading assignment. Section 2.5 is related to a later chapter about portfolio management.

2.1 The Time Value of Money and Interest Rates

The ***value*** of a financial product is a subjective measure of worth which could be derived from mathematical models. The ***price*** of a financial product is based on observations of bids, offerings, and/or transactions in financial markets. Technically speaking, they are different. A market price is based on supply and demand, and may differ markedly from opinions with regard to value. However, in equilibrium, market prices should converge to widely accepted opinions with regard to value. Because most financial derivatives trade in well-functioning markets and are valued using widely accepted derivative pricing models, there is less need to distinguish carefully between the terms *value, price, market value,* and *fair value.* For example, scholars often describe derivative valuation models as pricing models. Because of this, we may at times use the terms interchangeably.

Most modern money has no value other than what people will accept in exchange for it. The value of money changes through time due to changes in its supply and demand. The value of a dollar today is obviously \$1. But what about the value today of a dollar that won't be received for 5, 10, or 20 years – how much is it worth? The ***time value of money*** refers to the idea that the exchange ratio between the values of money to be received at different points in time is usually not exactly one. Money that won't be received for very long periods of time is typically only worth a few cents on each dollar. Note that a ***perpetuity*** – a promise of periodic payments forever – is worth a finite amount even though the sum of the promised cash flows is infinite.

Money such as U.S. dollars at different points in time should be viewed as different commodities (e.g., apples vs. oranges) rather than as the same asset. Just like in the case of apples vs. oranges, decisions about money at different points in time are based on the ratios of their market prices. The price ratios of money at different points in time are expressed using ***interest rates***. Thus, if \$1 due in one year is worth only \$0.80 today (i.e., its present value is \$0.80), the one-year interest rate would be 25% because \$0.80 grows to \$1.00 with an annual growth rate of 25%. In this case, the time value discount factor would be 0.80. ***Time value discount factors*** can be used to find the present value of a future cash flow by multiplying the future cash flow by the time value discount factor (e.g., 0.80) to get its present value.

Note that interest rates and discount factors are equivalent metrics for describing the time value of money – if we know the discount rate, we can derive the corresponding interest rate and vice versa.

2.1.1 Future Values and Interest Rates

A ***future value*** is the amount of money at a future point in time that has the same market value as a given amount of money available today. For example, market prices might indicate that the value in exactly one year of \$1.00 invested today is \$1.08, and that the value in two years of \$1.00 invested today is \$1.17. Under these conditions, the marketplace is telling us that \$1.00 today is equivalent to \$1.08 in one year, and that \$1.00 today is equivalent to \$1.17 in two years. We can utilize the financial marketplace to turn any present amount of dollars into future amounts of dollars. A student who borrows \$50 for a date as a freshman in college by charging the evening to a credit card might find that four years later that debt might grow to \$100 if unpaid due to interest. ***Interest*** is the amount of money a borrower promises to pay the lender to compensate them for the use of the money. If unpaid, and allowed to continue to grow on the credit card (with interest being charged on accrued interest), that debt could easily reach \$40,000 by retirement.

Interest rates are a type of price ratio that can serve as a useful metric for the time value of money. For example, credit card companies often charge an interest rate of about 18%. The student that charges \$50 today would owe \$50 × 1.18, or \$59 after one year. A student charging \$100 today and owing \$120 after one year would experience an interest rate of 20%. Note that *interest* is the amount of money charged by the lender and paid by the borrower as compensation for the time value of money, which is \$20 in this example.

It is important to view interest rates as a way of expressing price ratios rather than as the underlying basis for making economic decisions. People make decisions on prices. A person deciding whether or not to spend money using a credit card should compare the price of paying now versus the price of paying later. The interest rate serves merely as a convenient metric for making that comparison. Note that given today's price, and future value, we can derive its corresponding interest rate. Conversely, given an interest rate and a future value, we can find today's price.

2.1.2 Interest Rates, Future Values, and Compounding Methods

The previous discussion made interest rates sound easy: the interest rate that equates \$100 today with \$120 in one year was reported as being 20%. But interest rate methods are made complex by compounding. ***Compounding*** is the practice of interest being assessed on interest. It makes economic sense that the dollar compensation received for lending or paid for borrowing should be adjusted through time as the size of the accrued money grows. Ignoring compounding is known as using ***simple interest*** and is sometimes used for low interest rates and short periods of time. Simple interest is illustrated in Equation 2.1, defining T as the number of years between the present and future values:

$$\boldsymbol{FV_T = PV \times (1 + rT)} \tag{2.1}$$

where FV_T is the future value in T years, PV is the present or current value (year 0), and r is the annual interest rate. For longer-term time horizons (T), it makes sense for interest to be compounded periodically such as ***annual compounding*** which is depicted in Equation 2.2:

$$\boldsymbol{FV_T = PV \times (1 + r)^T} \tag{2.2}$$

Example 2.1: If T is 2 and r is 8%, find the future value of \$1,000 assuming annual compounding.

$$FV_2 = PV(1 + r)^2 = \$1{,}000(1.08)^2 = \$1{,}000(1.1664) = \$1{,}166.40$$

Financial calculators and spreadsheets make these computations and the ones that follow very easy.

Interest is such an important expression of the value of money through time that in most financial practices compounding is performed more often than annually. Equation 2.3 expresses a general formula for *discrete* compounding called ***m periods per year compounding.***

$$\boldsymbol{FV_T = PV \times [1 + (r/m)]^{mT}} \tag{2.3}$$

where m is the number of compounding periods per year (e.g., $m = 12$ denotes monthly compounding which is common in mortgage computations). Banks often use daily compounding ($m = 365$).

Example 2.2: If T is 2, r is 8% and $m = 365$, find the future value of $1,000.

$FV_2 = PV[1 + (r/m)]^{m\times 2} = \$1{,}000(1 + (0.08/365))^{2\times 365} = \$1{,}000 \times (1.17349) = \$1{,}173.49$

Continuous compounding lets the number of periods of compounding, m, go to infinity:

$$FV_T = PV \times \lim_{m\to\infty}\left[1 + \frac{r}{m}\right]^{mT}$$

In calculus, there is a famous limit:

$$\lim_{x\to\infty}\left[1 + \frac{1}{x}\right]^{x} = e$$

It is easy to derive the above limit using the following:

$$\lim_{m\to\infty}\left[1 + \frac{r}{m}\right]^{mT} = \lim_{m\to\infty}\left[1 + \frac{r}{m}\right]^{\frac{m}{r}\times rT} = \left(\lim_{m\to\infty}\left[1 + \frac{r}{m}\right]^{\frac{m}{r}}\right)^{rT} = \boldsymbol{e^{rT}}$$

here we let $x = \frac{m}{r}$.

Equation 2.4 uses that limit in the formula for continuous compounding. It is often used in academic research because of its simplicity and is the primary compounding assumption of this book.

$$\boldsymbol{FV_T = PV \times [e^{rT}]} \tag{2.4}$$

Example 2.3: If T is 2, r is 8%, and m approaches infinity (denoting continuous compounding), then:

$FV_2 = \$1{,}000e^{0.08\times 2} = \$1{,}173.51.$

Note that continuous compounding in this example only increases the future value by two cents relative to the use of daily compounding in the previous example.

Example 2.4: Consider the following examples of compounded interest each over a period of one year:
\$100 compounded annually at 10% = $100 \times (1 + 0.1)^1 = 110.00$
\$100 compounded semi-annually at 10% = $100 \times (1 + 0.1/2)^2 = 110.25$
\$100 compounded quarterly at 10% = $100 \times (1 + 0.1/4)^4 = 110.38$
\$100 compounded monthly at 10% = $100 \times (1 + 0.1/12)^{12} = 110.47$
\$100 compounded weekly at 10% = $100 \times (1 + 0.1/52)^{52} = 110.51$
\$100 compounded daily at 10% = $100 \times (1 + 0.1/365)^{365} = 110.52$
\$100 compounded continuously at 10% = $100 \times e^{0.10} = 110.52$

Note again that the compounding assumption used in research models (continuous) is quite similar in results to the daily compounding used in most bank accounts and other institutional arrangements.

2.1.3 Discounting and Present Values

Future values were used in the previous section to introduce the time value of money because the idea of an investment value growing through time is intuitively easy. But in finance in general and in this book the more common application is in the computation of present values, called *discounting*. A ***present value*** is the market value today of one or more cash flows to be received on a deferred basis when taking into account market interest rates.

A present value can be computed or discounted from a future value by simply reversing the computation of FV from the previous section into the computation of PV (given FV). The present value of a cash flow is equal to the future value multiplied by the time value discount factor. For example, using annual compounding and rearranging Equation 2.2:

$$PV = FV_T (1 + r)^{-T} \tag{2.5}$$

$(1 + r)^{-T}$ is called ***time value discount factor*** and is the multiplying factor to bring future cash flows to today.

Example 2.5: If T is 2, r is 8%, assuming annual compounding and FV_T = \$1,000, then:

$$PV = \$1{,}000(1.08)^{-2} = \$1{,}000/(1.1664) = \$857.34$$

The case of continuous compounding to calculate a present value (i.e., discount a cash flow) is shown in Equation 2.6:

$$PV = FV_T \times e^{-rT} \tag{2.6}$$

e^{-rT} is a time value discount factor.

Example 2.6: If T is 4, r is 14%, and assuming continuous compounding and FV_T is \$1,000, then:

$$PV = \$1{,}000\ e^{-0.14\times 4} = \$571.21$$

Note that this example uses an interest rate of 14% per year. It may seem to be unrealistic to a young generation, but in 1970–80 there were very high interest rates. In addition, some low-grade bonds can offer very high interest rates.

More generally, the present value P of a stream of cash flows using continuous compounding is described in Equation 2.7:

$$P = \sum_{i=1}^{n} C_{t_i} e^{-r_i t_i} \tag{2.7}$$

In Equation 2.7, $t_1, \ldots, t_n(=T)$ are the times between $[0, T]$ where cash flow C_{t_i} is received (if positive) or paid (if negative). The interest rate, r_i, is the continuously compounded market interest rate on the cash flow from that time period. Equation 2.7 is the general equation for valuing assets. It is the general equation from which numerous important financial equations are derived under various assumptions regarding the prospective cash flows. For instance, if the cash flows are all positive, occur with equal spacing or as a continuous stream, and persist through infinity then the underlying asset is a perpetuity and can be solved with the perpetuity formula ($P = C_1/r$). Similarly, if the cash flow stream persists for a finite length of time, the solution can be found using an annuity formula.

Note that when the interest rates used in discounting are market interest rates, the present value resulting from that computation can be described as a market value or price in the absence of arbitrage. As we mentioned before, interest rates and discount factors contain equivalent information; if we know the discount factor, we can derive the interest rate and vice visa.

2.1.4 Valuation of Pure Discount Bonds

Bonds represent specific claims to future cash flows. Like all financial securities, a ***bond's value*** is the sum of the present value of its potential cash flows. A ***pure discount bond,*** also known as a ***zero coupon bond,*** offers a single future cash flow, usually depicted in textbooks for simplicity in a denomination such as \$100, \$1,000, or \$10,000, which is often referred to as the ***face value, principal,*** or ***par value***. The time-to-receipt of a bond's final cash flow is known as the ***time-to-maturity*** of the bond. The value of a \$1,000 (face value) pure discount bond with a maturity of five years is equal to \$1,000 discounted for five years at a given or prevailing interest rate. Note that different issuers of bonds can have different levels of default risk, so the interest rates used to discount their promised cash flows to present value can be different.

Denote the price of a T-year bond today as $B(0, T)$. The ***value*** or ***price***, $B(0, T)$, of a pure discount bond given a continuously compounded market interest rate simply involves the application of Equation 2.6 (although the general case in Equation 2.7 also works) in which T represents the time-to-maturity in years and F is the face value or principal amount of the pure discount bond.

$$\boldsymbol{B(0, T) = F \times e^{-rT}} \tag{2.8}$$

Example 2.7: A two-year zero coupon bond pays a \$100 face value when two-year continuously compounded interest rates are 7.25%. What is its price today?

Applying Equation 2.8 generates:
$B(0, T) = F \times e^{-rT} = \$100\ e^{-2\times 0.0725} = \$86.502$

Alternatively, given the face value, F, and bond price $B(0,T)$ of a zero coupon bond, an analyst may compute the implied interest rate, which is called the ***T-year zero rate***.

Example 2.8: A two-year zero coupon bond pays \$100 face value and is priced at \$95 today. What is the two-year zero rate assuming continuous compounding?

Multiplying each side of Equation 2.8 by e^{rT} and inserting \$95 for $B(0, T)$ and \$100 for F produces:
$\$95e^{r\times 2} = \$100$
$r = ln(100/95)/2 = 2.5647\%$

2.1.5 Coupon Bond Yields and Prices

We first assume that all bonds do not have default risk (i.e., they are risk-free bonds). Later in this section defaultable bonds are considered. ***Coupon bonds*** represent fixed claims to two types of cash flows: a series of equal payments referred to as coupons, C, and a final payment referred to as the principal amount or face value, F, of the bond. A coupon bond has a stated ***coupon rate*** (e.g., 7.375%) that is distributed each year as a cash flow equal to the coupon rate times the bond's principal value. The coupon rate on a particular bond is often set equal to market interest rates when a bond is initially issued so that the initial price of the bond will approximate its face value. The periodic coupon payments can then be viewed as regular payments of interest, unlike a zero coupon bond wherein an economic sense the payment of interest is deferred until maturity (and is distributed at maturity as the difference between the bond's face value and its initial price).

The value of a T-year coupon bond, $B(0, T)$, can be viewed as the sum of the present values of an annuity (the stream of coupons) and the present value of the principal or face value (a single cash flow). The formula for the present value of a coupon bond is shown in Equation 2.9. Note that the formula is simplified from Equation 2.7 because all interest (coupon) payments (C) are identical in a fixed rate bond and the bond's last coupon, C, is combined with the principal or face value payment of F.

$$B(0,T) = C\sum_{i=1}^{n-1} e^{-r_{t_i}t_i} + (C+F)e^{-r_T T} \quad (2.9)$$

In practice, most coupon bonds pay coupons semiannually (at half the annual coupon rate). Equation 2.9 is used assuming that C represents a semiannual coupon payment, that t_i is expressed in half-year intervals, that there are $n = 2T$ periods, and with r_{t_i} denoting the corresponding market interest rate in each half-year period. The interest rate used for discount purposes is also called the ***discount rate***. If the discount rate for each cash flow is equal to the market discount rate for the time-to-receipt of that cash flow, then the sum of the present values of each cash flow should equal the market price.

In practice, bonds are often valued using a single discount rate, y, that is applied both to all coupon payments and the principal from the same bond rather than using the market interest rates corresponding to the longevity of each cash flow. This single discount rate is known as the ***bond's yield*** or ***yield-to-maturity***. Equation 2.10 is the formula for a bond price based on its yield-to-maturity, y:

$$B(0,T) = \sum_{i=1}^{n} C_{t_i} e^{-yt_i} \quad (2.10)$$

Or equivalently,

$$B(0,T) = C\sum_{i=1}^{n-1} e^{-yt_i} + (C+F)e^{-yT} \quad (2.11)$$

The discount rates in Equation 2.9, which allow each cash flow corresponding to a different longevity to have potentially different discount rates, r_{t_i}, are ***zero coupon rates*** or ***zero rates***. They reveal the time value of money specific to the particular longevity of a single cash flow. A zero coupon rate on a single cash flow is equal to the annualized rate of return that an investor will receive in a competitive market if investing only in that cash flow and holding that position until the cash flow is received.

The yield-to-maturity of a bond is a discount rate applied to all of the cash flows of a particular bond. In theory, it does not make sense to discount cash flows with different longevities using the same rate (i.e., yield). But yields can serve as a simple and useful metric to value bonds and to approximate the return that an investor will receive from holding the bond to maturity. An investor's final annualized return from holding a coupon bond to maturity will vary based on the rates at which the bond's coupons can be reinvested. If coupons can be reinvested at the bond's original yield, then the investor's final annualized rate of return will equal the bond's yield. If reinvestment rates are lower (higher), the final annualized return will be lower (higher) than the yield. Note that there is no reason to expect that coupons can be reinvested at the bond's original yield-to-maturity and so the use of yields is based primarily on simplicity.

The market value of bonds tends to vary inversely with interest rate levels. As indicated above, the final annualized return of a bond (or a bond portfolio) will vary based on reinvestment rates. Price risk is an important risk management concept that is discussed further in Section 2.2.

Comparison of Equations 2.9 and 2.11 indicates that the yield in Equation 2.11 may be viewed as a complex average of the zero coupon rates in Equation 2.9. Thus, a bond's yield-to-maturity is a function of the entire zero coupon rates corresponding to all of the cash flows of the bond. An investor in a coupon bond that holds the bond to maturity and reinvests the coupon payments will realize an annualized rate of return that depends on the interest rates at which the coupons are reinvested. Because these reinvestment rates are unknown and subject to randomness, ownership of a coupon bond (or other financial security with multiple future cash flows) in order to build value at some horizon point such as the time-to-maturity of the bond subjects the bond owner to reinvestment rate risk.

So far we have considered interest rates as a function of time, in particular, zero rates are a function of the longevity or time until the cash flow is received. However, we have ignored default risk (i.e., credit risk) as a determinant of interest rates. More risky borrowing or lending requires a higher interest rate. There are two types of interest rates in the market. ***Risk-free interest rates*** contain no default/credit risk. ***Risky interest rates*** are interest rates that include default/credit risk. They are a function of time and the risk of the underlying cash flow. We use risk-free interest rates to derive risk-free bond prices (such as Treasury bond prices) as well as the prices of derivatives because of the risk neutrality principal. We use risky interest rates to derive risky bond prices (such as corporate bond prices). In the next section, we will discuss major types of interest rates. There are different types of interest rates corresponding to different types of fixed-income products and different financial markets. It is important to understand different types of interest rates before we discuss derivative valuation.

2.1.6 Major Types of Fixed-Income Rates

Financial assets and financial interest rates are valued using a variety of interest rates that differ by the risks, currencies, and other factors such as jurisdiction. This section briefly reviews some of the major U.S. dollar rates with little or no default risk.

Inflation is a key issue in fixed-income analysis. ***Inflation*** is the rate at which the value of a particular currency declines through time. ***Nominal interest rates*** are the stated rates on fixed-income instruments without any adjustment for inflation. The ***real rate of interest*** is an inflation-adjusted measure of the return on short-term bonds with no default risk. Continuously compounded real interest rates are calculated by subtracting the anticipated continuously compounded inflation rate from the continuously compounded nominal interest rate.

Following Hull[1], here are several different interest rates and their descriptions:

U.S. Treasury rates*:** Treasury rates are often viewed as proxies of the riskless or risk-free interest rates corresponding to U.S. dollars. Treasury rates are derived from the prices of U.S. Treasury bills (short term), notes (middle term), and bonds (long term). Some other nations issue similar government bonds in their own currencies that are collectively referred to as ***sovereign debt. The nominal interest rates or yields on the sovereign debt of a nation depend on factors such as anticipated inflation, default risk, and real rates of interest rate.

SOFR (Secured Overnight Financing Rates) or ***LIBOR rates*** (LONDON InterBank Offered Rates): Both types of rates are used to specify the terms of variable-rate loans and to value those loans. Market participants are transitioning from using the previous standard, LIBOR, to a newer set of rates SOFR, based in part on concerns over manipulation of LIBOR rates by large banks. SOFR rates are based on observations of numerous transactions in the money markets while LIBOR is based on bank quotes. LIBOR for a particular maturity is the approximate rate of interest that major London banks are prepared to *offer* on large wholesale deposits. It is quoted every day for all major currencies for maturities up to 12 months. SOFR only quotes the daily (overnight) rates. Though SOFR is volatile from day to day, the payments of financial contracts based on SOFR use averaged rates; therefore, the volatilities of the daily rates can be averaged out. Due to the fact that SOFR is an overnight rate rather than a term structure and currently only one-month SOFR futures and three-month SOFR futures exist, it is hard to use it to derive the term structure for use in valuing interest rate derivatives. As time passes, more derivative products on SOFR will likely be available. For example, options on three-month SOFR futures were launched in early 2020. When more contracts and more liquidity of those derivative contracts on SOFR evolve, analysts will be better able to build a term structure of SOFR

---

[1] See *n*. 1 in Chapter 1.

rates and SOFR rates will be better able to be used in derivative valuation. The reminder of this book uses the LIBOR term structure. It is likely that the LIBOR term structure will be replaced by the SOFR term structure in the near future as the standard structure for derivative valuation.

Just as all assets are traded at two prices such as LIBOR and LIBID (respectively, the offer and bid), most interest rates and currency exchange rates have a bid side and an offer side in the marketplace – with the spread between the two representing compensation to market makers.

Repo rates: Repurchase agreements are short term, well-collateralized loans of liquid securities used to provide short-term financing to borrowers and short-term investing to lenders. In a repo, a financial institution that owns a security agrees to sell that security today for $X and buy the security back in the near future for a slightly higher price: $X + interest. A repo is like a well-collateralized loan except that the lender takes actual ownership of the security until it has been repurchased to settle the loan. A repo is safe way for a financial institution to arrange a short-term, cost-effective loan. The rate of interest is known as the repo rate.

Risky interest rates: In the case of corporate bonds, there is a risk that the corporation may default on its scheduled coupon and principal payments. So, investors demand a higher interest rate than the risk-free interest rate. The risky interest rates that discount scheduled principal and coupon payments to equal market prices must reflect rate premiums to reflect the default risk. The premium of these risky rates above the riskless rate of corresponding longevity is called a ***bond yield spread***, which compensates the bond holder for the default risk. The early chapters of this book focus on fixed-income securities with little or no default risk, namely on risk-free rates. Later chapters will discuss credit risk and risky interest rates.

The theoretic topics of interest rates are discussed next.

2.2 Term Structures, Yield Curves, and Forward Rates

Previous sections discussed present values, future values, time value discount factors (time discount factors), and interest rates. They are not independent concepts. Given a future value (for example the face value of a zero-coupon bond) and the present value (i.e., the bond's price), we can derive both the time value discount factor and its equivalent form, the interest rate. Note that there are actually a variety of interest rates corresponding to different compounding assumptions.

As discussed in the previous section, cash flows with different times-to-receipt are typically valued with different market discount rates. Specifically, cash flows with long times-to-receipt tend to have higher market discount rates than cash flows with short times-to-receipt, and cash flows with greater risk tend to have higher discount rates. In this section, we will derive various interest rates as a function of time using different bond prices. Perhaps the most straightforward way to value future cash flows is through the discount factor function – the relationship between discount factors and time-to-maturity. Using continuous compounding, the time value discount factor function is simply e^{-rT}.

There are three equivalent ways to represent the time value of money: the time value discount factor function, the term structure of interest rates, and forward interest rate structures (to be discussed later). Each can be derived from the other. Another concept introduced in this section is the yield curve. The yield curve is an ambiguous and vague concept. A ***yield curve*** is an expression of the relationship between bond yields and their times to maturity. Though it is used in practice, it has little mathematical or financial economic meaning.

2.2.1 The Bootstrap Method for Treasury Zero Coupon Interest Rates and the Term Structure of Interest Rates

The ***term structure of interest rates*** is the relationship between zero coupon interest rates and their corresponding terms to maturity. Henceforth, we will not distinguish the following synonymous terms: ***zero coupon interest rates, zero rates,*** and the ***zero coupon term structure***. The term structure is usually depicted with rates on the vertical axis and time to maturity on the horizontal axis, as shown in Figure 2.1.

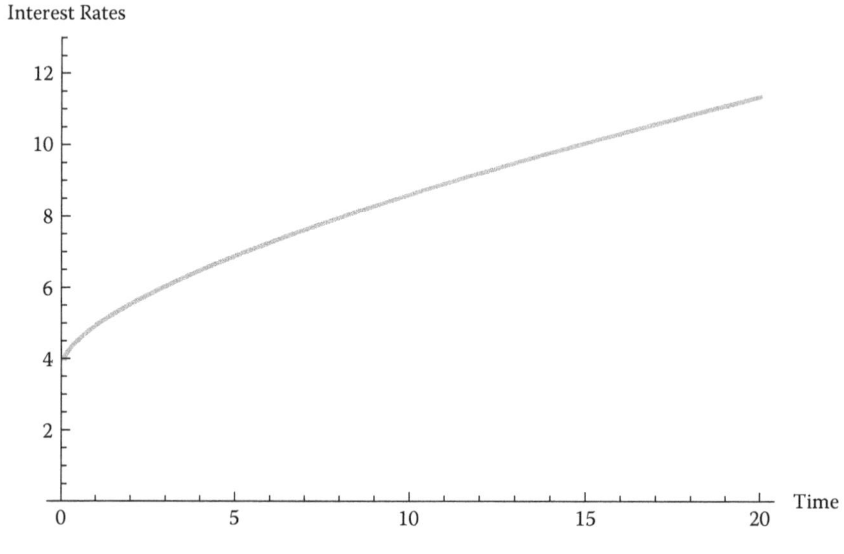

FIGURE 2.1
The Term Structure of Interest Rates.

While zero coupon debt securities are common in maturities of less than one year, most bonds with maturities beyond one year offer coupons. Some investors prefer zero coupon bonds in order to avoid the risk and inconvenience of reinvesting coupon payments at unknown future interest rates (reinvestment rate risk). Other investors prefer the regular income of coupon bonds.

When lenders are reluctant to issue long-term zero coupon bonds, financial engineering can be used to create zero coupon bonds from long-term coupon bonds. There is an active market for pure-discount fixed-income securities created by separating U.S. Treasury coupon bonds into their individual cash flows (each coupon payment and the principal payment), securitizing each promised cash flow (i.e., creating a security with a claim to an individual cash flow), and trading the securities in public markets. The resulting securities are entitled ***U.S. Treasury strips*** and are actively traded to the benefit of investors wishing to pure-discount bonds.

Treasury bills have maturities of a year or less. ***Treasury notes*** are issued with maturities from 2 to 10 years. ***Treasury bonds*** are long-term investments that have maturities of 10 to 30 years from their issue date. As zero coupon bonds, the prices of Treasury bills offer direct estimates of points along the term structure of interest rates since each price determines a rate. However, Treasury notes and bonds are coupon bonds with prices that are functions of the interest rates corresponding to interest rates of various longevities. This section introduces a method for determining points along the term structure of interest rates using coupon bond prices. The method to find risk-free zero rates as a function of time from Treasury bills/notes/bonds is called ***bootstrapping.***

In a previous section (see Example 2.8), we found the implied interest rate (i.e., the T-year zero rate) from a zero coupon bond price and its future cash flow. If we have multiple coupon bond prices with varying maturities and with coupon payment dates that align, we can derive a term structure of interest rates using a bootstrap method. Note that in order to apply the bootstrap method, the cash flows of the coupon bonds need to align within a number of dates equal to or less than the number of coupon bonds. If so, the Treasury bills/notes/bonds can be used to estimate or ***calibrate*** the Treasury term structure of interest rates. It is unusual to bootstrap corporate bonds since they usually have coupons and principal payments happening at too many different times (and their credit risks are not exactly equal).

Consider the following example to illustrate the bootstrap method. The bonds are Treasury bills (i.e., short-term risk-free zero coupon bonds), Treasury notes, and Treasury bonds (with the notes and bonds being risk-free coupon bonds). To the extent that markets are perfect, the risk-free zero rates derived from bootstrapping are the implied returns available from investing in corresponding U.S. Treasury strips.

Example 2.9: Data on Treasury bills/notes/bonds from the *Wall Street Journal*:
Three-month Treasury bill quote: 1.28
Six-month Treasury bill quote: 1.44
One-year Treasury notes: coupon 1.25%, price 99.5
One-and-a-half-year notes: coupon 0.875%, price 98.64
Two-year Treasury notes: coupon 1.35% price 99.0
Find risk-free zero rates as a function of time (Note that principal is \$100.)

Note that Treasury bill quotes are rates rather than prices: to obtain the price of a Treasury bill, multiply the quoted rate by the number of days until maturity, divide by 360, and then subtract that number from 100. For example, consider the bills with 90 and 180 days to maturity and bid rates of 1.28 and 1.44, respectively, that correspond to prices of \$99.68 and \$99.28 per \$100 of face value:

\$1.28 × 90/360 = \$0.32
\$100 − \$0.32 = \$99.68
\$1.44 × 180/360 = \$0.72
\$100 − \$0.72 = \$99.28

Next, the two 3-month and 6-month Treasury bill prices above (\$99.68 and \$99.28) can be used one step at a time to derive the zero coupon rates of three-month, six-month, other zero coupon rates from one-year, one-and-a-half-year, …, three-year coupon prices using the method of bootstrapping:

Step 1: Calculate the six-month rate using Equation 2.8.

$$99.28 = 100e^{-r\times\frac{1}{2}}$$

So, the six-month rate is 1.445%.

Step 2: Calculate the one-year rate r_1 using Equation 2.9, the data on a one-year coupon bond, and the rate calculated in step 1.

$$99.5 = 0.625e^{-0.01445\times0.5} + (0.625 + 100)e^{-r_1\times1}$$

Solving for the unknown rate produces r_1 = 1.75%. (Note that 0.625 is the semiannual coupon payment of the 1.25% coupon rate.)

Step 3: Calculate the one-and-a-half-year rate using Equation 2.9 and the rates calculated in steps 1 and 2.

$$98.64 = 0.4375e^{-0.01445\times 0.5} + 0.4375e^{-0.0175\times 1} + (0.4375 + 100)e^{-r_{1.5}\times 1.5}$$

Solving for the unknown rate produces $r_{1.5} = 1.79\%$.

Step 4: Calculate the two-year rate using Equation 2.9 and the rates calculated in steps 1, 2, and 3.

$$99 = 0.675 \times e^{-0.01445\times 0.5} + 0.675 \times e^{-0.0175\times 1} + 0.675 \times e^{-0.0179\times 1.5} + (0.675 + 100)e^{-r_2\times 2}$$

Solving for the unknown rate produces $r_2 = 1.85\%$.

In the second, third, and fourth steps, zero rates calculated in previous steps are used to discount the bond's cash flows (coupons) prior to maturity. Then, the only unknown zero rate (the zero rate for the last period) can be found. With bonds of different maturities and coupon payment dates that align, the bootstrap method can be used to infer the term structure of zero coupon interest rates.

2.2.2 Coupon Bond Yield Curves

In the previous section, we bootstrapped risk-free zero rates from the prices of Treasury bills, notes, and bonds. In this section, we will discuss yield curves. Note that "yield" itself is a flawed concept in economics because implicitly the yield of a coupon bond is based on discounting cash flows at different times using the same rate – the yield. The "yield curve" is even worse. Yield curves are estimates of the relationship between the yields and times-to-maturity of coupon bonds using bonds with different times to maturities. These different yields have no economic coherence. So, "yield curve" is a purely heuristic instrument. However, it is used in the industry and is worthy of understanding.

There is little precise economic meaning to a bond's yield other than that it is the single rate that equates a bond's price to its present value and that it is an estimate of realized return under the highly unrealistic assumption that coupon payments can be reinvested at the bond's previous yield-to-maturity. Nevertheless, yields are easy to calculate and therefore yield curves can represent convenient approximations to implied zero coupon rates.

Bonds tend to be issued with coupons that approximate current interest rates and therefore most newly issued bonds sell at prices near their par, face values, or principal amounts. As time passes, market interest rates often rise or fall. Falling rates tend to cause bond prices to rise, while rising rates tend to cause bonds prices to fall. The longer term a bond, the more its price is volatile and sensitive to interest rate changes. These issues are explored later in more detail in a discussion of duration.

Coupon bonds often trade at prices above their par value (***premium bonds***) or below their par value (***discount bonds***). The total return of a coupon bond during a particular time period is based on the sum of any coupon received and the change in the bond's price. A bond with a coupon rate that exceeds current zero coupon rates must sell at a premium to its face value in order to offer a yield-to-maturity that prevents arbitrage. Conversely, bonds with low coupons relative to current interest rates must sell at discounts.

The market price of all bonds must converge to the bond's principal amount as maturity approaches in the absence of default concerns. The premium on a high coupon bond tends to decline as the bond approaches maturity. Thus the higher-than-market coupon rate available with a premium bond is offset by the anticipated decline in its premium to zero at maturity. Conversely, low coupon bonds offer anticipated price gains through time as their prices converge toward par.

2.2.3 Forward Interest Rates

In finance, the term ***spot rate*** or ***spot price*** usually refers to rates and prices in cash markets (i.e., markets for immediate or nearly immediate delivery and payment such as most stock and bond markets). Forward contracts on securities and commodities are agreements focused on deferred exchange (e.g., a forward contract to buy a share of stock in three months) and were introduced in Chapter 1. In this section, forward interest rates are discussed. A ***forward interest rate*** is the market interest rate implied from spot interest rates (or observed in a forward contract) for a loan or deposit that will commence at some pre-specified future date and terminate at a subsequent pre-specified date. A forward interest rate, f_{T_1,T_2}, is the interest rate observed today that will apply on a deposit or loan beginning at time T_1 and ending at time T_2. A spot interest rate has $T_1 = 0$ (the current time). For example, a bond fund may contract with an investment bank to receive a rate of return that will be credited on a deposit to be made in three years and that will be in effect for two years (i.e., paying interest from the end of the second year to the end of the third year) and can be expressed as $f_{2,3}$. This notation is different from Chapter 7's notation. In this chapter, forward rates are continuously compounded.

The equilibrium forward rate can often be observed in financial markets containing forward contracts. But forward rates can also be implied using a no arbitrage model of spot rates. The ***no arbitrage model of forward rates*** uses two bond strategies with identical maturities, one that uses a hypothetical forward contract and one that does not, and solves for the forward rate that sets their final rates of return equal. For example, an investor can select the following two equivalent investment strategies (assuming continuous compounding for simplicity): (1) Invest in a two-year zero coupon bond (receiving the return r_2 for two years), or (2) invest in a one-year zero coupon bond (receiving r_1 for one year) and enter a forward contract that guarantees receiving the forward rate $f_{1,2}$ from the end of year 1 to the end of year 2 on the proceeds from the one-year bond. The final returns of the two alternatives are set equal to each other:

$$e^{r_2 \times 2} = e^{r_1 \times 1} e^{f_{1,2}}$$

Taking the natural logarithm of each side and rearranging produces the formula for the forward rate in terms of the spot rates:

$$f_{1,2} = (2 \times r_2) - r_1$$

More generally, we can prove

$$f_{T_1,T_2} = \frac{r_2 T_2 - r_1 T_1}{T_2 - T_1} \tag{2.12}$$

Note that if $r_1 < r_2$ then $-r_1 T_1 > -r_2 T_1$ and so:

$$f_{T_1,T_2} = \frac{r_2 T_2 - r_1 T_1}{T_2 - T_1} > \frac{r_2 T_2 - r_2 T_1}{T_2 - T_1} = r_2 > r_1$$

The above equation shows that forward rates will *exceed* spot rates of identical times to maturity when the term structure is *upward* sloping. The converse is also true. And when the term structure is level (i.e., flat), spot rates will equal forward rates.

2.2.4 Forward Interest Rate Structures

A forward rate structure is the graph of forward rates against starting time T_1. Typically, the forward rates used here are for an instantaneous forward rate f_T or a single period forward rate $f_{T,T+1}$ (i.e., the forward interest rate time period approaches 0 or is equal to 1). Note that in a perfect market (i.e., with no transaction costs or trading restrictions) forward rates that are implied from spot rates would represent opportunities at which agreements could be made today for money to be borrowed or lent over some future period of time (i.e., forward contracts).

An implied single period forward rate $f_{T,T+1}$ is often estimated from zero coupon term structures $r(T)$ using Equation 2.12:

$$f_{T,T+1} = \frac{r(T+1) \times (T+1) - r(T) \times T}{1} \tag{2.13}$$

An implied instantaneous forward rate f_T can be derived as follows:

Assume that the zero coupon term structure $r(T)$ is a function of time.

$$f_T = \lim_{\Delta t \to 0} f_{T,T+\Delta t} = \lim_{\Delta t \to 0} \frac{r(T+\Delta t)(T+\Delta t) - r(T)T}{\Delta t} = r'(T)T + r(T) \tag{2.14}$$

Example 2.10: Consider five zero coupon rates with maturities from one to five years: 5%, 6%, 6.5%, 6.8%, and 7.0%. Calculate the four forward rates comprising a four-period term structure of single-period forward rates.

Inserting the four pairs of adjoining rates into Equation 2.12 (with the denominator set equal to one): (2 × 6% – 5%) = 7%; (3 × 6.5% – 2 × 6%) = 7.5%; (4 × 6.8% – 3 × 6.5%) = 7.7%; and (5 × 7.0% – 4 × 6.8%) = 7.8%. Therefore the single period forward rate structure from $T_1 = 1$ to $T_2 = 5$ contains: 7%, 7.5%, 7.7%, and 7.8%.

A key concept is that forward rates are not generally equal to ***expected future spot interest rates*** unless investors are risk neutral. In other words, in Example 2.10 the implied forward rate of 7.8% between periods four and five is a *biased* estimate of the market's expected one-period spot rate at the end of period 4 to the extent that investors are risk averse. The explanation is straightforward. Forward contracts contain ***interest rate risk*** (changing value when interest rates change since the "locked-in" forward rate becomes more attractive or less attractive). In the real world, where investors are risk averse, forward rates must therefore reflect a risk premium because they share the general tendency of fixed-income instruments to change in value when interest rate levels change. This risk premium forces expected spot prices and rates to diverge from forward prices and rates.

The higher interest rate risk of long-term bonds causes the term structure and yield curve to be upward sloping (except when the market has a strong expectation that interest rates will fall). Upward-sloping term structures are consistent with forward interest rates exceeding spot rates as previously discussed. Therefore, in theory, forward interest rates (both implied forward rates and observed forward rate agreements and contracts) tend to exceed expected spot interest rates (although expected spot rates are not visible) due to risk aversion. However, expectations of future interest rate expectations also drive both the slope of the term structure and forward curve.

2.3 Fixed-Income Risk Measurement and Management

The primary two risks in fixed-income investment management are ***interest rate risk*** and ***credit risk***. As previously indicated, interest rate risk is often divided into ***price risk*** and ***reinvestment rate risk***. This section begins with interest rate risk and then concludes with credit risk.

2.3.1 Duration

Duration is the premier measure of the interest rate risk of a fixed-income security. Duration, D, has several interpretations. The most common interpretation is that it is the weighted average of the longevities of a fixed-income asset or portfolio. The weights of each cash flow's time-to-maturity are defined as the portion of the bond's value attributable to the cash flow as expressed in Equation 2.15:

$$D = \sum_{i=1}^{n} w_i \times t_i \tag{2.15}$$

where $w_i = C_{t_i}e^{-yt_i}/\sum_{i=1}^{n} C_{t_i}e^{-yt_i}$ and t_i is the time when cash flow is to be received, and $t_n = T$.

Note that Equation 2.15 uses a bond's yield to maturity for simplicity as the discount rate in calculating all of the weights. In theory, each cash flow should be discounted by a zero coupon rate corresponding to its longevity. Using zero coupon rates raises three problems: (1) Zero coupon rates are somewhat difficult to estimate, (2) estimates of zero coupon rates can vary based on the method used, and (3) if each cash flow is discounted with estimated zero coupon rates, the sum of the discounted cash flows may not exactly equal the bond's price and the weights may not sum to one. Accordingly, Equation 2.15 provides the convenient and popular yield-based method of estimating the duration of a bond with fixed promised cash flows.

A second important interpretation of duration (D) is that it is a measure of price risk from interest rate shifts. Specifically, ***duration*** may be defined and interpreted as the elasticity (percentage change) of a fixed-income value (B) with respect to the underlying yields (y). If the duration is estimated using the term structure of spot rates rather than yields then duration may be defined and interpreted as the elasticity of a fixed-income value (B) with respect to a parallel shift in the term structure of interest rates.

$$D = -\frac{\frac{\partial B}{\partial y}}{B} \tag{2.16}$$

where $B = B(0, T) = \sum_{i=1}^{n} C_{t_i}e^{-yt_i}$ as defined in Equation 2.10.

Simple calculus tells us that $\frac{\partial B}{\partial y} = \sum_{i=1}^{n} (-t_i) C_{t_i} e^{-yt_i}$ so:

$$D = -\frac{\partial B/\partial y}{B} = \frac{\sum_{i=1}^{n} (t_i) C_{t_i} e^{-yt_i}}{B} = \sum_{i=1}^{n} t_i \times \frac{C_{t_i} e^{-yt_i}}{B} = \sum_{i=1}^{n} t_i w_i$$

So, the two definitions of duration are consistent. The first definition is helpful for us to conceptualize the determinants and calculation of duration. Given a bond yield (or an entire term structure of zero coupon rates) we can calculate the value of a bond using each of its cash flows. Using these weights, we can find the duration of a bond. The duration in Equation 2.15 is often termed the ***Macaulay duration***.

The second definition of duration is helpful for us to understand price risk and price movements in a fixed-income asset or position due to interest rate changes. We can use duration, D, to estimate bond gains or losses very quickly given a change in its yield or a general shift in the term structure of interest rates. The duration formula above, $D = -\frac{\frac{\partial B}{\partial y}}{B}$, using $\frac{\Delta B}{\Delta y}$ to approximate $\frac{\partial B}{\partial y}$ when Δy is small, can be written as

$$\frac{\Delta B}{\Delta y} \approx -DB$$

$$\frac{\boldsymbol{\Delta B}}{\boldsymbol{B}} \approx -\boldsymbol{D\Delta y} \tag{2.17}$$

or

$$\boldsymbol{\Delta B} \approx -\boldsymbol{BD\Delta y} \tag{2.18}$$

Note that the left-hand side of Equation 2.17 is the formula for the percentage rate of return. Equation 2.17 approximates the instantaneous bond return from a change in its yield as the simple product of the bond's duration and the change in the yield. For example, a bond with a duration of 5.0 should fall approximately 2.5% in price if its yield rises 0.5% (or general interest rate levels rise 0.5%).

Duration as a measure of interest rate sensitivity is particularly useful for understanding the interest rate risk of a portfolio of fixed income instruments. In Equations 2.15 and 2.16, C_{t_i} does not have to be the coupon of a bond. It could be the cash flows of multiple bonds in a bond portfolio. In the following, we use two-bond portfolios as an example. We will show that the portfolio duration is a weighted average of the durations of the portfolio's constituent bonds where the weight of each bond equals the percentage of the bond portfolio's total current value attributable to the bond.

Assume portfolio B contains two bonds, B_1 and B_2, B = $B_1 + B_2$

$$D = -\frac{\partial B/\partial y}{B} = -\frac{\frac{\partial B_1}{\partial y} + \frac{\partial B_2}{\partial y}}{B} = -\frac{\frac{\partial B_1}{\partial y}}{B} - \frac{\frac{\partial B_2}{\partial y}}{B} = -\frac{\frac{B_1}{B_1}\frac{\partial B_1}{\partial y}}{B} - \frac{\frac{B_2}{B_2}\frac{\partial B_2}{\partial y}}{B}$$
$$= \frac{B_1}{B}D_1 + \frac{B_2}{B}D_2.$$

We can quickly approximate the portfolio's interest rate risk by tracking the portfolio duration.

Note that the above approximation for the percentage price change of a bond portfolio as a result of interest rate changes is only accurate when Δy is small. Also, the approximation's accuracy is better in the case of a portfolio of bonds when the term structure of interest rates has ***parallel shift*** (i.e., different yields shift the same amount).

Finally, it should also be noted that Equations 2.15 and 2.16 are based on continuously compounded yields and interest rates. If annual compounding is assumed then a 1% shift in yields or interest rates causes a smaller change in the bond's price (compared to a 1% shift in continuously compounded rates). Analysts using discrete compounding often adjust the bond's duration to its ***modified duration***, as shown in Equation 2.19.

$$\boldsymbol{D_{\text{mod}} = D/[1 + (y/m)]} \tag{2.19}$$

where m is the number of compounding intervals per year being used to measure yields (or zero coupon rates) and used to estimate ΔB in Equations 2.17 and 2.18. Note that if continuous compounding ($m \to \infty$) is used to calculate ΔB then $D_{\text{mod}} = D$ (modified duration equals Macaulay duration).

Equations 2.17 and 2.18 are true for modified duration, too. Let's look at an example.

Example 2.11: Consider a zero coupon bond with a time-to-maturity of seven years and a yield-to-maturity of 8%. If its yield shifts up by 0.4%, what is the approximated instantaneous percentage price change in the bond if: (1) continuous compounding is being used and (2) if semiannual compounding is used?

The Macaulay duration of the bond is 7.0 by Equation 2.15 because there is only one future payment. The modified duration is 6.73 (i.e., 7.0/1.04) found by Equation 2.19. The approximated bond price change ($\Delta B/B$) is found as $-D\Delta y$ (Equation 2.17). For continuous compounding the approximation is -7×0.004, or -2.8% (which does not depend on its yield). For semiannual compounding it is -6.73×0.004, or -2.69%.

2.3.2 Convexity

The use of duration in the approximation of bond returns can be viewed as a first-order Taylor expansion. In order to model the potential effects of large interest rate shifts, a second-order Taylor expansion may be used. Equations 2.20 and 2.21 provide the second-order risk measure (convexity) and the second-order approximation formula.

$$C = \frac{1}{B}\frac{\partial^2 B}{\partial y^2} \tag{2.20}$$

$$\Delta B \approx -BD\Delta y + B\frac{1}{2}C(\Delta y)^2 \tag{2.21}$$

Convexity approximates the curvature of a bond's price-yield relationship. The formula for convexity is based on squaring the times-to-receipt of a bond's cash flows. Roughly speaking, the convexity of a T year zero coupon bond is approximately T^2. For two portfolios with equal durations, the portfolio with higher dispersion in the times-to-receipt of cash flows will tend to have greater convexity.

Duration is limited in its ability to approximate accurately in the case of: (1) large interest rate shifts and (2) non-parallel interest rate shifts. It should be noted that convexity not only provides a second-order approximation of the effects of large parallel interest rate shifts, it also approximates a first-order effect of slope changes in the term structure. Therefore, convexity and higher-order duration terms can be used in the case of managing the interest rate risk of fixed-income portfolios when greater accuracy is sought.

2.3.3 Interest Rate Risk Management

Fixed-income portfolio management is usually focused on managing the interest rate risk of a portfolio relative to a benchmark or a related set of liabilities. Benchmarking is the process of identifying an index or other source of observable returns and using those returns as a point of reference for evaluating the success of an investment strategy. For example, a fixed-income manager for a mutual fund might desire that the effect of interest rate shifts on the manager's portfolio be very close to the effects of the same interest rate shifts on the value of the manager's benchmark such as a published index of bond returns. A fixed-income manager of a pension fund may be more concerned about making sure that the bond portfolio's ability to provide funds to meet anticipated liabilities (pension benefits) is protected from interest rate shifts.

The process of ensuring that the goals for a fixed-income portfolio (benchmarking or liability funding) will be met despite interest rate changes is known as ***immunization***. One of the premier methods of obtaining immunization is ***duration matching***. In the case of a mutual fund manager attempting to track a benchmark such as a major bond index, the manager will tend to make sure that the duration of her portfolio is very close to the duration of the benchmark. In the case of a pension fund manager attempting to ensure that the fund's investments will be able to meet the pension obligations that manager will tend to make sure that the duration of her portfolio is very close to the duration of the plan's anticipated liabilities (assuming the plan is fully funded with the value of the investments equaling the present value of the liabilities).

Another method of interest rate risk management is ***cash flow matching*** (i.e., matching the various cash flow longevities of the portfolio to the cash flow longevities of the benchmark or liability stream). The problem with cash flow matching is that it can be very inflexible with regard to the investment opportunities that the manager can select.

Duration provides a first-order solution that permits flexibility. The inclusion of convexity matching tends to improve the accuracy of the immunization but also tends to restrict the flexibility of the manager to include diverse investment opportunities.

In Chapter 5, we will discuss hedging strategies to manage interest rate risk using duration in detail. In the next subsection, we will discuss other risks associated with fixed-income security: credit risk and default risk. The management of credit risk and default risk will be discussed in detail in Chapter 8.

2.3.4 Introduction to Credit Risk and Default Risk

Credit risk refers to uncertainty regarding a borrower's ability to make timely interest and principal payments. Credit risk includes: (1) volatility of market prices caused by changes in a borrower's perceived ability to meet debt obligations, (2) fluctuations in the risk premiums (credit spreads) throughout the credit markets, and (3) losses due to defaults. ***Default risk*** is the risk that the borrower will make only partial payments, no payments, and/or delayed payments.

In the case of fixed income assets that have no risk of default, the present value or market value of the asset can be found by discounting the promised cash flows at the riskless interest rates corresponding to the longevity of each cash flow. In the case of fixed-income investments (not derivatives) with default risk, the present value or market value of the asset is found by discounting the promised cash flows with a risky discount rate that is equal to the sum of a riskless interest rate and a ***credit spread***, s. Equation 2.22 modifies the general asset valuation formula (Equation 2.10) by including a discount factor that captures the effect of potential default through the inclusion of a credit spread

$$\tilde{B}(0, T) = \sum_{i=1}^{n} C_{t_i} e^{-(r_i+s)t_i} \tag{2.22}$$

where $\tilde{B}(0, T)$ is the T year risky bond price today.

Equation 2.22 is based on a term-structure approach in which cash flows of different longevity are discounted at interest rates corresponding to the cash flow's longevity. Equation 2.23 illustrates the use of a credit spread in a yield-based valuation model in which all cash flows are discounted at the same rate (y) in which y contains the credit spread

$$\tilde{B}(0, T) = \sum_{i=1}^{n} C_{t_i} e^{-yt_i} \tag{2.23}$$

where y is the risky bond yield.

If the term structure of risk-free interest rates is flat, then Equation 2.22 becomes:

$$\tilde{B}(0, T) = \sum_{i=1}^{n} C_{t_i} e^{-(r+s)t_i} \tag{2.24}$$

Comparing Equations (2.23) and (2.24), note that we use $y = r + s$ in Equation 2.24 in which s is the ***yield spread*** or ***credit spread***.

Example 2.12: Consider a \$1,000 zero coupon bond with a maturity of five years. If current five-year riskless interest rates are 6% and the credit spreads on comparable bonds are 2%, what is the value of the bond? Then repeat the problem assuming that the credit spread (2%) was not provided but instead assume that the bond's market price is \$600. What is the implied credit spread?

Since there is a single cash flow due from the bond, Equations 2.23 and 2.24 are effectively equal. The value of the bond is \$1,000 $e^{-(0.06+s)5}$. Inserting $s = 2\%$ generates \$670.32. For the second question, solve this equation for s: \$1,000 $e^{-(0.06+s)5}$ = \$600. The implied credit spread is 4.2165%.

Higher perceived risk of default should increase credit spreads. The credit risk of most large debt securities is rated by various agencies. There are numerous rating levels with, for example, a rating of AAA indicating the smallest risk of default. The credit spread for a particular bond tends to be near the average credit spread for bonds with the same credit rating. Credit ratings change through time with changes in the financial prospects of the borrower. Overall bond ratings tend to move in tandem with economic cycles.

Financial credit derivatives, such as credit default swaps, discussed in Chapter 8, have played an enormous role in facilitating better credit risk management. Diversification of bond portfolios across a multitude of borrowers is one effective risk-reducing strategy. However, in many cases credit derivatives such as credit default swaps can provide a lower-cost means of achieving appropriate levels of diversification. Credit derivatives can provide highly liquid and cost-effective tools for ongoing management of specific credit risks as well as the total credit risk of a fixed-income portfolio.

Equities and portfolio of equities are discussed in the next two sections.

2.4 Equities and Equity Valuation

The assets of most operating firms are financed by a combination of major security types, especially debt (bonds) and equity (common stock) as illustrated in the following simplified balance sheet:

$$\text{Assets} = \text{Debt} + \text{Equity}$$

The capital structure of a firm refers to the various sources of funds that support the firm's assets.

This section provides an introduction to the valuation of a corporation's equity.

2.4.1 Equity as a Residual Claimant in a Firm's Capital Structure

The ***equity*** or ***common stock*** of a corporation is a residual claim, meaning that the owners of the equity (i.e., equity holders or stockholders) are entitled to cash flows from the firm's assets after the other major claimants have received value promised to them. The two primary claimants ahead of stockholders in a firm's capital structure are the debt holders and the holders of preferred stock. For simplicity, the remainder of the section assumes that there is no preferred stock in the firm's capital structure. The idea of equity holders being the residual claimants of the firm is illustrated by rearranging the previous equation:

$$\text{Equity} = \text{Assets} - \text{Debt}$$

The assets of the firm may be viewed as generating operating income that gets distributed first to the firm's debt holders for interest (coupon) and principal payments. Income taxes are then paid on the profits (i.e., operating income minus interest expense). The after-tax earnings of the firm are then divided into the portion that is distributed on a pro rata basis to the firm's stockholders in the form of dividends and a portion that will be retained inside the firm for acquisition of new assets (retained earnings). A ***dividend*** is a per-share distribution from the firm to its stockholders.

The flow is illustrated as follows:

Assets => Operating Income
– Interest =========> Debt Holders
– Income Taxes =====> Governments
– Dividends =======> Stockholders
= Retained Earnings => Assets (and the cycle repeats)

Note that the firm's earnings that are retained and reinvested in the firm's assets will generally be deployed to the benefit of the firm's shareholders because reinvested earnings increase the firm's assets and the value of the firm's equity is equal to the value of the assets minus the debt.

2.4.2 The Dividend Discount Model

Though modern financial modeling treats stock prices as random variables, it is important to understand the dividend discount model. The value of a common stock, like the value of any financial asset, emanates from the value of its future dividends. While in practice there are stock valuation models based on earnings or book values, in theory the ***dividend discount model*** is the most direct method of valuing stock.

Dividend amounts are set by the board of directors of a corporation who are selected by equity holders. Dividend policy has been a subject of great discussion. Modern corporate finance theory argues that dividend policy is irrelevant to the value of the firm in the absence of taxes and market imperfections. The argument is quite simple: the net earnings of the firm belong to the shareholders and it is irrelevant whether the earnings are distributed immediately to the shareholders in the form of dividends or reinvested in the firm's assets on behalf of the shareholders.

On average, U.S. operating firms pay very roughly one-third to one-half of their earnings to shareholders in the form of dividends. The remainder gets reinvested in the firm causing the firm's assets and equity to grow. Therefore, stockholders derive their investment return in the form of both *dividends* and *capital gains*. Most firms pay dividends – per-share distributions of the firm's earnings – on a regular schedule such as quarterly to their shareholders, but some firms pay no dividends.

To the shareholders, ***capital gains*** are realized when shares are sold at a price above their original purchasing price. An investor may buy a share of stock for $40, collect $1.20 in dividends during the year, and sell the stock at the end of the year for $44. The shareholder will have received a 3% annual dividend yield and a capital gain of 10% for a total annual return of 13%. Unlike the situation of bondholders who received fixed payments according to a specified schedule, neither dividends nor capital gains are fixed obligations of the firm. Thus, the process of valuing common stock is difficult because the key sources of return from stock – dividends and capital gains – must be estimated.

The ***dividend discount model*** specifies the price of a share of stock at time zero, P_0, as the present value of all estimated future dividends. Let's assume that dividends are paid annually. Equation 2.25 depicts the value of a share of stock as the present value of all future expected dividends.

$$P_0 = \sum_{i=1}^{\infty} \frac{DIV_i}{(1 + r)^i} \tag{2.25}$$

The discount rate on the dividends, r, is the market-based rate of return that investors expect and require on investments of similar risk to the stock being valued. It may be viewed as the sum of the risk-free rate and a risk premium for bearing the risk of the stock. In order to simplify the dividend discount model, the future expected dividends are often modeled as growing at a fixed rate g (or a rate that changes through time in a deterministic path).

$$DIV_{i+1} = DIV_i \times (1 + g)$$

$$\text{In general, } DIV_{i+1} = DIV_1 \times (1 + g)^i$$

$$\text{So, } P_0 = \sum_{i=1}^{\infty} \frac{DIV_1 \times (1 + g)^{i-1}}{(1 + r)^i}$$

The growth rate, g, is often modeled as the product of the percentage of earnings (i.e., ***the plowback ratio***) that the firm reinvests or plows back into operations and the firm's profitability as measured by the ***ROE***, or return on equity:

$$g = \text{Plowback ratio} \times ROE$$

Thus, if a firm earns 10% on its equity and reinvests (or plows back) 60% of its annual earnings (and distributes the remainder of the earnings to shareholders in the form of dividends), then both the firm's dividend and equity should grow at 6% per year.

Assuming that the firm's ROE and dividend policy stay constant, the expected value of g is equal for all future years, which enables the expected future dividends to be modeled as a growth perpetuity.

$$P_0 = DIV_1/(r - g) \tag{2.26}$$

Here is the derivation of Equation 2.26

P_0 is a geometric series with initial term $\frac{DIV_1}{1+r}$ and ratio $\frac{1+g}{1+r}$. By the geometric series formula, the sum is $\frac{\frac{DIV_1}{1+r}}{1-\frac{1+g}{1+r}} = DIV_1/(r-g)$.

Note that the numerator in Equation 2.26 is the expected level of next year's dividend, assumed to be equal to the most recently paid dividend grossed up by the growth rate.

Equation 2.26 can be factored into Equation 2.27 to illustrate the role of dividend yield and capital gains in forming the total return to a stockholder.

$$r = (DIV_1/P_0) + g \tag{2.27}$$

Example 2.13: Consider a stock that recently paid a $1 annual dividend and is expected to pay steadily increasing annual dividends forever. The firm's ROE is 10% and its dividend payout ratio is 25%. What is the value of the stock if investors expect and require an annual rate of return of 12% on similar investments? For a second challenge, ignore the 12% rate of return and determine what rate of return the stock offers if its price is $40.

First, estimate the annual growth rates in dividends using g = Plowback ratio × *ROE*. The plowback ratio is one minus the dividend payout ratio or 0.75, generating a growth rate of 7.5%. The annual growth rate can then be used to calculate the next annual dividend: $1 × 1.075 = $1.075. Equation 2.26 solves the first question with g = 7.5%, DIV_1 = $1.075, and r = 12%. The solution is P_0 = $23.89.

Equation 2.27 solves the second question using the known values for DIV_1, P_0, and g: ($1.075/$40) + 0.075 = 10.19%.

2.4.3 Technical and Fundamental and Analysis

One of the first techniques of security selection, popular centuries ago, and still utilized by some investors today, is ***technical analysis***. In technical analysis, information concerning past prices and trading volume is used to try to predict future price movements to enhance investment returns. Some technical analysts use charts of past prices and trading volume in an attempt to find favorable trades. Others use computer models.

For instance, a technical analyst might study patterns of stock price movements and conclude that stock prices tend to move in distinct waves, with the result that brief periods of price movements in one direction are often followed by reversals in the other direction.

Technical analysts might chart the movements of stocks over some previous period of trading in order to identify those that are prime candidates for reversal. A key to many technical trading strategies (e.g., the reversal strategy) is that past price patterns are assumed to be useful in predicting future stock price changes.

A more recent type of investment technique, called ***fundamental analysis,*** became popular after the Great Depression. Fundamental analysis attempts to find the true value of a security using publicly available information about the firm's underlying economic value such as financial statements. Fundamental analysts look for securities whose market price differs from their true price.

For example, fundament analysts might examine a company's financial statements to identify information such as financial ratios and to compare them with ratios of similar firms. One popular ratio, for example, divides the price of the firm's stock by its earnings per share. Fundamental analysis includes any financial analysis that utilizes underlying economics information about a firm, an industry, or even an entire economy.

2.4.4 Informational Market Efficiency

In the last 60+ years, there has been a movement toward *modern portfolio theory* as a major driver of investing decisions. Often modern portfolio theory is not used in an attempt to choose underpriced securities using publicly available information; rather, it attempts to control risk and form well-diversified portfolios. Many modern portfolio theoreticians believe that the only way to beat the market is to have information not available to the general public. According to modern portfolio theory, investors who do not possess such information should not expect to outperform other investors consistently.

In the broadest sense, the concept that the price of an asset is equal to its value is known as ***informational market efficiency***. An *efficient market* exists when the assets in the market are traded at prices that equal their values, based upon all available information. In a perfectly efficient market technical and fundamental analysis cannot be used to generate consistently superior returns.

However, if any market were perfectly informationally efficient there would be no incentive for technical or fundamental analysts to gather, process, and utilize information in their investment decisions. If no investors analyze information for investment purposes, then market prices could not possibly reflect the information. Accordingly, markets should not be labeled as being completely efficient or completely inefficient. Perhaps it is more useful to describe the degree of efficiency in various markets – which will tend to be higher when the markets are large, the analysts are many, information is highly available, and transactions costs and other impediments to trading are minimal. There is reasonable evidence that most major U.S. financial markets – for example, the stock markets and bond markets – are reasonably informationally efficient. More specifically, it is quite unlikely that common forms of analysis and the use of widely available information will generate consistent and superior risk-adjusted returns.

2.4.5 Dividend Policy and Theory

Dividend policy is a corporation's strategy with regard to the long-term decisions as to which portion of earnings to distribute directly to stockholders and which portion to retain inside the firm. Either way, the shareholder is entitled to the wealth – as a current dividend or as a higher future dividend.

Given that Equation 2.25 describes the value of a firm's stock as the present value of its anticipated dividends, it would seem that dividend policy is very important. But in the absence of market imperfections, modern corporate finance theory argues that, typically, the dividend decision is irrelevant. Shareholders desiring large dividends can satisfy their desire for cash by selling off a portion of their stock holdings. Shareholders who have no immediate preference to receive cash dividends can reinvest the cash in additional shares of stock. From the corporation's perspective, dividend policy should typically have little or no effect on the operations of the firm. Corporations paying large dividends can replenish their cash balances by issuing new shares to the public or by borrowing. Corporations that pay little or no dividends and are flush with cash can repurchase their own shares in the marketplace, pay down debt, or invest the excess cash in marketable securities such as the preferred stock of other corporations.

Some analysts express concerns that large dividends can increase the probability that a firm experiencing low earnings could increase its level of financial distress and ultimately even force the firm into bankruptcy. After all, it is true that dividends remove equity and cash from the firm. However, modern corporate finance theory includes rationale that even the threat of potential bankruptcy should not make dividend policy an *important* issue from the perspective of shareholders for reasonably healthy corporations.

However, there is full agreement on one issue: dividend policy changes appear to signal important information to the market. This explains why stock prices often experience large changes when an announcement is made with regard to a large change in dividends or dividend policy. Typically, a large dividend increase is viewed as indicating that a corporation's management team is optimistic about the financial future of the firm, while a dividend cut reveals the opposite. However, even that tendency has exceptions. Research indicates that sometimes dividend increases can be a sign that a corporation has few attractive uses for its cash and sometimes dividend decreases can be a sign that corporations have highly attractive uses for its cash.

Here is a summary of the implications of dividends and dividend policy with regard to financial derivative valuation. When a stock pays a dividend, it transfers shareholder wealth from inside the firm directly to the shareholder. As a result, and as is obvious from inspection of Equation 2.25, the price of a share of stock should drop by the amount of the dividend on the day after the legal right to the cash transfers from the corporation to the shareholder (the ex dividend date). The price of a financial derivative prior to that date should reflect the anticipated dividend (because dividends are announced well in advance of their distribution). Accordingly, the price of a financial derivative on a share of stock should not systematically rise or fall on a dividend distribution date (or the coupon distribution date of a bond) because the distribution has already been anticipated in the derivative's price. This means that a firm's dividend *policy* should be included in the valuation of a financial derivative on a stock such as an option.

The next section discusses equity risk management.

2.5 Asset Pricing and Equity Risk Management

Since the 1950s, enormous advancements have been made in the field of asset pricing. Asset pricing models generally describe ***asset returns*** (i.e., percentage price changes). The foundation for the field of asset pricing was laid by Harry Markowitz in the 1950s. The primary breakthrough, the capital asset pricing model, was made in the 1960s. The remainder of this chapter summarizes those insights. We leave the derivation of the formulas and the details to Chapter 6.

2.5.1 Two-Asset Portfolios and Markowitz Diversification

Nobel laureate Harry Markowitz provided a seminal contribution to portfolio theory by modeling equity returns as normally distributed random variables and focusing on mean and variance as measures of return and risk.

The ***mean*** or ***expected return of a portfolio*** is simply the weighted average of the mean or expected returns of the portfolio's constituent assets (where weights are defined as the percentage or proportion of the portfolio's total value attributable to the market value of each asset). Equation 2.28 depicts a portfolio's expected return as a function of the expected returns of its constituent assets. Markowitz demonstrated that the ***variance of the returns of a portfolio*** depends both on the variances of the returns of the portfolio's constituent assets and the *correlation coefficients* between the returns of each pair of assets. Consider the simple case of two assets such as stocks, X and Y. Equation 2.29 applies simple statistics to describe the effects of the correlation coefficient between the assets, ρ_{xy}, on the variance of a portfolio, $\sigma^2{}_p$.

$$E(p) = w_X \times E(X) + w_Y \times E(Y) \tag{2.28}$$

$$\sigma_P^2 = w_X^2\sigma_X^2 + w_Y^2\sigma_Y^2 + 2w_X w_Y \sigma_X \sigma_Y \rho_{XY} \tag{2.29}$$

where $E(X)$ and $E(Y)$ are the expected returns of stocks X and Y, σ_x and σ_y are the standard deviations of the returns of stocks X and Y, respectively, ρ_{XY} is correlation coefficient between the returns of stocks X and Y, w_x is the portion of the portfolio's total value attributable to the value of asset X, and σ_x^2 is the variance of the returns of asset X. The standard deviation of the returns of an asset is often referred to as its ***volatility*** in finance. Note that Equation 2.29 is often expressed using covariance rather than correlation coefficient, namely replacing $\sigma_X\sigma_Y\rho_{XY}$, by covariance σ_{XY}. Correlation coefficient is used here because it offers a scaled indication of the linear relationship between two variables that is always greater than or equal to −1 and less than or equal to +1 with 0 indicating no linear relationship. Example 2.14 illustrates mean-variance analysis for a two-asset portfolio.

Example 2.14: Consider a portfolio with two stocks as indicated:

Stock	Weight	Expected Return	Standard Deviation
X	0.60	8%	30%
Y	0.40	6%	20%

Calculate the standard deviation of the returns of the portfolio of these two assets under the following three values for the correlation coefficient between X and Y: 0.0, 0.4, and 0.8.

Note that only the last term in Equation 2.29 contains $\rho_{xy.}$ This solution begins with the first two terms: $w_x^2\sigma_x^2 = 0.60 \times 0.60 \times 0.30 \times 0.30 = 0.0324$ and $w_y^2\sigma_y^2 = 0.40 \times 0.40 \times 0.20 \times 0.20 = 0.0064$. The last term is $2 \times 0.60 \times 0.40 \times 0.30 \times 0.20 \times \rho_{xy} = 0.0288\ \rho_{xy}$. The three cases of ρ_{xy} are 0.0, 0.4, and 0.8. For $\rho_{xy} = 0.0$, the portfolio variance is the sum of the first two terms: $0.0324 + 0.0064 = 0.0388$, which is a standard deviation of 0.197. For $\rho_{xy} = 0.40$, the portfolio variance is $0.0388 + 0.0288 \times 0.40 = 0.05032$ (standard deviation = 0.224) and for $\rho_{xy} = 0.80$ the portfolio variance is $0.0388 + 0.0288 \times 0.80 = 0.06184$ (standard deviation = 0.249). Note that as the correlation between the two stocks increases the portfolio's volatility increases from 0.197 to 0.224 to 0.249. With $\rho_{xy} = 1$ the standard deviation of the portfolio is simply a market-weighted average of the standard deviations of the two assets (0.26) because no diversification takes place.

Figure 2.2 demonstrates the expected returns and standard deviations of the returns of a portfolio of assets X and Y when varying the proportion w of the portfolio in assets X and Y under four scenarios: perfect correlation of the returns between assets X and Y ($\rho = 1$), zero correlation ($\rho = 0$), 0.5 correlation ($\rho = 0.5$), and perfect negative correlation ($\rho = -1$). The

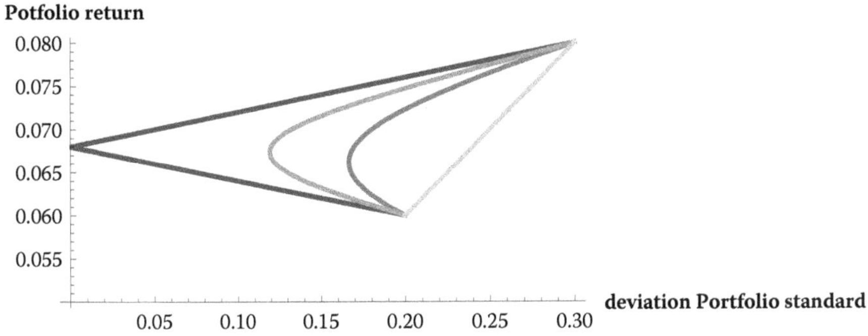

FIGURE 2.2
Risk-Return Opportunities for Two-Asset Portfolios with Four Potential Return Correlation Coefficients.

correlation coefficient, ρ, is the covariance divided by the product of the standard deviations: $\rho = \sigma_{XY}/\sigma_x\sigma_y$.

The curves from left to right correspond to four different correlations. For a fixed correlation, ρ, each point on a particular curve corresponds to different weight allocations (w_X, w_Y).

Note that $w_Y = 1 - w_X$. So, we can regard $E(p)$ as a function of w_X, and σ_P as a function of w_X.

So, the curve corresponds to the following parametric equation with $0 \leq w_X \leq 1$.

$$\begin{cases} E(p) = w_X \times E(X) + w_Y \times E(Y) \\ \sigma_P = \sqrt{w_X^2\sigma_X^2 + w_Y^2\sigma_Y^2 + 2w_Xw_Y\sigma_X\sigma_Y\rho_{XY}} \end{cases}$$

The following Mathematica codes can be used to generate the curves. Readers can generate their own curves by varying the correlation coefficients.

Mathematica Code 2.1 Generate two asset Markowitz efficient frontiers with different correlations

```
rho1=-1;
f1[w_]:=0.08*w+0.06*(1-w)
g1[w_]:=Sqrt[0.3^2*w^2+0.2^2*(1-w)^2+2*w*(1-
w)*0.2*0.3*rho1]
rho2=0;
f2[w_]:=0.08*w+0.06*(1-w)
g2[w_]:=Sqrt[0.3^2*w^2+0.2^2*(1-w)^2+2*w*(1-
w)*0.2*0.3*rho2]
rho3=-0.5;
f3[w_]:=0.08*w+0.06*(1-w)
g3[w_]:=Sqrt[0.3^2*w^2+0.2^2*(1-w)^2+2*w*(1-
w)*0.2*0.3*rho3]
rho4=1;
f4[w_]:=0.08*w+0.06*(1-w)
g4[w_]:=Sqrt[0.3^2*w^2+0.2^2*(1-w)^2+2*w*(1-
w)*0.2*0.3*rho4]
ParametricPlot[{{g1[w],f1[w]},{g2[w],f2[w]},
{g3[w],f3[w]},{g4[w],f4[w]}},{w,0,1},
AspectRatio->0.5,   AxesOrigin->{0,0.05},PlotStyle->
{Thick}]
```

In the case of stocks, the estimated correlation coefficients over various pairs should tend to range from small positive values to values somewhat near 1, depending on the length of the time interval spanned by the returns and the nature of the two stocks. The primary point is this: the relationship between the average of the risks of the two assets and the risk of the portfolio of the two assets tends to resemble the two middle curves in Figure 2.2. Specifically, when imperfectly correlated assets are combined into a portfolio, the portfolio's risk is reduced relative to the averaged risks of the two assets, as illustrated by the bending of the opportunity set from right to left. This is a visual representation of diversification. ***Diversification*** is a reduction in risk from combining assets into a portfolio that have return correlation coefficients less than 1.

Markowitz extended the analysis to a portfolio of many or even all available assets. The resulting optimal portfolios form a hyperbola (similar to the case of zero correlation illustrated in Figure 2.2) if allowing short sales and several pieces of hyperbolas (similar to the case illustrated in Figure 2.3) if disallowing short sales. The top portion of this curved line represents the highest expected return opportunities for each feasible level of risk, or, equivalently, the lowest

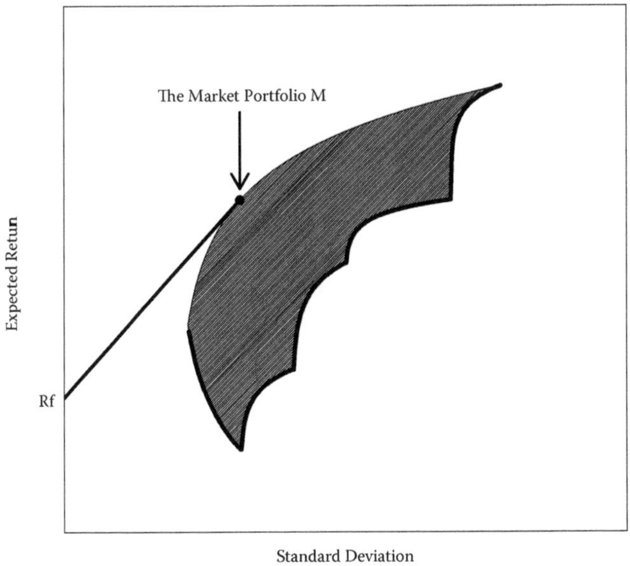

FIGURE 2.3
Markowitz Efficient Frontier.

total risk for each feasible level of expected return. The Markowitz Model's curved line is the set of all potentially optimal portfolios and is called the ***Markowitz efficient frontier***. Chapter 6 provides a thorough analysis of the Markowitz efficient frontier using linear algebra, the covariance matrix, and Lagrange multipliers for cases of more than two assets.

Markowitz modeled portfolio selection as the choice of where to invest on the efficient frontier. Each investor was viewed as considering the portfolios represented on the efficient frontier and selecting that portfolio that best suited the investor's preferences regarding risk and return. Markowitz's path-breaking work raised a major problem. In his model, various investors select different locations on the efficient frontier for investing – based on their different degrees of risk aversion driven for example by having different levels of initial wealth or time horizons. Each point on the efficient frontier has a different slope – indicating that investors selecting different risky portfolios (i.e., points on the efficiency frontier) experience a different trade-off between marginal risk (portfolio return standard deviation) and marginal expected return (portfolio return). This key issue is detailed in the next section as well as in Chapter 6. Note that the term *marginal* is an important concept in economics. A ***marginal*** quantity is the change that occurs over a very small change in its determinants. For example, a person's marginal utility (added happiness) from eating one piece of pizza may be quite positive. But the *total* change in his or her utility from eating an entire very large pizza might be quite different.

2.5.2 Investor Preferences and Market Equilibrium

One of the foundations of microeconomics is that there are enormous benefits when individuals exchange goods until each person's ratio of marginal preferences matches the corresponding ratio of market prices. For example, suppose that the market price per pound of apples is twice the market price of bananas per pound, and that all people consume both apples and bananas. In a perfect market, people will exchange until each person's marginal utility from consuming apples is twice his or her marginal utility from consuming bananas. Further, every producer of apples and bananas adjusts production such that the marginal cost of producing a pound of apples is twice that of producing a pound of bananas.

Financial markets are focused on exchanging capital involving time and risk. The lessons of microeconomics demonstrate that an optimal equilibrium occurs when every investor and every user of capital engages in exchange until everyone shares the same perception of the marginal tradeoff between risk and return. But in the Markowitz model, each investor can be on a different point along the efficient frontier and therefore can perceive a different trade-off between risk and return. Worse yet, each investor perceives the underlying risk differently because each portfolio on the efficient frontier is imperfectly correlated with every other portfolio. The result is an inefficient allocation of risk across society.

Consider a very wealthy young professional and an elderly retiree of modest means. If both invest within the Markowitz framework, it is reasonable to assume that the wealthy professional will select a portfolio of high risk and return (near the upper-right region of the efficient frontier) and that the retiree will select a portfolio of low risk and low return (near the lower-left region). In practice, the professional's portfolio would tend to contain high-risk stocks such as pharmaceuticals, high-tech ventures, and distressed firms. The retiree's portfolio would tend to contain large, established firms with safe earnings prospects. In that case, the two investors would perceive investment risk very differently – with the retiree more concerned about large, safe stocks and the professional more concerned about speculative stocks. The obvious concern is that different investors are missing opportunities to engage in mutually beneficial exchange. Additionally, the heterogeneity of their attitudes toward risks raises the practical dilemma that the financial manager of a corporation, who is charged with making decisions to benefit all shareholders, does not know whose risk preferences should dominate when the manager makes investment decisions and other important corporate decisions with regard to risk.

In the 1960s, the ***Capital Asset Pricing Model(CAPM)*** was developed, providing a solution to these problems. Unlike Markowitz's model, the

CAPM is a general equilibrium model – meaning that all economic decisions are determined internally within the model rather than externally imposed on the model. Derivative pricing models tend to be *partial equilibrium* models because the prices of underlying assets are exogenously determined (i.e., are imposed from outside rather than determined within the model). Very importantly, the CAPM generates a solution in which all investors are concerned about the same risk and all investors share the same risk-return trade-off – both of which address concerns regarding economic efficiency and corporate governance.

The next three sections provide a summary of this model and its important implications. Briefly, the CAPM derives a single measure of risk entitled *beta* and a single market risk premium (i.e., a single market price to risk).

2.5.3 Systematic and Diversifiable Risk

The key point to the diversification of risk in Markowitz's model is that when risky assets are combined into a portfolio, a portion of the portfolio's risk is diversified away. The risk that can be removed through diversification is called ***diversifiable risk***. Diversifiable risk should be viewed as harmless (and undeserving of added return in a competitive market) because diversifiable risk can be easily avoided by combining assets into large portfolios. Other terms used to describe diversifiable risk include nonsystematic risk, unique risk, non-market risk, and firm-specific risk.

Not all risk, however, can be diversified away. The risk that remains even after assets are combined into huge and well-diversified portfolio is called ***nondiversifiable risk, systematic risk, beta risk,*** or ***market risk***. Systematic risk is the tendency of an asset's returns to be correlated with the returns of all assets combined. The hypothetical portfolio containing all risky assets is called the market portfolio. The ***market portfolio*** not only includes all assets, it includes them in proportion to their size. The returns of most assets are positively correlated with the returns of the market portfolio.

The economics of equity risk is therefore conceptually simple. Most assets contain a combination of diversifiable and systematic risk. The diversifiable risk is not priced and vanishes as investors seek diversification. The systematic risk is priced: investors demand higher expected returns for bearing higher correlation with the market portfolio. This discussion of risk focuses on risks in equity markets and other investment markets wherein assets generate returns that are correlated with the performance of the overall economy. The risk management of fixed-income investments such as bonds and bond funds focuses on interest rate risk and credit risk as previously discussed in this chapter, and as further discussed in Chapters 5 and 7. The risk management of financial derivatives focuses on risk exposures known as the Greeks and on counterparty default risk which is discussed in Chapter 5 and Chapter 8, respectively.

The CAPM solves the challenges of the Markowitz model. In the CAPM, diversifiable risk is viewed as offering no risk premium and it is concluded that no rational investor would select a portfolio highly exposed to that risk. In the CAPM, the only risk that is widely borne and that offers a risk premium is systematic risk. In the CAPM, investors diversify until the only meaningful risk is systematic risk, and all of the benefits of exchange with regard to systematic risk are utilized. The result is that there is a consensus on the single risk that investors bear (systematic risk), there is a consensus on the single source of systematic risk (changes in total global wealth), there is a single risk premium (the market risk premium), and there is consensus among investors on the trade-off between risk and return. The next section explains the CAPM.

2.5.4 The Capital Asset Pricing Model (CAPM)

One of the most restrictive assumptions in the Markowitz model is that it ignores riskless investment opportunities such as short-term Treasuries. When the riskless asset is added to an investor's opportunity set, the optimal portfolios are indicated by a line formed by the combination of the riskless asset and the portfolio of risky assets that forms a point of tangency with the efficiency frontier as shown in Figure 2.3 in section 2.5.1. Figure 2.3 plots all investment opportunities with expected return on the vertical axis and total risk (standard deviation) on the horizontal axis. The hyperbola is the efficient frontier with no riskless asset allowed and the straight line is efficient frontier by adding a riskless asset.

When a riskless asset is added to Markowitz's model, investors can invest anywhere along the straight line between the riskless return, R_f, and the portfolio, M, that lies at the point of tangency (by varying the proportions of the riskless asset and the market portfolio). As discussed later, portfolio M is the ***market portfolio*** – the hypothetical portfolio containing all of the assets in the world. Conservative investors may opt for a portfolio dominated by the riskless asset (e.g., money market securities and short-term CDs at the intercept). Aggressive investors may opt to place most or all of their money in the market portfolio. In a perfect market, it is possible for investors to not only place all their funds in the market portfolio, but also to borrow at the riskless rate (pledging the total portfolio as collateral) to buy more stock. This financial leverage can be modeled as extending the straight line in Figure 2.3 farther up and to the right (not illustrated).

Note that all investors select portfolio M as their only holding of risky assets because it is the opportunity that offers the highest risk-return trade-off. There are several reasons that portfolio M must contain all available risky assets. First, if an asset were not in portfolio M, then no investor would select it and its price would fall until it was attractive enough to be included in the portfolio. In fact, each asset's price can be viewed as adjusting until the demand for that asset by investors in portfolio M exactly matches the supply of that asset. Second, portfolio M is the ultimately diversified portfolio – and it makes intuitive sense that ultimate diversification must include all risky assets. Finally, portfolio M contains the only risk that must be borne in a perfect market – the risk of fluctuations in the total wealth of the world.

The ***systematic risk*** of each investment is measured by beta. Beta is the key measure of risk in the CAPM. Beta is calculated as shown in Equation 2.30:

$$\beta_i = \sigma_{im}/\sigma_m{}^2 = \rho_{im}\sigma_i/\sigma_m \tag{2.30}$$

where σ_{im} is the covariance and ρ_{im} is the correlation coefficient between the returns of the asset and the returns of the market. Also, σ_m^2 is the variance of returns of the market, and σ_i and σ_m are the standard deviations of the returns of asset i and the market, respectively.

In the middle expression, the numerator measures the amount of risk that an individual stock adds to an already diversified portfolio. The denominator represents the total risk of the market portfolio. Thus, beta measures the systematic risk of an asset in relation to the systematic risk of the market. Note that the beta of the market portfolio must equal 1. Because the beta of the overall market is equal to 1, assets with betas greater than 1 signify a level of systematic risk greater than the overall market, and assets with betas less than 1 signify a level of systematic risk less than the overall market.

Example 2.15: Consider three stocks, all of which have a correlation coefficient of 0.50 with the market portfolio. The standard deviations of the three stocks are 0.20, 0.30, and 0.40. If the standard deviation of the returns of the market portfolio is 0.125, what is the beta of each stock?

The beta of each stock is found using $\rho_{im}\sigma_i/\sigma_m$, which in this case generates $0.50 \times \sigma_i/0.125$. Inserting the three standard deviations corresponding to the returns of the three stocks generates betas of 0.80, 1.20, and 1.60.

Given an asset's beta, we can use the following CAPM equation to determine the asset's expected or required rate of return (the derivation of the CAPM is in Chapter 6).

$$E(R_i) = R_f + \beta_i[E(R_m) - R_f] \tag{2.31}$$

Equation 2.31 depicts the key concept in portfolio theory, investments, and financial economics in general. It indicates that the expected return of any asset is a combination of compensation for the time value of money (R_f), and compensation for bearing the risk of general economic fluctuations: $\beta_i[E(Rm) - R_f]$. Economies inherently contain risk, and Equation 2.31 describes how assets will be priced with respect to that risk in a perfect market. The term β_i measures the quantity of risk for asset i, while the term $[E(Rm) - R_f]$ is the market price of each unit of risk. Note that for $\beta_i = 0$ the expected return is the riskless rate, and for $\beta_i = 1$ the expected return is the expected return of the market portfolio. Investors who contribute to society by bearing large amounts of systematic risk (high betas) will on average be rewarded with higher expected return. Investors who bear unnecessary risks (e.g., holding poorly diversified portfolios or engaging in casino gambling) will not receive higher expected returns from risk premiums.

The CAPM equation is illustrated in Figure 2.4, where the x-axis is systematic risk (beta) and the y-axis is expected return. The line in Figure 2.4 is called the ***security market line.*** Figure 2.4 differs from Figure 2.3 because the horizontal axis measures the *systematic risk*, or *beta*, of each opportunity rather than the total risk (standard deviation).

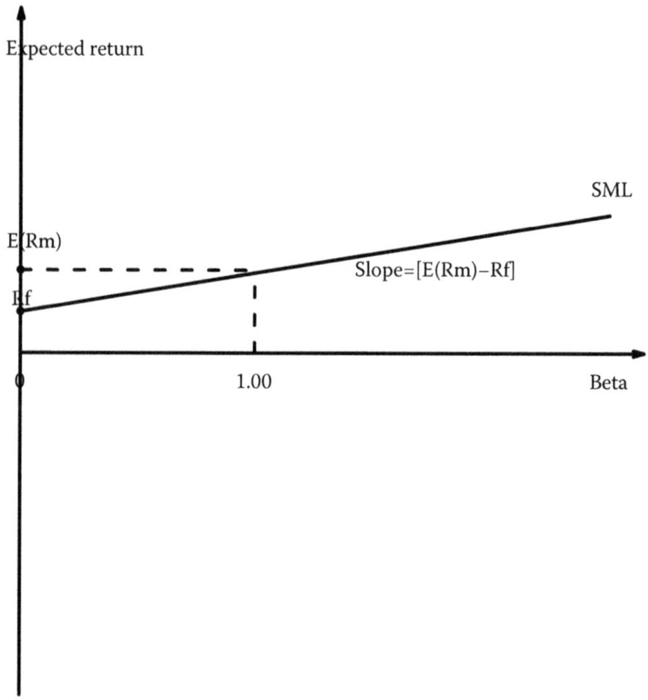

FIGURE 2.4
The Security Market Line of the CAPM.

Example 2.16 provides practice in applying the CAPM model.

Example 2.16: Complete the table by finding the missing values of the CAPM:

E(Ri)	R_f	β_i	E(Rm)
a	2%	0.80	10%
6%	b	0.50	8%
8%	1%	c	11%
12.5%	2%	1.5	d

Inserting the three known values in each case into $E(Ri) = R_f + \beta_i[E(Rm) - R_f]$ generates $a = 8.4\%$, $b = 4\%$, $c = 0.7$, and $d = 9\%$.

The CAPM should not be viewed as an extremely accurate and unquestioned method of describing the relationship between risk and return – rather, it should be viewed as a good tool for: (1) separating risks into those that do and do not require rewards and (2) for estimating the amount of return that the market requires for bearing a particular level of systematic risk. It is especially noteworthy that the CAPM asserts that the expected return of every asset depends on only one attribute unique to that asset: its beta.

2.5.5 Volatility and Beta

Especially in derivative valuation, it is important to have a firm understanding of the difference between ***total risk*** (measured by volatility) and ***systematic risk*** (measured by beta). Volatility (σ) is the standard deviation of the returns of an asset. Beta (β) is a volatility-scaled measure of the correlation of the returns of an asset to the returns of the market portfolio. Volatility is a key measure of stand-alone total risk. Beta is a key measure of the risk that is rewarded through a risk premium in a CAPM world.

Note that volatility is positively related to the time horizon. The volatility of the annual returns of the S&P 500 stock index in recent years has tended toward perhaps 12%–18%. Daily volatility has tended between 0.7% and 1.1%. In fact, if stock returns through time are not serially correlated, and if the single period volatility is σ, then the volatility of asset returns over T time periods is $\sigma\sqrt{T}$.

The beta of individual stocks is quite unpredictable. Beta estimates differ based on the granularity of the returns (daily returns, weekly returns, ...), the selection of the equity index serving as a market proxy, the specification of the riskless rate, the sample period, and the number of observations. However, in much of derivative valuation, the estimation of beta is unnecessary because of the risk-neutrality principle – the concept that in many cases the value of a derivative is independent of risk aversion.

Finally, one of the most important developments in financial markets in the last two decades has been the increasing volume and convenience of trading with regard to volatility levels. There are numerous methods by which investors and speculators can take positions that create or hedge exposure to the realized or anticipated volatility of major market indices. For example, there are numerous opportunities to place trades regarding the volatility of the S&P 500 – a premier index of major U.S. equities. Those trading opportunities have soared in popularity in recent years and include futures contracts and exchange traded products such as exchange traded notes and exchange traded funds.

The popularity of volatility trading opportunities extends far beyond interest from speculators. The importance of these opportunities emanates from the crux of the CAPM: it is the volatility of the overall market that captures the key risk of economic activity – systematic risk. Volatility products allow market participants to manage not just their systematic risk, but also their exposures to changes in the anticipated or realized levels of that systematic risk.

Chapter Summary

Chapter 2 introduces foundational concepts about interest rates, fixed-income securities, and equities. Time value of money is important in finance. Cash flows occurring at different times have different present values. Interest rates are used to provide an intuitive expression of the time value of money but they are quoted with different compounding methods. In addition, market interest rates are a function of time known as the term structure of interest rates. Each default-free zero coupon bond can be valued by discounting its promised cash flow with the market interest rate matching its maturity. Coupon bonds can also be valued using rates from the term structure of zero rates corresponding to each cash flow or heuristically by discounting all cash flows by a single artificial interest rate known as a bond yield. Bootstrapping is a technique to extract the term structure of zero rates from the market quotes of coupon bonds prices. Forward interest rates are discussed and compared with spot rates. Duration and convexity are introduced as important risk measures in the management of interest rate risk. Equities and equity valuation models are briefly discussed. Finally, the chapter introduces Markowitz portfolio theory and the capital asset pricing model (CAPM) as a foundation for later chapters.

End-of-Chapter Problems (Fundamental Problems)

1. \$100 is invested in a bank CD. How much will the money grow to at the end of five years if it is compounded
   a. annually with an annual interest rate of 3%?
   b. quarterly with an annual interest rate of 3%?
   c. monthly with an annual interest rate of 3%?
   d. weekly with an annual interest rate of 3%?
   e. daily with an annual interest rate of 3%?
   f. continuously with an annual interest rate of 3%?
2. \$100 is invested in a CD of Bank A for five years at a continuously compounded interest rate of 3% annually. \$100 is also invested in a CD of Bank B for five years at an annually compounded interest rate of 3%. Which bank offers more money after five years?
3. Bank A offers a continuously compounded interest rate of 3%. What annually compounded interest rate must Bank B offer to match Bank A's offer?
4. Bank A offers an annually compounded interest rate of 3%. What continuously compounded interest rate must Bank B offer to match Bank A's offer?
5. Risk-free annual interest rates are reported as follows (all interest rates are compounded continuously):

Half-year	3.0%
One year	3.5%
One and a half years	3.8%
Two years	4.0%

   a. What is the price of a two-year risk-free coupon bond that pays its coupon semiannually at an annual rate of 5%? (Assume that the bond's principal value is \$100.)
   b. What is the annualized yield, y, for this bond? (Use the trial-and-error method to find yield to the nearest 0.01%.)

6. Risk-free annual interest rates are reported as follows (all interest rates are compounded continuously):

Half year	3.0%
One year	3.5%
One and a half years	3.8%
Two years	4.0%

a. What is the price of a two-year risk-free coupon bond that pays its coupon semiannually at an annual rate of 9%? (Assume that the bond's principal value is $100.)

b. What is annualized yield, y, for this bond? (Use the trial-and-error method to find yield to the nearest 0.01%.)

c. What is the difference between the annualized yields in this problem and the previous problem? Explain why they differ.

7. Assume that a two-year bond with a face value of $100 pays a semiannual coupon at an annual rate of 6%. Use the following zero coupon rates to find the current bond price.

Maturity	Zero-Coupon Rates (Continuously Compounded)
0.5	5.0%
1.0	5.5%
1.5	6.5%
2.0	7.0%

8. Consider the following five bonds each with a face value of $100.

Maturity	Coupon per Year	Current Bond Price
0.25	0	97.5
0.5	0	95
1.0	0	90
1.5	8%	96
2.0	12%	101.5

a. Find the zero coupon rates for each maturity (assume continuous compounding).

b. Find the continuously compounded yields for the two coupon bonds.

9. The six-month zero coupon rate is 8% (with semiannual compounding) and the price of a $100 face value one-year bond that provides an annual coupon of 6% (paid semiannually) is $97. What is the one-year continuously compounded zero rate? Answer as a percent with two-decimal-place accuracy.

10. Compute the following annual interest rates assuming continuous compounding.

a. A six-month zero coupon bond with a \$100 principal amount has a current price of \$98. What is the zero coupon interest rate for a maturity of six months?
b. A one-year zero coupon bond with a \$100 principal amount has a current price of \$95. What is the zero coupon interest rate for a maturity of one year?
c. What is the forward annual interest rate for the period from 6 months to 12 months?

11. When the zero coupon interest rate curve is upward sloping, evaluate whether each of the following statements is true or false (assume all rates are continuously compounded). Explain why.
    a. The 1.5-year zero coupon rate must be greater than the forward rate for the period between 1 year and 1.5 years.
    b. The two-year zero coupon rate must be less than the bond yield on a two-year coupon bearing bond.
12. If the two-year zero coupon rate is 3% annually (compounded continuously) and the three-year zero coupon rate is 4% annually (compounded continuously), what is the continuously compounded forward rate for the period between years two and three? Is it higher or lower than the two-year zero coupon rate?
13. If the two-year zero coupon rate is 4% annually (compounded continuously) and the three-year zero coupon rate is 3% annually (compounded continuously), what is the continuously compounded forward rate for the period between years two and three? Is it higher or lower than the two-year zero coupon rate?
14. If the two-year zero coupon rate is r_2 annually (compounded continuously) and the three-year zero coupon rate is r_3 annually (compounded continuously), what is the formula for the continuously compounded forward rate for the period between years two and three, $f_{2,3}$?
15. If the two-year zero coupon rate is r_2 annually (compounded continuously) and the three-year rate is r_3 annually (compounded continuously), use the forward rate for the period between years two and three, $f_{2,3}$, to provide a sequential order of r_2,r_3 and $f_{2,3}$ for two cases regarding r_2 and r_3 other than the case where they are equal.
16. A pension fund's portfolio consists of a five-year zero coupon bond with a principal amount of \$5,000 and a 10-year zero coupon bond with a principal amount of \$10,000. The current continuously compounded yield on all bonds is 8% per annum.
    a. What is the duration of the portfolio?

b. Use duration to approximate the change in the portfolio value from a 0.1% per annum increase in the yields for all bonds. Compare that estimate to the actual price change based on discounting the bonds at the new yield.

c. Use duration to approximate the change in the portfolio value from a 1.0% per annum increase in yields for all bonds. Compare that estimate to the actual price change based on discounting the bonds at the new yield.

d. In which of the above scenarios is the duration-based approximation better: scenario (b) or (c)? Why is the approximation better for the scenario you chose?

17. Consider a three-year coupon bond with a 6% coupon rate paid annually and a face value of $100. Assume that the current continuously compounded yield on all bonds is 8% per annum.

a. Without calculating, guess the duration: Is it near 1, near 2, or near 3?

b. Calculate the duration.

c. Calculate Convexity.

18. The following table lists information for four stocks:

	Stock A	Stock B	Stock C	Stock D
Expected return	13%	8%	13%	8%
Standard deviation	20%	12%	20%	12%

Assume that there are two portfolios 1 and 2. Portfolio 1 is an equally weighted portfolio of Stocks A and B that have a return correlation of 0.30. Portfolio 2 is an equally weighted portfolio of Stocks C and D that have a return correlation of −0.30.

a. Based on Markowitz's portfolio theory, which portfolio (1 or 2) is more attractive?

b. Assume that you invest 40% into Stock C and 60% in Stock D, what is the expected return of your portfolio? What is standard deviation of the returns of your portfolio?

c. Assume that you invest 60% into Stock C and 40% in Stock D, what is the expected return of your portfolio? What is standard deviation of the returns of your portfolio?

d. Discuss your preference between investing in the portfolio from parts (b) or (c).

e. Use the Mathematica code offered in the book to draw the expected return and standard deviation graph with different weights for investing in Stocks A and B first, and then draw the graph for Stocks C and D. Compare the two graphs and discuss which portfolio is more attractive.

19. Consider six stocks: Stock A has $\beta = 1$, Stock B has $\beta = 1.2$, Stock C has $\beta = 0.8$, Stock D has $\beta = 0$, Stock E has $\beta = -0.5$, and Stock F has $\beta = -1$ (where β measures systematic risk). Assume that the risk-free rate is 5% and the expected return of the market portfolio is 9%.
    a. Use the CAPM to find the expected return for each stock.
    b. Explain why the expected return of a negative β stock should be lower than the risk-free rate. In other words, why would a risk averse investor select negative beta stocks when such stocks have positive total risk (standard deviations) and offer expected returns lower than the riskless rate?

20. Assume that firm's A and B have ROEs of 8% and each has a dividend payout ratio of 20%. Both stocks recently paid a $1 annual dividend and are expected to pay steadily increasing annual dividends forever.
    a. What is the value of the Stock A if investors expect and require an annual rate of return of 10% on similar investments?
    b. If we know that the current price of Stock B is $40, what is its anticipated rate of return?

3

Fundamentals of Financial Derivative Pricing

This chapter examines the value of a financial derivative prior to its expiration or settlement date, T. If we term the underlying asset price S_t, where t is any time between $[0, T]$, the value of a financial derivative may be described as a function of S_t and t: $f(S_t, t)$ or f_t. Note that $f(S_t, t)$ is its value at any time t, $f(S_0, 0)$ is the value today, and $f(S_T, T) = f_T$ is the value at expiration which is also known as the derivative's payoff at expiration T.

The term *price* is best used to represent a market price, while the terms *value* or *theoretical price* are the fair value derived from a model, which can differ from the market price. This chapter, like most of financial economics, focuses on *partial equilibrium models,* in which the observed market *prices* of some assets (e.g., the market price of a stock underlying an option) are used as model inputs to derive theoretical *values* of other assets such as financial derivatives (e.g., the theoretical value of the option). Unfortunately, most partial equilibrium models that estimate *values* for financial derivatives and other assets are termed *asset pricing models* even though the models are finding *values* that model market prices. To be consistent with these conventions, when discussing formal valuation models we use the terms *asset pricing model, call option pricing model,* and so forth.

In practice, arbitrageurs use model values to predict behavior of market prices. Further, practitioners often use model values to derive hedge ratios used in managing risk, a topic discussed in detail in Chapters 5 and 7. This chapter builds up the underlying theory and leaves the applications to subsequent chapters. For example, exotic options such as barrier options and Asian options are discussed in Chapter 4, while option risk measures known as "the Greeks" are discussed in Chapter 5.

In this chapter, we demonstrate the three primary derivative valuation approaches: (1) tree models, (2) simulation models, and (3) analytical models (e.g., the Black-Scholes and Merton models). We demonstrate these three approaches mostly using a stock as the derivative's underlying asset. In the discussions of Chapters 3, 4, and 5 (except in the duration section) regarding derivative pricing/hedging models, it is assumed that the interest rate is a single constant. When the underlying assets are sensitive to interest rate changes, those models fail. Chapter 7 discusses derivative pricing/hedging with fixed-income securities as underlying assets and with stochastic interest rates.

3.1 Overview of the Three Primary Derivative Pricing Approaches

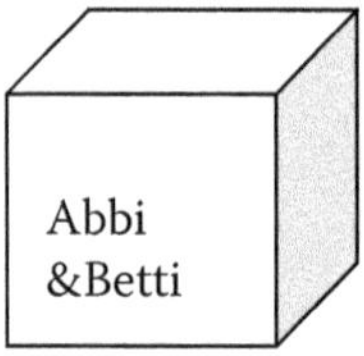

To illustrate the three derivative pricing approaches, let's examine a somewhat simple bet between Abbi and Betti. Abbi offers Betti to pick three distinct integers from 1 to 6 and then to roll three dice and receive payoffs based on the outcomes. Abbi offers Betti a chance to play for one ticket costing $2.80. Abbi promises to pay Betti as follows:

For all three numbers that Betti picked, she receives the following payoffs from comparing those picks with the outcomes of the three dice:

- $0 if there are no matches
- $2 if there is one single match
- $3 if there is one double match
- $4 if there are two single matches or one triple match
- $5 if there are one single match and one double match
- $6 if there are three single matches

For example, let's assume that Betti pays $2.80 and picks the numbers 2, 3, and 5. She rolls three dice. Here are the payoffs she would receive under just seven of the many scenarios:

Dice show 1, 1, 1: Abbi receives nothing
Dice show 4, 5, 6: Abbi receives $2 for a single match of 5
Dice show 1, 2, 3: Abbi receives $4 for two single matches of 2 and 3
Dice show 2, 3, 5: Abbi receives $6 for three single matches
Dice show 1, 5, 5: Abbi receives $3 for one double match
Dice show 2, 5, 5: Abbi receives $5 for one double match and one single match
Dice show 2, 2, 2: Abbi receives $4 for one triple match

How can Abbi determine if $2.80 is an attractive price for a ticket to play the game? Let's use each of the three primary approaches to derivative valuation to estimate an appropriate price.

3.1.1 Valuing the Dice Game with a Tree Model

A ***tree model*** projects the possible paths or outcomes of the underlying asset (usually with their probabilities) through time emanating from a node (i.e., the point from which the paths emanate). We illustrate the dice game as if the three dice rolls occur consecutively over three very brief time periods. Figure 3.1 illustrates the six possible outcomes of the first dice that emanate from the initial node. There is an identical node at the end of each of these six outcomes from which six outcomes emanate based on the second roll and so forth to the third and final roll. For simplicity, Figure 3.1 only illustrates the outcomes of the second and third rolls assuming that both the first and second rolls are 3s.

The solution using the tree model in this example involves valuing the outcome of each of the 216 nodes at the end of the three-period tree. Then, assuming that the dice are perfectly balanced so that each of the 216 outcomes is equally likely, the value of the gamble can be found as the average of the 216 payoffs. The solution may be found using a simple computer program with nested loops.

Most tree models in finance usually involve two paths from each node (***binomial tree models***) rather than six, although some models use three (***trinomial tree models***) or more paths emanating from each node.

3.1.2 Valuing the Dice Game with a Monte Carlo Simulation Model

A ***Monte Carlo simulation*** involves programming a computer to use a random number generator to project thousands or even millions of potential outcomes based on the probabilities and payoffs specified by the programmer.

In this case, the program is quite simple: (1) millions of dice rolls are simulated, (2) the payoff to each outcome is calculated, and (3) the payoffs are averaged to form an estimate of the expected future value of the gamble.

Here is a sample computer program in R (don't worry about this sample code, we will guide you step by step later in the model):

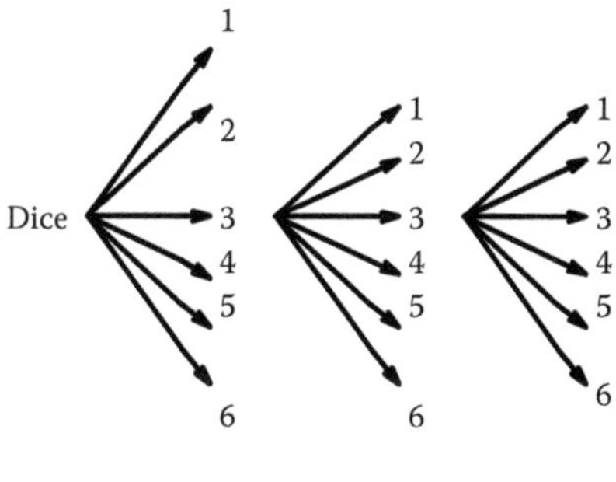

FIGURE 3.1
Three Dice Lead to 216 Outcomes.

R Code 3.1 Abbi and Betti dice game Monte Carlo simulation for expected payoff

```
bet=c(2,3,5)
n=1000000
totalreward=0
#run n trials
for(i in 1: n)
{
# simulate rolling three dice, saved as a vector x
x=sample(c(1,2,3,4,5,6),3,replace=TRUE)
#track corresponding matches for each bet number, saved
#as a vector count
count=c(0,0,0)
#reward for this trial
reward=0
# for each trial
for(j in x)
{
#check for matches
for(k in 1:3)
{
if(bet[k]==j)
{
count[k]=count[k]+1
}
}
}
# calculate reward based on matches
for(j in count)
{
#if there is a positive match number for a bet number
if(j>0)
{
reward=reward+1+j
}
}
totalreward=totalreward+reward
}
averagereward=totalreward/n
averagereward
```

Here is the estimate that we found on our first try with 1 million paths: 2.764421.

3.1.3 Valuing the Dice Game with an Analytical Model

An ***analytical model*** solves for the value of an uncertain outcome based on an assumed probability distribution of outcomes.

Although the dice game does not provide a perfect example, it does illustrate the basic approach. The researcher forms a payoff distribution through assumptions and modeling. The dice game between Abbi and Betti can be viewed as set of possible outcomes with probabilities based on modeling:

Payoff	State	Probability
$0	No matches	1/8
$1	Not possible	0
$2	One single match	3/8
$3	One double match	1/8
$4	Two single matches or one triple match	19/72
$5	One single match and one double match	1/12
$6	Three single matches	1/36

The sum of the probabilities times the outcomes (i.e., the expected payoff) is $2.7639 (= 199/72).

Note that for each roll, there is a 1/6 chance of matching a particular chosen number for a total probability of 1/2 of a single roll matching any one of the three chosen numbers. Therefore, there is a 1/2 chance of a particular roll not matching any of the three chosen numbers. These probabilities can be used to find the probability of each payoff level as follows:

$0 payoff (No match): $1/8 = 1/2 \times 1/2 \times 1/2$
$2 payoff (One single match): $3/8 = 3 \times 3 \times 1/6 \times 1/2 \times 1/2$; you have three possible match numbers and three possible locations.
$3 payoff (One double match): $1/8 = 3 \times 3 \times 1/6 \times 1/6 \times 1/2$; you have three possible match numbers and three possible locations.
$4 payoff (Two single matches or one triple match): $19/72 = 6 \times 3 \times 1/6 \times 1/6 \times 1/2 + 3 \times 1/6 \times 1/6 \times 1/6$, for triple match, there are three possible match numbers; for two single match, it has three possible match numbers and six possible locations.
$5 payoff (One single match and one double match): $1/12 = 3 \times 6 \times 1/6 \times 1/6 \times 1/6$, there are six possible ways to pick match numbers and three possible locations.
$6 payoff (Three single matches): $1/36 = 6 \times 1/6 \times 1/6 \times 1/6$, there are six possible locations.
Note that $1/8 + 3/8 + 1/8 + 19/72 + 1/12 + 1/36 = 1$.

3.1.4 Summary of Three Approaches to Valuing the Dice Game

The simulation model produces solutions that vary around the true solution (the analytical model) based on the randomness inherent in sampling methods. Generally, the greater the number of trials, the more accurate the estimate is. The tree model and analytical methods will generate exact solutions based on the payoffs and probabilities assumed. Note that in the case of the dice game there is no such thing as a perfectly balanced set of dice. In that sense all three methods are approximations because as models, the methods are based on an abstraction from reality. In securities markets most assets are priced to the nearest cent or so, indicating that their prices are not perfectly accurate. Further, the implementation of the approach requires assumptions regarding probabilities and outcomes that themselves depart from reality.

In the next three sections, we will describe these three approaches to valuing derivatives. The analytic Black-Scholes model was the first option pricing model based purely on observable market data, for which the authors received the Nobel Prize in Economics. The formula for valuing European calls and puts is simple, however, the theory behind it is deep. The Black-Scholes model does not have a closed form solution for European-style derivatives other than simple calls and puts. In addition, due to free boundary problems for the Black-Scholes partial differential equation (PDE for short), it is impossible to apply the same techniques to models of American calls and puts. Because of these reasons, we start with the binomial tree method because it is the most intuitive to understand and because it demonstrates the deep theory behind the theoretic Black-Scholes model. Also, in the proof of the Black-Scholes PDE in the analytic model, the approach is similar to the approach used in a binomial tree: constructing a risk-free portfolio. Finally, the discrete version of the risk-neutral principle is very intuitive and can be applied step by step, therefore the binomial tree approach can value American calls and puts quite easily.

After detailing the binomial tree method, we build on the theory for the analytic Black-Scholes model, which is a continuous model where the underlying asset price evolves over time. Since analytic models do not have a closed form formula for an arbitrary European-style derivative, we first introduce a Monte Carlo simulation model that is based on continuous models. Sample paths are simulated for the underlying asset price in the analytic model section and the Monte Carlo simulation approach is used to value European-style derivatives after the Black-Scholes PDE and the risk-neutral principle are used for the case of the continuous model.

All three approaches are described in detail in Chapter 4. Note that the binomial tree approach can value American-style derivatives quite easily. However, it is only an approximate method that depends on the number of periods used in the model. The Black-Scholes formula can be used to value European calls and puts easily, but it can't price American calls and puts. Also, there is no closed form formula for arbitrary European-style derivatives. Finally, Monte Carlo simulation can be directly used to approximate the value of European-style derivatives, but not American-style derivatives.

In general, all of these methods have advantages and disadvantages that are discussed in detail in Chapter 4. In this chapter, we lay a foundation and build up all machineries needed later.

3.2 The Binomial Tree Model Approach to Derivative Pricing

This section discusses the first major approach to derivative valuation: the binomial tree. A one-period binomial tree model was used in Chapter 1 to solve for the no-arbitrage value for a call option and to introduce risk-neutral valuation. You may recall that the option value was found by setting up two portfolios with identical potential payouts and solving for the call option value that caused the portfolios to have equal current values (thus avoiding arbitrage opportunities).

This section derives a similar binomial option pricing model with three modest differences: (1) the notation is made more flexible, (2) the solution to the derivative value is found by creating a single riskless portfolio and then finding the derivative value that ensures that the riskless portfolio will earn the riskless rate of return, and (3) the solution that is obtained is a more general formula for the derivative value that facilitates additional modeling.

3.2.1 The Assumptions and Framework of Derivative Valuation with a Binomial Tree

Models require simplifying assumptions. At the core of a tree model to value a financial derivative is the specification of the behavior of the derivative's underlying asset through time. We initially assume that the security underlying the derivative is a share of common stock. Consider a binomial tree model with the following eight assumptions:

Binomial Tree Model (Discrete Model) Assumptions:

1. A two-state economy where the initial price of a stock (today), S_0, could go up to $S_1 = S_0 \times u$ or down to $S_1 = S_0 \times d$ one period later. Note that u and d are the up and down factors. We assume that we know the values of u and d but in this particular model we do not know the probability of the stock going up and down.
2. u, d, and the risk free interest rate, r_f, are constants.
3. All securities are fully divisible such that we can trade fractional shares.
4. The underlying stock does not pay a dividend.
5. Short selling is permitted where market participants can borrow securities, sell them, and establish negative economic exposures to the securities.
6. There are no transaction costs such as commissions or bid-asked spreads and no taxes.
7. The payoff function of the derivative at expiration is f_T, which in the case of a one-period model is known as f_u and f_d where f_u and f_d are functions of $S_0 \times u$ and $S_0 \times d$ respectively.
8. Arbitrage opportunities do not exist.

Graphically, the underlying stock is assumed to form a tree in which it moves up or down one period later as follows:

One Period in the Stock Tree

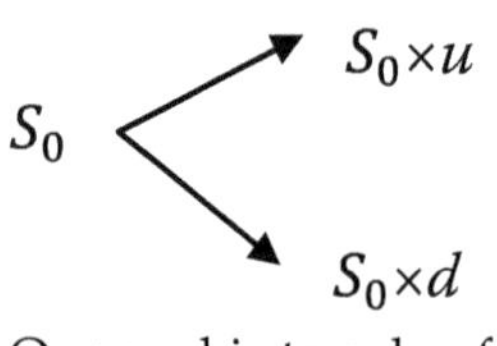

Our goal is to solve for the current value of the financial derivative, f_0, as shown in the tree of derivative values:

One Period in the Derivative Tree

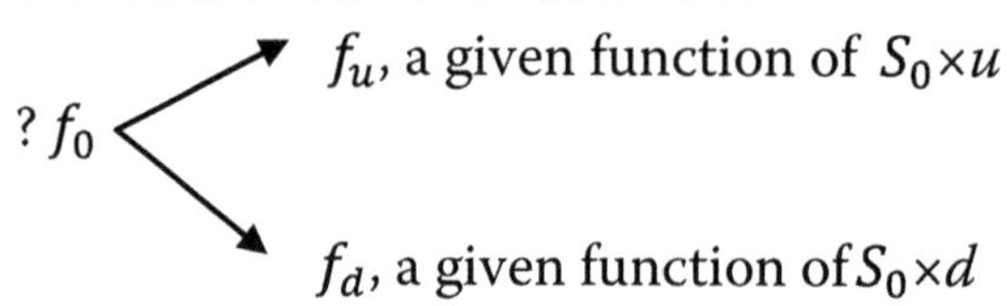

Note that at time 0 (today), by assumptions, we know values of the following: S_0, $S_0 \times u$, $S_0 \times d$, f_u, and f_d. Our objective is to find the value of f_0 that is consistent with the assumption that arbitrage opportunities do not exist. The next section solves a one-period tree.

3.2.2 No Arbitrage Tree Method to Value a Derivative with a One-Period Binomial Tree

The ***No-Arbitrage Tree Method***:

To find f_0, the value of the derivative today, we construct a riskless portfolio between the derivative f and its underlying stock S. Because the change in value of the derivative is directly tied to the price change of its underlying stock, the risk of a portfolio containing both assets can be eliminated by establishing the correctly sized positions in each asset. The key is finding the relative weights of the stock and derivative that form a riskless portfolio. Since this method is based on the idea that competitive markets will not allow values that permit arbitrage, it is called a no-arbitrage tree method.

We use the convention that money to be received is a positive value and money to be paid is a negative value. Consider a strategy to buy one unit of the derivative, f, and short Δ shares of the underlying stock, S, with cash flows as follows (Table 3.1):

TABLE 3.1

Construct Risk-Free Portfolio

Strategy/Payoff	Payoff Today	Payoff in the Up State	Payoff in the Down State
Buy one unit of derivative	$-f_0$	f_u	f_d
Short Δ shares of stock	ΔS_0	$-\Delta S_0 \times u$	$-\Delta S_0 \times d$
Form a risk-free portfolio	$\Delta S_0 - f_0$	$f_u - \Delta S_0 \times u$	$f_d - \Delta S_0 \times d$

In order to construct a riskless portfolio, Δ must satisfy the following equation, namely that the payoffs are same in up and down states:

$$f_u - \Delta S_0 \times u = f_d - \Delta S_0 \times d$$

By algebra:

$$f_u - f_d = \Delta S_0 \times u - \Delta S_0 \times d$$

$$f_u - f_d = \Delta(S_0 \times u - S_0 \times d)$$

The above equation can be used to derive a formula for the hedge ratio Δ:

$$\Delta = \frac{f_u - f_d}{(S_0 u - S_0 d)} \tag{3.1}$$

Note that as long as Δ is equal to the *change of the derivative value, f, over the change of underlying asset price S,* namely $\Delta = \frac{\Delta f}{\Delta S}$, the portfolio is hedged with regard to changes in S and is riskless. The riskless portfolio is formed by selling short Δ shares of the stock, which is accomplished by borrowing the shares from a holder of S, and selling the shares in the market. Eventually, the short position is closed by buying the shares in the market and returning them to the lender. Since S does not pay a dividend (by assumption), there is no cost to the security lender from lending the security.

By the ***no arbitrage assumption,*** the riskless portfolio should grow at the risk-free rate r_f because it contains no risk (i.e., its future payoff is fixed):

$$(f_0 - \Delta S_0)e^{r_f T} = f_u - \Delta S_0 \times u \tag{3.2}$$

Note that Equation 3.2 sets the future value of the initial cost of the portfolio (the left side) equal to the payoff to the portfolio (the right side). The payoff to the portfolio could have been expressed as the payoff in the down state since it is the same as the payoff in the up state. Since f_u, f_d, S_0, u, d, r_f, and T are given, the only unknown is f_0, and so the solution is easily calculated.

Note that in the model's derivation and construction of a riskless portfolio, numerous assumptions were used: we assumed that the stock goes up and down with fixed factors u and d, the interest rate is constant during the period, we can short fractional shares of stock, there are no transaction costs, bid-asked spreads or taxes, the stock does not pay a dividend, the derivative's payoff is given, and most importantly, there is no arbitrage so that the riskless portfolio must grow at the risk-free rate.

3.2.3 A Numerical Example of Derivative Pricing with a One-Period Binomial Model

This section provides an example of derivative pricing with a one-period binomial tree.

Example 3.1: Assume that the current price of a stock is \$30 and that it will go up by a factor of 1.1 (u) or down by a factor of 0.9 (d) in one period (e.g., one year). The risk-free rate is 0.05 compounded continuously. There is a call option on the stock with a strike price of \$32 that expires in one period. Find the value of the call option today.

The stock price tree is formed with S_0, u, and d.

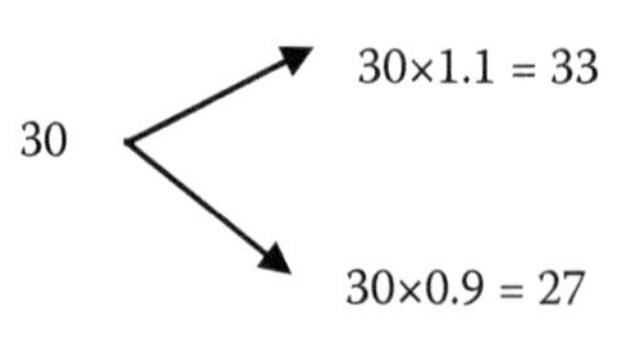

The payoffs in the tree for the call option are found using the formula for the payoffs to a call option in Chapter 1. Therefore, the stock price tree and the strike price (i.e., exercise price) can be used to form the payoff tree:

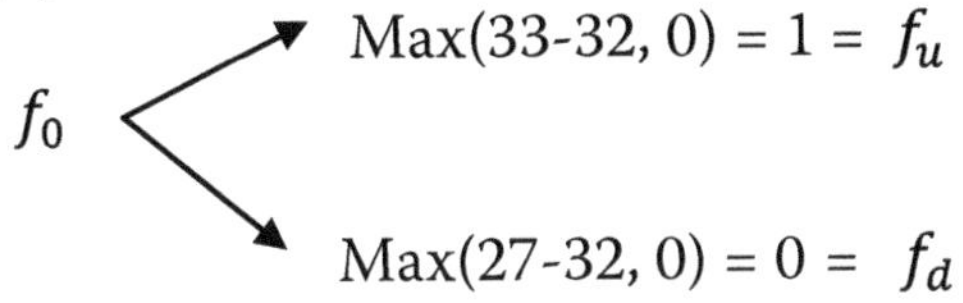

The hedge ratio between one call option and a short position in the underlying stock is found using the formula from the previous section, the potential stock prices, and the option payoff values:

$$\Delta = \frac{1-0}{33-27} = 1/6$$

So, buying one call and shorting 1/6 shares of the underlying stock forms a riskless portfolio that must grow at the risk-free rate in order to prevent arbitrage. Using the general formula:

$$(f_0 - \Delta S_0)e^{r_f T} = f_u - \Delta S_0 \times u$$

and inserting the values in this example produces:

$$\left(f_0 - \frac{1}{6} \times 30\right)e^{0.05\times 1} = 1 - \frac{1}{6} \times 33$$

which can be easily solved as:

$$f_0 = 0.719$$

This approach can be used to find the value of any derivative (e.g., put options) as long as the potential payoffs, f_u and f_d, are known and are a given function of the underlying stock price.

3.2.4 The Risk-Neutral Principle Derived in a One-Period Binomial Tree

The concept of risk-neutral derivative valuation was introduced in Chapter 1. In this section we will use the results of the previous section to derive a *general formula* that demonstrates the ***risk-neutral tree method*** to value derivatives.

In the previous sections two important equations were derived: the hedge ratio and the valuation formula for a derivative, f_0:

$$\Delta = \frac{f_u - f_d}{(S_0 \times u - S_0 \times d)}$$

and

$$(f_0 - \Delta S_0)e^{r_f T} = f_u - \Delta S_0 \times u$$

If we substitute the formula for Δ into the second equation:

$$\left(f_0 - \frac{f_u - f_d}{(S_0 \times u - S_0 \times d)} S_0\right)e^{r_f T} = f_u - \frac{f_u - f_d}{(S_0 \times u - S_0 \times d)} S_0 \times u$$

Canceling out S_0 from numerators and denominators:

$$\left(f_0 - \frac{f_u - f_d}{(u - d)}\right)e^{r_f T} = f_u - \frac{f_u - f_d}{(u - d)}u$$

$$f_0 = \left(f_u - \frac{f_u - f_d}{(u - d)}u\right)e^{-r_f T} + \frac{f_u - f_d}{(u - d)}$$

$$f_0 = e^{-r_f T}\left(\left(f_u - \frac{f_u - f_d}{(u - d)}u\right) + \frac{f_u - f_d}{(u - d)}e^{r_f T}\right)$$

$$f_0 = e^{-r_f T}\left(\frac{uf_u - df_u - uf_u + uf_d + e^{r_f T}f_u - e^{r_f T}f_d}{u - d}\right)$$

$$f_0 = e^{-r_f T}\left(\frac{-df_u + uf_d + e^{r_f T}f_u - e^{r_f T}f_d}{u - d}\right)$$

$$f_0 = e^{-r_f T}\left(\frac{e^{r_f T} - d}{u - d}f_u + \frac{u - e^{r_f T}}{u - d}f_d\right)$$

The above equation shows that the derivative value, f_0, depends on u, d, T, f_u, f_d, and r_f, but not explicitly on any other variables such as the probabilities of the up and down states or the expected rate of return on the underlying asset, S. This means that the derivative's value depends on the riskless rate (which is observable) but not on risk premiums (which are not observable).

Next, we recast this equation in the context of a Bernoulli distribution so that the components of the equation can be viewed as being probabilities or *pseudo probabilities*. Before proceeding, a review of some mathematics is provided:

Mathematics Review 3.1

Assume $X \sim$ Bernoulli (p), as illustrated below.

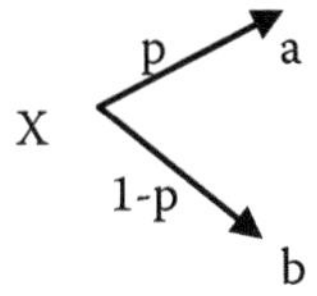

Find E(X) and Var(X).

$E(X) = pa + (1 - p)b$

$Var(X) = E(X^2) - (EX)^2 = pa^2 + (1 - p)b^2 - (pa + (1 - p)b)^2$

To cast the problem in the terms of risk-neutral derivative pricing, replace X with f, p with $Q = \frac{e^{r_f T} - d}{u - d}$, a with f_u, and b with f_d.

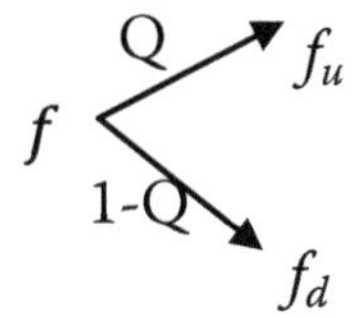

Finding E(f) and Var(f):

$$E(f) = Qf_u + (1 - Q)f_d$$

$$Var(f) = E(f^2) - (Ef)^2 = Qf_u^2 + (1 - Q)f_d^2 - (Qf_u + (1 - Q)f_d)^2$$

Return to the previously derived equation for the value of a derivative reproduced here:

$$f_0 = e^{-r_f T}\left(\frac{e^{r_f T} - d}{u - d} f_u + \frac{u - e^{r_f T}}{u - d} f_d\right)$$

The terms inside the parentheses can be viewed in the context of the formula for the expected value of a variable distributed Bernoulli (p) (derived in the mathematic review above).

Note that in the above equation, $\frac{e^{r_f T} - d}{u - d}$ and $\frac{u - e^{r_f T}}{u - d}$ sum to one, so that if we call $Q = \frac{e^{r_f T} - d}{u - d}$, then $1 - Q = \frac{u - e^{r_f T}}{u - d}$ and the formula for the value of the derivative can be expressed in terms of Q as:

$$f_0 = e^{-r_f T}(Qf_u + (1 - Q)f_d).$$

If we regard Q as a probability (assuming Q is between 0 and 1), then f_0 can be viewed as a probability-weighted average (i.e., an expected payoff) discounted back to today by the factor $e^{-r_f T}$. Q is not necessarily the statistical probability of the stock or derivative going up or down in the real world. Q and $(1 - Q)$ are known as Q-measures, and have very special properties in the context of derivative pricing that is discussed in the next section. What is important here is that there exists a Q that can be easily calculated (given u, d, r_f, and T) and can be used to value the derivative independent of the risk preferences of investors.

Let's calculate the expected value of the stock and the derivative using Q-measures as if they are true probabilities:

$$E_Q(S_T) = Q \times S_0 \times u + (1 - Q) \times S_0 \times d = \frac{e^{r_f T} - d}{u - d} \times S_0 \times u + \frac{u - e^{r_f T}}{u - d} S_0 \times d = S_0 e^{r_f T}$$

$$E_Q(f) = Qf_u + (1 - Q)f_d = f_0 e^{r_f T}.$$

Using Q-measures to compute expected values causes not only riskless bonds to have expected growth rates equal to the risk-free rates, but also risky stocks and their derivatives have expected growth rates equal to the risk-free rate. This imaginary world is termed a ***risk-neutral world***, because all assets have expected returns and growth rates equal to the risk-free rate and all expected values generated with Q-measures can be discounted to their current markets values by using the risk-free rate as the discount rate. In a risk-neutral world, there are no risk premiums (i.e., all risk premiums are set to zero).

Clearly this risk-neutral model would accurately describe the real world if all investors were risk neutral. The key breakthrough in derivative valuation is understanding the conditions under which the current values derived using risk neutrality are the prices that must exist in a risk averse world (i.e., the real world). The importance of this breakthrough is because risk-neutral valuation is relatively easy because it does not require knowledge of risk premiums which is the extra return above the risk-free rate demanded by risk-averse investors. The above derivation is a discrete version of the risk-neutral principle:

The Risk-Neutral Principle (Discrete Version):

The values of all assets today are equal to their discounted expected values of payoffs where all assets grow at the risk-free rate and are discounted at the risk-free rate as if we live in a risk-neutral world. This risk-neutral principle utilizes Q-measures in place of statistical probabilities (P-measures) so that for all assets in a two-state world:

$$f_0 = e^{-r_f T} E_Q(\mathrm{f}) = e^{-r_f T}\left(Qf_u + (1 - Q)f_d\right) \tag{3.3}$$

The risk-neutral principle gives us the risk-neutral tree method, which can be used to derive a derivative value. Even though the no-arbitrage method and risk-neutral method seem very different, they are equivalent from the above arguments. In derivative pricing, the no-arbitrage method is useful from a theoretical perspective, while the risk-neutral method is useful for performing calculations.

Next, Example 3.2 shows that the risk-neutral tree method discussed above works for the example introduced earlier in the chapter of a call option on a share of stock.

Example 3.2: Consider the parameters from Example 3.1: $S_0 = 30$, $u = 1.1$, $d = 0.9$, $T = 1$, $r_f = 0.05$, and $K = 32$. Find the risk-neutral probability Q and the call value using the risk-neutral tree method.

As previously defined:

$$Q = \frac{e^{r_f T} - d}{u - d} = \frac{e^{0.05\times 1} - 0.9}{1.1 - 0.9} = 0.75636, \text{ and } 1 - Q = \frac{u - e^{r_f T}}{u - d} = 0.24364$$

The call value is therefore $f_0 = e^{-0.05\times 1}(0.75636 \times 1 + 0.2436 \times 0) = 0.719.$

This modeling approach is very robust. For example, consider the following derivative with payoffs that are nonlinear in the underlying stock price.

Example 3.3: Given the parameters: $S_0 = 30$, $u = 1.1$, $d = 0.9$, $T = 1$, and $r_f = 0.05$, find the value of a derivative that has a payoff based on S_T^3. Note that there is no strike price needed.

The derivative's value follows the previous example except that the up-state and down-state payoffs are changed: $f_u = (S_0 u)^3 = 33^3$, $f_d = (S_0 d)^3 = 27^3$

$$f_0 = e^{-0.05\times 1}(0.75636 \times 33^3 + 0.24364 \times 27^3) = 30{,}417.34.$$

The current price of the stock (S_0) given u, d, and r_f reflects the degree of risk aversion in the economy. The Q-measures that generated S_0 can therefore be used to compute the value of a derivative on that stock, f_0, by simply assuming that arbitrage opportunities cannot exist.

3.2.5 P-Measures, Q-Measures, and Risk Aversion

The use of ***Q-measures*** and risk-neutral modeling is central to derivative valuation. The simplicity of risk-neutral modeling facilitates expanded use of derivatives for risk management purposes which in turn leads to improved allocation of risk across market participants. Here are five important concepts regarding ***P-measures*** (the real-life probability measures), Q-measures (the risk-neutral probability measures) and risk aversion.

Five Important Concepts Regarding P-measures and Q-measures:

1. *P-measures* are based on actual statistical probabilities of future outcomes. These statistical probabilities can be used to forecast actual expected future values.
2. P-measures are generally unobservable. The relationship between the true expected value of a risky asset and the asset's current value is driven by risk aversion (in the form of a risk premium added to the risk-free rate). The greater the degree of risk aversion in the market and the more undesirable the risk of a particular asset, the higher the expected return that the marketplace will demand. The higher the return demanded, the higher the expected value of the asset in the future. However, it is generally not possible to observe expected returns or statistical probabilities of risky assets. We cannot generally separate the effects of probabilities and risk aversion. For example, in the case of a risky bond selling at a low price relative to its promised payoff it is generally not possible to determine the extent by which the low price is driven by investor risk aversion rather than by pessimistic predictions of the likelihood of the bond defaulting.
3. *Q-measures* are based on probability measures that are only equal to actual statistical probabilities in two cases: when investors are risk neutral or when the asset being considered contains neither undesirable risk nor desirable risk (e.g., negative systematic risk that can be used to hedge positive systematic risk). In reality, since investors are not risk neutral, these two measures differ (when the risk is at least partially systematic). When Q-measures are used to forecast expected values, the resulting expected values are *biased* to the extent that

risk-neutrality does not hold. Thus, the forecasted expected values are unbiased only in the case of risk neutrality.

4. Q-measures are generally observable. By ignoring risk aversion, financial engineers can often determine risk-neutral probabilities and risk-neutral expected prices using observed prices). In other words, financial derivatives such as forward contracts can be precisely valued using risk-neutral modeling techniques that are based on easily observable market values such as the risk-free interest rate and the spot price of the derivative's underlying asset (as shown in Chapter 1).
5. Q-measures allow financial derivatives in a risk-averse world to be correctly valued using risk-neutral modeling. This is because, as demonstrated previously, the current value of the financial derivative relative to the current value of the asset underlying the derivative is unaffected by risk aversion. No matter what level of risk aversion exists in the market, the effects of that risk aversion are fully captured in the price of the underlying asset. Therefore, there is no need to adjust the derivative pricing model for risk aversion since the effects of risk aversion are already included in the value of the derivative's underlying asset.

3.2.6 The Risk-Neutral Tree Method to Value a Derivative with a Multi-Period Binomial Tree

The previous sections provided a simplified introduction to tree models by making the very unrealistic assumption that a stock evolves for only one period and with only two potential states of the economy. A natural extension is to generalize the results to a ***multi-period model***. Note that a multi-period Bernoulli distribution approaches a normal distribution for large numbers of periods. By dividing the total time period $[0,T]$ into a large number of subperiods, a tree model of log stock prices approaches a model based on the normal distribution, which is the commonly assumed distribution of log stock prices in finance. In a single period model, European-style and American-style options are identical. We begin the multi-period model assuming derivatives are European style (i.e., no early exercise is allowed).

Consider a ***two-period (n-period) model*** as an example: Assume that the total time period, $[0, T]$, is cut into 2 periods (n periods in general) with $\Delta t = T/2$ ($\Delta t = T/n$). The risk-neutral principle is then used to calculate the derivative value for each node starting at period $n - 1$ using the payoffs of period n. The values found for period $n - 1$ are then used to calculate the values at period $n - 2$ and so forth (a total of n times) until the derivative is valued in period 0. The process is known as ***backward induction.***

A n-period tree is easy to build when the tree recombines. A tree is ***recombining*** when the price after an up movement followed by a down movement matches the price after a downward movement followed by an up movement (i.e., when u and d are constant such that $S \times u \times d = S \times d \times u$).

In a two-state recombining tree, the number of nodes (outcomes) grows by one node per time period. Thus, in an n time period model there are only $n + 1$ outcomes in the final stage. For a fixed time period $[0, T]$, the larger the n, the closer the final outcomes will tend to be to matching risky assets in the real world, assuming that the asset log returns are normally distributed. The values of u and d must be a function related to n since modeled up and down price changes become smaller as n becomes larger. We begin with a two-period tree of stock prices as an illustration.

Two-Period Re-Combining Stock Tree:

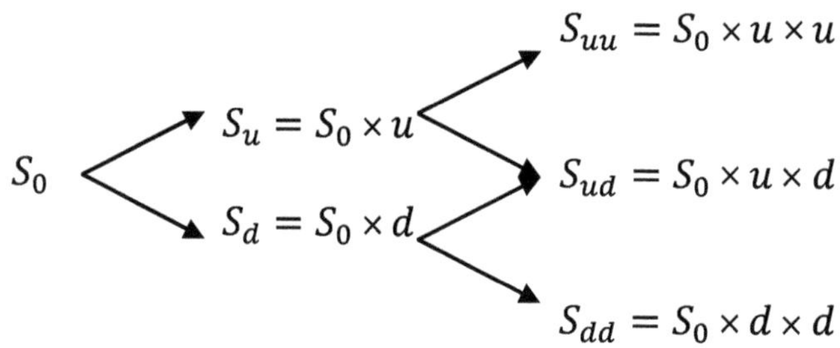

Two-Period Derivative Tree:

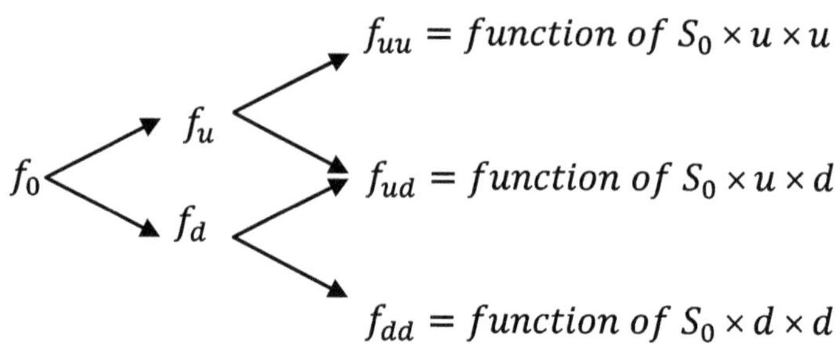

The solution to the current value of a derivative, f_0, is found using backward induction. Each "triangle" in the tree is solved separately, starting from the triangles that form the last time period. Starting from the final payoff of the derivatives, which are given as f_{uu}, f_{ud}, and f_{dd}, the derivative values one period before final payoff can be found as:

$$f_u = e^{-r_f \Delta t} E_Q(f_T) = e^{-r_f \Delta t}\left(Qf_{uu} + (1 - Q)f_{ud}\right) \quad (3.4)$$

$$f_d = e^{-r_f \Delta t} E_Q(f_T) = e^{-r_f \Delta t}\left(Qf_{ud} + (1 - Q)f_{dd}\right) \quad (3.5)$$

where $Q = \frac{e^{r_f \Delta t} - d}{u - d}$, $\Delta t = T/2$.

Next, the derivative value at the initial node is found as:

$$f_0 = e^{-r_f \Delta t} E_Q(f_{T/2}) = e^{-r_f \Delta t}(Qf_u + (1 - Q)f_d) \tag{3.6}$$

In general, any value can be selected for the number of time periods, n. The greater the n, the more the final outcomes resemble a continuous distribution.

The following example demonstrates a two-period model for an American and European call and put.

Example 3.4: Assume that the stock price is \$30 today and will go up by a factor of 1.05 or go down by a factor of 0.95 in every half a year. The risk-free rate is 0.05, compounded continuously. Consider a European put option with a strike of \$32 that expires in one year. Find the put value today using a two-period binomial tree. Then find the value of an American put option.

Here are the data needed to compute the option values:

$S_0 = 30$, $u = 1.05$, $d = 0.95$, $T = 1$, $n = 2$, $\Delta t = 0.5$, $r_f = 0.05$, $K = 32$

For a European put:
Stock Tree:

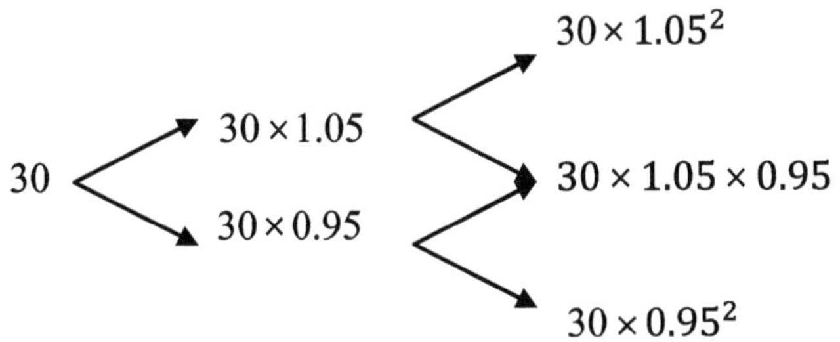

The risk-neutral probability can be calculated based on d, u, and r_f.

$$Q = \frac{e^{r_f \Delta t} - d}{u - d} = \frac{e^{0.05 \times 0.5} - 0.95}{1.05 - 0.95} = 0.75315,$$

$1 - Q = 0.24685$.

The solution given Q can now be used with the put payoff tree. To solve for the option price, start at the rightmost side and use the values from the stock tree to estimate payoffs and then work backwards (right to left) while solving each triangle as if it were a single period model.

Derivative Tree:

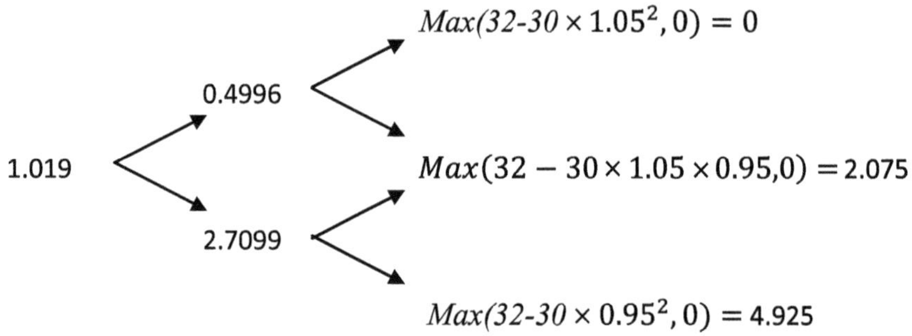

The three rightmost values are the payoffs to the put option at expiration. These three values are used to solve for the put values at the two nodes at the end of the first period:

$$0.4996 = e^{-0.05\times0.5}(0.75315 \times 0 + 0.24685 \times 2.075)$$

$$2.7099 = e^{-0.05\times0.5}(0.75315 \times 2.075 + 0.24865 \times 4.925)$$

Finally, the current value is found using the two values at the end of the first period:

$$1.019 = e^{-0.05\times0.5}(0.75315 \times 0.4996 + 0.24685 \times 2.7099).$$

Next, the same stock tree is used to find payoffs and values for an American put option.

Note that in a multi-period tree model the user can model future decisions made by the investor such as whether or not to exercise an American put prior to its expiration. This easy ability to model potentially intricate options, illustrated next, is a major reason to use a tree model.

Derivative Tree:

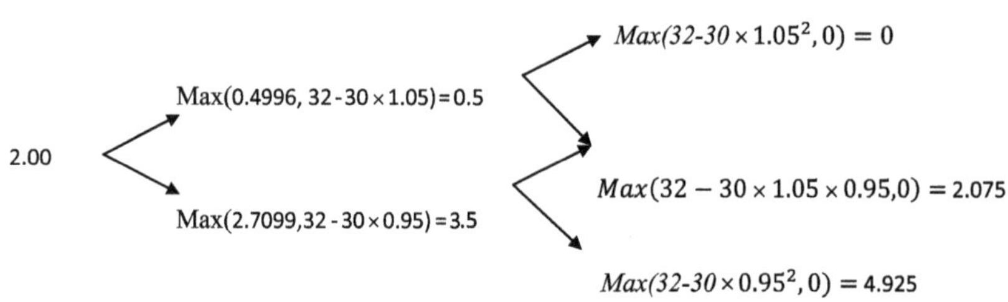

where Max(0.4996, 32 − 30 × 1.05) is Max(f_u, $K - S_u$) and Max(2.7099, 32 − 30 × 0.95) is Max(f_d, $K - S_d$), found by considering the immediate exercise value at period 1. The current put value 2.00 is from Max(32 − 30, $e^{-0.05\times0.5}$(0.75315 × 0.5 + 0.24685 × 3.5)).

In summary, in the down node at the end of the first period, the investor finds greater value from exercising the American put for $3.50 rather than waiting one more period to receive the two potential final payoffs of $2.075 and $4.925. So, American puts have higher values than an otherwise identical European put whenever the American option's early exercise has potential value. This is not a surprise since the option to exercise a put option prior to expiration can allow the option buyer to receive the strike price earlier, which can then be invested to earn a positive return when $r_f > 0$.

In general, f_{uu}, f_{ud}, and f_{dd} can be any deterministic function $f(S)$ of S_{uu}, S_{ud}, and S_{dd}, respectively, and backward induction can still be used to calculate no arbitrage values for any derivatives including calls. To value different derivatives of S simply involves inserting the corresponding payoff as the last column of the derivative tree.

3.2.7 Determination of Parameters *u* and *d* Using Volatility Matching

In previous sections, the selection of u and d has been arbitrary. In practice, the values of u and d can be linked to the standard deviation of the returns of the underlying asset (synonymous with ***volatility*** in finance). This section discusses the use of historical data to estimate volatility and the use of that volatility estimate to form a binomial tree. In practice, historical data and historical volatility are not used to calibrate the binomial tree. Instead, forecasted volatility is used to form a binomial tree. In fact, in most cases, market prices of derivatives are used to estimate the implied volatility.

Finance Review 3.1 Using YAHOO! FINANCE Historical Data to estimate volatility:

Download one year of daily adjusted stock prices (253 price observations) from YAHOO! Finance's "historical data" and "export" links. Form 252 daily (i.e., one year of data) price ratios using the adjusted stock price (which adjusts for dividends) to calculate daily total returns: $r_i = \frac{S_i - S_{i-1}}{S_{i-1}}$ where r_i is the daily return, and S_i and S_{i-1} are the adjusted stock prices separated by one day. Calculate the standard deviation of r_i and multiply by $\sqrt{252}$ to get an annualized estimate of the volatility of the returns, σ, which is an estimate of the dispersion in the annual stock returns. Note that there are many critical issues in the use of historical statistics to predict future behavior.

In the binomial tree model in the previous section, u and d are given explicitly (for example, 1.1 and 0.9, respectively). Those estimates of u and d were arbitrary – not formed based on estimated stock volatility σ. To estimate u and d using historical data, assume that $u = e^{\sigma\sqrt{\Delta t}}$ and $d = e^{-\sigma\sqrt{\Delta t}}$. The reason to select u and d as $u = e^{\sigma\sqrt{\Delta t}}$ and $d = e^{-\sigma\sqrt{\Delta t}}$ is to calibrate the binomial tree with a log normal distribution that has the same mean and variance as that viewed in markets. Next, estimate the risk-neutral probability as: $Q = \frac{e^{r_f\Delta t} - d}{u - d} = \frac{e^{r_f\Delta t} - e^{-\sigma\sqrt{\Delta t}}}{e^{\sigma\sqrt{\Delta t}} - e^{-\sigma\sqrt{\Delta t}}}$. The rest of this section is to justify this selection of u and d.

The selection of n (the number of steps in the tree) can be related to its implication for the distribution of the underlying asset. It is clear that u and d are functions of n. As n becomes large, Δt becomes small, and both u and d are close to 1 with $u > 1$ and $d < 1$. As n (the number of steps) approaches infinity, the ending stock distribution S_T approaches a lognormal distribution and therefore LnS_T approaches a normal distribution. The following two-step tree illustrates the issues, but the proof is for an n-step tree.

First, modify the stock tree in the previous section by taking the natural log of the stock's prices to form the log stock price tree:

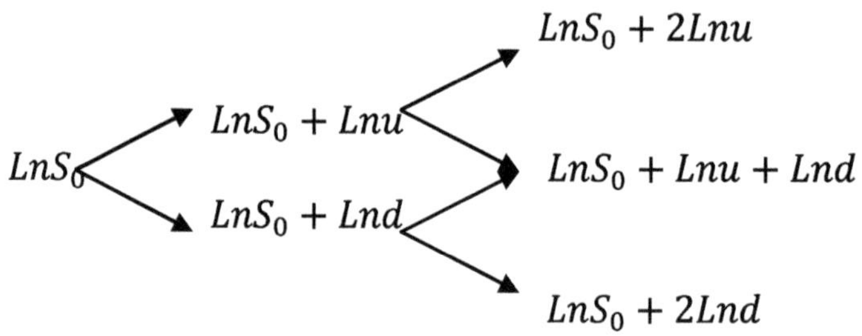

The final column is LnS_T. Define $Y = LnS_T$. We need to show that $Y = LnS_T$ converges to $N\left(LnS_0 + \left(r - \frac{1}{2}\sigma^2\right)T, \sigma^2 T\right)$ as n approaches infinity with the assumption that Q is the Bernoulli distribution probability. Equivalently, we need to show the followings as n goes to infinity:

- Y converges to a normal distribution
- $E(Y) = LnS_0 + \left(r - \frac{1}{2}\sigma^2\right)T$
- $Var(Y) = \sigma^2 T$

Define X_i as a Bernoulli random variable with a given probability Q defined as above:

X_i	$Ln\ u$	$Ln\ d$
Probability	Q	1 − Q

Assume that there are n periods in $[0, T]$. It is not difficult to see that $Y = LnS_0 + X_1 + X_2 + \ldots + X_{n-1} + X_n$. In particular, $Y = LnS_0 + X_1 + X_2$ for the above two-period model. We know that X_i is independent because Q is a constant and is not determined by the stock price at each node. So, $\sum_{i=1}^{n} X_i$ is the sum of independent identical Bernoulli distributions which itself follows the binomial distribution. As the period n approaches infinity, the binomial distribution approaches the normal distribution. So, Y converges to a normal distribution.

Now calculate $E(Y)$, the expected value of Y:

$$E(Y) = LnS_0 + E\left(\sum_{i=1}^{n} X_i\right),$$

which is equal to $LnS_0 + \sum_{i=1}^{n} EX_i$, which in turn equals $LnS_0 + n\ EX_i$.

First, calculate EX_i:

$$EX_i = QLnu + (1 - Q)Lnd$$

Substituting $u = e^{\sigma\sqrt{\Delta t}}$, $d = e^{-\sigma\sqrt{\Delta t}}$, and $Q = \frac{e^{r\Delta t} - d}{u - d}$ into EX_i:

$$EX_i = \frac{e^{r\Delta t} - e^{-\sigma\sqrt{\Delta t}}}{e^{\sigma\sqrt{\Delta t}} - e^{-\sigma\sqrt{\Delta t}}} Ln\left(e^{\sigma\sqrt{\Delta t}}\right) + \frac{e^{\sigma\sqrt{\Delta t}} - e^{r\Delta t}}{e^{\sigma\sqrt{\Delta t}} - e^{-\sigma\sqrt{\Delta t}}} Ln\left(e^{-\sigma\sqrt{\Delta t}}\right)$$

Use a Taylor expansion to simplify EX_i. We know that the Taylor expansion $e^x \approx 1 + x + \frac{x^2}{2}$ (to the second order) is a good approximation of e^x for small values of x. Here, Δt is small and all higher powers of Δt are small as n becomes large. Using the first-/second-order Taylor series approximation by ignoring high powers of Δt:

$$EX_i = \frac{1 + r \times \Delta t - \left(1 - \sigma\sqrt{\Delta t} + \frac{1}{2}\sigma^2\Delta t\right)}{\left(1 + \sigma\sqrt{\Delta t} + \frac{1}{2}\sigma^2\Delta t\right) - \left(1 - \sigma\sqrt{\Delta t} + \frac{1}{2}\sigma^2\Delta t\right)} \sigma\sqrt{\Delta t}$$
$$+ \frac{\left(1 + \sigma\sqrt{\Delta t} + \frac{1}{2}\sigma^2\Delta t\right) - (1 + r \times \Delta t)}{\left(1 + \sigma\sqrt{\Delta t} + \frac{1}{2}\sigma^2\Delta t\right) - \left(1 - \sigma\sqrt{\Delta t} + \frac{1}{2}\sigma^2\Delta t\right)} (-\sigma\sqrt{\Delta t})$$

By simplification, $EX_i = r \times \Delta t - \frac{1}{2}\sigma^2\Delta t$.

Recall that $n\Delta t = T$, which produces:

$$E(Y) = LnS_0 + n\ EX_i = LnS_0 + r \times n\Delta t - \frac{1}{2}\sigma^2 n\Delta t = LnS_0 + r \times T - \frac{1}{2}\sigma^2 T.$$

Since $Y = LnS_0 + X_1 + X_2 + \ldots + X_{n-1} + X_n$, and S_0 is a constant:

$$Var(Y) = \sum_{i=1}^{n} VarX_i$$

We know that $VarX_i = Q \times (Lnu)^2 + (1 - Q) \times (Lnd)^2 - (EX_i)^2$.

Substituting $u = e^{\sigma\sqrt{\Delta t}}$, $d = e^{-\sigma\sqrt{\Delta t}}$, and $Q = \frac{e^{r\Delta t} - d}{u - d}$ into $VarX_i$, while omitting technical details and ignoring higher powers of Δt, shows that $VarX_i = \sigma^2 \Delta t$. Hence, $Var(Y) = n\sigma^2 \Delta t = \sigma^2 T$. We leave this as an exercise. Note that:

$Y = LnS_T$ converges to $N(LnS_0 + r \times T - \frac{1}{2}\sigma^2 T,\ \sigma^2 T)$, as n goes to infinity.

In summary, binomial tree modeling of S_T is equivalent to a ***lognormal distribution model*** as n goes to infinity:

$$Ln\ S_T \sim N(Ln\ S_0 + r \times T - \frac{1}{2}\sigma^2 T,\ \ \sigma^2 T) \tag{3.7}$$

The formula for the binomial tree to value the derivative involves the volatility parameter σ and can be used to determine u and d:

$$f_{parent\ node} = e^{-r_f T}(Qf_{up\ child\ node} + (1 - Q)f_{down\ child\ node}) \tag{3.8}$$

which uses $u = e^{\sigma\sqrt{\Delta t}}$ and $d = e^{-\sigma\sqrt{\Delta t}}$ to calculate the stock tree. The risk-neutral probability $Q = \frac{e^{r_f \Delta t} - e^{-\sigma\sqrt{\Delta t}}}{e^{\sigma\sqrt{\Delta t}} - e^{-\sigma\sqrt{\Delta t}}}$ is used to do backward induction.

This section demonstrated that if u, d, and Q are selected as indicated above, the underlying asset S_T follows a distribution as in (3.7). The resulting derivative value at time 0 is $f_0 = e^{-r_f T} \times E_Q(f_T)$, where f_T is the derivative payoff at time T (which is a given function of S_T).

The binomial tree in this section is also called the ***Cox-Ross-Rubinstein binomial tree*** and is so important that it is often described using the initials: a CRR tree.

3.2.8 Using Spreadsheets to Value European and American Calls and Puts

The previous section demonstrates how derivative values can be found using binomial trees with many periods or states (i.e., by choosing a large value of n). A binomial tree with a large number of steps (n) can easily be constructed using an EXCEL spreadsheet or in other spreadsheet software such as Google Sheets. This section discusses issues involved and provides step-by-step guidance on constructing large trees. First, consider these technical issues in selecting parameter values for binomial trees:

Technical Issues of Binomial Tree Parameters:

Using σ to specify u and d generates $Q = \frac{e^{r_f \Delta t} - e^{-\sigma\sqrt{\Delta t}}}{e^{\sigma\sqrt{\Delta t}} - e^{-\sigma\sqrt{\Delta t}}}$.

In theory, the risk-neutral probability, Q, should be between 0 and 1: $0 \le \frac{e^{r_f \Delta t} - e^{-\sigma\sqrt{\Delta t}}}{e^{\sigma\sqrt{\Delta t}} - e^{-\sigma\sqrt{\Delta t}}} \le 1$.

This implies $0 \le e^{r_f \Delta t} - e^{-\sigma\sqrt{\Delta t}}$ (i.e., $r_f \Delta t \ge -\sigma\sqrt{\Delta t}$), which must be true.

Also, $e^{r_f \Delta t} - e^{-\sigma\sqrt{\Delta t}} \le e^{\sigma\sqrt{\Delta t}} - e^{-\sigma\sqrt{\Delta t}}$ (i.e., $r_f \Delta t \le \sigma\sqrt{\Delta t}$), which is $r_f \sqrt{\Delta t} \le \sigma$.

Thus, the only condition for $0 \le Q \le 1$ is that $r_f \sqrt{\Delta t} \le \sigma$. As long as we select an adequately small Δt (which is tantamount to selecting a large enough n) that condition holds.

For the example, assume that the stock price today is \$30 with an annualized return volatility of 20% and that the risk-free rate is 0.05, compounded continuously. We wish to value a European put option with a strike price of \$32 that expires in one year.

The Steps in Building a Tree in a Spreadsheet

We select $n = 7$ and create the tree using functions found in an EXCEL spreadsheet and generally available with other spreadsheet programs, especially when add-ins are utilized.

In order to make a spreadsheet clear, create a row of labels as shown in the second row:

	A	B	C	D	E	F	G	H	I	J
1	S_0	σ	r	T	n	Δt	u	d	Q	K
2	30	0.2	0.05	1	7	0.142857	1.0785	0.9272	0.5284	32

The value Δt in column F is found as T/n. All other values are defined in previous sections with values or formulas as given below addresses:

```
A2: 30 (the expression to the left indicates that 30
should be placed in cell A2)
B2: 0.2
C2:0.05
D2:1
E2:7
F2: =D2/E2
G2: =EXP(B2*SQRT(F2))
H2: =1/G2
I2: =(EXP(C2*F2)-H2)/(G2-H2)
J2:32
```

To create the tree of stock values (note that only three cells, A4, B4, and B5, need to be created and the rest cells are all copied and pasted):

```
Start in cell A4
A4: =A2 (the expression to the left indicates that
=A2 should be placed in cell A4)
B4: =A4*$G$2
B5: =A4*$H$2
```

	Col. A	Col. B
Row 4	30	32.3557024
Row 5		27.8158079

Next, ***copy*** B4 and ***extend it*** to the right for a whole row of size 8:

Col. A	Col. B	Col. C	Col. D	Col. E	Col. F	Col. G	Col. H
30	32.3557024	34.8963825	37.6365655	40.5919171	43.7793329	47.2170355	50.9246782

Then ***copy*** B5 and ***extend it*** to a rectangle box of size 7 × 6 to complete the stock price tree:

30	32.3557024	34.8963825	37.6365655	40.5919171	43.7793329	47.2170355	50.9246782
	27.8158079	30	32.3557024	34.8963825	37.6365655	40.5919171	43.7793329
		25.7906389	27.8158079	30	32.3557024	34.8963825	37.6365655
		0	23.9129152	25.7906389	27.8158079	30	32.3557024
		0	0	22.1719018	23.9129152	25.7906389	27.8158079
		0	0	0	20.5576454	22.1719018	23.9129152
		0	0	0	0	19.0609171	20.5576454
		0	0	0	0	0	17.6731603

Next, to create the option tree, start with the last column, cell H13: =MAX (J2-H4,0), and ***copy*** H13 to the whole column of size 8 (note that cell H13 needs to be created manually):

Col. H
0
0
0
4.18419214
8.0870848
11.4423546
14.3268397

Next, generate the second to the last column. Start with cell G13: =EXP (−C2 × F2) ×(I2 × H13 + (1 − I2) × H14) (note that cell G13 needs to be created manually).

Copy and paste the cell to generate column G:

G	H
0	0
0	0
0	0
1.95888672	0
5.98160407	4.18419214
9.60034114	8.0870848
12.7113258	11.4423546
7.51756401	14.3268397

Extend the second to the last column to a rectangle box backward by using ***copy and paste*** commands, to obtain the following:

Spreadsheet 3.1

2.58110822	1.52321772	0.7092284	0.20100295	0	0	0	0
5.14552778	3.80603449	2.45869605	1.28963095	0.42934334	0	0	0
7.61018155	6.72506153	5.37399702	3.80636606	2.27344936	0.91707958	0	0
8.36207188	8.71792235	8.34157769	7.21270138	5.58233018	3.82823461	1.95888672	0
6.59436732	8.09036092	9.27226732	9.73364039	9.14967864	7.63319176	5.98160407	4.18419214
3.49422842	5.01790717	6.88867376	8.89612391	10.5361249	10.9884616	9.60034114	8.0870848
1.10527076	1.83960784	2.99743395	4.74344619	7.19325316	10.1893236	12.7113258	11.4423546
0.15690527	0.29902727	0.56988085	1.08606878	2.06981056	3.94460814	7.51756401	14.3268397

Deleting the lower-left triangle (which contains useless data), produces the final derivative price tree:

2.58110822	1.52321772	0.7092284	0.20100295	0	0	0	0
	3.80603449	2.45869605	1.28963095	0.42934334	0	0	0
		5.37399702	3.80636606	2.27344936	0.91707958	0	0
			7.21270138	5.58233018	3.82823461	1.95888672	0
				9.14967864	7.63319176	5.98160407	4.18419214
					10.9884616	9.60034114	8.0870848
						12.7113258	11.4423546
							14.3268397

The tree of derivative prices (above) indicates four states with positive payoffs (the bottom four rows of the last – or rightmost – column). Those values are recursively brought to previous time periods – columns – until the leftmost column indicates the time 0 value. Thus, the European put value, is located in the upper-left corner of the derivative value tree and is $2.58.

Note that in the above tree, other than the input parameters, only five cells, A4, B4, B5, H13, and G13, are created manually and all the rest are copied and pasted. Both trees can be created in a couple of minutes.

This example highlights a potentially huge advantage of using spreadsheets and binomial option pricing models as part of an analysis of a financial derivative: spreadsheets are easily understood visual depictions of the behavior of the underlying asset's price, the derivative's price, and decisions such as early exercise. This allows experts (other than the computer programmer) to understand the role of each parameter and the extent to which the model may or may not be capturing the key economic realities such as various risks. The goal is to avoid the danger of relying solely on black-box methods that obfuscate the inner workings of a model.

It is very easy to generalize the above binomial model to value an American put: each node can recognize any early payment options by setting the applicable cells to: MAX(immediate exercise value, European option roll back value). So only one cell, G13, needs to be changed (and copied), taking less than one minute. In addition, if we need to switch from a call to a put or to another derivative, we only need to change (and copy) the payoff cell: H13.

Exercise 3.1: Use a 30-period tree to value the following European put: A stock price is currently \$50. Over the next six months, the stock price is expected to follow a binomial tree of 30 periods. The stock's annual volatility is 20%. The risk-free rate is 5% per annum with continuous compounding. What is the value of a six-month *European* put option with a strike price of \$52?

Exercise 3.2: Use a 30-period tree to value the following American put: A stock price is currently \$50. Over the next six months, the stock price is expected to follow a binomial tree of 30 steps. The stock's annual volatility is 20%. The risk-free rate is 5% per annum, continuously compounded. The put option may be exercised at any time. What is the value of this six-month *American* put option with a strike price of \$52?

The binomial model can be expanded to find European or American put values for any number of time periods (i.e., any value of n). A fascinating exercise is to graph the European put values as a function of the number of periods, n. This allows analysis of the ***convergence of the put values***.

Figure 3.2 illustrates a sample convergence analysis. Using tools available from the internet,[1] we derived a graph with the x-axis as the number of periods (steps), and the y-axis as the value of a European put for a binomial model of steps $n = 30$ to 170. A horizontal line indicates the analytic (Black-Scholes) option value of \$3.213.

The graph indicates three important properties of binomial models with regard to convergence: (1) that the put values generally trend toward convergence; (2) that lengthening n by one step at a time causes a see-saw

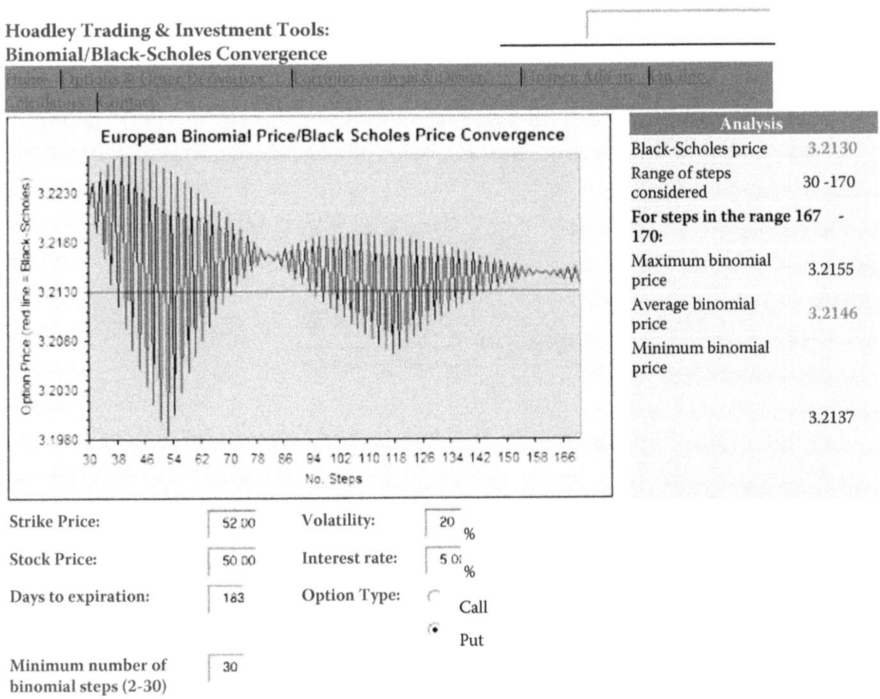

FIGURE 3.2
Convergence Analysis.

motion in the option price, and, perhaps most importantly; (3) that convergence is not monotonic even over somewhat long ranges. With regard to the third property, note that the general range of values actually widens over the interval from $n = 80$ to $n = 118$. Nevertheless, the key takeaway is that binomial values generally converge to near \$3.21 (the Black-Scholes value).

While the effort to construct a binomial valuation model (and its non-monotonic convergence) is unnecessary for valuing European options, the tree model is particularly useful for American-style options.

Section Summary

This section provides a foundation with which to:

1. Use a no-arbitrage tree method to value a derivative in a two-state economy by constructing a riskless portfolio consisting of one unit of derivative f (long) and $\Delta\left(=\frac{\Delta f}{\Delta S}\right)$ shares of the underlying asset (short).

2. Use a risk-neutral tree method to value a derivative. Note that the method is based on the risk-neutral principle where the derivative value today is the discounted expected value of the future payoff in a risk-neutral world in which all assets grow at the risk-free rate and can be discounted at the risk-free rate. The key valuation equation is as follows:

$$f_0 = e^{-r_f \Delta t} E_Q(f_{end}) = e^{-r_f \Delta t}(Qf_{up\ state} + (1 - Q)f_{down\ state})$$

$$Where\ Q = \frac{e^{r_f \Delta t} - e^{-\sigma\sqrt{\Delta t}}}{e^{\sigma\sqrt{\Delta t}} - e^{-\sigma\sqrt{\Delta t}}}$$

3. Understand that the results of a one-period tree can be generalized to an n-period tree in which convergence can be analyzed. Spreadsheets can be used to model large trees. In a multiple-period model, it is usually assumed that an estimate of σ is used to specify $u = e^{\sigma\sqrt{\Delta t}}$ and $d = e^{-\sigma\sqrt{\Delta t}}$. Backward induction is used to discount final payoffs backwards through the tree into an initial value.

Demonstration Exercises

Assume that the stock price is \$40 today and will go up by a factor of 1.1 or go down by a factor of 0.9 in the next four-month period. The risk-free rate is 0.05, compounded continuously.

a. Use the no-arbitrage tree method to find the value of a derivative that pays off $[Max(40 - S_T, 0)]^{1/2}$, where S_T is the stock price in four months.

b. Use the risk-neutral tree method to find the value of a derivative that pays off $[Max(40 - S_T, 0)]^{1/2}$, where S_T is the stock price in four months.

Solution to (a):
The Stock Tree:

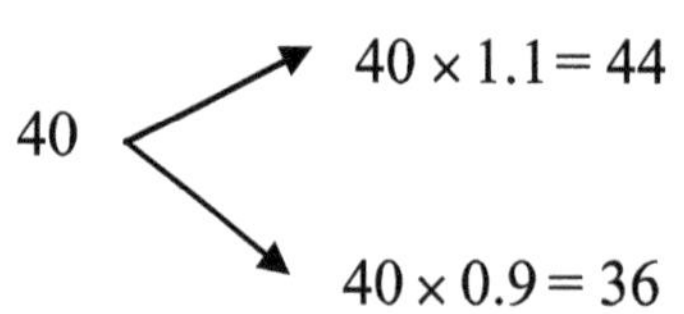

The Derivative Tree:

$$[Max(40-44,0)]^{1/2}=0=f_u$$

$$f_0$$

$$[Max(40-36,0)]^{1/2}=2=f_d$$

The riskless portfolio involves buying one unit of the derivative and short selling Δ shares of the stock. In order to form a perfect hedge, Δ needs to satisfy:

$$\Delta = \frac{0-2}{44-36} = -1/4$$

Also,

$$f_0 - \Delta S_0 = e^{-r_f T}(f_u - \Delta S_u)$$

$$f_0 - (-1/4)\times 40 = e^{-0.05\times\frac{1}{3}}(0-(-1/4)44)$$

$$f_0 = 0.8182.$$

Solution to (b): In order to apply the risk-neutral principle, first find the risk-neutral Q-measure, the probability Q:

$$Q = \frac{e^{r_f \Delta t} - d}{u-d} = \frac{e^{0.05\times 1/3} - 0.9}{1.1-0.9} = 0.584$$

By the risk-neutral principle, the value of the derivative is the expected payoff using Q-measures discounted by the riskless rate:

$$f_0 = e^{-0.05\times\frac{1}{3}}(0.584\times 0 + 0.416\times 2) = 0.8182.$$

3.3 The Geometric Brownian Motion and Monte Carlo Simulation

This section introduces Wiener processes, Ito processes, and in particular, Geometric Brownian motion, all of which are continuous models for an underlying asset price. The section develops theoretical tools such as Ito's lemma and the application of moment generating functions as well as practical tools such as Monte Carlo simulation of underlying asset prices. This section also discusses how to obtain a distribution of underlying asset prices at the maturity of financial derivatives by Ito's lemma without using a simulation. Finally, the question of which measure (the P-measure or Q-measure) should be used in valuing financial derivatives is explored.

Modeling the process of the underlying asset is the first step in modeling financial derivative values. The Black-Scholes formula is then derived to value European calls and puts in the next section. Simulation approaches to value derivatives are discussed in subsequent sections and are used for problems that may be too complex for an analytical solution or a simple tree model.

Financial simulation models often assume that certain underlying values such as equity prices and interest rates evolve as specified processes. The simulated paths of these underlying values are then used to specify the paths of securities that depend on those underlying values such as financial derivatives. In valuing financial derivatives, the Monte Carlo simulation approach generates estimates of derivative values that tend to increase in accuracy as the number of simulated paths is increased.

3.3.1 Introduction to Generalized Wiener Processes and Ito Processes

In this section, the focus is on the process of the underlying asset price and how it can be simulated. A ***process*** describes how something such as an asset price evolves non-deterministically.

Let's return to the case of a financial derivative such as a put option on a share of stock. As with the tree approach, we know S_0, the stock price today, and we want to model S_T, the stock price T years later, where T is the time to expiration or maturity of the put. In a discrete model such as binomial tree, the Q-measure expected value of S_T can be used in determining the derivative's value. This section moves to a *continuous* model and discusses the simulation approach which models the evolvement of S_t. ***Evolvement*** refers to how the change of t is related to the change of S_t.

We use a differential equation (DE) to model this evolvement. As a very simple example, we may assume that $dS = 0.14Sdt$. Namely, over a short period of time, dt, the change of S, dS, is proportional to S and dt. By basic calculus knowledge, this differential equation can be solved: $S_T = S_0e^{0.14T}$, where 14% is the stock price growth rate.

Note that for this DE, the value S_T is deterministic, specifying that the stock price grows at a continuous and constant annual rate of 14%. The stock acts like a bank account rather than containing the random price changes observed in the stock market. To add a random component, we add a random variable to the evolvement: $dS = 0.14Sdt + 0.2SdZ$, where the variable dZ is normally distributed with a mean of 0 and a variance of dt. The process dZ is called a Wiener process and is defined later. This kind of equation is called a ***stochastic differential equation (SDE)*** because it adds a random component to a differential equation. In particular, the process of dS is called Geometric Brownian motion, which is a special case of Ito processes. All three of these terms are defined in detail later. From now on, the words *process* and *evolvement* are used interchangeably.

Note that the value of S_T, the ending price of the stock after T years, is crucial in derivative valuation. A specific process can be used to simulate

$$f_0 \nearrow [Max(40-44,0)]^{1/2} = 0 = f_u$$

values of S_T. Each simulated path represents a single potential outcome from an infinite number of potential random outcomes. We can estimate the probability distribution of S_T by simulating many examples of S_T. Alternatively, using Ito's lemma, which is discussed later, we can determine the probability distribution of S_T for the given SDE. Whether or not this probability distribution is in P-measure or Q-measure is left for discussion in the next section.

In fact, the distribution of S_T is usually hard to derive analytically even when given the process (SDE) by which S_T evolves. We first analyze a relatively simple process $dX = 0.14dt + 0.2dZ$, which is a "generalized Wiener process." Note the difference between the process dX and Geometric Brownian motion dS: in this simple process, there are constants before dt and dZ, while in the process for a stock price there is an extra S prior to dt and dZ, which makes price changes proportional to the price (S). In the generalized Wiener process, the distribution of X_T is easily found but it is not good for modeling stock prices (obviously a stock trading near $400 per share experiences larger absolute price changes than a stock trading at $4 per share). To be useful, the process used to model market prices takes this form: $dS = 0.14Sdt + 0.2SdZ$. While this process (an Ito process) provides far superior modeling of the behavior of stock prices, it raises challenges in identifying the distribution of S_T due to the extra variable S before dt and dZ. A powerful tool called Ito's lemma can be used to identify the distribution of S_T.

For a generalized Wiener process such as $dX = 0.14dt + 0.2dZ$, the ***local result*** of an ending price after a short period of time, such as $X_{0.01}$ and the ***global result*** of an ending price after a long period of time, such as X_1, are similar. For a process similar to S_T (an Ito process), the local result of an ending price after a short period of time, such as $S_{0.01}$, and the global result of an ending price after a long period of time, such as S_1, are very different.

The next few sections analyze generalized Wiener processes and Ito processes in detail through examples.

3.3.2 Wiener Processes and the Simulation of Paths

Wiener processes are often used to describe the random variation that occurs in a stochastic process through time. To understand a Wiener process, begin by cutting the time interval $[0, T]$ into n equal time intervals $[0, t_1]$, $[t_1, t_2]$, $[t_2, t_3] \dots [t_{n-1}, t_n = T]$ with each interval length $\Delta t = T/n$.

Wiener processes describe how a function, say $Z(t)$ or Z_t, evolves between $[0, T]$ as follows.

Define Z_{t_i} as a function of time t_i at the discrete times $0, t_1, t_2, \dots t_{n-1}, t_n = T$.

$$
\begin{aligned}
&Z_0 = 0\\
&Z_{t_1} - Z_{t_0} = \sqrt{\Delta t} \quad \times \text{“}N(0, 1)\text{”}\\
&Z_{t_2} - Z_{t_1} = \sqrt{\Delta t} \quad \times \text{“}N(0, 1)\text{”}\\
&Z_{t_3} - Z_{t_2} = \sqrt{\Delta t} \quad \times \text{“}N(0, 1)\text{”}\\
&\ldots\\
&Z_T - Z_{t_{n-1}} = \sqrt{\Delta t} \quad \times \text{“}N(0, 1)\text{”}
\end{aligned}
$$

The above values indicated as "N(0, 1)" represent n independent standard normally distributed random variables. Note that Z_t begins as 0 and then evolves as an expanding cumulative sum of independent normally distributed random variables. As n goes to infinity, it generates a continuous path for Z_t as a function of time t. This process is called a ***Wiener process*** $\boldsymbol{Z(t)}$. We use dZ to represent how $Z(t)$ evolves at time interval dt. Note that the ending stage, Z_T, is a random variable. We cannot find a deterministic value of Z_T. We can only simulate paths that represent possible outcomes of Z_T or identify the distribution of Z_T.

Note that the change of Z, which is ΔZ, follows a normal distribution with mean 0 and variance Δt, namely $\Delta Z \sim N(0, \Delta t)$. Intuitively, a Wiener process is a sum of independent normal distributions accumulated through time with each normal distribution having a variance equal to the time interval Δt. Finally, note that Z_t is a continuous function of t, but it is not differentiable at any point.

Despite the shortcomings of a simple Wiener process in modeling the long-term path of stock price, each N(0, 1) nicely models the effect of the arrival of new information on the price of a stock. The function Z_t is a function of time t for discrete times $0, t_1, t_2, \ldots t_n = T$, that represents the cumulative effect of new information on the price of a given stock.

The easiest way to understand a Wiener process is through simulations.

Let's start with examples of simulating the path of a Wiener process at four discrete points. Note that the simulation is a discretization of the true Wiener process, so it is only an approximation of a Wiener process. The higher the number of periods selected, the closer it is approaching the Wiener process.

Example 3.5: Consider a four-period model. Select four random numbers from N(0, 1) = {1.40, 0.37, −0.64, −0.16}. Use these random numbers to generate the path for Z_1, Z_2, Z_3, Z_4 starting from $Z_0 = 0$.

Solution: The time intervals are assumed to be [0, 1], [1, 2], [2, 3], [3,4] so that $\Delta t = 1$ in this case.

$$Z_1 - Z_0 = \text{"}N(0,1)\text{"} = 1.40 \; \textit{which implies} \; Z_1 = Z_0 + 1.40 = 1.40$$
$$Z_2 - Z_1 = \text{"}N(0,1)\text{"} = 0.37 \; \textit{which implies} \; Z_2 = Z_1 + 0.37 = 1.40 + 0.37$$
$$Z_3 - Z_2 = \text{"}N(0,1)\text{"} = -0.64 \; \textit{which implies} \; Z_3 = Z_2 - 0.64$$
$$= 1.40 + 0.37 - 0.64$$
$$Z_4 - Z_3 = \text{"}N(0,1)\text{"} = -0.16 \; \textit{which implies} \; Z_4 = Z_3 - 0.16$$
$$= 1.40 + 0.37 - 0.64 - 0.16$$

After calculating Z_1, Z_2, Z_3, *and* Z_4, the path can be plotted with the x coordinates as time (0, 1, 2, 3, and 4) and the y coordinates as the Z-values. Additional potential paths can be simulated using new sets of random numbers.

Example 3.6: Again consider a four-period model. Using the same random numbers from N(0, 1) = {1.40, 0.37, −0.64, −0.16} but using a smaller time interval (0.01), generate the path for $Z_{0.01}$, $Z_{0.02}$, $Z_{0.03}$, and $Z_{0.04}$ starting from $Z_0 = 0$.

Solution: The time intervals are now [0, 0.01], [0.01, 0.02], [0.02, 0.03], [0.03, 0.04], so $\Delta t = 0.01$. In this case:

$$Z_{0.01} - Z_0 = \sqrt{\Delta t}\,\text{"}N(0,1)\text{"} = \sqrt{0.01} \times 1.40 = 0.140, \; \textit{which implies} \; Z_{0.01}$$
$$= Z_0 + 0.140 = 0.14$$
$$Z_{0.02} - Z_{0.01} = \sqrt{\Delta t}\,\text{"}N(0,1)\text{"} = \sqrt{0.01} \times 0.37 = 0.037, \; \textit{which implies} \; Z_{0.02}$$
$$= Z_{0.01} + 0.037 = 0.140 + 0.037$$
$$Z_{0.03} - Z_{0.02} = \sqrt{\Delta t}\,\text{"}N(0,1)\text{"} = \sqrt{0.01} \times (-0.64), \; \textit{which implies} \; Z_{0.03}$$
$$= Z_{0.02} - 0.064 = 0.140 + 0.037 - 0.064$$
$$Z_{0.04} - Z_{0.03} = \sqrt{\Delta t}\,\text{"}N(0,1)\text{"} = \sqrt{0.01} \times (-0.16), \; \textit{which implies} \; Z_{0.04}$$
$$= Z_{0.03} - 0.016 = 0.140 + 0.037 - 0.064 - 0.016$$

$Z_{0.01}$, $Z_{0.02}$, $Z_{0.03}$,$Z_{0.04}$, can also be plotted as one illustrated path of Z.

Based on a Wiener process, the next section investigates a generalized Wiener process.

3.3.3 The Generalized Wiener Process and the Simulation of Its Paths

The ***Generalized Wiener process*** is

$$d\mathbf{X} = \mathbf{a}d\mathbf{t} + \mathbf{b}d\mathbf{Z} \tag{3.9}$$

where a and b are constants.

In finance, the first term on the right side, adt, is a linear component, proportional to Δt and is often referred to as the ***drift*** or ***growth of the process*** because it represents a deterministic (usually positive) effect on the price or return being modeled with the parameter a indicating the magnitude of the drift. The second term on the right side, bdZ, is a Wiener process which adds a normal distribution with mean 0 and variance $b^2\Delta t$ and it is often referred to as the ***diffusion of the process*** because it represents the stochastic changes in the price or return through time with the parameter b indicating the magnitude of the volatility.

Example 3.7: Consider the Generalized Wiener process $dX = 5dt + 6dZ$ in a four-period model. Use these four random numbers from N(0,1): {1.40, 0.37, −0.64, −0.16} to simulate a path for $X_{0.01}$, $X_{0.02}$, $X_{0.03}$, and $X_{0.04}$, $X_{0.02}$, $X_{0.03}$, $X_{0.04}$, assuming X_0 =30.

Solution: The time intervals are [0, 0.01], [0.01, 0.02], [0.02, 0.03], [0.03, 0.04] with Δt = 0.01 in this case.

$$X_{0.01} - X_0 = 5\ \Delta t + 6\sqrt{\Delta t}\ \text{“}N(0,1)\text{”} = 5 \times 0.01 + 6 \times \sqrt{0.01} \times 1.40$$
$$= 0.05 + 0.6 \times 1.40 = 0.89\ \textit{which implies}\ X_{0.01} = 30 + 0.89 = 30.89$$
$$X_{0.02} - X_{0.01} = 5\ \Delta t + 6\sqrt{\Delta t}\ \text{“}N(0,1)\text{”} = 5 \times 0.01 + 6 \times \sqrt{0.01} \times 0.37$$
$$= 0.272\ \textit{which implies}\ X_{0.02} = X_{0.01} + 0.272 = 30.89 + 0.272$$
$$= 31.162$$
$$X_{0.03} - X_{0.02} = 5\ \Delta t + 6\sqrt{\Delta t}\ \text{“}N(0,1)\text{”} = 5 \times 0.01 + 6 \times \sqrt{0.01} \times (-0.64)$$
$$= -0.334\ \textit{which implies}\ X_{0.03} = X_{0.02} - 0.334 = 31.162 - 0.334$$
$$= 30.828$$
$$X_{0.04} - X_{0.03} = 5\ \Delta t + 6\sqrt{\Delta t}\ \text{“}N(0,1)\text{”} = 5 \times 0.01 + 6 \times \sqrt{0.01} \times (-0.16)$$
$$= -0.046\ \textit{which implies}\ X_{0.04} = X_{0.03} - 0.046 = 30.828 - 0.046$$
$$= 30.782.$$

Exercise 3.3: Use R to generate a path of $X_{0.01}$, $X_{0.02}$, $X_{0.03}$,$X_{0.04}$,…, $X_{0.99}$, $X_{1.00}$ where X_0 = 20 and $dX = 5dt + 6dZ$.

Solution: Here is a sample R-code for generating one path for this Wiener process where the x-axis is the index for 100 time points 0.01, 0.02, …, 1.00.

R Code 3.2 Generating one path for generalized Wiener process

```
#input initial value and parameters
X0=20
a=5
b=6
T=1
#specify a n period model
n=100
deltaT=T/n
#generate n random number from N(0,1)
W=rnorm(n,0,1)
#record the path at discrete n values as a vector X using a
for loop
X=rep(X0,n+1)
for(i in 1:n)
{
X[i+1]=X[i]+a*deltaT+ b*sqrt(deltaT)*W[i]
}
plot(X, main="Generalized Wiener Process",type="l")
```

The following figure graphs a sample path (Figure 3.3):

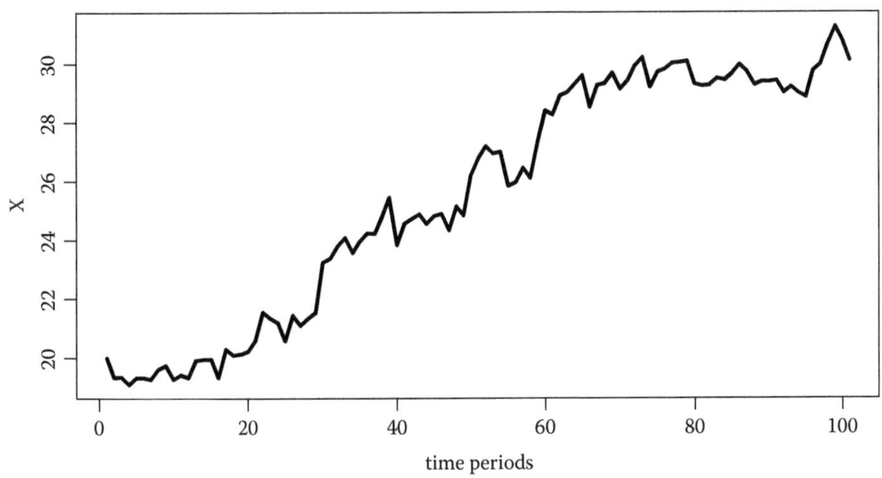

FIGURE 3.3
A Sample Path for a Generalized Wiener Process.

Note that X_t is a continuous function of t, but not differentiable at any point. The picture is a discretization of the process, so it is differentiable at all points except corner points.

Exercise 3.4: Repeat the previous exercise but change the number of cuts to n = 10,000 and draw the path.

Exercise 3.5: Write R-code to generate 50 paths assuming that it is only necessary to record the ending price of each path.

Solution: Here is a sample R-code (note that we need to nest an inside loop mentioned above into an outside loop of 50 paths):

R-Code 3.3 Generating many paths for generalized Wiener process

```
#input initial value and parameters
X0=20
a=5
b=6
T=1
# specify a n period model for m paths
m=50
n=100
deltaT=T/n
#record each path at discrete n values as a vector X and m
path's ending price as a vector Y
X=rep(X0,n+1)
Y=rep(0,m)
#double loops, outside loop for m paths and inside loop
for n periods, we overwrite X values
for(j in 1:m)
{
W=rnorm(n,0,1)
for(i in 1:n)
{
X[i+1]=X[i]+a*deltaT+ b*sqrt(deltaT)*W[i]
}
Y[j]=X[n+1]
}
#print Y
Y
```

Note that Y is the ending price (S_T) for 50 paths in the above code.

Exercise 3.6: Change the number of cuts to n = 1,000, and the number of paths to m = 500. Then find the mean of Y and the standard deviation of Y using the commands mean(Y) and $sd(Y)$.

Compare the resulting mean and standard deviation estimates with $X_0 + aT$ and $b\sqrt{T}$, respectively. Are they equal? In a later section, we will prove that $E(X_T) = X_0 + aT$ and $Var(X_T) = b^2T$ theoretically.

Exercise 3.7: Keep the number of cuts at n = 1,000 and the number of paths m = 500. Draw the distribution of X_T using command hist(Y). What distribution do you observe? In a later section, we will prove that X_T follows a normal distribution theoretically.

The next section discusses Ito processes, which are more challenging to understand than generalized Wiener processes.

3.3.4 Ito Processes, Geometric Brownian Motion, and the Simulation of Its Paths

An ***Ito process*** has the following SDE:

$$dX = \mathbf{a}(X, t)dt + \mathbf{b}(X, t)dZ \tag{3.10}$$

where a(X, t) and b(X, t) are functions of X and t.

This process has a small time interval with a length of Δt. The change of X has two components, one is Δt multiplying a function of X and t, and the other is a Wiener process times a function of X and t.

Consider the following special case, called ***Geometric Brownian motion***, that is the standard way to model a stock price:

$$dS = \mu Sdt + \sigma SdZ \tag{3.11}$$

where μ and σ are constants, and S is the price of the stock. Recall that μ is the ***stock's expected return*** or ***annual growth rate***, and σ is the ***stock return's annual volatility*** or ***standard deviation***. This Ito process contains a growth part, μSdt, and a random part, σSdZ. Next, some examples of simulating paths for Geometric Brownian motion are provided.

Example 3.8: Consider a four-period model with four random numbers from N(0, 1) = {1.40, 0.37, −0.64, −0.16}. Assume an Ito process $dS = 0.14Sdt + 0.20SdZ$ to generate one path ($S_{0.01}$, $S_{0.02}$, $S_{0.03}$, $S_{0.04}$) and assume that S_0 = 30.

Solution: S_0 grows from 30 to 30.7432 through four periods as shown:

$$S_{0.01} - S_0 = 0.14\ S_0 \Delta t + 0.20 S_0 \sqrt{\Delta t}\ \text{"}N(0,1)\text{"} = 0.14 \times 30 \times 0.01$$
$$+ 0.20 \times 30 \times \sqrt{0.01}\ 1.40 = 0.042 + 0.84$$
$$= 0.882 \ \textit{which implies}\ S_{0.01} = 30 + 0.882 = 30.882$$
$$S_{0.02} - S_{0.01} = 0.14\ S_{0.01}\ \Delta t + 0.20 S_{0.01} \sqrt{\Delta t}\ \text{"}N(0,1)\text{"}$$
$$= 0.14 \times 30.882 \times 0.01 + 0.20 \times 30.882 \times \sqrt{0.01} \times 0.37$$
$$= 0.0432 + 0.2285 = 0.2717\ \textit{which implies}\ S_{0.02}$$
$$= 30.882 + 0.2717 = 31.1537$$
$$S_{0.03} - S_{0.02} = 0.14\ S_{0.02}\ \Delta t + 0.20 S_{0.02} \sqrt{\Delta t}\ \text{"}N(0,1)\text{"}$$
$$= 0.14 \times 31.1537 \times 0.01 + 0.20 \times 31.1537 \times \sqrt{0.01} \times (-0.64)$$
$$= 0.04362 - 0.3988 = -0.3551\ \textit{which implies}\ S_{0.02}$$
$$= 31.1537 - 0.3551 = 30.7986$$
$$S_{0.04} - S_{0.03} = 0.14\ S_{0.03}\ \Delta t + 0.20 S_{0.03} \sqrt{\Delta t}\ \text{"}N(0,1)\text{"} = 0.14 \times 30.7986 \times 0.01$$
$$+ 0.20 \times 30.7986 \times \sqrt{0.01} \times (-0.16) = 0.04311 - 0.09856$$
$$= -0.0554\ \textit{which implies}\ S_{0.04} = 30.7986 - 0.0554 = 30.7432$$

The above results can be used to form a stock price plot through time.

Exercise 3.8: Use R to generate a path of $S_{0.01}, S_{0.02}, S_{0.03}, S_{0.04}, \ldots, S_{0.99}, S_1$ where $S_0 = 20$ and $dS = 0.14Sdt + 0.20SdZ$ and then plot the path where the x-axis is the index for time points 0.01, 0.02, ..., 1.00.

R-Code 3.4 Generating one path for Geometric Brownian motion

```
S0=20
a=0.14
b=0.2
T=1
n=100
deltaT=T/n
S=rep(S0,n+1)
W=rnorm(n,0,1)
for(i in 1:n)
{
S[i+1]=S[i]+a*S[i]*deltaT+ b*S[i]*sqrt(deltaT)*W[i]
}
plot(S, main="Geometric Brownian Motion",type="l")
```

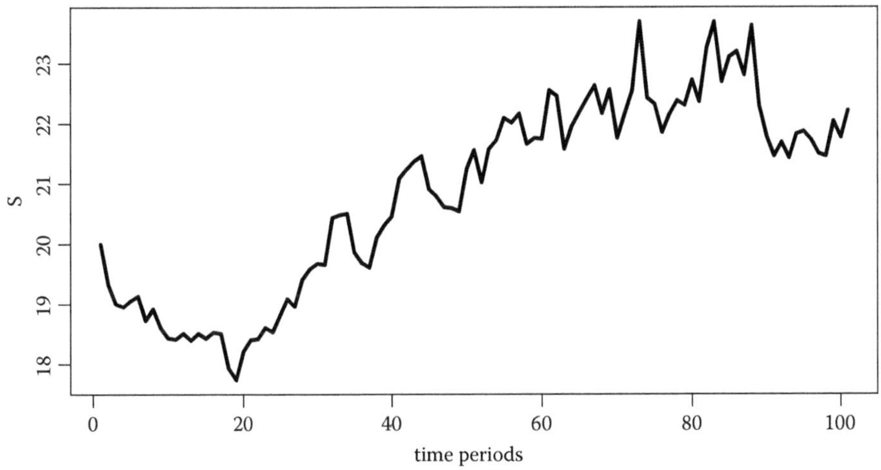

FIGURE 3.4
A Sample Path for Geometric Brownian Motion.

Figure 3.4 is a sample graph of one outcome.

Exercise 3.9: Repeat Exercise 3.8 with the number of cuts, n, increasing from 100 to 10,000 and then draw the path.

Solution: Repeat the program from Exercise 3.8 but change n from 100 to 10,000 in the R-code.

Exercise 3.10: Generate 50 paths with $n = 100$ using R-code but only record the ending prices.

Solution:

R-Code 3.5 Generating many paths for Geometric Brownian motion

```
S0=20
a=0.14
b=0.2
T=1
m=50
n=100
deltaT=T/n
S=rep(S0,n+1)
Y=rep(0,m)
for(j in 1:m)
{
W=rnorm(n,0,1)
for(i in 1:n)
{
S[i+1]=S[i]+a*S[i]*deltaT+ b*S[i]*sqrt(deltaT)*W[i]
}
Y[j]=S[n+1]
}
Y
```

Note that Y is the ending price S_T for 50 paths in the above codes.

Exercise 3.11: Repeat Exercise 3.10 changing the number of cuts to n = 1,000 and the number of paths to m = 500. Find the mean of S_T and the standard deviation of S_T using the commands: mean(Y) and *sd* (Y). Then compare the estimates of the expectation and standard deviation with $S_0e^{\mu t}$ and $\sqrt{S_0^2(e^{2\mu T+\sigma^2 T} - e^{2\mu T})}$, respectively. Are they equal? In a later section, we will prove that $E(S_T) = S_0e^{\mu t}$ and that $Var(S_T) = S_0^2(e^{2\mu T+\sigma^2 T} - e^{2\mu T})$ theoretically.

Exercise 3.12: Repeat Exercise 3.11 keeping the number of cuts at n = 1,000 and the number of paths at m = 500. Draw the distribution of LnS_T using the command: hist(log(Y, base = exp(1))). What distribution do you observe? In a later section, we will prove LnS_T follows the normal distribution theoretically, or equivalently, S_T follows a lognormal distribution.

Exercise 3.13: Use YAHOO! historical stock data from a recent year to get an estimate of μ and σ from the returns of a stock of your choice such as IBM. Use those estimates to generate a path of the one-year stock price with $\Delta t = 1/252$. If Δt is measured in units of a year, then μ and σ must be annualized estimates.

Note that if the growth/drift rate and volatility are used in the real world, then the process is regarded as a P-measure process. The P-measure process is not used to find the price of derivatives.

The above exercises simulate paths for a generalized Wiener process and Geometric Brownian motion. The next section finds the distribution of the processes without using simulation.

3.3.5 Finding the Distribution of a Generalized Wiener Process without Simulation

Given a process $dX = adt + bdZ$, how can the distribution of X_T be identified without using simulation?

First, cut the time into n intervals: $[t_0, t_1], [t_1, t_2], [t_2, t_3], \ldots, [t_{n-1}, t_n]$. Note that $t_0 = 0$ and $t_n = T$, and the length is $\Delta t = t_i - t_{i-1}$

$$X_{t_1} - X_0 = a\Delta t + b\ \sqrt{\Delta t}\ \text{“}N(0, 1)\text{”}$$
$$X_{t_2} - X_{t_1} = a\Delta t + b\ \sqrt{\Delta t}\ \text{“}N(0, 1)\text{”}$$
$$X_{t_3} - X_{t_2} = a\Delta t + b\ \sqrt{\Delta t}\ \text{“}N(0, 1)\text{”}$$
$$X_{t_4} - X_{t_3} = a\Delta t + b\ \sqrt{\Delta t}\ \text{“}N(0, 1)\text{”}$$
$$\cdots$$
$$X_T - X_{t_{n-1}} = a\Delta t + b\ \sqrt{\Delta t}\ \text{“}N(0, 1)\text{”}$$

Note that “N(0, 1)” in the above equations denotes independent standard normal distributions. Summing all of the above left-hand sides and right-hand sides, respectively:

$$X_T - X_0 = n\ a\Delta t + b\sqrt{\Delta t}\ \textstyle\sum_{i=1}^{n}\ N(0, 1).$$

Note that the sum of n independent N(0, 1) terms is N(0, n) or $\sqrt{n}$N(0, 1) and that $n\Delta t = T$. So, for a generalized Wiener process $dX = adt + bdZ$, we have:

$$\mathbf{X_T = X_0 + aT + b\sqrt{T}N(0, 1)} \tag{3.12}$$

So, it is not necessary to simulate the whole path to estimate the distribution of X_T because it has been proved that $X_T \sim N(X_0 + aT,\ b^2T)$. The equation $X_{t_1} - X_0 = a\Delta t + b\sqrt{\Delta t}N(0, 1)$ is termed a local formula because it only describes what happens in the next moment, t_1. However, $X_T = X_0 + aT + b\sqrt{T}N(0, 1)$ is a global formula since T can be arbitrary, big or small. The global formula describes the distribution of the end variable, X_T. Note that X_T can be generated by one simulation, namely generating one random variable from N(0,1) and applying it to the formula $X_T = X_0 + aT + b\sqrt{T}N(0, 1)$. It is not necessary to simulate the whole path with $X_{t_1}, X_{t_2}, \ldots, X_{t_{n-1}}, X_T$.

The above method cannot be generalized to Geometric Brownian motion directly; Ito's lemma is necessary. Before discussing advanced mathematical machinery, a review of relevant Mathematics is needed.

3.3.6 Review of Mathematics Regarding the Normal Distribution, Moment Generating Functions, and Taylor Expansions

This section begins by reviewing the properties of the normal distribution. Consider a normal random variable X with mean $E(X)=\mu$ and variance $Var(X)= \sigma^2$, namely $X\sim N(\mu,\sigma^2)$.

Mathematics Review 3.2

Warm-up 3.1

If $X = LnY_T$ follows a normal distribution $N(\tilde{\mu}, \tilde{\sigma}^2)$ with p.d.f.:

$$f_X(x) = \frac{1}{\sqrt{2\pi\tilde{\sigma}^2}}e^{-\frac{(x-\tilde{\mu})^2}{2\tilde{\sigma}^2}}$$

then Y_T has a lognormal distribution with p.d.f. of Y_T:

$$f_{Y_T}(y) = \frac{1}{\sqrt{2\pi\tilde{\sigma}^2}}e^{-\frac{(lny-\tilde{\mu})^2}{2\tilde{\sigma}^2}}\frac{1}{y}$$

Warm-up 3.2

If the random variable X has a p.d.f. $f_X(x)$, then the probability that $X > x$ has the following formula:

$$P(X > x) = \int_x^{+\infty} f_X(x)dx$$

Given $g(x)$, a function of x, the expected value of $g(X)$ can be expressed as follows:

$$E(g(X)) = \int_{-\infty}^{+\infty} g(x)f_X(x)dx$$

Warm-up 3.3

If $X\sim N(\mu, \sigma^2)$, then the probabilities have the following formulas:

$$P(X < x) = N(z) = N\left(\frac{x - \mu}{\sigma}\right)$$

$$P(X > x) = N(-z) = N\left(-\frac{x - \mu}{\sigma}\right) = N\left(\frac{\mu - x}{\sigma}\right)$$

where $N(\cdot)$ is the cumulative normal distribution function of the standard normal distribution which can be found in a z-table.

Warm-up 3.4

The moment generating function for $X \sim N(\mu, \sigma^2)$ is

$$\Phi(t) = e^{\mu t + \frac{1}{2}\sigma^2 t^2} = E(e^{Xt}) = \int_{-\infty}^{\infty} e^{xt} \frac{1}{\sqrt{2\pi\sigma^2}} e^{-\frac{(x-\mu)^2}{2\sigma^2}} dx$$

The moment generating function $\Phi(t)$ is defined as $E(e^{Xt})$. Note that the expected value of e^{Xt} is deterministic and is a function of t. It is well known that the moments of the normal distribution can be calculated by taking derivatives of the moment generating function:

$$EX = \Phi'(0)$$

$$E(X^2) = \Phi''(0)$$

$$E(X^3) = \Phi'''(0)$$

$$E(X^4) = \Phi^{(4)}(0).$$

Using Mathematica, the higher moments for a normal random variable can easily be found:

Mathematica Code 3.1 Finding first four moments for normal distribution

```
f[t_]:=Exp[mu*t+1/2*sigma^2*t^2]
f1[t_]:=f'[t]
f2[t_]:=f1'[t]
f3[t_]:=f2'[t]
f4[t_]:=f3'[t]
f1[0]
f2[0]
f3[0]
f4[0]
```

The formulas for all of the first four moments of $N(\mu, \sigma^2)$ are as follows:

$$EX = \mu$$

$$E(X^2) = \mu^2 + \sigma^2$$

$$E(X^3) = \mu^3 + 3\mu\sigma^2$$

$$E(X^4) = \mu^4 + 6\mu^2\sigma^2 + 3\sigma^4$$

For the normal distribution, all higher moments are functions of the mean and variance, which is not generally true for other distributions.

In addition to calculating moments, moment generating functions can be used for another purpose. After learning the risk-neutral principle for a continuous model, various expected payoffs will be calculated. For the normal (lognormal) distribution, the moment generating function is often used to calculate the expected values of random variables (payoffs). So, we will review how to use the moment generating function to calculate expected values of random variables related to the normal (lognormal) distribution in Review 3.3.

Mathematics Review 3.3

Warm-up 3.5

Recall: $\Phi(t) = E(e^{Xt}) = e^{\mu t + \frac{1}{2}\sigma^2 t^2}$ for $X \sim N(\mu, \sigma^2)$

If $X \sim N(5, 3^2)$, verify that $E(e^{2X}) = e^{5\times 2 + \frac{1}{2}3^2 2^2}$ using the moment generating function with $t = 2$.

If $Ln(X) \sim N(5, 3^2)$, verify that $E(X^2) = E(e^{2LnX}) = e^{5\times 2 + \frac{1}{2}3^2 2^2}$ using the moment generating function with $t = 2$.

Warm-up 3.6

Recall: $EX = \mu$, $E(X^2) = \mu^2 + \sigma^2$, $E(X^3) = \mu^3 + 3\mu\sigma^2$, and $E(X^4) = \mu^4 + 6\mu^2\sigma^2 + 3\sigma^4$ for $X \sim N(\mu, \sigma^2)$.

Verify that if $Y \sim N(0, \Delta t)$ where Δt is a constant, then $E(Y^2) = \Delta t$, $E(Y^4) = 3\Delta t^2$ using formulas for all of the first four moments.

Now we will review Taylor expansions, which are important for Ito's lemma.

Mathematics Review 3.4

Mathematical approximation methods are used in mathematical finance in general and in a very important lemma known as Ito's lemma. Here are three warm-ups to prepare for material that follows:

Warm-up 3.7 Review of Maclaurin series applied to e^x

$$e^x = 1 + \frac{x}{1!} + \frac{x^2}{2!} + \dots$$

$$e^x \approx 1 + x \ \textit{if} \ x \ \textit{is very small.}$$

$$e^x \approx 1 + x + \frac{x^2}{2} \ \textit{if} \ x \ \textit{is small.}$$

Warm-up 3.8

More generally, many functions have a Taylor series expansion:

$$f(x) = f(a) + \frac{f'(a)}{1!}(x-a) + \frac{f''(a)}{2!}(x-a)^2 + \dots$$

Note that $L(x) = f(a) + f'(a)(x-a)$ is a tangent line approximation of $f(x)$ at a.

$Q(x) = f(a) + f'(a)(x-a) + \frac{f''(a)}{2}(x-a)^2$ is a second-degree approximation of $f(x)$ at a.

Warm-up 3.9

Let $z=f(x, y)$ from R^2 to R^1 represent a surface in R^3.

Consider a Taylor series in two-variable cases:

Note that $L(x, y) = f(a, b) + f_x(a, b)(x-a) + f_y(a, b)(y-b)$ is a tangent plane approximation of $f(x, y)$ at point (a,b).

$Q(x, y) = f(a, b) + f_x(a, b)(x-a) + f_y(a, b)(y-b) + \frac{1}{2}f_{xx}(a, b)(x-a)^2$ is a

$$+ f_{xy}(a, b)(x-a)(y-b) + \frac{1}{2}f_{yy}(a, b)(y-b)^2$$

second-degree approximation of the surface $f(x, y)$ at point (a,b).

We know that $z - f(a, b)$ is the change of f, df, while $x - a$ is the change of x, dx, and $y - b$ is the change of y. Here are the two above equations in a total differentials format:

Degree 1:

$$df = \frac{\partial f}{\partial x}dx + \frac{\partial f}{\partial y}dy$$

Degree 2:

$$df = \frac{\partial f}{\partial x}dx + \frac{\partial f}{\partial y}dy + \frac{1}{2}\frac{\partial^2 f}{\partial x^2}(dx)^2 + \frac{\partial^2 f}{\partial x \partial y}dxdy + \frac{1}{2}\frac{\partial^2 f}{\partial y^2}(dy)^2$$

Now we are ready for Ito's lemma!

3.3.7 Finding the Distribution of Geometric Brownian Motion without Simulation Using Ito's Lemma

Recall the Geometric Brownian motion commonly used in finance and introduced earlier in this chapter: $dS = \mu S dt + \sigma S dZ$.

The distribution of S_T cannot be found for an Ito process using the strategy demonstrated in the previous section. To see this, assume the time intervals are $[t_0, t_1]$, $[t_1, t_2]$, $[t_2, t_3]$,…,$[t_{n-1}, t_n]$ as before.

$$
\begin{aligned}
&S_{t_1} - S_0 = \mu S_0 \Delta t + \sigma S_0 \sqrt{\Delta t} \ \text{"}N(0,1)\text{"} \\
&S_{t_2} - S_{t_1} = \mu S_{t_1} \Delta t + \sigma S_{t_1} \sqrt{\Delta t} \ \text{"}N(0,1)\text{"} \\
&S_{t_3} - S_{t_2} = \mu S_{t_2} \Delta t + \sigma S_{t_2} \sqrt{\Delta t} \ \text{"}N(0,1)\text{"} \\
&S_{t_4} - S_{t_3} = \mu S_{t_3} \Delta t + \sigma S_{t_3} \sqrt{\Delta t} \ \text{"}N(0,1)\text{"} \\
&\ldots \\
&S_{t_n} - S_{t_{n-1}} = \mu S_{t_{n-1}} \Delta t + \sigma S_{t_{n-1}} \sqrt{\Delta t} \ \text{"}N(0,1)\text{"}
\end{aligned}
$$

Summing both sides, as before, does not provide a common factor since the RHS (right-hand side) has different values of S_{t_i}. In fact, S_T does not follow a normal distribution. To find the distribution of S_T involves using a new and very important tool: Ito's lemma.

Ito's lemma is a tool for modeling the process df, where $f(X, t)$ is a function of X and t, and X is an Ito process. Ito's lemma starts with the total differential df in the second degree. If $f(X, t)$ is a function of X and t, then:

$$
df = \frac{\partial f}{\partial X} dX + \frac{\partial f}{\partial t} dt + \frac{1}{2}\frac{\partial^2 f}{\partial X^2}(dX)^2 + \frac{\partial^2 f}{\partial X \partial t} dX dt + \frac{1}{2}\frac{\partial^2 f}{\partial t^2}(dt)^2 \quad (3.13)
$$

where dX is an Ito process, which follows some SDE.

Ito's lemma

If $f(X,\ t)$ is a function of X and t, and X follows a process:

$$dX = \mu(X,\ t)dt + \sigma(X,\ t)dZ$$

then $f(X,\ t)$ follows a process:

$$df = \frac{\partial f}{\partial X}dX + \frac{\partial f}{\partial t}dt + \frac{1}{2}\frac{\partial^2 f}{\partial X^2}(dX)^2 \tag{3.14}$$

We can use either of the following equations as Ito's lemma:

$$df = \frac{\partial f}{\partial X}(\mu(X,\ t)dt + \sigma(X,\ t)dZ) + \frac{\partial f}{\partial t}dt + \frac{1}{2}\frac{\partial^2 f}{\partial X^2}\sigma(X,\ t)^2 dt \tag{3.15}$$

$$df = \left(\frac{\partial f}{\partial X}\mu(X,\ t) + \frac{\partial f}{\partial t} + \frac{1}{2}\frac{\partial^2 f}{\partial X^2}\sigma(X,\ t)^2\right)dt + \frac{\partial f}{\partial X}\sigma(X,\ t)dZ \tag{3.16}$$

Remarks

Adopting Hull's approach, this book provides some intuition regarding Ito's lemma but does not demonstrate a formal proof. Because dt is small (for example, $dt = 1/252$), $(dt)^2$ *and* $dt^{3/2}$ are very small. Therefore, for very small time intervals, both terms can be ignored. In fact, all higher powers of dt can be ignored while the linear term dt is retained. The next step is to determine which terms in Equation 3.13 have a high power of dt in a second-order Taylor expansion.

Note that the random component of dX is dZ. We analyze dZ first. The key observation is that $(dZ)^2 = dt$ when ignoring the high powers of dt. Here is the key intuition. Since $\Delta Z \sim N(0,\ \Delta t)$, we will use the notation $dZ \sim N(0,\ dt)$. $E((dZ)^2)$ is the second moment of the normal distribution (as shown in Warm-up 3.6) and is equal to dt. By definition of variance and Warm-up 3.6:

$$Var(((dZ)^2) = E((dZ)^4) - (E((dZ)^2))^2 = 3dt^2 - dt^2 = 2dt^2.$$

We know that $E((dZ)^2) = dt$ and $Var((dZ)^2) = a\ high\ order\ of\ dt$ (which can be ignored for small values of dt). The random variable $(dZ)^2$ has a constant mean and a negligible variance, so we know $(dZ)^2$ is deterministic and it is equal to the expected value dt. Therefore:

$$(dZ)^2 = dt$$

$$dZ = \sqrt{dt}$$

$$dZdt = dt^{3/2}$$

With this key observation and ignoring high powers of dt such as $(dt)^2$ and $dt^{3/2}$:

$dXdt$ can be ignored
$(dX)^2 = \sigma(X, t)^2 dt$ and cannot be ignored.

By inserting $(dX)^2 = (\mu(X, t)dt + \sigma(X, t)dZ)(\mu(X, t)dt + \sigma(X, t)dZ) = \sigma(X, t)^2 dt$

and $dXdt = (\mu(X, t)dt + \sigma(X, t)dZ)dt = \mu(X, t)dt^2 + \sigma(X, t)dt^{3/2}$, which can be ignored.

With the above equations it is easy to see that Equations 3.14 to 3.16 are equivalent.

It is very important to understand the key contribution of Ito's lemma. For a given process, dX, and a function of X and t, $f(X, t)$, Ito's lemma states how to find the process of $f(X, t)$. The results do not give us a distribution of the underlying X_T or a distribution of $f(X, t)$. The results only provide the evolvement of $f(X, t)$ as shown in previous equations. Note that (3.16) of Ito's lemma is a SDE, a process of f:

$$df = \left(\frac{\partial f}{\partial X}\mu(X, t) + \frac{\partial f}{\partial t} + \frac{1}{2}\frac{\partial^2 f}{\partial X^2}\sigma(X, t)^2\right)dt + \frac{\partial f}{\partial X}\sigma(X, t)dZ$$

In other words, a financial derivative follows an Ito process if its underlying asset follows an Ito process. It is important that both processes receive their randomness from the same source: dZ.

Now let's consider a financial derivative on an underlying asset that follows Geometric Brownian motion:

Corollary 1

Given a stock that follows Geometric Brownian motion, $dS = \mu Sdt + \sigma SdZ$, and where $f(S, t)$ is a function of S (such as a financial derivative with a known payoff $f(S_T, T)$ at time T), the process of the derivative f is:

$$df = \left(\frac{\partial f}{\partial S}\mu S + \frac{\partial f}{\partial t} + \frac{1}{2}\frac{\partial^2 f}{\partial S^2}\sigma^2 S^2\right)dt + \frac{\partial f}{\partial s}\sigma SdZ \quad (3.17)$$

By Ito's lemma (3.16):

$$df = \left(\frac{\partial f}{\partial X}\mu(X, t) + \frac{\partial f}{\partial t} + \frac{1}{2}\frac{\partial^2 f}{\partial X^2}\sigma(X, t)^2\right)dt + \frac{\partial f}{\partial X}\sigma(X, t)dZ$$

Here: $\frac{\partial f}{\partial X} = \frac{\partial f}{\partial S}$, $\mu(X, t) = \mu S$, $\frac{\partial^2 f}{\partial X^2} = \frac{\partial^2 f}{\partial S^2}$, *and* $\sigma(X, t)^2 = \sigma^2 S^2$.

Inserting from all four of the above equations (for $\frac{\partial f}{\partial X}$, $\mu(X, t)$, $\frac{\partial^2 f}{\partial X^2}$ *and* $\sigma(X, t)^2$) generates:

$$df = \left(\frac{\partial f}{\partial S}\mu S + \frac{\partial f}{\partial t} + \frac{1}{2}\frac{\partial^2 f}{\partial S^2}\sigma^2 S^2\right)dt + \frac{\partial f}{\partial S}\sigma S dZ$$

Ito's lemma only produces a process of how f evolves. What is the importance of knowing the process? If the process of f is easier to work with than the process of X, there may be valuable applications.

Next, Ito's lemma is applied to a model with Geometric Brownian motion. Recall that Geometric Brownian motion is not a generalized Wiener process because there is an extra S in the equation:

$$dS = \mu S dt + \sigma S dZ$$

In a generalized Wiener process:

$$dX = adt + bdZ$$

The distribution of X_T is known: $N(X_0 + aT,\ b^2T)$

So, dividing by S on both sides:

$$\frac{1}{S}dS = \mu dt + \sigma dZ$$

In calculus, $dLnS = \frac{1}{S}dS$, but this is not true when S is a random variable. Next, we find the distribution of S using Ito's lemma.

Corollary 2

Assuming that a stock follows Geometric Brownian motion, then assuming $dS = \mu S dt + \sigma S dZ$ is equivalent to assuming that:

$$\mathbf{Ln}\ \boldsymbol{S_T} \sim \mathbf{N}\left(\mathbf{Ln}\ \boldsymbol{S_0} + \boldsymbol{\mu} \times \boldsymbol{T} - \frac{\mathbf{1}}{\mathbf{2}}\boldsymbol{\sigma^2 T},\ \boldsymbol{\sigma^2 T}\right) \tag{3.18}$$

In other words, if stock price S_T follows Geometric Brownian motion, then S_T follows the above lognormal distribution.

We can define $f(S, t) = LnS$ as a function of S and t in Corollary 1:

$$dLnS = \left(\frac{\partial LnS}{\partial S}\mu S + \frac{\partial f}{\partial t} + \frac{1}{2}\frac{\partial^2 LnS}{\partial S^2}(\sigma S)^2\right)dt + \frac{\partial LnS}{\partial S}\sigma S dZ$$

It simplifies to:

$$dLnS = \left(\frac{1}{S}\mu S + 0 + \frac{1}{2}\frac{-1}{S^2}(\sigma S)^2\right)dt + \frac{1}{S}\sigma S dZ$$

$$dLnS = \left(\mu + 0 + \frac{-1}{2}(\sigma)^2\right)dt + \sigma dZ$$

$$dLnS = \left(\mu - \frac{1}{2}(\sigma)^2\right)dt + \sigma dZ$$

The process of LnS is much easier to work with than the process of S. LnS is a generalized Wiener process in which there is no variable before dt and dZ. For any generalized Wiener process, the global formula will give us the distribution of the ending variable LnS_T.

$$LnS_T - LnS_0 = \left(\mu - \frac{1}{2}(\sigma)^2\right)T + \sigma\sqrt{T}N(0, 1)$$

So, LnS_T's distribution is known: $N(LnS_0 + (\mu - \frac{1}{2}(\sigma)^2)T, \sigma^2 T)$. It has a mean of LnS_T, $E(LnS_T) = LnS_0 + \left(\mu - \frac{1}{2}(\sigma)^2\right)T$ and a variance of LnS_T, $\text{Var}(LnS_T) = \sigma^2 T$.

Also, if our starting point is t instead of 0, then LnS_T's distribution is $N(LnS_t + (\mu - \frac{1}{2}(\sigma)^2)(T - t), \sigma^2(T - t))$.

Note that Geometric Brownian motion and the binomial tree discussed earlier in this chapter have the same underlying process if n is large and if we assume a risk-neutral world where all assets grow at the risk-free rate ($\mu = r$). In this case, it is assumed that a stock price follows Geometric Brownian motion, then all probability questions about S_T can be solved since we have a global solution for S_T. Some applications of Corollary 2 are discussed next.

Example 3.9: A stock grows by Geometric Brownian motion with an annual (continuously compounded) growth rate of 14% and a volatility of 20%. Find the probability that the stock price is less than \$60 in one year when the initial price is \$50.

From Corollary 2, LnS_T follows

$$N\left(Ln50 + \left(0.14 - \frac{1}{2}0.2^2\right) \times 1,\ \ 0.2^2 \times 1\right).$$

$P(S_T < 60) = P(LnS_T < Ln60)$, by Warm-up 3.3 in the review of Mathematics:

$$= N\left(\frac{Ln60 - \left(Ln50 + \left(0.14 - \frac{1}{2}0.2^2\right) \times 1\right)}{0.2\sqrt{1}}\right)$$

where $N\left(\frac{X-\mu}{\sigma}\right)$ can be found from the common z-score table.

Note that in applying corollary 1, it can be assumed that $f(S, t)$ is any function of S and t. For example, if S^2 is inserted in Corollary 1 of Ito's lemma, we can obtain a random process for S^2:

$$dS^2 = (2S\mu S + \sigma^2 S^2)dt + 2S\sigma SdZ.$$

Most of the processes are not generalized Wiener processes or Geometric Brownian motion, but the process can be simulated.

We found the distribution of $Ln(S_T)$ as well as the mean and the variance of $Ln(S_T)$. A very common mistake is that people think that the expected value $E(\ )$ and $Ln()$ are commutative. In fact,

$$E(Ln(S_T)) \neq Ln(E(S_T)).$$

So, effort is needed to find $E(S_T)$ and $Var(S_T)$. In fact, moment generating functions are used.

3.3.8 Finding the Mean and Variance of S_T for Geometric Brownian Motion

This section demonstrates the local and global solutions for the mean and variance of S_T. Consider the local solution for S_T when T (i.e., Δt) is small (the local solution only works well for a time horizon up to a couple of days):

The Geometric Brownian motion local solution for S_T when Δt is very small:

$$dS = \mu Sdt + \sigma SdZ$$

$$\Delta S = \mu S\Delta t + \sigma S\Delta Z$$

$$S_{\Delta t} - S_0 = \mu S_0\Delta t + \sigma S_0 N(0, \Delta t)$$

So,

$$S_{\Delta t} = S_0 + \mu S_0 \Delta t + \sigma S_0 N(0, \Delta t)$$

So,

$$\boldsymbol{E(S_{\Delta t}) = S_0 + \mu S_0 \Delta t} \tag{3.19}$$

$$\boldsymbol{Var(S_{\Delta t}) = \sigma^2 S_0{}^2 \Delta t} \tag{3.20}$$

Example 3.10: Given $S_0 = 60$, $\mu = 0.10$, $\sigma = 0.15$, and $\Delta t = \frac{1}{252}$, find $E(S_{1/252})$ and $Var(S_{1/252})$.

By the above equations, $E(S_{1/252}) = 60 + 0.10 \times 60 \times 1/252$ and $Var(S_{1/252}) = 0.15^2 \times 60^2 \times 1/252$.

The Geometric Brownian motion *global* solution for S_T where T is large or small:

The solution for when T is large works for both large T and small T. By Ito's lemma (this is a key formula that needs to be remembered):

$$LnS_T - LnS_0 \sim N\left(\left(\mu - \frac{1}{2}\sigma^2\right)T, \sigma^2 T\right)$$

or

$$LnS_T \sim N(LnS_0 + (\mu - \frac{1}{2}(\sigma)^2)T, \ \sigma^2 T)$$

So, $E(LnS_T) = LnS_0 + (\mu - \frac{1}{2}(\sigma)^2)T$, $Var(LnS_T) = \sigma^2 T$.

Ln and E cannot be commuted; a moment generating function is necessary.

Recall Warm-up 3.5 regarding the moment generating function with $t = 1$:

$$E(S_T) = E(e^{LnS_T}) = e^{E(LnS_T)\times 1 + \frac{1}{2}\times Var(LnS_T)\times 1^2}$$
$$= e^{LnS_0 + \left(\mu - \frac{1}{2}(\sigma)^2\right)T + \frac{1}{2}\times\sigma^2 T} = e^{LnS_0 + \mu T} = S_0 e^{\mu T}$$

So, $E(S_T) = S_0 e^{\mu T}$.

Since $Var(S_T) = E(S_T^2) - E(S_T)^2$, first calculate $E(S_T^2)$.

From the moment generating function with $t = 2$ in Warm-up 3.5:

$$E(S_T^2) = \mathrm{E}(e^{2\mathrm{Ln}S_T}) = e^{\mathrm{E}(LnS_T)\times 2+\frac{1}{2}\times Var(LnS_T)\times 2^2} = e^{(LnS_0+\left(\mu-\frac{1}{2}(\sigma)^2\right)T)\times 2+\frac{1}{2}\times\sigma^2 T\times 4}.$$
$$= e^{2LnS_0+2\mu T+\sigma^2 T}$$

$$Var(S_T) = e^{2LnS_0+2\mu T+\sigma^2 T} - (S_0 e^{\mu T})^2$$

So, the global mean and variance of S_T *is:*

$$\mathbf{E}(S_T) = S_0 e^{\mu T} \tag{3.21}$$

$$\mathbf{Var}(S_T) = S_0^2(e^{2\mu T+\sigma^2 T} - e^{2\mu T}) \tag{3.22}$$

Example 3.11: Given $S_0 = 60$, $\mu = 0.10$, $\sigma = 0.15$, and $\Delta t = 1/252$:

$$E(S_{1/252}) = S_0 e^{\mu T} = 60 \times e^{0.10\times 1/252}$$

$$Var(S_{1/252}) = S_0^2(e^{2\mu T+\sigma^2 T} - e^{2\mu T}) = 60^2 \times \left(e^{2\times 0.10\times\frac{1}{252}+0.15^2\times 1/252} - e^{2\times 0.10\times 1/252}\right)$$

A comparison of these results with the results in Example 3.10 should indicate that they are very close. The local results in Example 3.10 are approximate results of the accurate results of the global solution in Example 3.11 while T is small. If T is large, only the method in Example 3.11 should be used.

So far, whether the process/distribution is P-measure or Q-measure has not been discussed. The stock process involves two parameters, the growth rate (μ) and the volatility parameter (σ). The next section discusses which Q-measure parameters are important to the derivative value.

3.3.9 Risk Aversion vs. Risk Neutrality

Using Ito's lemma, for a given process, dS_t, we can find the process of df, where $f_t = f(S_t, t)$ is a function of underlying S_t and t. Once the process df is known, $E(f_T)$ can be solved either analytically or by a simulation. It seems the derivative value formula is $f_0 = e^{-rT}E(f_T)$. However, there is an important economic concept that is being ignored: risk aversion. Consider two simple offers, one is to receive \$3.5 in three months, the other is rolling dice and receiving the dice number as the dollar payment in three months. Both offers have an expected value of 3.5. However, risk-averse people will prefer \$3.5 with certainty to rolling dice. So, the value of the rolling dice offer should be reduced to reflect a risk premium demanded by the risk-takers.

Generally, the risk premium can be avoided in derivative valuation. Derivative prices can be derived through two approaches using expected values: P-measures (which use the real-world probabilities and in which it is necessary to include a risk premium) and Q-measures (which use risk-neutral probabilities and in which the risk-free rate is used for discounting). Since risk premiums are unobservable, it is hard to value derivatives accurately using P-measures. Therefore, Q-measures are used. In particular, paths for various stochastic processes are used such as a generalized Wiener process, an Ito process, and Geometric Brownian motion – all in a risk-neutral framework.

Simulation allows researchers to model many possible paths of a financial derivative (approaching infinity) given a process for the underlying asset in a risk-neutral world. These paths can then be used to calculate payoffs to securities that in turn can be used to estimate expected future values of those securities as well as expected payoffs of their financial derivatives. Two fundamental questions remain:

- What is the growth rate and volatility for an underlying asset in a risk-neutral world? In other words, in simulating a path what should be the average rate of growth, μ, in the price of each asset, what is volatility?
- What is the interest rate that should be used as a discount rate in computing present values?

The answer to these questions has been found in discrete cases (e.g., a tree model) by applying the risk-neutral principle (which allows both the growth rates and discount rates to be set equal to the riskless rate). At this stage, we have not yet proved whether the risk-neutral principle holds for a continuous model, where the underlying asset follows a stochastic process such as the generalized Wiener process or an Ito process. Is it valid to assume that (for derivative valuation purposes) investors live in a risk-neutral world where risk aversion and risk premiums do not exist? The next section shows the continuous version of the risk-neutral principle through the Black-Scholes model. At this stage, we have finished most but not all of the important machinery of the Monte Carlo simulation model approach to derivative valuation.

Section Summary

This section discusses the theoretical properties of different processes starting with a review of the processes discussed so far.

A generalized Wiener process is $dX = adt + bdZ$, and is equivalent to:

$$X_T \sim N(X_0 + aT, b^2T)$$

An Ito process is $dX = \mu(X, t)dt + \sigma(X, t)dZ$, usually does not have a closed form distribution for X_T. But, Monte Carlo simulation can be used to find the distribution for X_T.

A special case of an Ito process is Geometric Brownian motion:

$$dS = \mu Sdt + \sigma SdZ$$

The above process is equivalent to a log normal distribution:

$$LnS_T - LnS_0 \sim N\left(\left(\mu - \frac{1}{2}\sigma^2\right)T, \sigma^2 T\right)$$

The expectation and variance of S in the case of Geometric Brownian motion ($dS = \mu Sdt + \sigma SdZ$) is:

$$E(S_T) = S_0 e^{\mu T}$$

$$Var(S_T) = S_0^2 (e^{2\mu T + \sigma^2 T} - e^{2\mu T})$$

The following three exercises reinforce the previous material. After that, a new section introduces the famous Black-Scholes model, the risk-neutral principle, and Monte Carlo simulation to value derivatives.

Demonstration Exercises

1. An asset follows Geometric Brownian motion: $dX = \mu Xdt + \sigma XdZ$ while a financial derivative, Y, has payoff $Y = \frac{1}{X}$.
   a. Using Ito's lemma find the process dY and verify whether Y follows a Geometric Brownian motion.
   b. Find $E(Y_T)$ and $Var(Y_T)$.
   c. If $\sigma^2 = 2\mu$ *and* $X_0 = Y_0 = 1$, prove $E(X) = E(Y)$.

   Solution:

   a. By Ito's lemma Corollary 1, $dY = \left(\frac{\partial Y}{\partial X}\mu X + \frac{\partial Y}{\partial t} + \frac{1}{2}\frac{\partial^2 Y}{\partial X^2}(\sigma X)^2\right)dt + \frac{\partial Y}{\partial X}\sigma XdZ$

$$dY = \left(-1X^{-2}\mu X + 0 + \frac{1}{2}2X^{-3}(\sigma X)^2\right)dt + (-1X^{-2})\sigma XdZ$$

$$dY = (-\mu X^{-1} + \sigma^2 X^{-1})dt + (-\sigma X^{-1})dZ$$

$$dY = (-\mu + \sigma^2)X^{-1}dt + (-\sigma)X^{-1}dZ$$

$$dY = (-\mu + \sigma^2)Ydt + \sigma Y(-dZ)$$

Therefore, Y follows a Geometric Brownian motion with $\hat{\mu} = -\mu + \sigma^2$, $\hat{\sigma} = \sigma$.

b. Let $\hat{\mu} = -\mu + \sigma^2$ *and* $\hat{\sigma} = \sigma$

$$E(Y_T) = Y_0 e^{\hat{\mu}T} = Y_0 e^{(-\mu+\sigma^2)T}$$

$$\text{var}(Y_T) = Y_0^2\left(e^{2\hat{\mu}T+\hat{\sigma}^2T} - e^{2\hat{\mu}T}\right) = Y_0^2\left(e^{-2\mu T+3\sigma^2T} - e^{-2\mu T+2\sigma^2T}\right)$$

c. Since $E(X_T) = X_0 e^{\mu T}$, $E(Y_T) = Y_0 e^{(-\mu+\sigma^2)T}$. If $\sigma^2 = 2\mu$ *and* $X_0 = Y_0 = 1$, then

$$E(X_T) = E(Y_T) = e^{\mu T}$$

2. Return to the previous problem. Assume that X and Y are exchange rates on the same pair of currencies differing by which currency is in the numerator and which is in the denominator. If $\sigma^2 = 2\mu$ *and* $X_0 = Y_0 = 1$, then as previously found, $E(X) = E(Y)$. However, this contradicts our intuition. Let's look at the contradiction intuitively. Assume that one unit of foreign currency in country A is equal to one unit of foreign currency in country B initially so that current foreign currency exchange rates are $X_0 = Y_0 = 1$. Assume that the exchange rate X follows a Geometric Brownian motion with a growth rate of 10% and a volatility of $\sqrt{0.2}$. The previous problem demonstrated that the expected values of both exchange rates are equal at time T: $E(X_T) = E(Y_T) = e^{0.10T}$. It seems that both exchange rates are growing; however, they are the reciprocal of each other, so how can they both grow? This paradox is known as ***Siegel's paradox***. Mathematics has shown that both exchange rates are expected to grow, so our intuition must be tricking us. Here is an economic example to see why intuition can be wrong.

Suppose that one gold coin can buy one share of the S&P 500. Assume that this value of the S&P 500 in terms of gold coins is going to either double to 2 or drop in half to 0.50. Assuming that there is a 50/50 chance of both paths, then the expected value of the S&P 500 in terms of gold coins is 1.25. Conversely, view the S&P 500 as a better measure of wealth than gold coins. Viewing the value of gold coins in terms of the S&P 500 (currently priced 1 to 1) the "price" of the gold coins in terms of S&P 500 units is equally likely to rise to 2 S&P 500 shares or fall to 0.50 S&P 500 shares with an expected value of 1.25. Thus, gold coins and the S&P 500 have an expected value of 1.25 when measured in terms of each other. Therefore, both ratios can have higher future expected values than their current values even though they are the reciprocal of each other.

3. Assume that X follows a ***mean reversion process***: $dX = k(\theta - X)dt + \sigma dZ$. Mean reversion processes can be used to model interest rates and other common financial values. Find $E(X_T)$ and $Var(X_T)$ using the following two steps:
   a. Construct a financial derivative $f(X, t) = e^{kt}X$, and find the process df.
   b. Find $E(X_T)$ and $Var(X_T)$ using the process of df.

Solution:

a. By Ito's lemma, $df = \left(\frac{\partial f}{\partial X}k(\theta - X) + \frac{\partial f}{\partial t} + \frac{1}{2}\frac{\partial^2 f}{\partial X^2}(\sigma)^2\right)dt + \frac{\partial f}{\partial X}\sigma dZ$

$$df = (e^{kt}k(\theta - X) + ke^{kt}X + 0)dt + \sigma e^{kt}dZ$$

$$df = (e^{kt}k\theta)dt + \sigma e^{kt}dZ$$

Next, discretize it:

$$\begin{aligned}
f_{t_1} - f_0 &= e^{k0}k\theta \ \Delta t + \sigma e^{k0}\Delta Z_1 \\
f_{t_2} - f_{t_1} &= e^{kt_1}k\theta \ \Delta t + \sigma e^{kt_1}\Delta Z_2 \\
f_{t_3} - f_{t_2} &= e^{kt_2}k\theta \ \Delta t + \sigma e^{kt_2}\Delta Z_3 \\
f_{t_4} - f_{t_3} &= e^{kt_3}k\theta \ \Delta t + \sigma e^{kt_3}\Delta Z_4 \\
&\cdots \\
f_{t_n=T} - f_{t_{n-1}} &= e^{kt_{n-1}}k\theta \ \Delta t + \sigma e^{kt_{n-1}}\Delta Z_n
\end{aligned}$$

$$f_T = f_0 + \sum_{i=0}^{n-1} e^{kt_i}k\theta\,\Delta t + \sigma\sum_{i=0}^{n-1} e^{kt_i}\,\Delta Z_i$$

Where ΔZ_i are independent $\sqrt{\Delta t}$ "N(0, 1)"
Even though the process of $f = e^{kt}X$ is not Geometric Brownian motion, $\mu(X, t)$ and $\sigma(X, t)$ are deterministic. So, it is still a linear combination of independent normal distributions.

b. Calculate the mean of $e^{kt}X$ at time t first. In order to calculate $E(e^{kT}X_T) = E(f_T)$, find the mean of the first component in the above f_T equation, which is deterministic. As n goes to infinity, solve the differential equation:

$$E(e^{kT}X_T - e^{k0}X_0) = E\left(f_T - f_0\right) = \int_0^T (e^{kt}k\theta)dt = \theta(e^{kT} - e^{k0})$$

$$E(X_T) = e^{-kT}X_0 + \theta(1 - e^{-kT})$$

Now find $Var(f_T) = E(f_T^2) - E(f_T)^2$. First find $E(f_T^2)$.

$$E(f_T^2) = \left(f_0 + \sum_{i=0}^{n-1} e^{kt_i}\, k\theta\, \Delta t\right)^2 + \sigma^2\left(\sum_{i=0}^{n-1} e^{2kt_i}\right)E(\Delta Z_i^2)$$

This is because $E(\Delta Z_i) = 0,\ E(\Delta Z_i \Delta Z_j) = 0 \textit{ for } i \neq j$

$$E(f_T^2) = \left(f_0 + \sum_{i=0}^{n-1} e^{kt_i}\ k\theta\ \Delta t\right)^2 + \sigma^2\left(\sum_{i=0}^{n-1} e^{2kt_i}\right)\Delta t$$

Because $E(f_T) = (f_0 + \sum_{i=0}^{n-1} e^{kt_i}\ k\theta\ \Delta t)$

$$(E(f_T))^2 = \left(f_0 + \sum_{i=0}^{n-1} e^{kt_i}\ k\theta\ \Delta t\right)^2$$

$$Var(f_T) = \sigma^2\left(\sum_{i=0}^{n-1} e^{2kt_i}\right)\Delta t$$

As n goes to infinity:

$$Var(f_T) = \sigma^2 \int_0^T e^{2kt}dt = \frac{\sigma^2}{2k}(e^{2kT} - 1)$$

$$Var(e^{kT}X_T) = \frac{\sigma^2}{2k}(e^{2kT} - 1)$$

So, $Var(X_T) = \frac{\sigma^2}{2k}(1 - e^{-2kT})$

3.4 Analytical Model Approaches to Derivative Pricing

This section derives the Black-Scholes PDE and risk-neutral principle. This section also describes why only Q-measure processes are used for a continuous model of derivative pricing.

The seminal work of Black and Scholes, as well as Merton, revolutionized derivative valuation in general and option pricing in particular. This section reviews the assumptions for the Black-Scholes PDE, the derivation of the Black-Scholes PDE, and how the no-arbitrage argument applies.

Similarly with derivations earlier in this chapter in which a riskless portfolio is formed using a binomial tree model, this section designs a riskless portfolio in continuous time. The no-arbitrage principle is used (along with the assumption that any riskless portfolio must grow at the risk-free rate) to derive the Black-Scholes PDE. There is an important difference relative to the binomial model. In the binomial model, the same riskless portfolio lasts for an entire time period while here the riskless portfolio is instantaneous and remains so only for a short period of time unless continuously adjusted.

We will show that the stock price growth rate, μ, has no effect on an option's value through the Black-Scholes PDE. Thus, the risk-neutral principle holds: the derivative value today is independent of the underlying asset's expected return and is independent of any risk premium in the expected return of the underlying asset. Further, we will confirm that the real-world value of the derivative is identical to that determined in a risk-neutral world where all assets grow at the risk-free rate and the discount rate is the risk-free rate. So the fair value of a derivative is the discounted expected value of the future payoff using risk-free rates as discount rates. At the end of this section, a simulation algorithm is given for valuing any European-style derivative.

3.4.1 The Black-Scholes PDE

Here are the ***assumptions of the Black-Scholes model (also known as the Black-Scholes-Merton model)***:

Black-Scholes Model Assumptions

1. The stock price follows Geometric Brownian motion: $dS = \mu S dt + \sigma S dZ$.
2. The stock volatility, σ, and the risk-free interest rate, r_f, are constants. The subindex f of r_f is dropped for convenience in the rest of the book by replacing r_f with r.
3. Securities are fully divisible.
4. The underlying stock pays no dividend.
5. Short selling has no restrictions.
6. There are no transaction costs or bid-asked spreads or taxes.
7. Continuous trading is possible.
8. Arbitrage opportunities do not exist.

Note that the assumptions of the Black-Scholes model are very similar to assumptions of the binomial tree model. Here is the ***Black-Scholes PDE***.

Black-Scholes PDE

If the above assumptions are satisfied, the derivative value $f(S_t, t)$ as a function of underlying asset S_t and time t follows the following partial differential equation:

$$\frac{\partial f(S_t, t)}{\partial S_t} rS_t + \frac{\partial f(S_t, t)}{\partial t} + \frac{1}{2}\frac{\partial^2 f(S_t, t)}{\partial S_t{}^2}\sigma^2 S_t^2 = rf(S_t, t)$$

or, where dropping the subscript t for convenience:

$$\frac{\partial f}{\partial S} rS + \frac{\partial f}{\partial t} + \frac{1}{2}\frac{\partial^2 f}{\partial S^2}\sigma^2 S^2 = rf \qquad (3.23)$$

There are several important observations about this model:

1. The resulting equation is a PDE, which is deterministic, not a SDE. This means that all of the randomness of f is driven by the randomness of the underlying asset (S), so f_0 can be solved as a deterministic function of S_0.

2. If the assumptions of the model hold, all financial derivatives satisfy the Black-Scholes PDE including both European-style and American-style options, and call options and put options. In the PDE approach, boundary conditions are required to establish the uniqueness of the solution of the PDE. However, the boundary issues can be tricky sometimes. For financial derivatives, we may impose some economically justified constraints on the solution of the PDE. For example, for a European-style derivative the payoff is at expiration, generating the terminal condition: $f_T = f(S_T, T)$ as a known function of S_T *and* T. With economically justified boundaries and terminal conditions, the PDE can be solved.
   For American-style derivatives (e.g., an American put or a call with dividends) there is an early-exercise boundary, which isn't known in advance. These kinds of challenges can be reduced to solving a certain free boundary problem for the Black-Scholes equation. Thus, valuing American options using the PDE is actually quite different from valuing European options using the PDE method. The general methods of solving any PDE are beyond the scope of this book. However, Chapter 4 discusses a numerical approximation method (a finite different method) that can be used.
3. The PDE does not contain μ, the stock's growth rate. The stock's growth rate is irrelevant in valuing derivatives since only the risk-free rate r matters. This implies that the risk-neutral principle holds for derivatives with an underlying asset that follows a Geometric Brownian motion (given that the model's assumptions hold). In addition, the interest rate that should be used as a discount rate in computing present values is the risk-free rate.
4. It is equivalent to describe the stock price as following the Geometric Brownian motion: $dS = rSdt + \sigma SdZ$, where r is a risk-free interest rate instead of μ.

The next section provides an intuitive derivation of the Black-Scholes PDE that is not a strict proof in Mathematics. However, an intuitive derivation is good enough for introductory purposes and it shares a lot of common ideas found in the binomial tree approach which was previously analyzed.

3.4.2 The Intuition Behind the Black-Scholes PDE: Comparison of the PDE Approach and a Binomial Tree Approach

Our approach of understanding the Black-Scholes PDE is through a comparison of the PDE approach and a binomial tree approach. Based on the Black-Scholes assumptions, the stock price has the following process:

$$dS = \mu Sdt + \sigma SdZ$$

The derivative follows a process from Corollary 1:

$$df = \left(\frac{\partial f}{\partial S}\mu S + \frac{\partial f}{\partial t} + \frac{1}{2}\frac{\partial^2 f}{\partial S^2}\sigma^2 S^2\right)dt + \frac{\partial f}{\partial s}\sigma S dZ$$

Our objective is to find f_0. From a mathematical point of view, the above equations can be used to cancel dZ, namely by using the elimination trick in solving a system of equations. First, multiply the first equation by $\frac{\partial f}{\partial s}$ and then subtract the second equation. Having canceled dZ, there is no randomness.

Now, consider the financial intuition. Recall that in a binomial tree model, the hedging strategy is to buy one unit of the derivative, f, and short $\Delta = \frac{change\ of\ f}{change\ of\ S} = \frac{\Delta f}{\Delta S}$ shares of stock S. An analogous approach is applied here. First, let's clarify some mathematical notation. Note that dS is the change in the stock price and df is the change in the derivative value. As time changes, Δt is infinitesimally small, so we do not distinguish the notation dS vs. ΔS, df vs. Δf. For convenience, the differential operator is d. So, $\Delta = \frac{\Delta f}{\Delta s} = \frac{\partial f}{\partial s}$. This is true only if Δt is infinitesimally small (i.e., just like dt).

The financial intuition is that the elimination trick to solve a system of equations discussed above is analogous to designing a portfolio. Specifically, the portfolio is selected at time t by buying one unit of derivative f, short selling Δ shares of stock, and keeping this portfolio until $t + dt$. The value of the initial portfolio (π) is:

$$\begin{aligned}\pi &= f(S_t, t) - \Delta \times S_t \\ &= f(S_t, t) - \frac{\partial f}{\partial S} \times S_t \\ &= f - \frac{\partial f}{\partial S} S\end{aligned}$$

where subscripts are dropped for convenience.

The change of the portfolio value at the end of this small time period $[t, t+dt]$ is:

$$d\pi = df - \frac{\partial f}{\partial S} \times dS$$

Now, insert our processes df and dS in order to obtain:

$$d\pi = \left(\frac{\partial f}{\partial S}\mu S + \frac{\partial f}{\partial t} + \frac{1}{2}\frac{\partial^2 f}{\partial S^2}\sigma^2 S^2\right)dt + \frac{\partial f}{\partial S}\sigma S dZ - \frac{\partial f}{\partial S} \times (\mu S dt + \sigma S dZ)$$

$$d\pi = \left(\frac{\partial f}{\partial S}\mu S + \frac{\partial f}{\partial t} + \frac{1}{2}\frac{\partial^2 f}{\partial S^2}\sigma^2 S^2\right)dt + \frac{\partial f}{\partial S}\sigma S dZ - \frac{\partial f}{\partial S} \times \mu S dt - \frac{\partial f}{\partial S}\sigma S dZ$$

$$d\pi = \left(\frac{\partial f}{\partial t} + \frac{1}{2}\frac{\partial^2 f}{\partial S^2}\sigma^2 S^2\right)dt$$

Note that the change of the portfolio, π, has no randomness over the time period $[t, t + dt]$, so it is a riskless portfolio during this small time period $[t, t + dt]$. The riskless portfolio should earn the risk-free rate, so $d\pi = r\pi dt$. Therefore,

$$d\pi = r\pi dt = r\left(f - \frac{\partial f}{\partial S} \times S\right)dt = \left(rf - \frac{\partial f}{\partial S} rS\right)dt.$$

Setting the equations of $d\pi$ equal,

$$\left(\frac{\partial f}{\partial t} + \frac{1}{2}\frac{\partial^2 f}{\partial S^2}\sigma^2 S^2\right)dt = \left(rf - \frac{\partial f}{\partial S} rS\right)dt.$$

Canceling out dt,

$$\left(\frac{\partial f}{\partial t} + \frac{1}{2}\frac{\partial^2 f}{\partial S^2}\sigma^2 S^2\right) = \left(rf - \frac{\partial f}{\partial S} rS\right).$$

After re-arranging,

$$\frac{\partial f}{\partial S} rS + \frac{\partial f}{\partial t} + \frac{1}{2}\frac{\partial^2 f}{\partial S^2}\sigma^2 S^2 = rf.$$

This completes the derivation of the Black-Scholes PDE.

The above equations did not contain the subscript t for convenience. Here is the final equation with the subscript t:

$$\frac{\partial f(S_t, t)}{\partial S} rS_t + \frac{\partial f(S_t, t)}{\partial t} + \frac{1}{2}\frac{\partial^2 f(S_t, t)}{\partial S^2}\sigma^2 S_t^{\,2} = rf(S_t, t)$$

Note that if it is assumed that the underlying stock price has Geometric Brownian motion, the Black-Scholes PDE can be derived as above. In general, if the underlying stock price is an Ito process, we can derive a similar PDE using the previous approach and can solve for the derivative value in a general framework (which is beyond the scope of this book).

3.4.3 Boundary Conditions for the Black-Scholes PDE and the Black-Scholes Formula

In order to specify a unique solution to the Black-Scholes PDE, conditions must be imposed on the solution. These conditions are the: ***boundary (or asymptotic) conditions*** or ***initial (terminal) conditions of the Black-Scholes PDE***.

For the price (i.e., value) of a European call, we set the following conditions: (1) At expiration, the terminal condition for the call price is the call's payoff, (2) as the stock price approaches 0, so does the call price, and (3) as the stock price becomes very large the ratio of the call price to the stock price approaches one. These conditions are summarized next in equation form:

Boundary/terminal conditions for Black-Scholes PDE for European call option

$$f(S_T, T) = Max(S_T - K, 0) \tag{3.24}$$

$$\lim_{S_t \to 0^+} f(S_t, t) = 0 \text{ for } t \in [0, T] \tag{3.25}$$

$$\lim_{S_t \to \infty} f(S_t, t)/S_t = 1 \text{ for } t \in [0, T] \tag{3.26}$$

Advanced mathematics of PDE analysis beyond this introduction can solve the call value at any time t. The solution is perhaps the most famous equation in financial economics and is called the ***Black-Scholes European call price formula***. The equation is detailed below:

The Black-Scholes European Call Price Formula

A European call value, C_t, at time t can be solved as a function of the underlying asset price at time t, S_t, the strike price, K, the risk-free interest rate, r, the volatility of the returns of the underlying asset, σ, and the option expiration time, T.

$$C_t = f(S_t, t) = S_t\mathbf{N}(d_1) - Ke^{-r(T-t)}\mathbf{N}(d_2) \tag{3.27}$$

where $d_1 = \dfrac{Ln(S_t/K) + \left(r + \frac{\sigma^2}{2}\right)(T-t)}{\sigma\sqrt{T-t}}$ and $d_2 = \dfrac{Ln(S_t/K) + \left(r - \frac{\sigma^2}{2}\right)(T-t)}{\sigma\sqrt{T-t}} = d_1 - \sigma\sqrt{T-t}$.

Setting $t = 0$ generates today's call value, which is the most common application:

$$C_0 = S_0\mathbf{N}(d_1) - Ke^{-rT}\mathbf{N}(d_2) \tag{3.28}$$

where $d_1 = \dfrac{Ln(S_0/K) + \left(r + \frac{\sigma^2}{2}\right)T}{\sigma\sqrt{T}}$ and $d_2 = \dfrac{Ln(S_0/K) + \left(r - \frac{\sigma^2}{2}\right)T}{\sigma\sqrt{T}} = d_1 - \sigma\sqrt{T}$.

Note: Setting $B = Ke^{-rT}$ (the present value of the strike price) and $v = \sigma\sqrt{T}$ (the non-annualized volatility over T) allows simplification of the Black-Scholes call option pricing model to:

$$C_0 = S_0\mathbf{N}(d_1) - B\mathbf{N}(d_2) \tag{3.29}$$

where $d_1 = \frac{Ln(S_0/B)}{v} + (v/2)$ and $d_2 = \frac{Ln(S_0/B)}{v} - (v/2) = d_1 - v$.

Further, for an at-the-time-value-of-money call option (i.e., $S_0 = B = Ke^{-rT}$) the equation simplifies to:

$$C_0 = S_0[\mathbf{N}(v/2) - N(-v/2)].$$

Remarks:

1. N() is the cumulative normal distribution function of N(0,1), which returns a constant between 0 and 1 (i.e., the value from a Z table value).
2. C is a deterministic solution to a call option and therefore does not contain any randomness.
3. If t is set close to T, the call value at the expiration is equal to the payoff at expiration T:

$$C_T = \lim_{t\to T} f(S_t, t) = S_T \lim_{t\to T} N(d_1) - K \lim_{t\to T} N(d_2)$$

Note that:

$$\lim_{t\to T} N(d_1) = \lim_{t\to T} N(d_2) = N(\infty) = 1 \ \textit{if} \ S_T > K$$

$$\lim_{t\to T} N(d_1) = \lim_{t\to T} N(d_2) = N(-\infty) = 0 \ \textit{if} \ S_T < K$$

So, $C_T = f(S_T, T) = Max(S_T - K, 0)$.

If S_0 *is set close to* 0, the call value at any time is 0.

$$C_0 = \lim_{S_0\to 0^+} f(S_0, t) = 0 \times N(-\infty) - Ke^{-rT}N(-\infty) = 0$$

$$\lim_{S_t\to 0^+} f(S_t, t) = 0$$

It is easy to check that:

$$\lim_{S_t\to \infty} f(S_t, t)/S_t = 1$$

Analogously to the approach for a European call option, for the price of (i.e., value) a European put, we set the following conditions: (1) At expiration, the terminal condition for the put price is the put's payoff; (2) as the stock price approaches 0, the put price approaches the present value of K; and (3) as the stock price becomes very large the put value approaches 0. These conditions are summarized next in equation form:

Boundary/Terminal Conditions for Black-Scholes PDE for European Put Option

$$f(S_T, T) = \mathbf{Max}(K - S_T, 0) \tag{3.30}$$

$$\mathbf{lim}_{S_t\to 0^+} f(S_t, t) = Ke^{-r(T-t)} \ \mathbf{for} \ t \in [0, T] \tag{3.31}$$

$$\mathbf{lim}_{S_t\to \infty} f(S_t, t) = 0 \ \mathbf{for} \ t \in [0, T] \tag{3.32}$$

The European put option value is called the ***Black-Scholes European put price formula*** and is summarized below:

Black-Scholes European Put Price Formula

A European put value P_t, at time t can be solved as a function of underlying asset price at time t, S_t, the strike price, K, the risk-free interest rate, r, the volatility of the returns of the underlying asset, σ, and the option expiration time, T.

$$P_t = Ke^{-r(T-t)}N(-d_2) - S_t N(-d_1) \tag{3.33}$$

where $d_1 = \dfrac{Ln(S_t / K) + \left(r + \frac{\sigma^2}{2}\right)(T-t)}{\sigma\sqrt{T-t}}$ and $d_2 = \dfrac{Ln(S_t / K) + \left(r - \frac{\sigma^2}{2}\right)(T-t)}{\sigma\sqrt{T-t}} = d_1 - \sigma\sqrt{T-t}$.

In particular,

$$P_0 = Ke^{-rT}N(-d_2) - S_0 N(-d_1) \tag{3.34}$$

where $d_1 = \dfrac{Ln(S_0 / K) + \left(r + \frac{\sigma^2}{2}\right)T}{\sigma\sqrt{T}}$ and $d_2 = \dfrac{Ln(S_0 / K) + \left(r - \frac{\sigma^2}{2}\right)T}{\sigma\sqrt{T}} = d_1 - \sigma\sqrt{T}$.

Even though solving the PDE is beyond the scope of this book, we will derive the above formulas in a later section by discounting risk-neutral expectations. Having checked already that the call solution satisfies the terminal and boundary conditions, next we develop some machinery to show that the Black-Scholes formula satisfies Black-Scholes PDE for a European call:

$$C_t = f(S_t, t) = S_t N(d_1) - Ke^{-r(T-t)}N(d_2)$$

satisfies the Black-Scholes PDE:

$$\frac{\partial f(S_t, t)}{\partial S} rS_t + \frac{\partial f(S_t, t)}{\partial t} + \frac{1}{2}\frac{\partial^2 f(S_t, t)}{\partial S^2}\sigma^2 S_t^2 = rf(S_t, t).$$

The details are left as an exercise. However, below is an important property that is frequently used in forming the hedge ratio that maintains a riskless portfolio.

Property 3.1

If a European call value, C_t, at time t satisfies the Black-Scholes formula, the hedge ratio (the shares needed to hedge one call) at time t is

$$\frac{\partial C}{\partial S} = \mathbf{N}(d_1) \text{ where } d_1 = \frac{\text{Ln}(S_t/K) + \left(r + \frac{\sigma^2}{2}\right)(T-t)}{\sigma\sqrt{T-t}} \tag{3.35}$$

If a European put value, P_t, at time t satisfies the Black-Scholes formula, the hedge ratio (the shares needed to hedge one call) at time t is

$$\frac{\partial P}{\partial S} = \mathbf{N}(d_1) - 1 \text{ where } d_1 = \frac{\text{Ln}(S_t/K) + \left(r + \frac{\sigma^2}{2}\right)(T-t)}{\sigma\sqrt{T-t}} \tag{3.36}$$

Now we show $\frac{\partial C}{\partial S} = N(d_1)$. Note that $N(d_1)$ and $N(d_2)$ both contain S_t, so a partial derivative can be found using the product rule along with other aids. The following steps solve for the following derivative one part at a time.

$$\frac{\partial C}{\partial S} = N(d_1) + S_t N'(d_1)\frac{\partial d_1}{\partial S_t} - Ke^{-r(T-t)}N'(d_2)\frac{\partial d_2}{\partial S_t}$$

It is easy to find: $S_t\frac{\partial d_1}{\partial S_t} = S_t\frac{\partial\left(\frac{LnS_t}{\sigma\sqrt{T-t}}\right)}{\partial S_t} = \frac{1}{\sigma\sqrt{T-t}}$ and $\frac{\partial d_2}{\partial S_t} = \frac{1}{S_t}\frac{1}{\sigma\sqrt{T-t}}$.

To show $\frac{\partial C}{\partial S} = N(d_1)$, it is necessary to show that

$S_t N'(d_1)\frac{\partial d_1}{\partial S_t} - Ke^{-r(T-t)}N'(d_2)\frac{\partial d_2}{\partial S_t} = 0$

i.e., $N'(d_1)\frac{1}{\sigma\sqrt{T-t}} - N'(d_2)Ke^{-r(T-t)}\frac{1}{S_t}\frac{1}{\sigma\sqrt{T-t}} = 0$

i.e., $N'(d_1) - N'(d_2)Ke^{-r(T-t)}\frac{1}{S_t} = 0$

By definition, $N(x) = \int_{-\infty}^{x} \frac{1}{\sqrt{2\pi}}e^{-\frac{1}{2}\times x^2}dx$.

So, $N'(x) = \frac{1}{\sqrt{2\pi}}e^{-\frac{1}{2}\times x^2}$

$$N'(d_1) = \frac{1}{\sqrt{2\pi}}e^{-\frac{1}{2}\times d_1^2}$$

$$N'(d_2) = \frac{1}{\sqrt{2\pi}}e^{-\frac{1}{2}\times d_2^2} = \frac{1}{\sqrt{2\pi}}e^{-\frac{1}{2}\times(d_1-\sigma\sqrt{T-t})^2} = \frac{1}{\sqrt{2\pi}}e^{-\frac{1}{2}\times d_1^2}e^{\sigma d_1\sqrt{T-t}}e^{-1/2\times\sigma^2(T-t)}$$

Now substitute d_1 *into* $e^{\sigma d_1\sqrt{T-t}}$:

$$e^{\sigma d_1\sqrt{T-t}} = e^{Ln\left(\frac{S_t}{K}\right)+\left(r+\frac{1}{2}\times\sigma^2\right)(T-t)} = \left(\frac{S_t}{K}\right)e^{r(T-t)}e^{1/2\times\sigma^2(T-t)}$$

Inserting the results generates:

$$N'(d_2) = \frac{1}{\sqrt{2\pi}}e^{-\frac{1}{2}\times d_1{}^2}\left(\frac{S_t}{K}\right)e^{r(T-t)}e^{1/2\times\sigma^2(T-t)}e^{-1/2\times\sigma^2(T-t)}$$
$$= \frac{1}{\sqrt{2\pi}}e^{-\frac{1}{2}\times d_1{}^2}\left(\frac{S_t}{K}\right)e^{r(T-t)}$$
$$= N'(d_1)\left(\frac{S_t}{K}\right)e^{r(T-t)}$$

Now that $N'(d_1) - N'(d_2)Ke^{-r(T-t)}\frac{1}{S_t} = 0$ has been proved, it is clear that $\frac{\partial C}{\partial S} = N(d_1)$.

A similar method can be used to find $\frac{\partial P}{\partial S}$.

Or, simply use put call parity: $P = Ke^{-(T-t)} + C - S$

Taking the derivative of each of the three components of P:
$\frac{\partial P}{\partial S} = 0 + \frac{\partial C}{\partial S} - 1 = N(d_1) - 1.$

We can also find $\frac{\partial C}{\partial t}$ and $\frac{\partial^2 C}{\partial S^2}$ and can show that the Black-Scholes formula for a European call satisfies the Black-Scholes PDE: $\frac{\partial C}{\partial S}rS + \frac{\partial C}{\partial t} + \frac{1}{2}\frac{\partial^2 C}{\partial S^2}\sigma^2 S^2 = rC$, which is left for the end-of-chapter problems.

3.4.4 Conclusions Concerning the Black-Scholes Formula

1. The Black-Scholes formula provides call or put values given five parameters: S_0, r, T, σ, K. Four of them are directly observable in the market. The volatility of the returns of the underlying assets, σ, must be estimated. The observed historical σ is often used as a basis for forecasting σ.
2. The Black-Scholes formula is based on all the assumptions of the Black-Scholes PDE. Any violation of the assumptions could cause the formula to explain derivative values poorly.
3. The Black-Scholes formula may not work for American options when early exercise is possible (and when dividends are included) since there is no fixed boundary condition.
4. The main purpose to learn the Black-Scholes formula is not to value the European call and put. Due to a very liquid market for standard European calls and puts, market prices are likely fair values. The volatility parameter is not directly observed, it is inferred reversely from the market price of the option using the Black-Scholes model. The inferred volatility is termed the ***implied volatility***. Another use of the Black-Scholes formula is to find a hedge ratio to hedge away

the risk of the option's underlying asset. These perspectives are discussed in Chapters 4 and 5.

Now let's practice some examples using the Black-Scholes formula.

Example 3.12: We know that the stock price is \$30 today with a volatility of 20%. The risk-free rate is 0.05, compounded continuously. Consider a put option with a strike of 32 and an expiration date of one year:

$$S_0 = 30, \ \sigma = 0.2, \ r = 0.05, \ T = 1, K = 32$$

The put value today can be found using the Black-Scholes formula.

$$P = 32 \times e^{-0.05\times 1} N(-d_2) - 30 \times N(-d_1)$$

where $d_1 = \dfrac{Ln\left(\frac{30}{32}\right) + \left(0.05 + \frac{0.2^2}{2}\right) \times 1}{0.2\sqrt{1}}$ and

$$d_2 = \frac{Ln(30/32) + \left(0.05 - \frac{0.2^2}{2}\right)1}{0.2\sqrt{1}} = d_1 - 0.2\sqrt{1}.$$

We can use a Z-table to find N(·) or use the cumulative normal distribution formulas in software such as EXCEL: The solution is $P = 2.633$.

A natural corollary from the Black-Scholes PDE is the risk-neutral principle in which underlying assets follow a stochastic process with a continuous sample space (infinitely many outcomes) and is discussed in the next subsection.

3.4.5 Risk-Neutral Evaluation and the Black-Scholes Formula

In the Black-Scholes PDE, the actual stock price growth rate μ is irrelevant to the derivative's value. Therefore, a risk-neutral world can be assumed in which all asset values are expected to grow at the risk-free rate and all future cash flows can be discounted at the risk-free rate. In other words, given our assumptions we can utilize the ***risk-neutral principle (continuous version)*** when underlying assets follow a stochastic process with a continuous sample space (infinite many outcomes).

The Risk-Neutral Principle (Continuous Version):

The value of a European style financial derivative today, f_0, is equal to its discounted expected values of its payoffs in a risk-neutral world where all assets grow and are discounted at the risk-free rate r. Assuming that a derivative's underlying asset is S_t which follows a Geometric brownian motion, $f_T = f(S_T, T)$ is the payoff function at expiration T, and $h(S_T)$ is the Q-measure p.d.f. of S_T:

$$f_0 = e^{-rT}E_Q(f_T) = e^{-rT}\int_{-\infty}^{\infty} f(S_T, T)h(S_T)dS_T \tag{3.37}$$

where the expected value $E_Q()$ is with respect to random variable S_T in a risk-neutral world. Note that this Q-measure expectation is only an expectation under the risk-neutral principle – it is not an actual statistical expectation. The Q-measure expected payoff is calculated in a risk-neutral framework assuming that the expected growth rate of all assets is also r.

If there is no closed form solution for the above integral, it can be estimated using numerical approximation. In fact, Mathematica has a relative good built-in numerical integration package. This book will not extend further here. Note that the above formula does not work for American-style derivatives since the payoff function is not precisely known. In the cases of European calls and puts, the integral can be solved analytically, generating a solution that matches the Black-Scholes formula as discussed previously. The next section solves this integral analytically.

3.4.6 The Risk-Neutral Principle and Derivation of the Black-Scholes Formula

In this section, similar to Hull, we derive the Black-Scholes formula for a European call using basic calculus and the risk-neutral principle while avoiding the need to solve the PDE, which is beyond this introduction.

Under the risk-neutral principle, the value of a European call option is the present value of its expected cash flows (assuming all asset values grow at an expected rate equal to the risk-free rate) discounted at the risk-free rate.

$$C_0 = e^{-rT}E_Q(Max(S_T - K, 0))$$

The call value is under the Q-measure (i.e., in risk-neutral world). The expected payoff is discounted back to today by the risk-free interest where the stock follows a Geometric Brownian motion with risk-free growth rate r.

In order to calculate $E_Q(Max(S_T - K, 0))$, it is necessary to have the p.d.f. of S_T in a risk-neutral world.

Recall that, in a risk-neutral world, stock price follows the Geometric Brownian motion with drift r:

$$dS = rSdt + \sigma SdZ$$

Geometric Brownian motion is equivalent to LnS_T having the following distribution:

$$LnS_T \sim N\left(LnS_0 + \left(r - \frac{\sigma^2}{2}\right)T,\ \sigma^2 T\right) = N(\tilde{\mu}, \tilde{\sigma}^2)$$

The above equation denotes $LnS_0 + \left(r - \frac{\sigma^2}{2}\right)T$ *by* $\tilde{\mu}$, and $\sigma^2 T$ by $\tilde{\sigma}^2$.

The objective is to calculate $e^{-rT}E_Q(Max(S_T - K, 0))$.

By Warm-up 3.2, the option's expected value can be calculated:

$$\begin{aligned}
&= e^{-rT}\int_K^{+\infty}(s - K)f_{S_T}(s)ds \\
&= e^{-rT}\int_K^{+\infty} sf_{S_T}(s)ds - e^{-rT}\int_K^{+\infty} Kf_{S_T}(s)ds \\
&= Part\ I - Part\ II
\end{aligned}$$

Solving Part II first:

$$\begin{aligned}
&e^{-rT}\int_K^{+\infty} Kf_{S_T}(s)ds \\
&= Ke^{-rT}\int_K^{+\infty} 1f_{S_T}(s)ds \\
&= Ke^{-rT}P(S_T > K)\ (Warm\ Up\ 3.2) \\
&= Ke^{-rT}P(lnS_T > lnK) \\
&= Ke^{-rT}N\left(\frac{\tilde{\mu} - LnK}{\tilde{\sigma}}\right)\ (Warm\ Up\ 3.3) \\
&= Ke^{-rT}N\left(\frac{LnS_0 + \left(r - \frac{\sigma^2}{2}\right)T - lnK}{\sigma\sqrt{T}}\right) \\
&= Ke^{-rT}N\left(\frac{Ln(S_0 / K) + \left(r - \frac{\sigma^2}{2}\right)T}{\sigma\sqrt{T}}\right) \\
&= Ke^{-rT}N(d_2)
\end{aligned}$$

Next, Part I:

$$e^{-rT}\int_K^{+\infty} sf_{S_T}(s)ds$$
$$= e^{-rT}\int_K^{\infty} s\frac{1}{\sqrt{2\pi\tilde{\sigma}^2}}e^{-\frac{(lns-\tilde{\mu})^2}{2\tilde{\sigma}^2}}\frac{1}{s}ds \;\; (Warm\;\; Up\;\; 3.1)$$

Now let $y = \frac{lns - \tilde{\mu}}{\tilde{\sigma}}$. Using a substitution trick in the integral:

$$dy = \frac{1}{\tilde{\sigma}}\frac{1}{s}ds \quad implies \quad \tilde{\sigma}dy = \frac{1}{s}ds$$

And $y\tilde{\sigma} = lns - \tilde{\mu}$ which implies $s = e^{y\tilde{\sigma}+\tilde{\mu}}$.

Inserting, $= e^{-rT}\int_{\frac{lnK-\tilde{\mu}}{\tilde{\sigma}}}^{+\infty} e^{y\tilde{\sigma}+\tilde{\mu}}\frac{1}{\sqrt{2\pi\tilde{\sigma}^2}}e^{-\frac{y^2}{2}}\tilde{\sigma}dy$.

(Note that it is necessary to change the limit for substitution too.)

$= e^{-rT}e^{\tilde{\mu}}\int_{\frac{lnK-\tilde{\mu}}{\tilde{\sigma}}}^{+\infty} e^{y\tilde{\sigma}}\frac{1}{\sqrt{2\pi}}e^{-\frac{y^2}{2}}dy$ (move $e^{\tilde{\mu}}$ out and cancel $\tilde{\sigma}$ with the denominator)

$$= e^{-rT}e^{\tilde{\mu}}\int_{\frac{lnK-\tilde{\mu}}{\tilde{\sigma}}}^{+\infty}\frac{1}{\sqrt{2\pi}}e^{-\frac{y^2}{2}+y\tilde{\sigma}}dy = e^{-rT}e^{\tilde{\mu}}\int_{\frac{lnK-\tilde{\mu}}{\tilde{\sigma}}}^{+\infty}\frac{1}{\sqrt{2\pi}}e^{-\frac{y^2-2y\tilde{\sigma}}{2}}dy$$

$$= e^{-rT}e^{\tilde{\mu}}\int_{\frac{lnK-\tilde{\mu}}{\tilde{\sigma}}}^{+\infty}\frac{1}{\sqrt{2\pi}}e^{-\frac{(y-\tilde{\sigma})^2}{2}+\frac{\hat{\sigma}^2}{2}}dy \;\; (complete\;\; the\;\; square)$$

$$= e^{-rT}e^{\tilde{\mu}}\int_{\frac{lnK-\tilde{\mu}}{\tilde{\sigma}}}^{+\infty}\frac{1}{\sqrt{2\pi}}e^{-\frac{(y-\tilde{\sigma})^2}{2}}e^{+\frac{\hat{\sigma}^2}{2}}dy$$

$$= e^{-rT}e^{\tilde{\mu}}\int_{\frac{lnK-\tilde{\mu}}{\tilde{\sigma}}}^{+\infty}\frac{1}{\sqrt{2\pi}}e^{-\frac{(y-\tilde{\sigma})^2}{2}}dye^{+\frac{\tilde{\sigma}^2}{2}}$$

$$= e^{-rT}e^{\tilde{\mu}+\frac{\tilde{\sigma}^2}{2}}\int_{\frac{lnK-\tilde{\mu}}{\tilde{\sigma}}}^{+\infty}\frac{1}{\sqrt{2\pi}}e^{-\frac{(y-\tilde{\sigma})^2}{2}}dy$$

Now, with another substitution $z = y - \tilde{\sigma}$.

$$= e^{-rT}e^{\tilde{\mu}+\frac{\tilde{\sigma}^2}{2}}\int_{\frac{lnK-\tilde{\mu}}{\tilde{\sigma}}-\tilde{\sigma}}^{+\infty}\frac{1}{\sqrt{2\pi}}e^{-\frac{z^2}{2}}dz$$

$$= e^{-rT}e^{\tilde{\mu}+\frac{\tilde{\sigma}^2}{2}}N\left(-\frac{lnK-\tilde{\mu}}{\tilde{\sigma}}+\tilde{\sigma}\right)$$

$$= e^{-rT} e^{LnS_0 + \left(r - \frac{\sigma^2}{2}\right)T + \frac{\sigma^2 T}{2}} N\left(-\frac{lnK - (LnS_0 + \left(r - \frac{\sigma^2}{2}\right)T)}{\sigma\sqrt{T}} + \sigma\sqrt{T} \right)$$

$$= e^{LnS_0} N\left(\frac{-lnK + \left(LnS_0 + \left(r - \frac{\sigma^2}{2}\right)T\right) + \sigma^2 T}{\sigma\sqrt{T}} \right)$$

(*multiply negative and common denominator*)

$$= S_0 N\left(\frac{Ln(S_0/K) + \left(r + \frac{\sigma^2}{2}\right)T}{\sigma\sqrt{T}} \right)$$

$$= S_0 N(d_1)$$

So, this proves the call value formula using the risk-neutral expected value approach:

$$C = S_0 N(d_1) - Ke^{-rT} N(d_2)$$

where $d_1 = \frac{Ln(S_0/K) + \left(r + \frac{\sigma^2}{2}\right)T}{\sigma\sqrt{T}}$ and $d_2 = \frac{Ln(S_0/K) + \left(r - \frac{\sigma^2}{2}\right)T}{\sigma\sqrt{T}} = d_1 - \sigma\sqrt{T}$.

In the proof, an interesting property is obtained:

Property 3.2

If S_t follows Geometric Brownian motion, then:

$$\boldsymbol{P}(S_T > K) = \mathbf{N}(d_2) \tag{3.38}$$

$N(d_2)$ is the probability that call will be exercised in a risk-neutral world.

Property 3.1 provide a practical meaning of $N(d_1)$, which is the hedge ratio for a European call, and Property 3.2 provides the practical meaning of $N(d_2)$, which is the probability that a call is exercised in a risk-neutral world. Note that the derivation also works for a European put. But it does not work for American options since there is an early-exercise boundary, which isn't known in advance. So, the American option value does not have an analytic formula in general through we can prove the value of an American call without a dividend is equal to the value of European call as a very special case. Usually the value of the American put is larger than the value of European put which we exposed in the binomial tree method.

This completes the analytic method of valuing derivatives. Note that only a few derivatives can be valued this way. We will do more review on that in Chapter 4. Now we return back to using the Monte Carlo simulation method to value derivatives.

3.4.7 Monte Carlo Simulation and Derivative Valuation

The risk-neutral principle is true for any derivative with any underlying asset that follows an Ito process. If the Ito process has no closed form solution of the p.d.f. $h(S_T)$, Monte Carlo simulation can be used to obtain an estimate of the initial value of the financial derivative. The following is the ***risk-neutral principle (simulation version)***:

The Risk-Neutral Principle (Simulation Version):

The value of a European style financial derivative today, f_0, is equal to the discounted expected values of its potential payoffs at expiration in a risk-neutral world where all assets grow and are discounted at the risk-free rate r.

$$f_0 = e^{-rT}\frac{\sum_{j=1}^{n} f(S_{T,j}, T)}{n} \tag{3.39}$$

Where $S_{T,j}$ is the underlying asset price at expiration T for the jth path. Note that in this model, the underlying asset S_t must grow at an expected rate of r. In particular, if the underlying asset is a stock price which follows Geometric Brownian motion, then the process is: $dS = rSdt + \sigma SdZ$.

Exercise 3.14: Assume that a stock price follows Geometric Brownian motion with $S_0 = 60$ and volatility equal to 15%. The strike price is 62 and the risk-free rate is 5%. Find the value of a six-month European put using simulation and R-code.

R-Code 3.6 Price European put using simulation

```
#input initial value and parameters, specify a n period
model for m paths
S0=60
r=0.05
b=0.15
T=0.5
m=5000
n=10000
K=rep(62,m)
deltaT=T/n
#record the path at discrete n values as a vector S and m
#path's ending price as a vector Y
S=rep(S0,n+1)
Y=rep(0,m)
#record m path's payoff function as a vector D
D=rep(0,m)
for(j in 1:m)
{
W=rnorm(n,0,1)
for(i in 1:n)
{
S[i+1]=S[i]+r*S[i]*deltaT+ b*S[i]*sqrt(deltaT)*W[i]
}
Y[j]=S[n+1]
D[j]=max(K[j]-Y[j],0)
}
p=exp(-r*T)*mean(D)
p
```

Note that the Monte Carlo simulation method does not apply to an American option directly because the payoff is not a known function of the underlying asset at expiration.

Summary of Formulas in Derivative Valuation

Analytic models for valuing derivatives are important. There are nine key concepts regarding this introduction to analytical models for valuing derivatives.

1. Stock prices may be well described by Geometric Brownian motion:

$$dS = \mu S dt + \sigma S dZ$$

In a risk-neutral world, a stock price follows Geometric Brownian motion with an expected growth rate equal to the riskless rate: $dS = rSdt + \sigma SdZ$.

2. In a risk-neutral world, Geometric Brownian motion is equivalent to having a log normal distribution:

$$LnS_T \sim N\left(LnS_0 + \left(r - \frac{\sigma^2}{2}\right)T,\ \sigma^2 T\right) = N(\tilde{\mu}, \tilde{\sigma}^2).$$

3. Ito's lemma is used to derive a derivative process, df, with the assumption that there is an underlying process, dX, and that the financial derivative is a deterministic function $f(X, t)$:

   If $f(X, t)$ is a function of X and t, and X follows a process:

$$dX = \mu(X, t)dt + \sigma(X, t)dZ$$

   then $f(X, t)$ follows a process:

$$df = \left(\frac{\partial f}{\partial X}\mu(X, t) + \frac{\partial f}{\partial t} + \frac{1}{2}\frac{\partial^2 f}{\partial X^2}\sigma(X, t)^2\right)dt + \frac{\partial f}{\partial X}\sigma(X, t)dZ$$

4. The Black-Scholes PDE is (ignoring subscripts): $\frac{\partial f}{\partial S}rS + \frac{\partial f}{\partial t} + \frac{1}{2}\frac{\partial^2 f}{\partial S^2}\sigma^2 S^2 = rf$.
   The derivation is similar to that of the binomial tree approach: buy one unit of derivative f and short sell $\Delta\left(=\frac{\partial f}{\partial S}\right)$ shares to construct a riskless portfolio. Note that this Δ hedge is risk-free for a very short time period (dt). The portfolio needs to be continuously rebalanced to maintain the perfect (riskless) hedge.
5. A European call value is calculated as:

$$C_0 = S_0 N(d_1) - Ke^{-rT}N(d_2)$$

where $d_1 = \frac{Ln(S_0 / K) + \left(r + \frac{\sigma^2}{2}\right)T}{\sigma\sqrt{T}}$ and $d_2 = \frac{Ln(S_0 / K) + \left(r - \frac{\sigma^2}{2}\right)T}{\sigma\sqrt{T}} = d_1 - \sigma\sqrt{T}$.

6. A European put value is calculated as:

$$P_0 = Ke^{-rT}N(-d_2) - S_0 N(-d_1)$$

7. The moment generating function for $X \sim N(\mu, \sigma^2)$ is:

$$\Phi(t) = e^{\mu t+\frac{1}{2}\sigma^2 t^2} = E\left(e^{Xt}\right) = \int_{-\infty}^{\infty} e^{xt}\frac{1}{\sqrt{2\pi\sigma^2}}e^{-\frac{(x-\mu)^2}{2\sigma^2}}dx$$

8. Using a no-arbitrage argument, the same option pricing model can be derived in both the binomial tree model approach and the Black-Scholes PDE.
9. The risk-neutral principle: A European style derivative value in the real world can be calculated as the risk-neutral expected payoff discounted back to today at the risk-free interest rate. The risk-neutral expected payoff is calculated by assuming that all asset values grow at an expected rate equal to the risk-free rate, r.

$$f_0 = e^{-rT}E_Q(f_T)$$

where f_T is a known function of S_T.

Compare the above formula for a derivative value with the formula derived using the risk-neutral principle in Section 3.2. Note that the only difference is that in the earlier section the expected value was calculated with respect to a discrete random variable as a probability weighted average. In the analytical model approach, the expected value is calculated with respect to a continuous random variable. The risk-neutral approach comes from the fundamental principle of asset pricing in financial economics: that any riskless portfolio must grow at the risk-free rate.

This chapter has introduced the three primary approaches to valuing financial derivatives. Chapter 4 compares the effectiveness of the three methods in valuing a financial derivative. Chapter 4 shows the advantages and disadvantages of different methods to value derivatives as well as their convergence. Finally, in Chapter 4, five methods to value a call or put are demonstrated: the Black-Scholes (analytical) formula, Monte Carlo simulation, two binomial tree approaches, and the finite difference method. All of these methods will be shown to converge.

Chapter Summary

Chapter 3 begins the mathematical foundation of the book. Three primary derivative pricing approaches (tree models, simulation models, and analytic models) are introduced through a simple dice game. After that, the chapter details the crucial concept of risk-neutral pricing through a binomial tree model. P-measures, Q-measures, and risk aversion are introduced and discussed in a tree setting. After the tree model, Geometric Brownian motion and simulation are introduced as a foundation to generating random paths. Ito's lemma is discussed next. Then, an instantaneous risk-free portfolio is constructed to provide an intuitive derivation of the Black-Scholes PDE. The risk-neutral principle in a continuous model is implied by the Black-Scholes PDE. Finally, applying the risk-neutral principle, the analytic Black-Scholes European call and put price formulas are derived through integration. At the end of the chapter, Monte Carlo simulation is used to value calls and puts. Sample codes of simulation in R are offered so that readers can mimic, practice, and experiment. Sample spreadsheets with a small numbers of steps are offered so that readers can visualize the underlying economics and mathematics and extend it. The book offers a project of comparing the three general derivative valuation methods (analytic, binomial tree by spreadsheet, and simulation by R) and analyzing convergence. Convergence is shown to be a potentially important issue.

End-of-Chapter Demonstration Exercises

1. Assume that a stock price follows Geometric Brownian motion:

$$dS = \mu S dt + \sigma S dZ$$

and that a financial derivative on that stock price will pay S_T^2 at time T.

a. Find the current derivative value, f_0.
b. Find the derivative value f_t as a function of $f(S_t, t)$ at any time t.
c. Check that the solution, f_t, satisfies the Black-Scholes PDE and its terminal condition.

Solutions:

a. By the risk-neutral principle, $f_0 = e^{-rT} E_Q\left(f_T\right) = e^{-rT}\left(E_Q(S_T^2)\right)$.

Method 1: Calculate $E_Q(S_T^2)$, using integration:

$$E_Q(S_T^2) = \int_{-\infty}^{\infty} s^2 \frac{1}{\sqrt{2\pi\tilde{\sigma}^2}} e^{-\frac{(lns-\tilde{\mu})^2}{2\tilde{\sigma}^2}} \frac{1}{s} ds$$

where $\tilde{\mu} = LnS_0 + \left(r - \frac{\sigma^2}{2}\right)T$, and $\tilde{\sigma}^2 = \sigma^2 T$.
Method 2: Use the moment generating function:

$$E_Q(S_T^2) = E_Q(e^{2LnS_T})$$

Note that $LnS_T \sim N(\tilde{\mu}, \tilde{\sigma}^2)$,

$$E_Q(e^{2LnS_T}) = e^{\tilde{\mu}\times 2 + 1/2\tilde{\sigma}^2\times 2^2} = e^{2\tilde{\mu}+2\tilde{\sigma}^2} = e^{2LnS_0+2\left(r-\frac{\sigma^2}{2}\right)T+2\sigma^2 T}$$
$$= e^{2LnS_0}e^{2rT+\sigma^2 T} = S_0^2 e^{2rT+\sigma^2 T}$$

So, $f_0 = e^{-rT}(E_Q(S_T^2)) = S_0^2 e^{rT+\sigma^2 T} = S_0^2 e^{(r+\sigma^2)T}$.
Method 3: Using Ito's lemma, derive the process for S_T^2:

$$df = \left(\frac{\partial f}{\partial S} rS + \frac{\partial f}{\partial t} + \frac{1}{2}\frac{\partial^2 f}{\partial S^2}\sigma^2 S^2\right)dt + \frac{\partial f}{\partial s}\sigma S dZ$$

$$dS^2 = (2SrS + \sigma^2 S^2)dt + 2S\sigma S dZ$$

$$dS^2 = (2r + \sigma^2)S^2 dt + 2\sigma S^2 dZ$$

$$df = (2r + \sigma^2)f dt + 2\sigma f dZ$$

So, $f = S^2$ follows Geometric Brownian motion.

It is easy to confirm that $E_Q(S_T^2) = E_Q(f_T) = f_0 e^{(2r+\sigma^2)T} = S_0^2 e^{(2r+\sigma^2)T}$

$$f_0 = e^{-rT}(E_Q(S_T^2)) = S_0^2 e^{(r+\sigma^2)T}$$

b. If the starting time is t, then the above equation changes to:

$$f_t = e^{-r(T-t)}(E_Q(S_T^2)) = S_t^2 e^{(r+\sigma^2)(T-t)}$$

It is a function of S_t and t, namely $f(S_t, t) = S_t^2 e^{(r+\sigma^2)(T-t)}$.
c. Check the Black-Scholes PDE terminal condition:

$$f(S_T, T) = S_T^2 e^{(r+\sigma^2)(T-T)} = S_T^2$$

Now check the PDE:

$$\frac{\partial f}{\partial S} = 2S_t e^{(r+\sigma^2)(T-t)}$$

$$\frac{\partial f}{\partial t} = -(r + \sigma^2) S_t^2 e^{(r+\sigma^2)(T-t)}$$

$$\frac{1}{2}\frac{\partial^2 f}{\partial S^2} = e^{(r+\sigma^2)(T-t)}$$

$$\begin{aligned}\frac{\partial f}{\partial S} rS + \frac{\partial f}{\partial t} + \frac{1}{2}\frac{\partial^2 f}{\partial S^2}\sigma^2 S^2 &= 2S_t e^{(r+\sigma^2)(T-t)} rS_t + \left(-(r + \sigma^2) S_t^2 e^{(r+\sigma^2)(T-t)}\right) \\ &\quad + \sigma^2 S_t{}^2 e^{(r+\sigma^2)(T-t)} \\ &= S_t^2 e^{(r+\sigma^2)(T-t)}(2r - r - \sigma^2 + \sigma^2) \\ &= S_t^2 e^{(r+\sigma^2)(T-t)} r = rf\end{aligned}$$

So, we have Black-Scholes PDE:

$$\frac{\partial f}{\partial S} rS + \frac{\partial f}{\partial t} + \frac{1}{2}\frac{\partial^2 f}{\partial S^2}\sigma^2 S^2 = rf$$

Chapter 3 Extensions and Further Reading

From a theoretic perspective, when Black-Scholes derived their model, they first derived the Black-Scholes partial different equation and then deduced the Black-Scholes formula. This is the approach used in this book. However, asset pricing theorems derive the Black-Scholes formula from the martingale property, a risk-neutral evaluation framework. The risk-neutral framework is very beautiful mathematically and is coherent with the Black-Scholes PDE. Stochastic calculus is the major tool in this framework. However, stochastic calculus is rarely covered adequately in an undergraduate course to make the material regarding the martingale property approach accessible. Additionally, it is very difficult to self-study the knowledge at the undergraduate level. Through hands-on programming implementation of the models, this book is designed to provide students with a gateway toward stochastic calculus-based modeling. Following are two good stochastic calculus-based books (although several others exist):

1. *Methods of Mathematical Finance* by Ioannis Karatzas and Steven Shreve
2. *The Concepts and Practice of Mathematical Finance* by Mark S. Joshi

On the modeling side, since the volatility in the Black-Scholes model is not observable, the Black-Scholes model is not able to explain the volatility smile implied by market data on option values. There are three main approaches to overcome this shortcoming. First, local volatility models are quite popular in the financial industry in which the volatility is modeled as a deterministic function of the underlying assets. The second approach is adding jumps to Geometric Brownian motion such as in using jump diffusion models and Levy models. The third approach is to use a stochastic volatility model. For example, in Heston's model the variance of the underlying asset follows a mean reversion process which is negatively correlated with the Geometric Brownian motion. In Bates' model, a jump is added to Heston's model resulting in a mixed the second and third approaches. Note that it is very hard to delta hedge any jump process because jump risk can only be completely hedged away by an infinity of hedging vehicles. However, it may well approximated by many finite hedging vehicles. Interested readers could explore the following books to go deeper into these issues:

3. *The Volatility Surface* by Jim Gathreal
4. *Applications of Fourier Transform to Smile Modeling* by Jianwei Zhu
5. *Mathematical Modeling and Computation in Finance: With Exercises and Python and MATLAB Computer Codes* by Cornelis W. Oosterlee and Lech A. Grzelak

There are also other volatility models such as ARCH and GARCH models. A good book with which to explore these models is related to data science:

6. *Analysis of Financial Time Series* by Ruey S. Tsay

Finally, another extension is two-factor models. For example, convertible bond prices are affected by both stock price and interest rate changes. A good book for two-factor models is:

7. *Options, Futures and other Derivatives* by John Hull

End-of-Chapter Project

Various Methods for Option Valuation

We have learned three approaches to value derivatives. This project uses a simple put option to illustrate that the three approaches converge. The binomial tree method is used to study the convergence pattern. For the Monte Carlo simulation, we study the statistical properties of the simulated prices. We use the analytic model (the Black-Scholes formula here) for a European put as a benchmark. We compare how fast the binomial tree method converges and how good the Monte Carlo simulation method is. Finally, we compare an American put valued using a binomial tree with the benchmark. All these methods can be generalized while learning the fundamental machineries.

A stock price is currently \$60 and is expected to follow Geometric Brownian motion. The stock's annual volatility is expected to be 15%. The risk-free rate is 5% per annum with continuous compounding. What is the value of a six-month European put option with a strike price of \$62?

a1. Use EXCEL to build a 30-period binomial tree to approximate the put value.

a2. Report the put values with different n's in a table by increasing the number of steps with $n = 30, 31, 32, \ldots, 39$. Place the put values on a graph with the x-axis as the number of steps and the y-axis as the values of the puts. Analyze the pattern for convergence.

a3. Perform a convergence analysis as in a2 but with $n = 300, 301, 302, \ldots, 309$. Graph the results as in a2. Compare the graph in a2 with the new graph in a3. What is similar about the two graphs and what is different? What does this comparison suggest regarding convergence?

a4. Show how to adjust the model if the put option is an American put.

b1. Use the Black-Scholes put value formula to find the solution. You may use Excel, or an add-in to Excel, or search the web for an online calculator (setting dividends to zero). Be careful to use continuous compounding with the riskless rate.

b2. Write a one-page story about the Black-Scholes model. (It could be history, background, importance to the emergence of derivative trading, or anything related.)

c1. Use Monte Carlo simulation to estimate the option value by projecting 10,000 trajectories, then averaging the 10,000 put payoffs, and then discount the average back by $e^{-0.05\times 0.5}$.

c2. Repeat the previous simulation 10 times. Use a graph (for example) of the histogram and report the mean and standard deviation to analyze the convergence.

c3. Do more simulation trials and a statistical analysis of the simulations, including the computing time.

End-of-Chapter Problems (Fundamental Problems)

1. A stock price is currently \$30. Over the next period (lasting three months), it is expected to go up by 10% or down by 10%. The risk-free interest rate is 5% per annum with continuous compounding.
   a. Use the no-arbitrage method to find the value of a European call option value with a strike of \$32.
   b. Is the hedge ratio Δ positive or negative? What is the practical meaning of Δ? Explain why the portfolio is riskless.
   c. Use the risk-neutral method to find the value of the same European call option value.

2. A stock price is currently \$30. Over the next period (lasting three months), it is expected to go up by 10% or down by 10%. The risk-free interest rate is 5% per annum with continuous compounding.
   a. Use the no-arbitrage method to find the value of a European put option value with a strike of \$32.

b. Is the hedge ratio Δ positive or negative? What is the practical meaning of Δ? Explain why the portfolio is riskless.
c. Use the risk-neutral method to find the value of the same European put option value.

3. A stock price is currently \$30. Over the next period (lasting three months), it is expected to go up by 10% or down by 10%. The risk-free interest rate is 5% per annum with continuous compounding.
   a. Use the no-arbitrage method to find the value of a derivative that pays off $[\,Max(32 - S_T, 0)]^{1/2}$, where S_T is the stock price in three months.
   b. Is the hedge ratio Δ positive or negative? What is the practical meaning of it? Explain why it is consistent with creating a riskless portfolio.
   c. Use the risk-neutral method to find the value of a derivative that pays off $[\,Max(32 - S_T, 0)]^{1/2}$, where S_T is the stock price in three months.

4. A stock price is currently \$30. Over the next period (lasting three months), it is expected to go up by 10% or down by 10%. The risk-free interest rate is 5% per annum with continuous compounding.
   a. Use the no-arbitrage method to find the value of a derivative that pays off $\sqrt{S_T}$, where S_T is the stock price in three months.
   b. Is the hedge ratio Δ positive or negative? What is the practical meaning of Δ? Explain why the portfolio is riskless.
   c. Use the risk-neutral method to find the value of a derivative that pays off $\sqrt{S_T}$, where S_T is the stock price in three months.

5. A stock price is currently \$30. Over the next period (lasting three months), it is expected to have a volatility of 15%. The risk-free interest rate is 5% per annum with continuous compounding.
   a. Use a two-period tree to find the value of a derivative that pays off $[\,Max(30 - S_T, 0)]^{1/2}$, where S_T is the stock price in three months (assume that early exercise is not allowed).
   b. Use a two-period tree to find the value of a derivative that pays off $[\,Max(30 - S_T, 0)]^{1/2}$, where S_T is the stock price in three months (assume that early exercise is allowed and the corresponding payoff is $[\,Max(30 - S_t, 0)]^{1/2}$ at any time t before T).

6. A stock price is currently \$30. Over the next period (lasting three months), it is expected to have a volatility of 15%. The risk-free interest rate is 5% per annum with continuous compounding.
   a. Use a two-period tree to find the value of a derivative that pays off $\sqrt{S_T}$, where S_T is the stock price in three months (assume that early exercise is not allowed).
   b. Use a two-period tree to find the value of a derivative that pays off $\sqrt{S_T}$, where S_T is the stock price in three months (assume that early exercise pays off $\sqrt{S_t}$ at any time t before T).
7. A stock price is currently \$30. Over the next period (lasting three months), it is expected to go up by 10% or down by 10%. The risk-free interest rate is 5% per annum with continuous compounding. Consider a European put with a payoff $Max(K - S_T, 0)$.
   a. What value of K will make the hedge ratio (Δ) equal to $-1/2$?
   b. What value of K will make the hedge ratio (Δ) equal to $-1/3$?
   c. What value of K will make the hedge ratio (Δ) equal to $-\frac{1}{n}$?
   d. What value of K will make the hedge ratio (Δ) equal to 0?
   e. What value of K will make the hedge ratio (Δ) equal to -1?
   f. Why must the hedge ratio in this example be between −1 and 0?
8. A stock price is currently \$30. Over the next period (lasting three months), it is expected to go up by 10% or down by 10%. The risk-free interest rate is 5% per annum with continuous compounding. Consider a European call with a payoff $Max(S_T - K, 0)$.
   a. What value of K will make the hedge ratio (Δ) equal to $1/2$?
   b. What value of K will make the hedge ratio (Δ) equal to $1/3$?
   c. What value of K will make the hedge ratio (Δ) equal to $\frac{1}{n}$?
   d. What value of K will make the hedge ratio (Δ) equal to 0?
   e. What value of K will make the hedge ratio (Δ) equal to 1?
   f. Why must the hedge ratio in this example be between 0 and 1?
9. If the risk-free rate is 10% and the annual stock volatility is 5%, what is the lowest steps, n, needed to make sure that the risk-neutral probability is between 0 and 1 for a derivative contract with maturity three months?
10. A stock price is currently \$30. Over the next period (lasting three months), it is expected to have an annual volatility of 15%. The risk-free interest rate is 5% per annum with continuous compounding.
    a. Use an EXCEL spreadsheet to form a 30-period tree to value a derivative that pays off $[Max(30 - S_T, 0)]^{1/2}$, where S_T is the

stock price in three months (assume that early exercise is not allowed).

b. Use an EXCEL spreadsheet to form a 30-period tree to value a derivative that pays off $[Max(30 - S_T, 0)]^{1/2}$, where S_T is the stock price in three months (assume that early exercise with a payoff of $[Max(30 - S_t, 0)]^{1/2}$ at any time t before t is allowed).

11. Assume that an underlying asset follows a generalized Wiener process:

$$dX = 3dt + 4dZ$$

$$X_0 = 20$$

a. Assume $T = 1$. Use R to generate a sample path.
b. Assume $T = 1$. Find 5,000 paths, calculate the expected value of X at $T = 1$, and calculate the variance of X at $T = 1$.
c. Derive $E(X_T)$ and Var (X_T) for any T.
d. Find the distribution of X_T.
e. What is the probability that X is greater than 15 at the end of five years? You can keep N(·), the cumulative distribution function, in your answer.

12. Assume that an underlying asset follows a generalized Wiener process: $dX = adt + bdz$
Assume that the asset starts at X_0.

a. What is the expected value of the asset at the end of t years?
b. What is the standard deviation of the asset value at the end of t years?

13. Assume that an underlying asset follows a generalized Wiener process: $dX = adt + bdz$. For the first two years, $a = 2$ and $b = 4$, but in years three to five $a = 3$ and $b = 5$. Assume $X_0 = 20$.

a. What is the expected value of the asset at the end of five years?
b. What is the standard deviation of the asset value at the end of five years?
c. What is the probability that X is greater than 50 at the end of five years? You can keep N(·), the cumulative distribution function, in your answer.

14. Assume that an underling asset has Geometric Brownian motion: $dS = \mu Sdt + \sigma SdZ$. During the first two years $\mu = 0.10$ and $\sigma = 0.20$.

During the following three years $\mu = 0.15$ and $\sigma = 0.25$. Assume $S_0 = 20$.

a. Assume $T = 1$. Use R to generate a sample path.

b. Assume $T = 1$. Find 5,000 paths, calculate the expected value of X at $T = 1$, and calculate the variance of X at $T = 1$.

c. Use simulation to find the expected value and the standard deviation of the asset at the end of five years.

15. Assume that an underling asset has Geometric Brownian motion, $dS = \mu S dt + \sigma S dZ$. During the first two years, $\mu = 0.10$ and $\sigma = 0.20$. During the following three years, $\mu = 0.15$ and $\sigma = 0.25$. Further assume $S_0 = 20$.

 a. Find the expected value of the asset at the end of five years analytically.

 b. Find the standard deviation of the asset at the end of five years analytically.

 c. What is the probability that S is greater than 50 at the end of five years? You can keep $N(\cdot)$, the cumulative distribution function in your answer.

16. Assume that a stock price S follows Geometric Brownian motion, $dS = \mu S dt + \sigma S dZ$. Find the process for df where $f(S, t)$ is:

 a. $\frac{1}{S}$

 b. e^{3S}

 c. $Ln(2S)$

 d. $S^{1/2}$

17. Assuming S follows a mean reversion process: $dX = k(\theta - X)dt + \sigma dZ$, find the process for df where $f(X, t)$ is:

 a. $\frac{1}{X}$

 b. e^X

 c. LnX

 d. X^2

18. Show that the Black-Scholes formula for European call value:

$$C_t = f(S_t, t) = S_t N(d_1) - Ke^{-r(T-t)} N(d_2)$$

satisfies the Black-Scholes PDE:

$$\frac{\partial f(S_t, t)}{\partial S} r S_t + \frac{\partial f(S_t, t)}{\partial t} + \frac{1}{2}\frac{\partial^2 f(S_t, t)}{\partial S^2}\sigma^2 S_t^2 = rf(S_t, t).$$

19. Show that the Black-Scholes formula for a European put value satisfies the Black-Scholes PDE: $\frac{\partial f(S_t,t)}{\partial S} rS_t + \frac{\partial f(S_t,t)}{\partial t} + \frac{1}{2}\frac{\partial^2 f(S_t,t)}{\partial S^2}\sigma^2 S_t^{\,2} = rf(S_t, t)$.
20. Assume that a stock price follows Geometric Brownian motion: $dS = \mu Sdt + \sigma SdZ$ and that a financial derivative will pay $\frac{1}{S_T}$ at time T.
    a. Find the derivative value today, f_0.
    b. Find the derivative value, f_t, as a function of $f(S_t, t)$ at any time t.
    c. Check that the solution, f_t, satisfies the Black-Scholes PDE and its terminal condition.
21. Assume that a stock price follows Geometric Brownian motion, $dS = \mu Sdt + \sigma SdZ$ and that a derivative will pay S_T^n at time T.
    a. Find the derivative value today, f_0.
    b. Find the derivative value, f_t, as a function of $f(S_t, t)$ at any time t.
    c. Check that the solution, f_t, satisfies the Black-Scholes PDE and check that its terminal condition is satisfied.

End-of-Chapter Problems (Challenging Problems)

This chapter discussed the fundamentals of derivative pricing. There are multiple ways to further our knowledge regarding these fundamentals. On the mathematical side, we can dig deeper into theory such as measure-theory-based derivative pricing. There are a lot of sources on that topic available elsewhere (especially mathematical books) so this book focuses on the application side to dig deeper into various financial derivative contracts and how they can be valued – a topic on which there are fewer materials available.

The CFA (Charted Financial Analyst) program discusses various derivative products and their valuation formulas – the fundamentals of which are the same. However, in the absence of a deep understanding of the fundamentals of the valuation formula, the various formulas are too often merely memorized. This section challenges readers to understand the fundamentals more deeply including how tools can be applied to various derivative products. This problem develops a deeper understanding of derivative pricing using the exercises that follow. Since some of the exercises that follow introduce derivatives not covered in previous sections, the exercises include some short lecture notes.

Four main kinds of derivatives on fixed-income securities are discussed: forwards/futures on bonds, forwards/futures on interest rates, options on bonds, and options on interest rates. They will be discussed in two topics. Note that the models here are elementary and Chapter 7 discusses advanced models.

In this chapter, the forward pricing formula is used to price futures. We will not distinguish forwards and futures for pricing purposes. In Chapter 7 they are distinguished.

Topic 1 Forwards/futures on fixed income securities (bonds and interest rates)

Previously the forward pricing formula was found to be: $F = S_0 e^{rT}$. Applying this pricing formula to a forward on a stock appeared straightforward since it was assumed that the underlying stock paid no dividends. However, in practice, all dividends paid prior to the delivery date of a forward contract should be included in the modeling of the forward price. The issue of cash flows being distributed by the underlying asset during the life of a forward contract is especially prevalent in the case of coupon-bearing bonds. Additionally, forward/futures contracts on bonds are complicated by there being two maturities: the forward/futures maturity and bond maturity. Further there are numerous conventions regarding bonds such as unique quotation methods and some other technical details that need to be understood.

For example, for short-term bonds such as Treasury bills and Eurodollars, there is a 360-day price quotation convention. The formula for forwards on these money-market securities uses 360 days per year as in either: $F = S_0 \times (1 + r_f)^{T/360}$ or $F = S_0\left(1 + r_f \times \frac{T}{360}\right)$ and where S_0 is today's price of the underlying short-term bond and T is the maturity in days for forward contract. For longer-term bonds there is 365-day convention. The formula for forwards on long-term bonds is $F = S_0(1 + r_f)^{T/365}$.

For futures contracts on long-term bonds such as Treasury bonds, the deliverable underlying bond is *any* Treasury bond with a maturity greater than 15 years. The holder of the short position has the option of which bond to deliver. There is a conversion factor corresponding to each bond that can be delivered, but even when delivery prices are adjusted by these conversion factors different bonds will tend to have different degrees of attractiveness for delivery. This complex option – and other related options such as options to delay identification of the deliverable bond (the "wild card option") – introduce very challenging issues into modeling financial derivative prices in practice. Table 3.2 summarizes quotation conventions and some other technical details for various fixed-income instruments.

TABLE 3.2

Quotation Conventions and Other Details of Fixed-Income Instruments

Name	Face Value and Quotation Method	What Does It Mean?
Treasury Bills, T-bills for short	Face value: $1,000. ***Quote method for T-bill: Annualized Discount Method (in percent)***	Assume that a T-bill has 100 days to maturity. It is quoted as 1.40: The non-annualized interest rate for 100 days is: 1.4% × 100/360 = 0.003889. Applying this non-annualized discount generates the price: (1 − 0.003889) × 1,000 = 996.111 So the T-bill offers $1,000 face value back in 100 days for paying $996.111 where $996.111 is called the settlement price. It is the bond's value, which is consistent with this book's use of the term underlying asset price S_0
LIBOR	Face value: Any amount ***Quote method for LIBOR: Annualized Add-on Method***	Assume a $1 million loan for 30 days at LIBOR 6%: This implies the interest is $1 million × 6% × 30/360 = $5,000. To borrow $1 million today requires paying back $1,005,000 at the end of 30 days.
Treasury Bond	Face value: $1,000 ***Quote method for Treasury bond***: Price per $1,000, coupon is paid semiannually, coupon rate is quoted as annual rate.	Assume a 7% (coupon) 15-year Treasury bond with a $1,000 face value and a $1,040 price: This bond pays $35 every half-year, returns the face value $1,000 in 15 years and its price is $1,040 today.
Eurodollar futures	Underlying asset: At the futures delivery date the deliverable is a $1,000,000 face value three-month loan based on LIBOR rate. ***Quote method for Eurodollar futures: 100 minus the Annualized Discount Method (in percent)***	Assume a Eurodollar future with a price quoted at 97.22 and a maturity of 90 days. It implies the LIBOR rate is 2.78%, and the interest rate is 2.78% × 90/360 = 0.00695. So, the futures price is: (1 − 0.00695) × 1 million = 993,050. This is the futures contract price consistent with our book's notation: F_0. In addition, if the Eurodollar futures price changes 0.01(1 basis point), the discount rate changes 0.01%, which changes the interest rate 0.01% × 90/360 = 0.0025%. For a $1 million face

(*continued*)

TABLE 3.2 (Continued)

Name	Face Value and Quotation Method	What Does It Mean?
		value contract, the futures contract value changes $25.
SOFR futures	Underlying asset: Secured Overnight Financing Rates. One-month SOFR futures with $5,000,000 face value and three-month SOFR futures with $1,000,000 face value. ***Quote method for SOFR futures: 100 minus the Annualized Discount Method (in percent)***	Assume that a one-month SOFR futures contract has a price quoted at $97.53 at its expiration date. If the final settlement in one month is $97.57. This implies the arithmetic average of overnight repo rates (quoted in annualized format) in this reference period between the expiration and settlement is 2.43%. If the one-month futures price changes 0.01 (1 basis point), the discount rate changes 0.01%, which changes the interest rate 0.01 % × 30/360 = 0.0008333%. For a $5 million face value contract, the futures contract value changes $41.67. Assume that a three-month SOFR futures contract has a price quoted at $97.53 at its expiration date. If the final settlement in three months is $97.57. This implies the daily compounding of overnight repo rates in this reference period between the expiration and settlement is 2.43%. With daily compounding, divide by 360 days since overnight repo rates are quoted as annual rates. If the three-month futures price changes 0.01 (1 basis point), the discount rate changes 0.01%, which changes the interest rate 0.01% × 90/360 = 0.0025%. For a $1 million face value contract, the futures contract value changes $25.

(*continued*)

TABLE 3.2 (Continued)

Name	Face Value and Quotation Method	What Does It Mean?
Treasury Notes/ Bond Futures (T-Notes and T-Bonds for short)	Sample underlying assets: For two- to three-year Treasury notes futures contracts, the underlying assets are two- to three-year Treasury notes with a face value of \$200,000. For 5-year (10-year) Treasury notes futures contracts, the underlying is 5-year to 10-year Treasury notes with a face value of \$100,000. For 15-25 year Treasury bond futures contracts, the underlying is any Treasury bond with 15–25 years to maturity. The holder of the short position decides which one to deliver. ***Quote method for Treasury notes/bond futures***: Prices are quoted per \$2,000 for the two-year and three-year contracts and per \$1,000 for the other U.S. Treasury notes/ bonds futures. Fractional points are expressed in1/32nds – in line with the convention used in the U.S. government bond market.	Assume that it is now January 2021. Consider a long position in a March 2021 five-year U.S. Treasury note futures contract at the quoted price 120'25. The future contract's value is:(120 + 25/32) × 1,000 = 120,781.25.
Stock Index Futures	Underlying asset: the stock index value times 250 ***Quote method for Stock Index Futures***: the futures price is 1/250th of the total forward contract value bringing it in line with the scale of the stock index.	If an index such as the S&P 500 index has a value of $S_0 = 3,000$, the contract is based on 250 × that value (i.e., \$750,000). The futures contract price, F_0, is 3,000 times the adjustment factor for the riskless rate and the dividend yield (i.e., $e^{(r-d)T}$): $F_0 = S_0 \times e^{(r-d)T}$.

Next, we provide exercises on forwards/futures on bonds and interest rates

Forwards/Futures on short-term and long-term bonds

1. Consider a long position in a T-bill forward contract deliverable in 70 days. The underlying asset is a 160-day T-bill. Assume that the 70-day T-bill contract is quoted at 2.00 and the 160-day T-bill is quoted at 3.00. Find the price formula and price for the T-bill forward.
   a. Using the T-bill quotation convention, find the 70-day T-bill price per \$100 of face value.
   Hint: $\boldsymbol{B_{70}} = \left(\mathbf{100} - q_{70} \times \frac{\mathbf{70}}{\mathbf{360}}\right)$, where q_{70} is the quotation discount.

b. Show that the interest rate for this time period (non-annualized and no compounding) is: $r_f = \frac{100}{B_{70}} - 1$.
Note this is the non-annualized interest rate that grows today's T-bill value B_{70} to the face value, $100.

c. Using the T-bill quotation convention, find the 160-day T-bill price. Hint: $\boldsymbol{B_{160}} = \left(\mathbf{100} - q_{160} \times \frac{\mathbf{160}}{\mathbf{360}}\right)$, where q_{160} is the quotation discount. We use $100 as a face value. Note that B_{160} is today's price of the underlying asset, S_0, per $100 of face value.

d. Show the price of the forward contract, $\boldsymbol{F} = \frac{B_{160}}{B_{70}} \times \mathbf{100}$.
Hint: Using the forward pricing formula $F = S_0(1 + r_f)$, where r_f is the interest rate for 70-day Treasury bill calculated in (b) and S_0 is calculated in (c).

2. Explain why no riskless arbitrage relation exists for Eurodollar futures contract. (Hint: The underlying asset's price will not change $25 for every basis point change in LIBOR since LIBOR is an add-on rate.)

3. A futures contract on T-notes will mature in 1.4 years. The underlying asset is a 4% coupon T-note with exactly 10 years to maturity (with coupon just paid) and with a price of $1,070. Assume that the risk-free rate is 3%. Find the price of the futures contract.

   a. Find the underlying asset price $\boldsymbol{S_0} = \boldsymbol{B} - \boldsymbol{PV}(\boldsymbol{D})$, which is the note's price B after subtracting the present value of its coupons (PV(D)). Hint: There are two coupons during the 1.4-year period, so $PV(D) = 20 \times e^{-0.05\times0.5} + 20 \times e^{-0.05\times1}$, where the $20 comes from 4% coupon rates paid semiannually with a face value of $1,000.

   b. Find the futures price using $\boldsymbol{F} = \boldsymbol{S_0}e^{rT}$, where $T = 1.4$.

   c. Note that the holder of the short position of the T-note futures contract has the option of delivering any of a number of eligible notes, each of which is assigned a ***conversion factor***. Adjusting prices for the conversion factor allows for computation of that note which is the ***cheapest-to-deliver***. The futures price for this cheapest-to-deliver note is therefore $\boldsymbol{FP} = \boldsymbol{F} \times \frac{\mathbf{1}}{\boldsymbol{CF}}$ where CF is the cheapest-to-deliver note's conversion factor. Assume that the cheapest-to-deliver note has a conversion factor of 1.1. Find the no arbitrage futures price for this 4% coupon 10-year T-bond relative to the cheapest-to-deliver note.

Forwards/Futures on interest rates

4. A $j\times(j + k)$ FRA (***forward rate agreement***) is a forward contract that matures j days from today. It is an over-the-counter contract between parties that determines the rate of interest to be paid or received on an obligation beginning at j days from today and lasting k days. Usually j and j+k are quoted as months in the market, here we make it days for

illustrating the pricing formula. Assume that the underlying asset is a $(j+k)$-day short-term zero coupon loan. Assume that $L_{(j+k)}$ is today's annualized risk-free interest rate for $j+k$ days loan contracts and that $L_{(j)}$ is today's annualized risk-free interest rate for j-days loan contracts. Find the price formula for the forward rate FRA(j,k) (i.e., the k-day forward rate j days from now) for the forward rate agreement, which is the rate quoted in the forward rate agreement contract.

a. Show the price of a \$1 face value $j+k$ days loan contract is $S_0 = \frac{1}{1 + L_{(j+k)} \times \frac{j+k}{360}}$.

b. Show the price of this forward at its settlement in j-days is: $F = S_0\left(1 + L_{(j)} \times \frac{j}{360}\right)$.

c. FRA(j,k) is the interest rate that makes the forward price in j-days grow to a \$1 face value in ($j+k$) days. So, $F \times \left(1 + FRA(j, k) \times \frac{k}{360}\right) = 1$. Show that $FRA(j, k) = \left(\frac{1}{F} - 1\right) \times \frac{360}{k}$.

d. Combine (a), (b), and (c), and show that $\mathbf{FRA}(j, k) = \left(\frac{1 + L_{(j+k)} \times \frac{j+k}{360}}{1 + L_{(j)} \times \frac{j}{360}} - 1\right) \times \frac{360}{k}$.

5. Calculate the price of a 1 × 4 FRA (90-day loan, 30 days from now). The current 30-day rate is 4% and the 120-day is 5% (both are annualized money market rates).

Now we will look at exercises on options on bonds and interest rates. Note that the Black-Scholes assumptions fail here since the underlying asset is sensitive to interest rates, so we can't assume r is constant. Instead, the interest rate has a tree. More machinery is discussed in Chapter 7. Here it is just briefly discussed.

Topic 2 Options on fixed-income securities (bonds and interest rates)

In order to value options on bonds or interest rates, it is necessary to set up a model for interest rate evolvement. It is clear Geometric Brownian motion does not work because interest rates usually have a mean reversion property and other properties not consistent with Geometric Brownian motion. Here, we assume a two-step interest rate tree to value the options on fixed-income securities. The generation of the interest rate tree is covered in Chapter 7.

Here is the two-period interest rate tree to be used for all the remaining exercises:

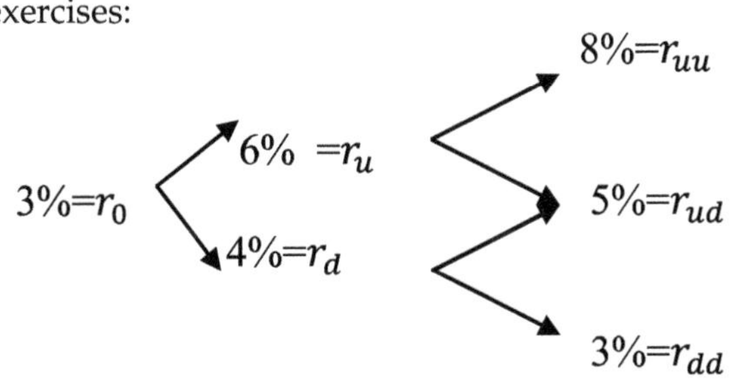

Where r_0 is the current one-year interest rate from year 0 to year 1; r_u and r_d are next year's possible interest rates from year 1 to year 2; and r_{uu}, r_{ud}, and r_{dd} are the one-year interest rates from year 2 to year 3 that are possible two years from now. (Note that in Chapter 7, the nodes are not recombined.)

Assume that the risk-neutral probability of up and down shifts is Q = ½ for each node. This assumption is explained in Chapter 7.

Options on bonds

6. Using the above tree, value a European call option and an American call option with two years to expiration and a strike price \$100. The underlying bond has a face value of \$100, an annual coupon of 6%, and three years to maturity.

Algorithm: Construct a bond tree and use the rates to value the possible future bond prices at various nodes where $B(2, 3)_{uu}$ refers to the three-year bond price at the end of year two, which matures at the end of year three if the years 2 to 3 interest rate is r_{uu}:

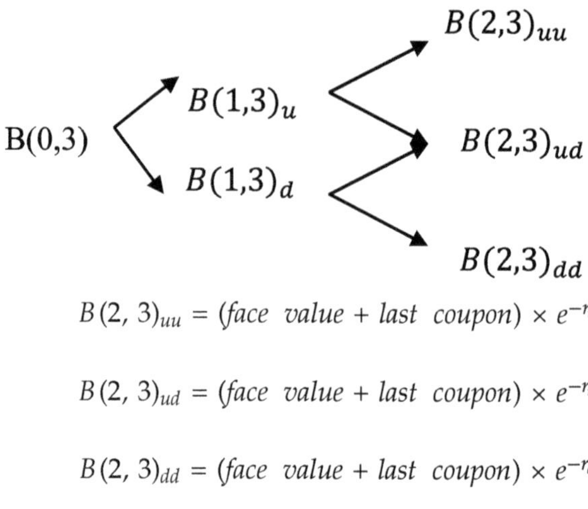

$$B(2, 3)_{uu} = (\textit{face value} + \textit{last coupon}) \times e^{-r_{uu}\times 1}$$

$$B(2, 3)_{ud} = (\textit{face value} + \textit{last coupon}) \times e^{-r_{ud}\times 1}$$

$$B(2, 3)_{dd} = (\textit{face value} + \textit{last coupon}) \times e^{-r_{dd}\times 1}$$

$$B(1, 3)_u = \left(\tfrac{1}{2} \times B(2, 3)_{uu} + \tfrac{1}{2} \times B(2, 3)_{ud} + \textit{2nd year coupon}\right) \times e^{-r_u\times 1}$$
$$B(1, 3)_d = \left(\tfrac{1}{2} \times B(2, 3)_{ud} + \tfrac{1}{2} \times B(2, 3)_{dd} + +\textit{2nd year coupon}\right) \times e^{-r_d\times 1}$$
$$B(0, 3) = \left(\tfrac{1}{2} \times B(1, 3)_u + \tfrac{1}{2} \times B(1, 3)_d + \textit{1st year coupon}\right) \times e^{-r_0\times 1}$$

Then construct the tree of option values:

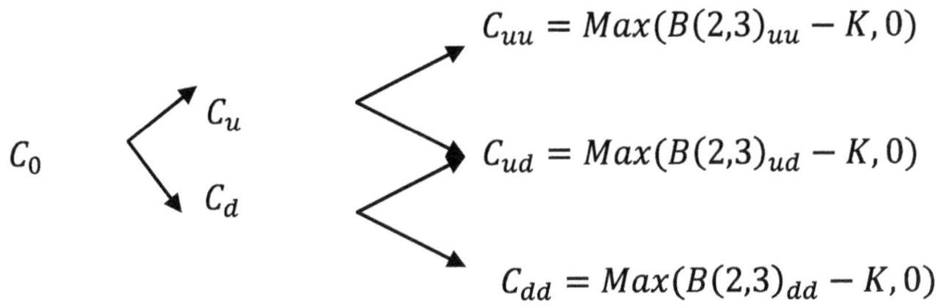

$$C_u = (\frac{1}{2} \times C_{uu} + \frac{1}{2} \times C_{ud}) \times e^{-r_u \times 1}$$

$$C_d = (\frac{1}{2} \times C_{ud} + \frac{1}{2} \times C_{dd}) \times e^{-r_d \times 1}$$

$$C_0 = (\frac{1}{2} \times C_u + \frac{1}{2} \times C_d) \times e^{-r_0 \times 1}$$

For an American option, at each node compare the value of the bond with its immediate exercise value and select the larger value to reflect the owner's choice of whether or not to exercise early:

$$C_u = Max\left\{B(1, 3)_u - K, \left(\frac{1}{2} \times C_{uu} + \frac{1}{2} \times C_{ud}\right) \times e^{-r_u \times 1}\right\}$$

$$C_d = Max\left\{B(1, 3)_d - K, \left(\frac{1}{2} \times C_{ud} + \frac{1}{2} \times C_{dd}\right) \times e^{-r_d \times 1}\right\}$$

$$C_0 = Max\left\{B(0, 3) - K, \left(\frac{1}{2} \times C_u + \frac{1}{2} \times C_d\right) \times e^{-r_0 \times 1}\right\}$$

Options on Interest Rates

7. Multiperiod ***interest rate caps and floors*** may be viewed as portfolios of European-style single period options on interest rates, called ***caplets*** and ***floorlets***, where a caplet is an n-year European call option with a strike r_K on notional principal and floorlet is an n-year European put option with a strike r_K on notional principal. Interest rate caps/floors are multiple caplets/floorlets with different maturities.

Value a two-year cap with an annual reset (containing a one-year caplet and a two-year caplet) and a strike rate of 4.5% using a notional principal value of $30 million dollars.

Algorithm: This solution contains a one-year caplet using a one-step interest rate tree and a two-year caplet using a two-step interest rate tree. Note that the two-year caplet is a two-year European call option with a strike r_K on notional principal. First, construct a two-year caplet tree:

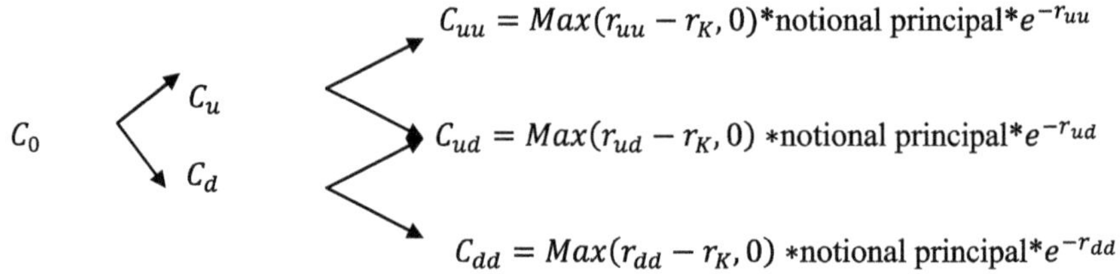

$$C_u = \left(\frac{1}{2} \times C_{uu} + \frac{1}{2} \times C_{ud}\right) \times e^{-r_u \times 1}$$

$$C_d = \left(\frac{1}{2} \times C_{ud} + \frac{1}{2} \times C_{dd}\right) \times e^{-r_d \times 1}$$

$$C_0 = \left(\frac{1}{2} \times C_u + \frac{1}{2} \times C_d\right) \times e^{-r_0 \times 1}$$

Next, construct a one-year caplet tree:

$$C'_u = Max(r_u - r_K, 0) \times notional\ \ principal \times e^{-r_u}$$

$$C'_d = Max(r_d - r_K, 0) \times notional\ \ principal \times e^{-r_d}$$

$$C'_0 = \left(\frac{1}{2} \times C_u + \frac{1}{2} \times C_d\right) \times e^{-r_0 \times 1}$$

The value of the cap is the sum of the two caplets: $C_0 + C'_0$. Note that this cap is slightly different from the caps in Chapter 7. In Chapter 7, a two-year cap uses the rates at the end of one year (parent node). If the tree is recombined, the parent node cannot be used since the parent node is not unique.

Note

1 http://www.hoadley.net/options/binomialvsbs.aspx (accessed 14.02.17.). NOTE: PERMISSION has been OBTAINED. Copyright © 2000 Peter Hoadley

4

More about Derivative Valuation

This chapter generalizes the previous chapter in four directions. First, it introduces more methods for valuing financial derivatives. However, it retains the assumption that a derivative's underlying asset price follows Geometric Brownian motion. Second, this chapter discusses various derivatives other than calls and puts. Third, this chapter considers an underlying asset that pays dividends and develops methods for valuing derivatives on that asset. Finally, the chapter discusses forward values as well as the relation between the futures price of an asset and the current forward contract price.

4.1 Various Methods to Value Derivatives

This section summarizes the tools that have been developed to value financial derivatives and it introduces two new valuation methods. Finally, the section compares the advantages and disadvantages of each model.

4.1.1 Overview of the Three Primary Derivative Valuation Approaches

We review three methods as followings:

4.1.1.1 Binomial Tree

We can use the binomial tree model, assuming the underlying asset price follows Geometric Brownian Motion, to find the value of a derivative. The first step is to build a binomial tree of the potential underlying asset prices forward through time, starting with the asset's initial price. Then we form a binomial tree of the derivative values calculating backward through time from its final payoffs to its current value.

Section 3.2 began with a one-step tree and showed that as the number of steps in the tree goes to infinity, the ending stock distribution, has a lognormal distribution. This statement is equivalent to the assumption that the underlying asset price follows Geometric Brownian Motion. Finally, Excel spreadsheets were demonstrated as a way to value financial derivatives using the tree method and to provide visualization of the process.

The valuation process is based on no-arbitrage valuation. Through constructing a risk-free portfolio for each step in the tree, the derivative's no-arbitrage value can be determined without specifying risk premiums. The process can be generalized to the Risk-neutral Principle in a discrete setting (n-step tree). *For the purpose of valuing financial derivatives, we can regard assets as being expected to grow at the risk-free rate and we can use risk-free interest rates to do all time value of money discounting.*

The advantages of the binomial tree method are its ease of computation and, more importantly, it can be used to value American-style derivatives. American-style derivatives can be exercised any time before maturity, unlike European style derivatives, which can only be exercised at the maturity date. Early exercise is modeled as follows: calculate the derivative values backward through the binomial tree while comparing the immediate exercise value with the discounted expected value of future cash flows at each node. The maximum of these two values represents the optimal decision of the option holder and therefore can be used to value American-style derivatives at each node. This chapter investigates the use of binomial tree models further by introducing an alternative binomial tree model for valuing derivatives.

4.1.1.2 Black-Scholes Partial Differential Equation (Black-Scholes PDE)

Assuming Geometric Brownian Motion for the underlying asset price, we derive the Black- Scholes partial differential equation, using a hedging argument and an instantaneous risk-free portfolio. The values of different financial derivatives can be derived with the same Black- Scholes PDE by imposing different boundary conditions. For example, European-style derivatives are valued with a boundary which corresponds to their payoffs at maturity while American-style derivatives are valued with an early exercise free boundary.

The Black-Scholes PDE does not require an assumption about the expected growth rate of the underlying asset; only the risk-free interest rate is needed. The Risk-neutral Principle is therefore applied in a continuous setting: *for derivative-valuation purposes, we can regard assets as having an expected growth rate equal to the risk-free rate and we can use risk-free interest rates to do all time value of money discounting.*

Recall that Section 3.4, sketches a proof of the Black-Scholes PDE. In the Black-Scholes PDE, the actual stock price growth rate, μ, is irrelevant to the derivative's value. Therefore, for simplicity, we can assume that we live in a risk-neutral world, where all asset prices are expected to grow at the risk-free rate and all future cash flows can be discounted at the risk-free rate.

From the Black-Scholes PDE and the Risk-Neutral Principle, there are two valuation methods derived. One is solving the PDE and the other is solving the integral directly. We first look at solving the integral. Based on the Risk-Neutral Principle the derivative's value is known to be the present value of the expected payoff where riskless rates are used both for projecting expected value and for discounting. For European-style derivatives, the expected value of the payoff at maturity, a continuous random variable, is an integral. So we can integrate the payoff with respect to the risk-neutral probability measure for European-style derivatives where the payoff is a deterministic function of the underlying asset price at maturity. If the integral has a closed form solution, an analytic method can be used to value the European-style derivative. If the integral does not have a closed form solution, a semi-analytical method can be used to value the European-style derivative. Semi-analytical integrals need to be solved by numerical approximation. Black-Scholes valuation of European put and call options can be derived by calculating the expected value of the derivative payoff. The integral, in this case, has a closed form solution.

Now let's look at the second valuation method: solving the PDE. Using the European option's potential payoffs as boundary conditions, one can derive the Black-Scholes formula for European call and put option values by solving the PDE. Chapter 3 did not discuss how to solve the PDE explicitly, since it is beyond the scope of this book. For arbitrary European-style financial derivatives with various payoffs, there may not be a closed form solution. Numerical methods are commonly used to solve derivative PDEs. In this chapter, the finite difference method is introduced. The finite difference method is a numerical method that can solve a PDE even for American-style options. The free boundary problem can be solved by working on grids.

4.1.1.3 Monte-Carlo Simulation

Recall that Chapter 3, begins by generating potential paths for the underlying asset price using Mont-Carlo simulation and assuming Geometric Brownian Motion. For European-style derivatives, we can then calculate the payoffs of the derivatives using the arithmetic average of the ending price for all paths. The payoffs are then discounted back by the risk-free interest rate to find the derivative value today. Note that in the process of deriving a binomial tree and a Black-Scholes PDE, we developed the Risk-Neutral Principle as an output. In Section 3.3, Monte-Carlo simulation was introduced for generating underlying asset paths. Derivative values were not estimated in Section 3.3 using simulation because the Risk-Neutral Principle, a necessary assumption, had not yet been derived in the section. Monte-Carlo simulation was used to find derivative prices at the end of Chapter 3 after the Risk-Neutral Principle had been established.

In this chapter, the Risk-Neutral Principle is used as an assumption for Monte-Carlo simulation. In other words, since we now know the Risk-Neutral Principle holds (because of the binomial tree method and the Black-Scholes PDE), this chapter can apply the Risk-Neutral Principle to perform a Monte-Carlo simulation for valuing *exotic* options.

Monte-Carlo simulation has three advantages in deriving derivative values: (1) it can model asset prices that do not follow Geometric Brownian Motion, (2) it can model derivative values with multiple underlying assets, and (3) it can value path-dependent financial derivatives since the whole paths are recorded. Binomial tree models and Black-Scholes PDEs cannot be used to value path-dependent derivatives. For example, in the binomial tree method, although there is a unique value at each node, there are multiple paths leading to that node.

Chapter 3 discusses three methods for valuing derivative securities. Important questions arise. First, do these methods yield the same values? We must consider what assumption is used to model the underlying asset and what valuation principle is used for the derivative. It is clear that Geometric Brownian Motion is assumed for the underlying asset in the Black-Scholes PDE. The binomial tree model appears different. However, through distribution-matching and mean-variance matching it can be shown that the price of the underlying asset in the binomial tree model follows the same process as Geometric Brownian Motion with the same parameters.

Finally, in the Monte-Carlo simulation method, Geometric Brownian Motion growing at the risk free rate can be used, which matches the other two methods. The second question is whether the valuation principles are consistent. Since all three methods are based on the Risk-Neutral Principle, it is not surprising that the estimated derivative values for each of the three different models, converge. This was shown in the Chapter 3 project.

In the next few sections, we introduce two new models. One is an alternative binomial tree which belongs to the Binomial tree approach, where the mean and variance are matched differently. The other is the finite difference method, which is a numerical method to solve a PDE and belongs to the Black-Scholes PDE approach.

4.1.2 Alternative Binomial Trees

Recall that in Section 3.2.7, as n, the number of steps in the tree, goes to infinity, the ending stock distribution S_T has a lognormal distribution, which is equivalent to Geometric Brownian Motion. The normal distribution has two parameters: mean and variance. We can solve for the parameters u (the tree up factor) and d (the tree down factor) such that the mean and variance of the tree's prices match the mean and variance of the assumed Geometric Brownian Motion. Section 3.2.7 set the up factor to $u = e^{\sigma\sqrt{\Delta t}}$, the down factor to $d = e^{-\sigma\sqrt{\Delta t}}$ and the risk-neutral probability to $Q = \frac{e^{r\Delta t} - d}{u - d}$, giving:

$$EX_i = \mathbf{Q} \times Lnu + (1 - \mathbf{Q}) \times Lnd = r \times \Delta t - \frac{1}{2}\sigma^2 \times \Delta t \tag{4.1}$$

$$\mathbf{Var}X_i = \mathbf{Q} \times (Lnu)^2 + (\mathbf{1} - \mathbf{Q}) \times (Lnd)^2 - (EX_i)^2 = \sigma^2 \Delta t \tag{4.2}$$

Therefore the binomial tree method with u and d specified as above generates a process for the derivative's underlying asset that matches the mean and variance in Geometric Brownian Motion.

Note that there are three unknown variables (u, d and Q) in Equations 4.1 and 4.2. Because two equations are insufficient to solve for three unknowns, a third equation must be identified. There are two commonly established approaches to identifying a third equation in order to find a unique solution.

One approach is to restrict the relation between u and d to satisfy the equation:

$$ud = \mathbf{1} \tag{4.3}$$

Including Equation 4.3, there are three equations in u, d and Q and therefore 4.1, 4.2 and 4.3 can generate a solution, namely:

$$u = e^{\sigma\sqrt{\Delta t}},\ d = e^{-\sigma\sqrt{\Delta t}},\ Q = \frac{e^{r\Delta t} - d}{u - d} \tag{4.4}$$

The other approach is to impose that the up and down shifts have equal risk-neutral probabilities as indicated in Equation 4.5.

$$\mathbf{Q} = \mathbf{1/2} \tag{4.5}$$

Note that as long as 4.1 and 4.2 are assumed, the tree has the same distribution as Geometric Brownian Motion. The third equation (i.e., the choice between the two approaches) can be based on intuition or convenience.

Imposing Q = 1/2 leads to an alternative binomial tree with different parameters. Each step of the tree, including the last column $Ln(S_T)$, follows a Bernoulli distribution. Normality comes from the Bernoulli distribution of the tree. As n goes to infinity, the binomial distribution approaches the normal distribution. The tree can be forced to yield the same mean and variance of the normal distribution and generate a lognormal distribution with mean $LnS_0 + r \times T - \frac{1}{2}\sigma^2 T$ and variance $\sigma^2 T$.

It is not hard to see that inserting from the following Equation 4.6 into the mean and variance equations 4.1 and 4.2 will satisfy the goals. This can be shown using a Taylor expansion "trick" in a problem at the end of this chapter.

$$u = e^{\left(r - \frac{1}{2}\sigma^2\right)\Delta t + \sigma\sqrt{\Delta t}},\ d = e^{\left(r - \frac{1}{2}\sigma^2\right)\Delta t - \sigma\sqrt{\Delta t}},\ Q = \mathbf{1/2} \tag{4.6}$$

We refer to this new way to construct a binomial tree as the ***alternative binomial tree*** method.

Consider an example.

Example 4.1: Use a 30-period alternative binomial tree to value a European put:
A stock price is currently \$50. Over the next six-months, the stock price is expected to follow a binomial tree of 30 steps. The stock's annual volatility is 20% and the risk-free rate is 5% per annum with continuous compounding. What is the value of a six-month *European* put option with a strike price of \$52?

The given parameters are used with the equations in Equation 4.6 to generate u, d and Q. Using a spreadsheet:

S	K	r	T	Sigma	n	delta t	u	d	Q
50	52	0.05	0.5	0.2	30	0.016667	1.026669	0.974998	0.5

The spreadsheet with $n = 30$ calculates the put value: $P = 3.23399052$, while the Black-Scholes value is 3.211157. Note that we only need to change three cells from the CRR tree we introduced in Chapter 3: u, d and Q.

Example 4.2: Analysis of the convergence for the same put over different n (steps):

The value of the above European put can be calculated for any number of periods (steps). By graphing the European put value as a function of n, we can analyze the convergence of the put values as shown in Figure 4.1.

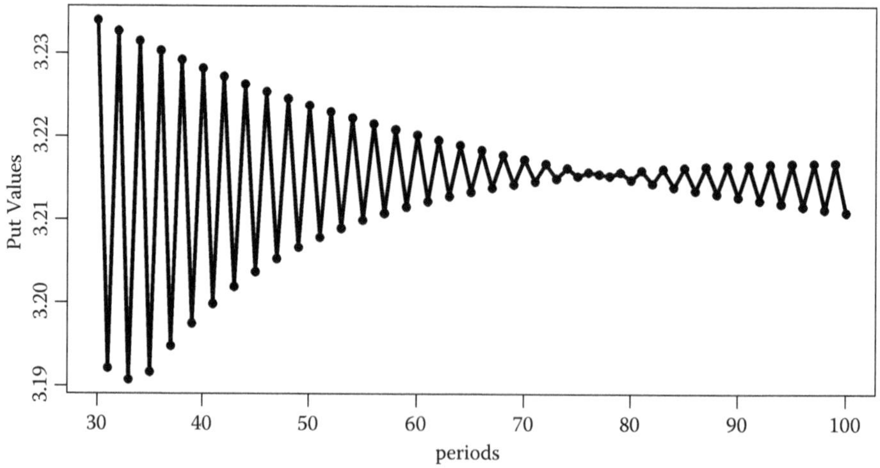

FIGURE 4.1
The Convergence Pattern of an Alternative Binomial Tree.

Note that the convergence is not monotonic. The pattern is similar to that for a Binomial tree shown in Section 3.2.8. Here are two convergence tables for two types of put options (American and European) over ranges of n from 30 to 39, and from 300 to 309:

n	**put value European**	**put value American**
30	3.23399052	3.42859438
31	3.19211283	3.39979894
32	3.23270131	3.42739284
33	3.19074562	3.40030208
34	3.23148391	3.4261053
35	3.19165219	3.4013562
36	3.23033294	3.4247776
37	3.19478627	3.4030779
38	3.22924332	3.42369841
39	3.19749523	3.4046457

	European	**American**
300	3.21344694	3.41126906
301	3.20902642	3.4083874
302	3.21343459	3.41125512
303	3.20895068	3.40836587
304	3.21342149	3.41124035
305	3.20887532	3.40836932
306	3.21340766	3.41122514
307	3.20881555	3.40837768
308	3.21339314	3.4112095
309	3.20891884	3.40844324

Note that within n from 30 to 39 and within n from 300 to 309 the estimated option values bounce up and down rather than quickly converging. However, comparing the results of n from 30 to 39 to the results of n from 300 to 309 indicates that the model is converging in the long-term.

The advantage of this alternative binomial tree is that there is no need to be concerned about probabilities outside of the (0,1) boundary. This is particularly useful when considering an underlying asset that pays a dividend (which is detailed later in this chapter). The binomial tree from Chapter 3 may have an invalid probability in the case of a dividend-paying underlying stock. It is interesting to investigate which trees or methods converge the fastest. However, this is beyond the scope of this book.

In theory, many kinds of tree methods can be generated by varying the third equation. These two tree methods are the popular ones. The next section discusses the numerical approximation method for the Black-Scholes PDE: the finite difference method.

4.1.3 Using the Finite Difference Method to Solve the Black-Scholes PDE

Recall that Section 3.4 derived and discussed the Black-Scholes PDE. The derivative price $f(S_t, t)$ as a function of the price of the underlying asset, S_t, and time t, follows this partial differential equation:

$$\frac{\partial f}{\partial S} rS + \frac{\partial f}{\partial t} + \frac{1}{2}\frac{\partial^2 f}{\partial S^2}\sigma^2 S^2 = rf.$$

A European call option has the following boundary conditions:

$$f(S_T, T) = \text{Max}(S_T - K, 0)$$

$$\lim_{S_t \to 0^+} f(S_t, t) = 0.$$

$$\lim_{S_t \to \infty} f(S_t, t)/S_t = 1$$

The above differential equation can be transformed to the standard heat equation and then solved, however, this method of solution is beyond the scope of this book. Rather, we solve this differential equation numerically, using the finite difference method as in Hull's book.

In mathematics, ***finite-difference methods*** (FDMs) are numerical methods for solving differential equations. The method approximates derivatives by approximating their finite differences. There are two different methods used to approximate: the ***explicit finite difference method*** and the ***implicit finite difference method***. We will discuss them separately.

- The ***Implicit finite-difference*** approximation method:

Consider a grid on the XY plane with the x-axis mapping the asset value and the y-axis mapping time. Assume that a financial derivative is a function of the underlying asset price and time: f(S, t). Begin by finding the value of $f(S_i, t_j)$ at each grid point (S_i, t_j).

Let $f_{i,j}$ represent $f(S_i, t_j)$ where S_i, $0 \le i \le m$ are equally spaced values between $[0, S_{\text{Max}}]$ such that $S_0 = 0$ *and* $S_m = S_{\text{Max}}$. S_{Max} is the largest expected S before maturity and $\Delta S = \frac{S_{Max}}{m}$. Similarly, t_j, $0 \le j \le n$ describes equally spaced values between [0, T] where $t_0 = 0$ *and* $t_n = T$, the maturity, and $\Delta t = \frac{T}{n}$.

Let $t_j = j\Delta t$ where $0 \le j \le n$ and $S_i = i\Delta S$ where $0 \le i \le m$. On the x-axis, m +1 points are indexed by 0 to m and on the y-axis, n+1 points are indexed by 0 to n. Note that it is necessary to pick S_{Max} as an integer multiple of today's underlying price, S, so that today's price is one of the S_i. Here the notation is altered by denoting today's underlying asset value as S_i for some i where $0 \le i \le m$ instead of S_0, as previously used.

The following diagram depicts the grids for f_{ij} (Figure 4.2):

If finite differences are used to replace the partial derivatives:

$$\frac{\partial f}{\partial t} = \frac{f_{i,j+1} - f_{i,j}}{t_{j+1} - t_j} \tag{4.7}$$

$$\frac{\partial f}{\partial S} = \frac{f_{i+1,j} - f_{i-1,j}}{S_{i+1} - S_{i-1}} \tag{4.8}$$

$$\frac{\partial^2 f}{\partial S^2} = \frac{f_{i+1,j} + f_{i-1,j} - 2 \times f_{i,j}}{(S_{i+1} - S_i)^2} \tag{4.9}$$

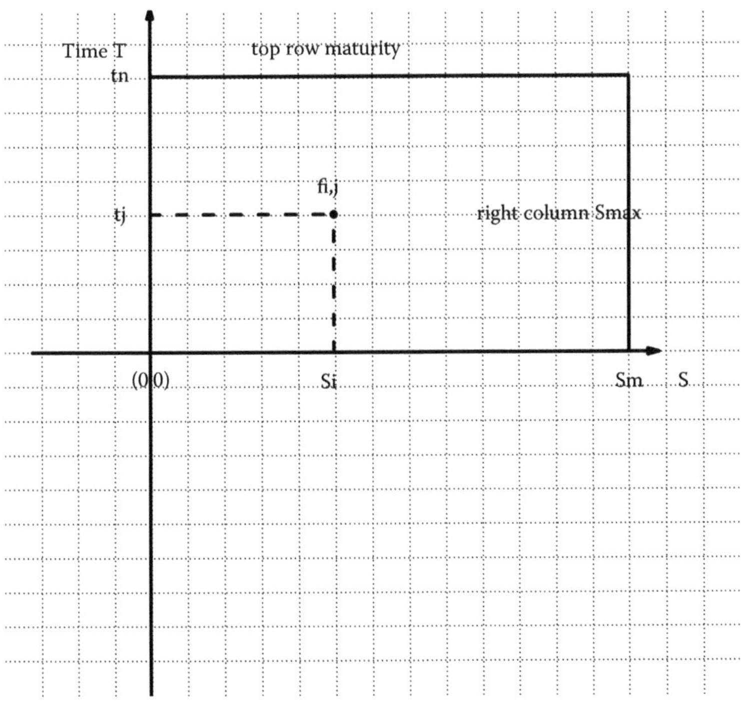

FIGURE 4.2
Grid of Finite Difference Method.

Then the above equations can be combined with the Black-Scholes PDE:

$$\frac{f_{i+1,j} - f_{i-1,j}}{S_{i+1} - S_{i-1}} \times r \times S_i + \frac{f_{i,j+1} - f_{i,j}}{t_{j+1} - t_j} + \frac{1}{2}\frac{f_{i+1,j} + f_{i-1,j} - 2 \times f_{i,j}}{(S_{i+1} - S_i)^2}\sigma^2 S_i^2 = rf_{i,j}$$

After inserting $t_j = j\Delta t$ and $S_i = i\Delta S$:

$$\left(f_{i+1,j} - f_{i-1,j}\right) \times \frac{1}{2} \times r \times i + \frac{f_{i,j+1} - f_{i,j}}{\Delta t}$$
$$+ \frac{1}{2}\left(f_{i+1,j} + f_{i-1,j} - 2 \times f_{i,j}\right) \times \sigma^2 \times i^2 = rf_{i,j}.$$

Expanding yields:

$$\frac{1}{2}ri\Delta tf_{i+1,j} - \frac{1}{2}ri\Delta tf_{i-1,j} + f_{i,j+1} - f_{i,j} + \frac{1}{2}\sigma^2 \times i^2\Delta tf_{i+1,j} + \frac{1}{2}\sigma^2 \times i^2\Delta tf_{i-1,j}$$
$$- \sigma^2 \times i^2\Delta tf_{i,j} = r\Delta tf_{i,j}.$$

Combining like terms and simplifying,

$$\left(\frac{1}{2}ri\Delta t - \frac{1}{2}\sigma^2 \times i^2\Delta t\right)f_{i-1,j} + (r\Delta t + 1 + \sigma^2 \times i^2\Delta t)f_{i,j}$$
$$+ \left(-\frac{1}{2}ri\Delta t - \frac{1}{2}\sigma^2 \times i^2\Delta t\right)f_{i+1,j} = f_{i,j+1}$$

If we define $a_i = \frac{1}{2}ri\Delta t - \frac{1}{2}\sigma^2 i^2\Delta t,$ $b_i = r\Delta t + 1 + \sigma^2 \times i^2\Delta t,$ and $c_i = -\frac{1}{2}ri\Delta t - \frac{1}{2}\sigma^2 i^2\Delta t,$ then $a_i f_{i-1,j} + b_i f_{i,j} + c_i f_{i+1,j} = f_{i,j+1}.$

The Implicit Finite Difference method:

In order to solve today's derivative values $f_{i,0}$, at $t_0 = 0$ we just need to solve the system of linear equations:

$$a_i f_{i-1,j} + b_i f_{i,j} + c_i f_{i+1,j} = f_{i,j+1} \text{ for } 1 \le i \le m - 1 \text{ and } 0 \le j \le n - 1 \quad (4.10)$$

where, $a_i = \frac{1}{2}ri\Delta t - \frac{1}{2}\sigma^2 i^2 \Delta t,\ b_i = 1 + \sigma^2 i^2 \Delta t + r\Delta t,\ c_i = -\frac{1}{2}ri\Delta t - \frac{1}{2}\sigma^2 i^2 \Delta t.$

Note that we treat the $j+1^{th}$ row values $f_{i,j+1}$ as known values and the jth row values $f_{i-1,j}$, $f_{i,j}$, $f_{i+1,j}$ as unknown values. We start with the top row (n^{th} row), which represents the values of the derivative payoffs at maturity and solve the system of equations, row by row, from top to bottom.

Next, make sure that the number of equations and the number of unknown variables match. Note from inspecting the values of the grid, $f_{i,j}$, that the $f_{i,j}$ on the three boundaries are known. A call option is used as an example to illustrate the method.

On the top horizontal line, t = T, so $f_{i,n} = f(S_i, T) = \text{Max}(S_i - K, 0)$ (Figure 4.3).

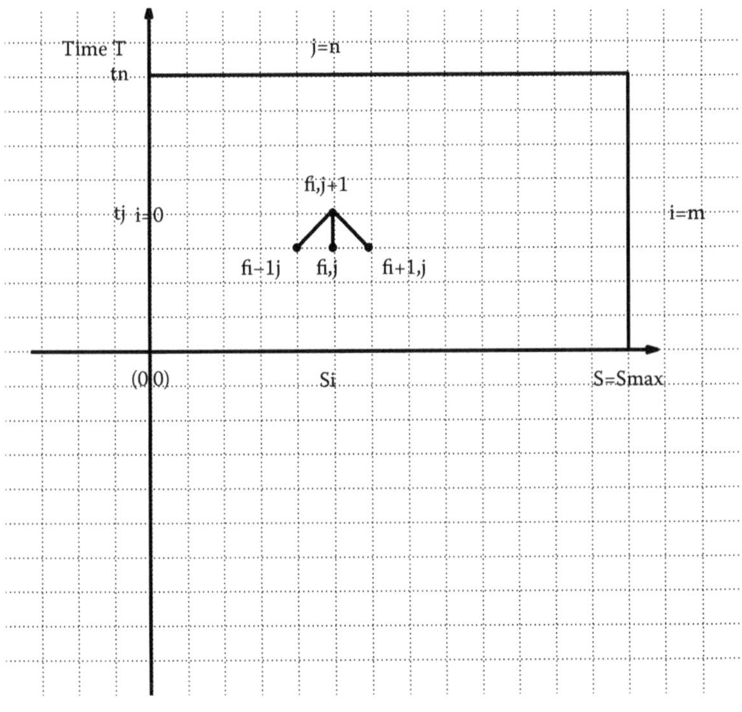

FIGURE 4.3
Implicit Finite Difference Method.

From the left vertical line (i = 0) it is clear that $f_{0,j} = 0$ because $S_t = 0$. There is no chance that S_T will be above K. Now consider the right vertical line i=m. For a call, since it is assumed that the upper bound for the stock is S_{max}, for large S_{max}: $f_{m,j} = \text{Max}(S_{max} - K, 0) = S_{max} - K$.

Now count the number of unknowns. There are three boundaries containing $(m + 1) + n + n$ elements of $f_{i,j}$. The unknowns are: $(m + 1)\times(n + 1) - m - 2n - 1 = mn - n = (m - 1)n$. Now consider the number of equations: $1 \le i \le m - 1$ and $0 \le j \le n - 1$. There are (m – 1)n equations. So, we will be able to solve all $f_{i,j}$, where $f_{i,0}$ is the call value at time 0 with different initial prices S_i. Since one of the S_i is the underlying price today, that corresponding $f_{i,0}$ is today's call value.

Note that in order to get option values, it is necessary to solve $(m - 1)n$ linear system of equations. This can be easily programmed in MATLAB or Mathematica. In fact, it is simple to work row by row starting from the second to the top row where $j = n - 1$ and solving each row as a system of linear equations in a matrix form. For example, the values are known for the top row (nth row). Our objective is to solve $f_{i,n-1}$. (For the 2nd to the top row, there are $m - 1$ unknowns since the left and right boundary elements are known):

$$
\begin{aligned}
&a_1 f_{0,n-1} + b_1 f_{1,n-1} + c_1 f_{2,n-1} = f_{1,n} \\
&a_2 f_{1,n-1} + b_2 f_{2,n-1} + c_2 f_{3,n-1} = f_{2,n} \\
&a_3 f_{2,n-1} + b_3 f_{3,n-1} + c_3 f_{4,n-1} = f_{3,n} \\
&\ldots \\
&a_{m-2} f_{m-3,n-1} + b_{m-2} f_{m-2,n-1} + c_{m-2} f_{m-1,n-1} = f_{m-2,n} \\
&a_{m-1} f_{m-2,n-1} + b_{m-1} f_{m-1,n-1} + c_{m-1} f_{m,n-1} = f_{m-1,n}
\end{aligned}
$$

Written in matrix form:

$$
\begin{pmatrix}
b_1 & c_1 & 0 & 0 & 0 & 0 & \ldots & 0 & 0 \\
a_2 & b_2 & c_2 & 0 & 0 & 0 & \ldots & 0 & 0 \\
0 & a_3 & b_3 & c_3 & 0 & 0 & \ldots & 0 & 0 \\
\ldots & \ldots & \ldots & \ldots & \ldots & \ldots & \ldots & \ldots & \ldots \\
0 & 0 & 0 & 0 & 0 & \ldots & a_{m-2} & b_{m-2} & c_{m-2} \\
0 & 0 & 0 & 0 & 0 & 0 & \ldots & a_{m-1} & b_{m-1}
\end{pmatrix}
\begin{pmatrix}
f_{1,n-1} \\ f_{2,n-1} \\ f_{3,n-1} \\ f_{4,n-1} \\ \ldots \\ f_{m-2,n-1} \\ f_{m-1,n-1}
\end{pmatrix}
=
\begin{pmatrix}
f_{1,n} - a_1 f_{0,n-1} \\ f_{2,n} \\ f_{3,n} \\ f_{4,n} \\ \ldots \\ f_{m-2,n} \\ f_{m-1,n} - c_{m-1} f_{m,n-1}
\end{pmatrix}.
$$

We define the matrix A = $\begin{pmatrix} b_1 & c_1 & 0 & 0 & 0 & 0 & \dots & 0 & 0 \\ a_2 & b_2 & c_2 & 0 & 0 & 0 & \dots & 0 & 0 \\ 0 & a_3 & b_3 & c_3 & 0 & 0 & \dots & 0 & 0 \\ \dots & \dots & \dots & \dots & \dots & \dots & \dots & \dots & \dots \\ 0 & 0 & 0 & 0 & 0 & \dots & a_{m-2} & b_{m-2} & c_{m-2} \\ 0 & 0 & 0 & 0 & 0 & 0 & \dots & a_{m-1} & b_{m-1} \end{pmatrix}$

$$\vec{f_j} = \begin{pmatrix} f_{1,j} \\ f_{2,j} \\ f_{3,j} \\ f_{4,j} \\ \dots \\ f_{m-2,j} \\ f_{m-1,j} \end{pmatrix}$$

$$\overrightarrow{s_{j-1}} = \begin{pmatrix} a_1 f_{0,j-1} \\ 0 \\ 0 \\ 0 \\ \dots \\ 0 \\ c_{m-1} f_{m,j-1} \end{pmatrix}$$

We have $\vec{f_j} = A^{-1}(\overrightarrow{f_{j+1}} - \vec{s_j})$. Note that $f_{0,j}$ and $f_{m,j}$ are two vertical boundaries that are known, so $\vec{s_j}$ is known for all j. Matrix A can also be calculated and it is a function of i, not j.

Starting with j = n − 1, we know $\vec{f_n}$, and we can solve $\overrightarrow{f_{n-1}}$. Recursively, we can find a formula for the price of any European-style financial derivatives as long as we know three boundaries:

Formula for Implicit Finite Difference Method to Value European-style Financial Derivatives

$$\overrightarrow{f_0} = (A^{-1})^n \overrightarrow{f_n} - (A^{-1})^n \overrightarrow{s_{n-1}} - (A^{-1})^{n-1} \overrightarrow{s_{n-2}} - \cdots - (A^{-1})^2 \overrightarrow{s_1} - A^{-1} \overrightarrow{s_0} \tag{4.11}$$

where

$$\text{matrix A} = \begin{pmatrix} b_1 & c_1 & 0 & 0 & 0 & 0 & \cdots & 0 & 0 \\ a_2 & b_2 & c_2 & 0 & 0 & 0 & \cdots & 0 & 0 \\ 0 & a_3 & b_3 & c_3 & 0 & 0 & \cdots & 0 & 0 \\ \cdots & \cdots & \cdots & \cdots & \cdots & \cdots & \cdots & \cdots & \cdots \\ 0 & 0 & 0 & 0 & 0 & \cdots & a_{m-2} & b_{m-2} & c_{m-2} \\ 0 & 0 & 0 & 0 & 0 & 0 & \cdots & a_{m-1} & b_{m-1} \end{pmatrix},$$

$$\text{the top boundary } \overrightarrow{f_n} = \begin{pmatrix} f_{1,n} \\ f_{2,n} \\ f_{3,n} \\ f_{4,n} \\ \cdots \\ f_{m-2,n} \\ f_{m-1,n} \end{pmatrix},$$

the left and right boundaries are included in:

$$\overrightarrow{s_j} = \begin{pmatrix} a_1 f_{0,j} \\ 0 \\ 0 \\ 0 \\ \cdots \\ 0 \\ c_{m-1} f_{m,j} \end{pmatrix},$$

$a_i = \frac{1}{2} ri\Delta t - \frac{1}{2}\sigma^2 i^2 \Delta t$, $b_i = 1 + \sigma^2 i^2 \Delta t + r\Delta t$, $c_i = -\frac{1}{2} ri\Delta t - \frac{1}{2}\sigma^2 i^2 \Delta t$ for $1 \leq i \leq m - 1$.

The implicit finite difference method converges faster than the explicit finite difference method as is introduced below - although the faster convergence is not established in detail since it is beyond the scope of this book.

For European puts, the only change is the top, left and right boundaries.

Alternatively, we can use various finite differences to replace the partial derivatives and in so doing utilize the explicit finite difference method, discussed next.

- The ***explicit finite difference*** method:

Like the implicit finite difference method, the explicit finite difference method requires construction of a grid. We have $f_{i,j}$ where $0 \le i \le m$, $0 \le j \le n$. If we use:

$$\frac{\partial f}{\partial t} = \frac{f_{i,j+1} - f_{i,j}}{t_{j+1} - t_j} \tag{4.12}$$

$$\frac{\partial f}{\partial S} = \frac{f_{i+1,j+1} - f_{i-1,j+1}}{S_{i+1} - S_{i-1}} \tag{4.13}$$

$$\frac{\partial^2 f}{\partial S^2} = \frac{f_{i+1,j+1} + f_{i-1,j+1} - 2f_{i,j+1}}{(S_{i+1} - S_i)^2} \tag{4.14}$$

by inserting them into the Black-Scholes PDE:

$$\frac{f_{i+1,j+1} - f_{i-1,j+1}}{S_{i+1} - S_{i-1}} \times r \times S_i + \frac{f_{i,j+1} - f_{i,j}}{t_{j+1} - t_j} + \frac{1}{2}\frac{f_{i+1,j+1} + f_{i-1,j+1} - 2 \times f_{i,j+1}}{(S_{i+1} - S_i)^2}\sigma^2 S_i^2$$
$$= rf_{i,j}.$$

Substituting $t_j = j\Delta t$ and $S_i = i\Delta S$ yields:

$$\left(f_{i+1,j+1} - f_{i-1,j+1}\right) \times \frac{1}{2} \times r \times i + \frac{f_{i,j+1} - f_{i,j}}{\Delta t}$$
$$+ \frac{1}{2}\left(f_{i+1,j+1} + f_{i-1,j+1} - 2 \times f_{i,j+1}\right) \times \sigma^2 \times i^2 = rf_{i,j}.$$

Expanding as follows:

$$\frac{1}{2}ri\Delta t f_{i+1,j+1} - \frac{1}{2}ri\Delta t f_{i-1,j+1} + f_{i,j+1} - f_{i,j} + \frac{1}{2}\sigma^2 \times i^2\Delta t f_{i+1,j+1}$$
$$+ \frac{1}{2}\sigma^2 \times i^2\Delta t f_{i-1,j+1} - \sigma^2 \times i^2\Delta t f_{i,j+1} = r\Delta t f_{i,j}.$$

and then combining like terms generates:

$$\left(-\frac{1}{2}ri\Delta t + \frac{1}{2}\sigma^2 \times i^2\Delta t\right)f_{i-1,j+1} + (1 - \sigma^2 \times i^2\Delta t)f_{i,j+1} + \left(\frac{1}{2}ri\Delta t + \frac{1}{2}\sigma^2 \times i^2\Delta t\right)f_{i+1,j+1} = (1 + r\Delta t)f_{i,j}.$$

Next, define:

$$a_i^* = \frac{-\frac{1}{2}ri\Delta t + \frac{1}{2}\sigma^2 i^2\Delta t}{1 + r\Delta t},\ b_i^* = \frac{1 - \sigma^2 i^2\Delta t}{1 + r\Delta t},\ c_i^* = \frac{\frac{1}{2}ri\Delta t + \frac{1}{2}\sigma^2 i^2\Delta t}{1 + r\Delta t}.$$

Thus, the simplified form is written as:

$$a_i^* f_{i-1,j+1} + b_i^* f_{i,j+1} + c_i^* f_{i+1,j+1} = f_{i,j}.$$

The Explicit finite difference method:

In order to calculate today's derivative values $f_{i,0}$, at $t_0 = 0$, it is necessary to recursively solve the following equation:

$$f_{i,j} = a_i^* f_{i-1,j+1} + b_i^* f_{i,j+1} + c_i^* f_{i+1,j+1} \tag{4.15}$$

where $1 \le i \le m - 1,\ 0 \le j \le n - 1$

and $a_i^* = \frac{-\frac{1}{2}ri\Delta t + \frac{1}{2}\sigma^2 i^2\Delta t}{1 + r\Delta t},\ b_i^* = \frac{1 - \sigma^2 i^2\Delta t}{1 + r\Delta t},\ c_i^* = \frac{\frac{1}{2}ri\Delta t + \frac{1}{2}\sigma^2 i^2\Delta t}{1 + r\Delta t}.$

Treat the j+1 th row values $f_{i-1,j+1}$, $f_{i,j+1}$ and $f_{i+1,j+1}$ as known values and the jth row values $f_{i,j}$ as unknown values. Starting with the top row, which has derivative payoffs at maturity, calculate back, row by row, from the top to the bottom as weighted sums.

Examining the values of the $f_{i,j}$ at each grid point for a call option, we recognize that $f_{i,j}$ on the three boundaries is known. A call option provides an example (Figure 4.4).

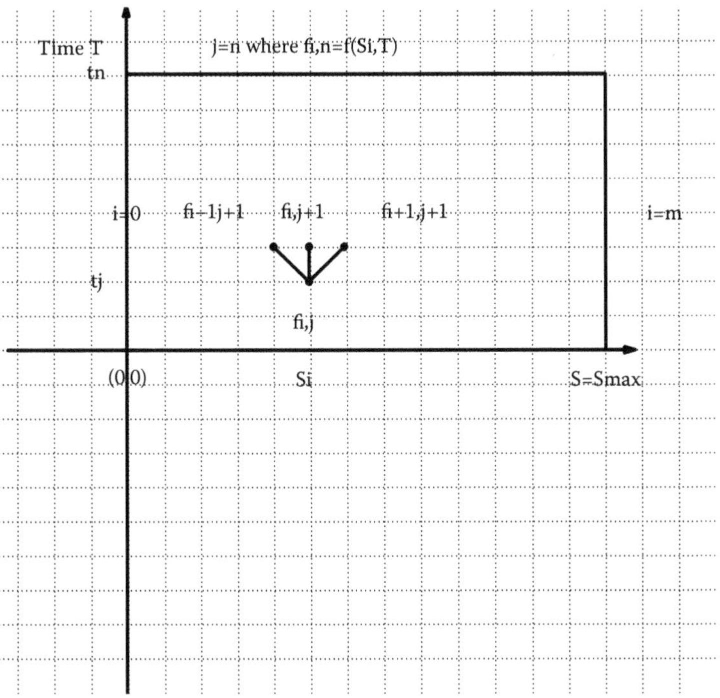

FIGURE 4.4
Explicit Finite Difference Method.

Since $f_{0,j} = 0$ because $S_t = 0$, there is no chance S_T will be above K. Assume an upper bound for the stock, $S_{\max}$. Thus, for large $S_{\max}$: $f_{m,j} = \text{Max}(S_{\max} - K, 0) = S_{\max} - K$.

We could interpret a_i^*, b_i^* and c_i^* as weights and then recursively solve the rows down until $f_{i,0}$ (the call value at time 0 for different initial prices S_i) is obtained.

The explicit finite difference method is easy to calculate; it does not require solving a linear system of equations. The explicit finite difference method is equivalent to a ***trinomial tree*** calculation where each parent node has three children nodes. However, the explicit method has slower convergence than the implicit method.

Consider an example that uses the explicit finite difference method:

Example 4.3: Use a 6x6 grid to value the following European put using the explicit difference method. The underlying stock is currently \$50 ($S_{\max} = 150$). Over the next six-months, the stock price is expected to follow a Geometric Brownian Motion. The stock's annual volatility is 20%. The risk-free rate is 5% per annum with continuous compounding. What is the value of a six-month *European* put option with a strike price of \$52?

Note that a_i^*, b_i^* and c_i^* are functions of i and $1 \le i \le 5$, and that:

$$T = 0.5, \Delta t = \frac{0.5}{6} = 0.08333, S_{\max} = 150, \text{ and } \Delta S = \frac{150}{6} = 25.$$

Using a spreadsheet, the value of the above put can be found:

Spreadsheet 4.1

T	K	S(0)	S(min)	S(max)	r	sigma	N	M	*dt*	dS
0.5	52	50	0	150	0.05	0.2	6	6	0.083	25
	1	2	3	4	5					
a*	-0.00041(c)	0.00249	0.008714	0.018257	0.03112					
b*	0.992531(c)	0.982573	0.965975	0.942739	0.912863					
c*	0.003734(c)	0.010788	0.021162	0.034855	0.051867					
T	in the money						out of money			
0.5	52(c)	27(c)	2	0	0	0	0(c)			
0.416667	52	26.78423	2.032365	0.017427	0	0(c)	0	j=5		
0.333333	52	26.5702	2.063817	0.034544	0.000318	0	0	j=4		
0.25	52	26.35788	2.094373	0.051359	0.000931	9.9E-06	0	j=3		
0.166667	52	26.14726	2.124049	0.067881	0.001815	3.8E-05	0	j=2		
0.083333	52	25.93832	2.152861	0.084118	0.002952	9.12E-05	0	j=1		
0	52	25.73106	2.180827	0.100077	0.004322	0.000175	0	j=0		
S	0	25	50	75	100	125	150			
	i=0	i=1	i=2	i=3	i=4	i=5	i=6			

The calculation of this spreadsheet is not as intensive as it appears. To generate a^*, b^*, c^*, we only need to calculate the first column to the right of the cells containing a^*, b^*, c^* (marked with a (c) in the top part of the table) and then ***copy*** the formulas across the remaining cells. Also, we can find three boundaries of the put value. In fact, the cell marked with a (c) and with a value of 52 is created and the column below the cell is ***copied and pasted***; the cell marked with a (c) and with a value of 27 is created and the row right to the cell is ***copied and pasted***; the cell marked with a (c) and with a value of 0 on the right boundary is created and the column below the cell is ***copied and pasted***. Then, calculate the 2nd row from the top of the grid, starting with the cell marked with a (c) and with a value of 0. Then ***copy*** the formulas across the columns in the second top row left to the cell and ***copy*** the formulas down the rows. It should only take a few minutes to build the whole table. Including boundaries, it is a 7 x 7 table. More generally, it is an $(m+1) \times (n+1)$ table. (Note that to be strict, the left boundary could be e to the power $Ke^{(-r(T-t))}$ instead of K.)

In generating the two tables, only 7 cells marked with (c) are calculated, the rest is copied and pasted.

The row near the bottom with t = 0 gives the current put value at t = 0 for different initial underlying asset prices, *S*. In the case where the initial underlying asset price, S_0, is equal to 50, the row shows that the value of the put option is 2.180827 using the explicit difference method.

The 6x6 grid does not converge to its Black-Scholes put price.

Consider another example.

Example 4.4: Analyze the convergence for the same put from Example 4.3. We can calculate the value for any number of grid dimensions. For the sake of simplicity, only the results for $m = n$ are shown. By graphing the European put value as a function of the grid dimension, *m*, the convergence of the put value can be analyzed. Assume that $S_{max} = 150$.

N	P
15	2.92655606
30	3.2074515
45	3.21434755
60	3.20102739
75	3.18830041
90	3.20731566
105	3.21306464

Figure 4.5 graphs the above data.

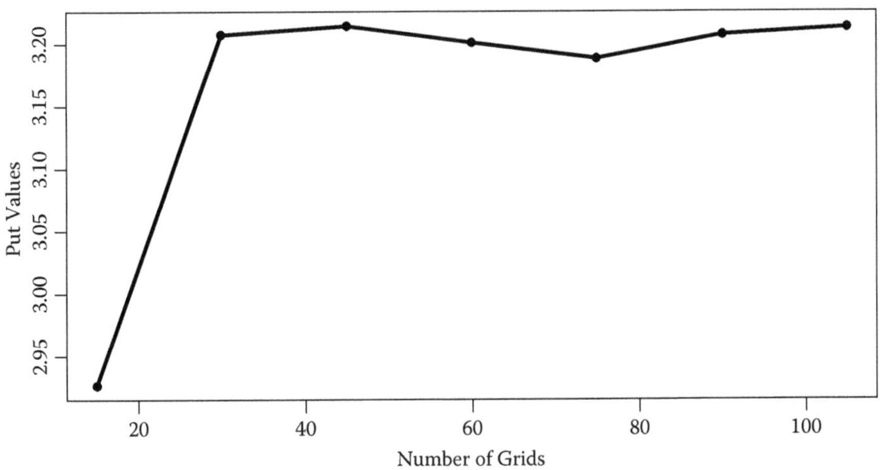

FIGURE 4.5
Convergence Pattern of the Explicit Finite Difference Method.

The Black-Scholes value for this put is 3.211157. The explicit difference method is quite accurate after $m = n = 30$. It is not surprising that the convergence is not monotonic, since the explicit finite difference method is equivalent to a trinomial tree.

Example 4.5: Use the implicit finite difference method to value a European put in a 6x6 grid. The underlying stock price is currently trading at \$50 (assume that $S_{max} = 150$). Over the next six-months, the stock price is expected to follow a geometric Brownian motion. The stock's annual volatility is 20%. The risk-free rate is 5% per annum with continuous compounding. What is the value of a six-month *European* put option with a strike price of \$52?

From the algorithm of the implicit finite difference method, the top row values of $f_{i,6}$ (where i = 1, 2, 3, 4, 5) are the payoffs at maturity for the put (which is known).

$$\begin{aligned}
a_1 f_{0,5} + b_1 f_{1,5} + c_1 f_{2,5} &= f_{1,6} \\
a_2 f_{1,5} + b_2 f_{2,5} + c_2 f_{3,5} &= f_{2,6} \\
a_3 f_{2,5} + b_3 f_{3,5} + c_3 f_{4,5} &= f_{3,6} \\
a_4 f_{3,5} + b_4 f_{4,5} + c_4 f_{5,5} &= f_{4,6} \\
a_5 f_{4,5} + b_5 f_{5,5} + c_5 f_{6,5} &= f_{5,6}
\end{aligned}$$

We know that:

$$\vec{f_6} = \begin{pmatrix} f_{1,6} \\ f_{2,6} \\ f_{3,6} \\ f_{4,6} \\ f_{5,6} \end{pmatrix} = \begin{pmatrix} 27 \\ 2 \\ 0 \\ 0 \\ 0 \end{pmatrix}$$

$$\vec{S_5} = \begin{pmatrix} a_1 f_{0,5} \\ 0 \\ 0 \\ 0 \\ c_5 f_{6,5} \end{pmatrix}$$

Calculating the value of a, b, and c in the spreadsheet:

	1	2	3	4	5
a	0.000417	−0.0025	−0.00875	−0.01833	−0.03125
b	1.0075	1.0175	1.034167	1.0575	1.0875
c	−0.00375	−0.01083	−0.02125	−0.035	−0.05208

The 5x5 matrix A is:

1.0075	−0.00375	0	0	0
−0.0025	1.0175	−0.01083	0	0
0	−0.00875	1.034167	−0.02125	0
0	0	−0.01833	1.0575	−0.035
0	0	0	−0.03125	1.0875

The inverse matrix A^{-1} is:

0.992565	0.003658	3.83373E-05	7.71E-07	2.48E-08
0.002439	0.982899	0.010299951	0.000207	6.67E-06
2.06E-05	0.008319	0.967394217	0.019458	0.000626
3.58E-07	0.000144	0.016787181	0.946864	0.030474
1.03E-08	4.15E-06	0.00048239	0.027209	0.920416

$a_1 = 0.000417$, $c_5 = -0.05208$, the left most column $f_{0,5} = f_{0,4} = f_{0,3} = f_{0,2} = f_{0,1} = f_{0,0} = 52$, the right most column $f_{6,5} = f_{6,4} = f_{6,3} = f_{6,2} = f_{6,1} = f_{6,0} = 0$.

$$\overrightarrow{S_5} = \overrightarrow{S_4} = \overrightarrow{S_3} = \overrightarrow{S_2} = \overrightarrow{S_1} = \overrightarrow{S_0} = \begin{pmatrix} a_1 \times 52 \\ 0 \\ 0 \\ 0 \\ c_5 \times 0 \end{pmatrix}$$

$$\overrightarrow{f_0} = (A^{-1})^6 \overrightarrow{f_6} - (A^{-1})^6 \overrightarrow{s_5} - (A^{-1})^5 \overrightarrow{s_4} - \cdots - (A^{-1})^2 \overrightarrow{s_1} - A^{-1} \overrightarrow{s_0}$$

$$\overrightarrow{f_0} = \begin{pmatrix} f_{1,0} \\ f_{2,0} \\ f_{3,0} \\ f_{4,0} \\ f_{5,0} \end{pmatrix}$$

Note that $f_{2,0}$, which is the put value, corresponds to S = 50. This is because $m = 6$ and $\Delta S = \frac{S_{max}}{m} = \frac{150}{6} = 25$. Thus, $i = \frac{50}{25} = 2$.

At the end of this section, we use the common substitution Z = ln(S) to obtain a PDE, which is equivalent to the Black-Scholes PDE.

$$\frac{\partial f}{\partial Z}\left(r - \frac{1}{2}\sigma^2\right) + \frac{\partial f}{\partial t} + \frac{1}{2}\frac{\partial^2 f}{\partial Z^2}\sigma^2 = rf$$

The finite difference method using the log of the underlying asset price is easier to solve than the Black-Scholes PDE.

These last two sections have shown that put values calculated by the alternative tree and finite difference methods both converge to the Black-Scholes value. So why use the alternative valuation methods instead of a simple Black-Scholes formula? There are two main reasons:

1. Early Exercise: The Black-Scholes formula does not allow early exercise. The alternative tree and finite difference method can be adjusted to value American options: The alternative methods adjust for early exercise by comparing the option value at each node (in a tree) or each grid (under the finite difference method) with its value if exercised immediately and selecting the larger value (modeling the option owner's decision to exercise any option that increases his or her wealth). In particular, only one cell in spreadsheet needs to update so that we can change the option from European to American in both the alternative tree and explicit finite difference settings.
2. Complex Payoffs: The alternative tree and finite difference methods can be used to calculate values of derivatives with complex payoffs as long as the derivative's payoff is a deterministic function of the underlying asset price at maturity S_T. Only one cell in spreadsheet needs to be updated. So we can change the payoff in a tree setting. Three boundary cells need to be updated to change the payoff of financial derivatives in the explicit finite difference setting.

The advantages and disadvantages of different valuation methods are discussed more in the next section.

4.1.4 Advantages vs. Disadvantages of Derivative Valuation Methods

There are three families of general methods that value derivatives: the Binomial Tree, the Black-Scholes PDE, and Monte-Carlo Simulation. There are also subcategories in the binomial tree and Black-Scholes PDE methods. The binomial tree method includes the Cox-Ross-Rubinstein (CRR) binomial tree discussed in Chapter 3 and the alternative binomial tree discussed in this chapter. The Black-Scholes PDE method can be applied to solve for the analytic solution, a semi-analytic solution, or a numerical solution.

Exhibit 4.1 categorizes and summarizes the approaches in five methods: binomial tree, analytic, ***semi-analytic*** (through numerical integration), finite difference, and Monte-Carlo Simulation.

Exhibit 4.1 Summary of valuation methods

Summary of Valuation Methods

	Advantages	Disadvantages
Monte-Carlo simulation	1. Allows underlying assets that do not follow GBM 2. Allows several random variables or assets 3. Allows path dependency 4. Error estimation	1. It is hard to value early exercise, such as American options.
Binomial Tree	1. Allows early exercise 2. Allows complex payoffs 3. It is intuitive and easy to calculate	1. It does not allow path dependent derivatives.
Finite difference	1. Allows early exercise 2. Implicit difference is robust which means that it convergences fast and the speed of convergence is not sensitive to the parameters. 3. Explicit difference is similar to trinomial tree. It is intuitive and easy to calculate.	1. Path dependency is not easily modeled 2. Implicit difference is computationally difficult. 3. Explicit difference method converges slowly.
Analytic	1. Easy to calculate	1. Restricted by the assumptions (such as GMB, payoff pattern) . 2. It does not allow early exercise. 3. It is very hard/impossible to value path dependent derivatives. 4. It does not work for complex European options (i.e., other than simple calls and puts).
Semi-analytic	1. Relatively easy to calculate complex European options other than calls and puts.	1. It does not allow early exercise. 2. Numerical approximation of integrals may be necessary.

Monte-Carlo simulation is the only method covered so far where the underlying asset is not assumed to follow geometric Brownian motion (GBM). The idea can be generalized to Heston's model and Bates' model, two additional methods where the underlying asset is not assumed to follow geometric Brownian motion. Those models that do not assume geometric Brownian motion are valued using the semi-analytic method and Monte-Carlo simulation. For the semi-analytic method, advanced integral packages in Mathematica or Matlab are needed. For simulation, R is the most efficient method.

As an application, the next section applies all five methods from Exhibit 4.1 to value various exotic derivatives and compare the results of the different methods.

4.2 Various Exotic Derivatives

Previous chapters have discussed several kinds of financial derivatives including futures contracts, forward contracts, European options and American options. These financial derivatives tend to be very popular and are usually exchange-traded. There is a large ***over-the-counter market*** (an electronic market where trading is done directly between two parties, without the supervision of a specialist) in which exotic derivatives are traded. Exotic derivatives are typically options. ***Exotic options*** are often distinguished from simple options by having complex exercise features or complex payoffs such as nonlinear payoffs. Valuation models for exotic derivatives are important for two primary reasons: (1) the Black-Scholes formula does not tend to value exotic options accurately, and (2) exotic derivatives tend not to be actively traded with visible market prices, therefore theoretic valuation becomes crucial for evaluating their values. This section introduces several exotic derivatives and compares the efficacy of different valuation methods – an analysis that is generally not shown in other books.

4.2.1 Gap Options

A ***gap call*** is a European option where the owner has the right to exercise the call at strike price K_1 at expiration only if the stock price exceeds the price K_2 at expiration. Consider the following example:

Example 4.6: A gap call has the following parameters: $S_0 = 50$, $K_1 = 35$, $K_2 = 40$, $r = 0.05$, $\sigma = 0.2$, $T = 1$. Thus this call's lowest non-zero payoff is \$5.01 for $S_T = \$40.01$ acting as a normal European call for values in excess of K_2. This gap call is valued next using the analytic method, tree method, explicit finite difference method and Monte-Carlo simulation method.

1. Analytic method as illustrated in Hull's book:

The Black-Scholes formula can be extended so that the exercise condition is $S_T > K_2$, and the payoff is $Max(S_T - K_1, 0)$. The integral can be solved in a similar method to that of Chapter 3.

$$
\begin{aligned}
C &= e^{-rT} \int_{K_2}^{+\infty} (s - K_1) f_{S_T}(s) ds \\
&= e^{-rT} \int_{K_2}^{+\infty} s f_{S_T}(s) ds + e^{-rT} \int_{K_2}^{+\infty} (K_1) f_{S_T}(s) \\
&= SN(d_1) - K_1 e^{-rT} N(d_2)
\end{aligned}
$$

where $d_1 = \dfrac{Ln(S_0 / K_2) + \left(r + \frac{\sigma^2}{2}\right)T}{\sigma\sqrt{T}}$ and $d_2 = \dfrac{Ln(S_0 / K_2) + \left(r - \frac{\sigma^2}{2}\right)T}{\sigma\sqrt{T}} = d_1 - \sigma\sqrt{T}$

Inserting the given values: C = 16.5616.

2. Tree methods:
   We use a 30-step CRR and alternative binomial trees in a spreadsheet.
   Binomial Tree: Keep $u = e^{\sigma\sqrt{\Delta t}}$, $d = e^{-\sigma\sqrt{\Delta t}}$, and $Q = \frac{e^{r\Delta t} - d}{u - d}$,
   The stock tree is the same as for a simple option. In the call tree, change the last column to (In fact, only one cell in EXCEL needs to update):

$$
\text{Payoff} = \begin{cases} S_T - K_1 & S_T \geq K_2 \\ 0 & S_T < K_2 \end{cases}
$$

   Rolling back produces C = 16.67746.
   For the alternative tree, use $u = e^{\left(r - \frac{1}{2}\sigma^2\right)\Delta t + \sigma\sqrt{\Delta t}}$, $d = e^{\left(r - \frac{1}{2}\sigma^2\right)\Delta t - \sigma\sqrt{\Delta t}}$, and Q = 1/2 to find C = 16.579329.
3. Explicit Finite Difference Method (30x30 grids using a spreadsheet):
   The formula for a^*, b^*, and c^* are the same as for a simple option; the only changes are the top, right, and left boundaries. (In fact, only three cells in EXCEL need to update) In the top row, the payoff formula is:

$$
\text{Payoff} = \begin{cases} S_T - K_1 & S_T \geq K_2 \\ 0 & S_T < K_2 \end{cases}
$$

   In the left column, $S_t = 0$ and Payoff = 0.
   In the right column, $S_{max} = 150$ and Payoff = 150-35 = 115.
   Discounting to the bottom row generates: C = 16.7619.
4. Simulation (In fact, there are just a few changes from European call or put codes): (R code)

R code 4.1 Price Gap call using simulation from Example 4.6

```
#parameters, we run m trials and cut the time interval
#into n pieces
S0=50
r=0.05
b=0.2
T=1
m=5000
n=10000
K1=rep(35,m)
K2=rep(40,m)
#Define d as an m-dimension vector to save the derivative
#value for m trials
D=rep(0,m)
#Define S as an n-dimension vector for stock price
#(geometric Brownian motion) paths
S=rep(S0,n+1)
#Define Y as an m-dimension vector for stock prices at
#maturity for m trials
Y=rep(0,m)
deltaT=T/n
#loop for m trials
for(j in 1:m)
{
W=rnorm(n, 0, 1)
#loop for each path
for(i in 1:n)
{
S[i+1]=S[i]+r*S[i]*deltaT+ b*S[i]*sqrt(deltaT)*W[i]
}
Y[j]=S[n+1]
if(Y[j]>=K2[j])
{
D[j]=Y[j]-K1[j]
}
}
c=exp(-r*T)*mean(D)
```

Our run of the above simulation yields C = 16.47342. Our several valuation methods for the gap option ranged from 16.47 to 16.76 with the analytic method at 16.56.

The second exotic option, analyzed in the next section, is a quite popular exotic option: an Asian option.

4.2.2 Asian Options

An ***Asian option*** has a payoff that depends on the *average* price of the underlying asset over a specified period of time as opposed to simply the price at maturity. A ***fixed strike Asian call*** has the payoff $\text{Max}(S_{\text{ave}} - K, 0)$, and a ***fixed strike Asian put*** has the payoff $\text{Max}(K - S_{\text{ave}}, 0)$ where S_{ave} is the average of the price of the underlying asset over the specified reference period. A ***floating strike Asian call*** has the payoff $\text{Max}(S_T - S_{\text{ave}})$, and a ***floating strike Asian put*** has the payoff $\text{Max}(S_{\text{ave}} - S_T, 0)$.

The average price, S_{ave}, may be obtained in many ways. Conventionally, an arithmetic average is used. In the *continuous* case, this is obtained by

$$S_{\text{ave}} = \frac{1}{T}\int_0^T S(t)dt$$

For the case of *discrete monitoring* (with monitoring at the times $0 \le t_i \le T$) the average is:

$$S_{\text{ave}} = \frac{1}{n}\sum_{i=1}^{n} S(t_i)$$

There also exist Asian options based on a geometric average. In the continuous case, this is:

$$S_{\text{ave}} = e^{\frac{1}{T}\int_0^T Ln(S(t))dt}$$

For the case of *discrete monitoring* based on a geometric average (with monitoring at the times $0 \le t_i \le T$):

$$S_{\text{ave}} = e^{\frac{1}{n}\sum_{i=1}^{n} Ln(S(t_i))}$$

Asian options can be attractive to firms that purchase or sell a commodity regularly (e.g., daily or weekly) and whose total revenues or costs are driven by the average price of the commodity over time. The following example values a fixed strike Asian call.

> **Example 4.7:** A fixed strike Asian call (continuous arithmetic average) has the following parameters: $S_0 = 50$, $K = 50$, $r = 0.1$, $\sigma = 0.4$, and $T = 1$. Select the proper method and value this Asian call. (This example purposely matches that in Hull's book[1] so that Hull's discussion of the analytic method can be compared to this book's use of simulation.)

In this case, with no further model assumptions on the distribution of S_{ave}, there is no analytic solution if we only have an assumption that S_t follows

Geometric Brownian motion. This is because the pdf of S_{ave} has no closed-form solution.

Both the tree and finite difference methods cannot trace the average price of the underlying asset, so the simulation method is the only method that works in general. The following R code values a fixed strike Asian call using simulation.

R code 4.2 Valuing a fixed strike Asian call using simulation

```
#parameters, we run m trials and cut the time interval
#into n pieces
S0=50
r=0.10
b=0.4
T=1
m=5000
n=10000
K=rep(50,m)
#Define d as an m-dimension vector to save the derivative
#value for m trials
D=rep(0,m)
deltaT=T/n
#Define S as an n-dimension vector for stock price
#(geometric Brownian motion) paths
S=rep(S0,n+1)
#Define Y as an m-dimension vector for stock prices at
#maturity for m trials
Y=rep(0,m)
#loop for m trials
for(j in 1:m)
{
W=rnorm(n,0,1)
#loop for each path
for(i in 1:n)
{
S[i+1]=S[i]+r*S[i]*deltaT+ b*S[i]*sqrt(deltaT)*W[i]
}
Y[j]=mean(S)
if(Y[j]>=K[j])
{
D[j]=Y[j]-K[j]
}
}
c=exp(-r*T)*mean(D)
```

The above simulation calculated C = 5.521735 on one try.

In the simulation method, only options that use discrete averages can be valued. Note that the problem asks for the arithmetic average. If the problem asks for the geometric average, only one line of code would need to be changed, from "Y[j] = mean(S)" to "Y[j] = exp(mean(log(S)))".

Hull's book adds an assumption that S_{ave} is lognormal with the analytic formulas given by including the first two moments M_1 and M_2 of S_{ave}. Two sets of formulas were given, one for the average calculated continuously and the other for the average calculated from observations at time T_i. Note that the assumption is based on the fact that *if the asset price follows geometric Brownian motion, the geometric average of the price is lognormal and the arithmetic average is approximately lognormal.* This fact is explored further as a subsequent exercise.

It is obvious that our example used the arithmetic average and intervals cut into n = 10,000 pieces, so it is close to the average calculated continuously. Also, our solution is close to the solution of 5.62 derived by Hull using an analytic formula.

The above simulation code values a fixed strike call, but the code can be easily changed to value a floating strike call, floating strike put, and a fixed strike put.

The third option analyzed here is a Barrier option.

4.2.3 Barrier Options

There are four main types of ***barrier options*** where the rights of the option to payoffs occur or expire only when the price of the underlying security reaches a specified barrier before the option matures:

Up-and-out: The spot price starts below the barrier level. If it reaches the barrier, the option is knocked out (become null and void).

Down-and-out: The spot price starts above the barrier level. If it reaches the barrier, the option is knocked out (become null and void).

Up-and-in: The spot price starts below the barrier level. Only if it reaches the barrier does the option become activated (i.e., knocked in as able to be exercised).

Down-and-in: The spot price starts above the barrier level. Only if it reaches the barrier does the option become activated (i.e., knocked in as able to be exercised).

These four types of barrier options can be based on either a call option or a put option as the type of option that gets knocked in or out. For example, consider a ***down-and-out call***. A down and out call is an otherwise simple call option that ceases to exist if the underlying asset price reaches the prespecified barrier H (with initially H < S_0) at any time prior to the option's expiration.

Example 4.8: A down-and-out call has the following parameters: H = 18, K = 20, S = 19, r = 0.05, q = 0.05, σ = 0.4, and T = 0.25, where H is the barrier level. Find the value of this call. We use (r-q) as the growth rate and r as discount rate for reasons discussed in a later section. One reason we have q (dividend yield) in the problem is to compare our results with those of Hull's analytic method. (This example purposely matches that in Hull's book[1] so that Hull's discussion of the analytic method can be compared to this book's use of simulation.)

The R code for the simulation method in Example 4.3 is shown below.

R code 4.3 Valuing a down-and-out call using simulation

```
S0=19
r=0.05
q=0.05
b=0.4
T=0.25
H=18
m=5000
n=10000
D=rep(0,m)
deltaT=T/n
S=rep(S0,n+1)
Y=rep(0,m)
K=rep(20,m)
Z=rep(0,m)
for(j in 1:m)
{
W=rnorm(n,0,1)
for(i in 1:n)
{
S[i+1]=S[i]+(r-q)*S[i]*deltaT+ b*S[i]*sqrt
(deltaT)*W[i]
}
Y[j]=S[n+1]
Z[j]=min(S)
if(Z[j]>H)
{
D[j]=max(Y[j]-K[j],0)
}
}
c=exp(-r*T)*mean(D)
```

In Hull's book, the analytic formula is shown to provide an analytic solution for the value equal to 0.626392578. A trial of the above simulation produced 0.6153175.

An important issue in barrier options is how often the price of the underlying asset is referenced. This issue can affect whether the underlying asset price has reached the barrier. The analytic formula assumes continuous observation of the price, while our simulation method assumed that all observations happened at the cutting points. In practice, the observations are typically performed daily.

There are other exotic options available, some of which are introduced as part of exercises.

In the above valuation model, the volatility of the underlying asset is given as a parameter. The volatility parameter is examined in more detail in the next section.

4.2.4 The Volatility Parameter

Market prices of simple calls and puts on common stocks are easily observed because many simple options are exchange-traded (i.e., liquid). Each option price can be used with the Black-Scholes formula to find an *implied volatility* of the underlying asset's returns. In the case of most exotic options whose prices are not readily observable, market participants use the implied volatility from simple liquid options as an input to value illiquid exotic options on the same asset. Further, arbitragers searching for relatively mispriced assets (e.g., market prices of simple options vs. market prices of exotic options, or within types of options) use implied volatilities from various options to identify potential arbitrage opportunities.

For example, differences in the implied volatilities of XYZ Corporation's common stock derived from different options may signal relative mispricing of the options (or it may signal that there are modeling misspecifications such as non-normality of the returns of the underlying asset). Another use of the Black-Scholes formula and other option valuation models is to estimate the price sensitivities (i.e., "Greeks") of various options for hedging purposes – as discussed in Chapter 5.

So far, we have not modeled dividends to be distributed by an option's underlying asset. Since most stocks pay dividends, option valuation models for stocks and stock indexes require modifications for dividends or dividend yields. These modifications are straight forward and are detailed in the next section.

4.3 Dividends

We have assumed so far that the underlying asset does not pay a dividend. If any option's underlying asset generates payments for the owner during the option's life, all methods of valuing the derivatives need to be adjusted for anticipated distributions. In reality, most stocks pay dividends quarterly, although often there are no dividends scheduled over the time-to-expiration of many short-term options. If the underlying asset is a bond, the same is true for anticipated coupons. Bonds have another complication of non-constant interest rates which is discussed Chapter 7. If an option's underlying asset is a broad stock index, the dividend issue can't be avoided. When an option's underlying asset is a foreign currency, the foreign interest rate can be viewed as income similar to a stock's dividend and must be included in the valuation model. Finally, if the option's underlying asset is a forward/futures contract, it can be regarded as an asset with a dividend yield equal to the risk-free rate, r, since the forward contract defers payment until settlement - so the time value of the deferment serves the same role as a dividend.

All derivative valuation methods must include anticipated distributions in constructing the risk-free portfolios used in their derivation. Therefore it is important to generalize option models to include dividend payments. In modeling stock behavior for derivative valuation there are two methods of specifying distributions: a ***cash dividend*** or a ***dividend yield***. The variable I is used to denote cash distributions (discretely paid and measured in dollars). The variable q is used to denote annualized distribution yields (assumed continuously paid and measured in percentages). The following section discusses the modification of valuation formulas for both types of dividends (discrete and continuous) and for different derivative pricing methods.

4.3.1 How Dividends Affect Various Valuation Formulas

In this section, we will see how dividends affect the valuation formulas we have already discussed.

1. Dividends and the Analytic method:

4.1 Principle for handling cash dividend I and dividend yield q in analytic method:

In all formulas, if the underlying asset S has a dividend, the formula deducts the anticipated intervening dividend from the asset's value as follows:

$$\text{Change } S \text{ into } S - \mathrm{PV}(I)$$

where PV(I) is present value of the cash dividend, I, or

$$\text{Change } S \text{ into } Se^{-qT}$$

where q is the dividend yield.

- Forward and Futures:
  The forward price of a stock, F_0, has the following analytic formula in the case of no dividends:

$$F_0 = S_0 e^{rT}.$$

where S_0 is underlying asset price today, r is risk-free interest rate and T is maturity time.
We will show in a later section that if the risk-free interest rate is constant r (uncorrelated to the underlying asset price), the futures price, f_0, has the same analytic formula:

$$f_0 = S_0 e^{rT}$$

where S_0 is the underlying asset price today, r is the risk-free interest rate and T is the time to maturity or expiration.
Now, assuming that an asset pays a continuous dividend yield at the rate q, the price of the forward/futures contract is:

$$\boldsymbol{F_0 = f_0 = S_0 e^{(r-q)T}} \tag{4.16}$$

Assuming that an asset pays a cash dividend with a present value of PV (I), the price of the forward/future contract is:

$$\boldsymbol{F_0 = f_0 = (S_0 - PV(I))e^{rT}} \tag{4.17}$$

The intuition of the above formula is that the current stock price, S_0, can be divided into claims on two future cash flows: a known and fixed dividend to be received prior to the option's expiration, I, and the remaining cash flows from the stock $(S_0 - PV(I))$. The option is a claim on only the later of the two cash flows. If an asset is modeled as having a continuous stream of dividends, then the size of the dividends occurring over an option's life depend on the price of the underlying asset over the option's life.

- Calls and Puts

A call value has the analytic formula (Black-Scholes):

$$C = S_0 \mathrm{N}(d_1) - Ke^{-rT}\mathrm{N}(d_2)$$

where $d_1 = \dfrac{Ln(S_0 / K) + \left(r + \frac{\sigma^2}{2}\right)T}{\sigma\sqrt{T}}$ and $d_2 = \dfrac{Ln(S_0 / K) + \left(r - \frac{\sigma^2}{2}\right)T}{\sigma\sqrt{T}} = d_1 - \sigma\sqrt{T}$

For an underlying asset, S, with a cash dividend, I, the call and put formulas can be adjusted so that:

$$C = (S_0 - PV(I))\mathbf{N}(d_1) - Ke^{-rT}\mathbf{N}(d_2) \tag{4.18}$$

Where, $d_1 = \dfrac{Ln((S_0 - PV(I))/K) + \left(r + \frac{\sigma^2}{2}\right)T}{\sigma\sqrt{T}}$ and $d_2 = \dfrac{Ln((S_0 - PV(I))/K) + \left(r - \frac{\sigma^2}{2}\right)T}{\sigma\sqrt{T}} = d_1 - \sigma\sqrt{T}$

$$P = Ke^{-rT}\mathbf{N}(-d_2) - (S_0 - PV(I))\mathbf{N}(-d_1) \tag{4.19}$$

Where, $d_1 = \dfrac{Ln((S_0 - PV(I))/K) + \left(r + \frac{\sigma^2}{2}\right)T}{\sigma\sqrt{T}}$ and $d_2 = \dfrac{Ln((S_0 - PV(I))/K) + \left(r - \frac{\sigma^2}{2}\right)T}{\sigma\sqrt{T}} = d_1 - \sigma\sqrt{T}$

For underlying asset S with a continuous dividend yield q:

$$C = S_0e^{-qT}\mathbf{N}(d_1) - Ke^{-rT}\mathbf{N}(d_2) \tag{4.20}$$

where $d_1 = \dfrac{Ln(S_0/K) + \left(r - q + \frac{\sigma^2}{2}\right)T}{\sigma\sqrt{T}}$ and $d_2 = \dfrac{Ln(S_0/K) + \left(r - q - \frac{\sigma^2}{2}\right)T}{\sigma\sqrt{T}} = d_1 - \sigma\sqrt{T}.$

Similarly, the put formula with a continuous dividend yield can be written as:

$$P = Ke^{-rT}\mathbf{N}(-d_2) - S_0e^{-qT}\mathbf{N}(-d_1) \tag{4.21}$$

where, $d_1 = \dfrac{Ln(S_0/K) + \left(r - q + \frac{\sigma^2}{2}\right)T}{\sigma\sqrt{T}}$ and $d_2 = \dfrac{Ln(S_0/K) + \left(r - q - \frac{\sigma^2}{2}\right)T}{\sigma\sqrt{T}} = d_1 - \sigma\sqrt{T}.$

The Black-Scholes (with dividend) formulas just derived can be simplified to:

$$C = e^{-rT}\left(S_0e^{(r-q)T}N(d_1) - KN(d_2)\right) \tag{4.22}$$

where, $d_1 = \dfrac{Ln(S_0/K) + \left(r - q + \frac{\sigma^2}{2}\right)T}{\sigma\sqrt{T}}$ and $d_2 = \dfrac{Ln(S_0/K) + \left(r - q - \frac{\sigma^2}{2}\right)T}{\sigma\sqrt{T}} = d_1 - \sigma\sqrt{T}$, and

$$P = e^{-rT}\left(KN(-d_2) - S_0e^{(r-q)T}\mathbf{N}(-d_1)\right) \tag{4.23}$$

where $d_1 = \dfrac{Ln(S_0/K) + \left(r - q + \frac{\sigma^2}{2}\right)T}{\sigma\sqrt{T}}$ and $d_2 = \dfrac{Ln(S_0/K) + \left(r - q - \frac{\sigma^2}{2}\right)T}{\sigma\sqrt{T}} = d_1 - \sigma\sqrt{T}.$

Consider the following principle for an underlying asset with a continuous dividend yield q in a different perspective.

4.2 Principle for handling dividend yield q (a second perspective):
If the underlying asset S pays a continuous dividend yield q, then adjust the expected growth rate of S from r to r-q in its option pricing model. The risk-free rate for discounting remains as r.

2. Dividends and the Binomial tree method for derivative valuation:

The principles regarding the modeling of dividends using the binomial tree method do not differ between options, forwards, futures, other financial derivatives. For an underlying asset that does not pay a dividend, recall that the risk-neutral valuation formula is:

$$f_0 = e^{-r\Delta t}(Qf_u + (1 - Q)f_d)$$

where, $Q = \frac{e^{r\Delta t} - d}{u - d}$, $u = e^{\sigma\sqrt{\Delta t}}$, and d = 1/u.
For a derivative with an underlying asset that has a dividend yield, q, the formula can be adjusted:

$$f_0 = e^{-r\Delta t}(Qf_u + (1 - Q)f_d) \tag{4.24}$$

where $Q = \frac{e^{(r-q)\Delta t} - d}{u - d}$, $u = e^{\sigma\sqrt{\Delta t}}$, and d = 1/u.
Note that the discount rate r (in the main equation) does not change while the stock growth rate (in the equation for Q), r, becomes r – q. For an underlying asset paying a discrete cash dividend, the tree does not recombine, and an adjustment is needed. The details are beyond the scope of this introductory book.

3. Dividends and the Black-Scholes PDE or Finite difference methods:

For all financial derivatives, a derivative without a dividend satisfies the Black-Scholes PDE:

$$\frac{\partial f}{\partial S}rS + \frac{\partial f}{\partial t} + \frac{1}{2}\frac{\partial^2 f}{\partial S^2}\sigma^2 S^2 = rf$$

A derivative with a dividend yield, q, satisfies this PDE:

$$\frac{\partial f}{\partial S}(r - q)S + \frac{\partial f}{\partial t} + \frac{1}{2}\frac{\partial^2 f}{\partial S^2}\sigma^2 S^2 = rf. \tag{4.25}$$

The principle for handling derivatives holds. In particular, in the finite different method to solve the PDE, we should use the above PDE with a dividend yield q. Adjust the formulas for a_i, b_i, c_i and a_i^*, b_i^*, c_i^* as follows.

$$a_i = \frac{1}{2}(r-q)i\Delta t - \frac{1}{2}\sigma^2 i^2 \Delta t,\ b_i = 1 + \sigma^2 i^2 \Delta t + r\Delta t,$$
$$c_i = -\frac{1}{2}(r-q)i\Delta t - \frac{1}{2}\sigma^2 i^2 \Delta t \tag{4.26}$$

$$a_i^* = \frac{-\frac{1}{2}(r-q)i\Delta t + \frac{1}{2}\sigma^2 i^2 \Delta t}{1 + r\Delta t},\ b_i^* = \frac{1 - \sigma^2 i^2 \Delta t}{1 + r\Delta t},$$
$$c_i^* = \frac{\frac{1}{2}(r-q)i\Delta t + \frac{1}{2}\sigma^2 i^2 \Delta t}{1 + r\Delta t} \tag{4.27}$$

4. Dividends and the Monte-Carlo Simulation method:

For a derivative without a dividend, the simulation path is given by:

$$dS = rSdt + \sigma SdZ$$

$$f = e^{-rT}E\,(\text{Payoff at expiration})$$

For a derivative with a dividend yield q, the simulation path is given by:

$$dS = (r-q)Sdt + \sigma SdZ \tag{4.28}$$

$$f = e^{-rT}E\,(\textbf{Payoff at expiration}) \tag{4.29}$$

The next section applies the formulas to specific assets.

4.3.2 Dividends and Options on Foreign Currency or Futures Contracts

If the underlying asset of the derivative is a foreign currency, then the underlying asset is modeled as a stock with a dividend yield equal to the risk-free interest rate, r_f, available in the market on the foreign currency and we use r to represent the risk-free interest rate available in the domestic market. If S is a foreign exchange rate, then $q = r_f$. Therefore, the forward formula is:

$$F_0 = f_0 = S_0 e^{(r-r_f)T} \tag{4.30}$$

The call value formula for a foreign currency is therefore:

$$C = S_0 e^{-r_f T}\mathbf{N}(d_1) - Ke^{-rT}\mathbf{N}(d_2) \tag{4.31}$$

where $d_1 = \dfrac{Ln(S_0 / K) + \left(r - r_f + \frac{\sigma^2}{2}\right)T}{\sigma\sqrt{T}}$ and $d_2 = \dfrac{Ln(S_0 / K) + \left(r - r_f - \frac{\sigma^2}{2}\right)T}{\sigma\sqrt{T}} = d_1 - \sigma\sqrt{T}.$

We must be careful about which country's interest rate is considered r_f since there are two interest rates between the two currencies. The risk free rate depends on how the currency is quoted. For example, if the underlying exchange rate is yen/dollar, then the dollar is viewed as the asset and the yen as cash, and the U.S. risk-free rate in dollars is the dividend yield (which is termed r_f), while the Japanese risk-free rate in yen is the riskless rate (which is termed r).

Another application of the stock-with-dividend-model is valuation of an option on a futures contract. If the underlying asset of an option is a futures contract with a longer maturity than the put/call contract, then the futures contract is treated as a stock with a dividend yield r where r is the risk-free interest rate.

If f_0 is the price of a futures contract, then $q = r$. So, the call on the futures contract has value $\hat{C}$:

$$\hat{C} = f_0 e^{-rT}\mathbf{N}(d_1) - Ke^{-rT}\mathbf{N}(d_2) \tag{4.32}$$

where $d_1 = \dfrac{Ln(f_0 / K) + \left(\frac{\sigma^2}{2}\right)T}{\sigma\sqrt{T}}$ and $d_2 = \dfrac{Ln(f_0 / K) + \left(-\frac{\sigma^2}{2}\right)T}{\sigma\sqrt{T}} = d_1 - \sigma\sqrt{T}.$

The above formula is called the ***Black formula*** for a call on a futures contract and can be written: $\hat{C} = e^{-rT}[f_0 N(d_1) - KN(d_2)]$.

Example 4.8: Find the formula for a call value where the underlying asset is a foreign exchange rate under the Binomial Tree method.

$$f_0 = e^{-r\Delta t}(Qf_u + (1 - Q)f_d)$$

where $Q = \dfrac{e^{(r-r_f)\Delta t} - d}{u - d}$, $u = e^{\sigma\sqrt{\Delta t}}$, d = 1/u

Example 4.9: Find the formula for a call value where the underlying asset is a futures contract under the Binomial Tree method.

$$f_0 = e^{-r\Delta t}(Qf_u + (1 - Q)f_d)$$

where $Q = \dfrac{e^{(r-r)\Delta t} - d}{u - d} = \dfrac{1 - d}{u - d}$, $u = e^{\sigma\sqrt{\Delta t}}$, d=1/u

Consider the following example that demonstrates all the methods to value a European call with a dividend yield.

Example 4.10: Consider a European call option on the S&P 500 stock index that is three months from expiration. The current value of the index is 2,430 and the exercise price is 2,400. Assume the risk-free interest rate is 5% per annum (continuously compounded) and the volatility of the index is 20% per annum. The dividend yield is 2.5% per annum paid continuously. Find the call value using the four main valuation methods: analytic, tree, explicit finite difference, and simulation.

1. Analytic Method:

Using the analytic solution:

$$C = S_0 e^{-qT} N(d_1) - K e^{-rT} N(d_2)$$

where $d_1 = \dfrac{Ln(S_0 / K) + \left(r - q + \frac{\sigma^2}{2}\right)T}{\sigma\sqrt{T}}$ and $d_2 = \dfrac{Ln(S_0 / K) + \left(r - q - \frac{\sigma^2}{2}\right)T}{\sigma\sqrt{T}} = d_1 - \sigma\sqrt{T}$,
Inserting the known values generates C = 119.4017.

2. Tree method (We use a 30 step CRR and alternative Binomial tree):
   Binomial Tree: As usual, $u = e^{\sigma\sqrt{\Delta t}}$, $d = e^{-\sigma\sqrt{\Delta t}}$
   Change Q so that: $Q = \dfrac{e^{(r-q)\Delta t} - d}{u - d}$,
   This yields C = 120.0111.

   For an alternative tree: $u = e^{\left(r-q-\frac{1}{2}\sigma^2\right)\Delta t + \sigma\sqrt{\Delta t}}$, $d = e^{\left(r-q-\frac{1}{2}\sigma^2\right)\Delta t - \sigma\sqrt{\Delta t}}$.
   Similarly this yields, C = 120.0741.
3. Explicit Finite Difference (30x30 grid)
   The formulas for a^*, b^*, c^* are adjusted so that

$$a_i^* = \frac{-\frac{1}{2}(r - q)i\Delta t + \frac{1}{2}\sigma^2 i^2 \Delta t}{1 + r\Delta t}, \quad b_i^* = \frac{1 - \sigma^2 i^2 \Delta t}{1 + r\Delta t}, \quad c_i^* = \frac{\frac{1}{2}(r - q)i\Delta t + \frac{1}{2}\sigma^2 i^2 \Delta t}{1 + r\Delta t}$$

C = 111.2529.

4. Simulation code:

R code 4.4 Price call on S&P 500 index using simulation

```
S0=2430
r=0.05
b=0.2
T=0.25
q=0.025
m=5000
n=10000
K=rep(2400,m)
D=rep(0,m)
deltaT=T/n
S=rep(S0,n+1)
Y=rep(0,m)
for(j in 1:m)
{
W=rnorm(n,0,1)
for(i in 1:n)
{
S[i+1]=S[i]+(r-q)*S[i]*deltaT+   b*S[i]*sqrt(deltaT)
*W[i]
}
Y[j]=S[n+1]
D[j]=max(-K[j]+Y[j],0)
}
C=exp(-r*T)*mean(D)
```

Our try of the above Monte Carlo simulation calculated the value C = 117.2191.

Having discussed valuation methods regarding a variety of derivatives, the next section covers one remaining issue: forward and future contract valuation methods.

4.4 More on Forward Values, Futures Contracts, and Futures Prices

The payoff for a forward is $S_T - F_0$ for the long position, where F_0 is the forward price. Forwards usually do not have upfront payments. This section begins with a discussion of the initial forward price and the changes in the value of a forward contract through time. Note that for most derivatives such as calls or puts, we abuse the notation of value and price. However, in the case of forward contracts, value and price are different.

4.4.1 Forward Value to the Buyer

Parties to a forward contract typically agree to a ***forward price (also called the forward delivering price)*** that does not require an upfront premium or payment between the parties. The forward price that sets the initial value of the forward contract to zero for both parties is $F_0 = S_0 e^{rT}$ and is sometimes termed as the "fair" price of the contract because it allows both parties to costlessly enter the contract and costlessly form a hedge that eliminates all risk (i.e., eliminates all subsequent net cash flows). As long as the forward price, F_0, is fair, neither the buyer nor seller should need additional compensation in the form of an upfront premium or payment. However, over time and as the underlying asset price changes, the "fair" forward price for new contracts, F_t, will change. This feature can be seen in the formula, $F_t = S_t e^{r(T-t)}$ in which variables will change through time. First, the underlying price, S_t, will vary through time. Second, the time to maturity, (T − t), will decrease. These changes will change the "fair" price of new forward contracts. This in turn causes contracts based on F_0 as the forward price to take on a positive value for one side of the contract and a negative value for the other side. The forward value at time t must be $(F_t - F_0)e^{-r(T-t)}$. This is shown in the following example. So the ***forward value*** is the price of the opposite forward postion that would need to be taken to close out the current forward position.

Assume that at today, t = 0, we enter into a long position in a forward contract with a maturity (i.e., settlement) at time T. Assume that there is no initial payment or premium at t = 0 for this contract. Further assume that at a later time, t, we short a new forward contract on the same asset with maturity T − t (i.e., matching the maturity date of our original contract). Assume that there is no initial payment at time t for this new contract. Note that at time t, we know for sure that at time T we will have a payoff equal to $F_t - F_0 = (S_T - F_0 - (S_T - F_t)) = F_t - F_0$ at maturity T. Since our final cash flow is known in advance, the portfolio will be neutral to changes in the price of the underlying asset. Therefore, the value of owning this portfolio at time t is:

$$\boldsymbol{V_t = (F_t - F_0)e^{-r(T-t)}} \tag{4.33}$$

where V_t is the forward value established for the long position at time t. Note that the example demonstrates that (ignoring counterparty risk) a long position in a forward contract at time t with value $V_t = (F_t - F_0)e^{-r(T-t)}$ can be closed out with a new short position with the same expiration date.

Alternatively:

$$V_t = S_t - F_0 e^{-r(T-t)} = S_t - S_0 e^{rt} \tag{4.34}$$

Consider a derivation of this equation.

$$F_t = S_t e^{r(T-t)}$$

Substitute the above equation for Equation 4.33:

$$V_t = (S_t e^{r(T-t)} - F_0)e^{-r(T-t)} = S_t - F_0 e^{-r(T-t)}.$$

Another substitution, $F_0 = S_0 e^{rT}$, yields:

$$V_t = S_t - S_0 e^{rT} \times e^{-r(T-t)} = S_t - S_0 e^{rt}.$$

The equation can be tested at initiation to make sure it is a viable valuation model: $V_0 = S_0 - S_0 = 0$. The forward value for the short position is $-V_t$. This is because owning a short position today can be closed out by a long position for a payoff $F_0 - F_t$. This is a logical result.

Depending on whether there is a cash dividend, I, or a dividend yield of q, there are two formulas for the forward price F_0.

$$F_0 = (S_0 - PV_t(I))e^{rT}$$

$$F_0 = S_0 e^{(r-q)T}$$

There are two formulas for the forward value V_t:

$$V_t = (S_t - PV_t(I)) - F_0 e^{-r(T-t)} \tag{4.35}$$

$$V_t = (S_t e^{-q(T-t)}) - F_0 e^{-r(T-t)} \tag{4.36}$$

In the case of a foreign exchange rate, the formula for the forward price is F_0:

$$F_0 = S_0 e^{(r-r_f)T}$$

The following formula is for the forward value V_t:

$$V_t = (S_t e^{-r_f(T-t)}) - F_0 e^{-r(T-t)} \tag{4.37}$$

It is very easy to memorize the forward value formula. The forward payoff for a long position is:

$$\text{Payoff} = S_T - F_0.$$

The forward value for a long position is found by discounting the formula for the payoff from time T to time t. Discounting S_T to time t gives S_t. Discounting the forward price due at time T, F_0, to time t gives $F_0 e^{-r(T-t)}$, so that:

$$V_t = S_t - F_0 e^{-r(T-t)}.$$

4.4.2 Forward Contracts vs. Futures Contracts

In the example in the previous section, the long position in a forward was closed out at time t by entering a short position with the same expiration date, thereby locking in the payoff at time t: $(F_t - F_0)e^{-r(T-t)}$.

A futures contract can be regarded as a modification of a forward contract whose gain is realized daily (i.e., marked-to-market) rather than at settlement. Consider a futures contract with price, f_0, that does not require or receive any immediate premium or payment but is marked to market on a daily basis. Suppose that at the end of the next day in the futures market, the futures price on new contracts is f_1. In the case of futures contracts, the futures exchange (i.e., its clearinghouse) requires an immediate payment between the counterparties with the long side receiving $f_1 - f_0$ from the short side if $f_1 - f_0 > 0$. If $f_1 < f_0$, then a payment is required from the long side to the short side in the amount of $f_0 - f_1$. In effect, all positions in futures contracts are settled at the end of each day using the end-of-day price and automatically rolled over into a new position with the new futures price. Therefore the delivery price contained in a futures position changes each day with the market price of the futures contract as the contract is *marked to market*.

The cumulative net cash flow to (or from if negative) a long position in both a futures contract and a forward contract held to settlement is equal to the market price of the underlying asset at settlement minus the original forward/previous day futures price. The difference is that the futures contracts call for daily settlements while the forward contracts are settled only at the delivery or settlement date. This difference in cash flow timing can affect market values when the time value of money is positive (as it almost always is).

One effect of the daily marking-to-market of a futures contract is obviously to accelerate realization of profits (cash inflows) and losses (cash outflows). Whether or not the marking-to-market of futures contracts causes futures prices and forward prices to differ is complex and is discussed more in Chapter 7. Note that since both futures contracts and forward contracts typically are formed with zero initial cash flows from both sides, it must be true that in a risk-neutral world, the forward price $F_T = E(S_T)$. Therefore, the total *expected* payoff to a forward contract in a risk-neutral world must be: $F_T - E(S_T) = 0$. A key question arises as to whether the same is true in a risk averse world for futures contracts in which cash flows are exchanged throughout the life of the contract rather than simply at settlement.

Let us consider three cases that drive potential difference between the valuation of futures and forward contracts.

Case 1:

If the anticipated correlation through time between the underlying asset price and short term interest rates is ***zero****, then the futures price is* ***equal*** *to the forward price,* $f_0 = F_0$.

Consider in detail the difference between the two contracts. Assume that a trader establishes long in positions both a forward contract and futures contract with a price F_0 and f_0 respectively. Assume that at time t_1, the forward price is F_1 and futures price is f_1. Suppose that at time t_1 the trader enters into a short position of the same forward contract at price F_1 but does nothing to hedge the futures contract. The consequence of the trade locks in a payoff $= F_1 - F_0$ at maturity T which is equivalent to the value $(F_1 - F_0)e^{-r(T-t_1)}$ at time t_1. The futures contract has a payoff $f_1 - f_0$ at time t_1.

At maturity, T, the forward has a payoff $S_T - F_0$, which is the only payoff.

On the other hand, at maturity T, the futures contract has a payoff $S_T - f_{n-1}$. During the process, the payoff each day is $f_1 - f_0, f_2 - f_1, \ldots, f_{n-1} - f_{n-2}, f_n - f_{n-1}$. Where $f_n = S_T$. Without the time value of money, the combined value of the payoff is $S_T - f_0$.

The difference is that the futures contract realizes the return (positive or negative) throughout the life of the contract while the forward contract receives or pays the cash flows at the end of the contract. However, if there is no correlation between the time value of money and the price of the underlying asset, then changes in the interest rate do not systematically favor the long or short side because the expected payoff in a risk-neutral world is zero.

Case 2:

If the anticipated ***correlation*** *through time between the underlying asset price and short term interest rates is* ***positive****, then the futures contract price is* ***higher*** *than the forward price prior to settlement.*

This futures price is higher because in those scenarios where the price of the underlying asset increases, both the long position in a forward and future contract earns positive returns. However, the future contract's profits are realized immediately *and when interest rates are more likely to be higher*. In those scenarios where the price of the underlying asset decreases, both the long positions in a forward and future contract earn negative returns. However, the future contract's losses are realized immediately *when interest rates are more likely to be lower*. The positive correlation of the interest rate and the underlying asset causes the futures contract to enjoy accelerated profits in high interest rate scenarios (earning reinvestment profits) and suffer accelerated losses in low interest rate scenarios (lowering the costs of financing the losses). The net effect is to make long positions in futures contracts more desirable than long positions in otherwise equal forward contracts *if their initial prices are the same.* Of course in a competitive market with positively correlated interest rates and underlying asset prices, futures prices will adjust to reflect the advantages to futures contracts and the price of futures contracts will be higher than the prices of forward contracts.

Case 3:

If the anticipated ***correlation*** *through time between the underlying asset price and short term interest rates is* ***negative****, then the future contract price is* ***lower*** *than the forward price prior to settlement.*

The argument follows Case 2 except that the futures contract suffers from a tendency to generate profits in low interest rate environments and losses in high interest rate environments. Therefore, in a competitive market with negatively correlated interest rates and underlying asset prices, futures prices will adjust to reflect the disadvantages to futures contracts and the price of futures contracts will be lower than the prices of forward contracts. Note that this is the reason that bond futures prices tend to be less than bond forward prices, as discussed further in Chapter 7.

Finally, consider an example of value and price of a forward contract in which the underlying asset pays dividends.

Example 4.11: Assume the underlying asset price is \$30 today, the asset pays a continuous dividend yield of 1%, and the risk-free rate for all maturities is 5% (continuously compounded).

a. What is the current two-year forward price?

$$
\begin{aligned}
F_0 &= S_0 e^{(r-q)T} \\
&= 30 \times e^{(0.05-0.01)\times 2} \\
&= 32.50
\end{aligned}
$$

b. Assume that instead of paying a continuous dividend yield of 1%, the asset provides a cash distribution of \$2 at the end of the first year. What is the current two-year forward price?

$$
\begin{aligned}
F_0 &= (S_0 - PV(I))e^{rT} \\
&= (30 - 2 \times e^{-0.05\times 1}) \times e^{(0.05)\times 2} \\
&= 31.05
\end{aligned}
$$

c. What is the value today of the long position in a two-year forward contract with a delivery price of $30?

$$\begin{aligned} V_t &= (S_t e^{-q(T-t)}) - F_0 e^{-r(T-t)} \\ &= 30 \times e^{-0.01\times 2} - 30 \times e^{-0.05\times 2} \\ &= 2.26 \end{aligned}$$

This completes Chapter 4 as a supplement to Chapter 3. In this chapter, more derivative valuation methods were introduced and they were applied to various exotic derivatives. We also generalized the geometric Brownian motion model to involve dividend payments. Finally, forward prices and futures prices were discussed. The next chapter examines a very important application of all valuation models: the hedging of risks.

Chapter Summary

Chapter 4 first reviews the three valuation methods from Chapter 3. It then explains and demonstrates the application of an alternative binomial tree (other than the CRR binomial tree mentioned in Chapter 3) and finite difference methods to solve the Black-Scholes PDE. After two new mathematical machineries are explained, various exotic options are introduced including gap options, Asian options and barrier options. Through those financial products, readers clearly understand why multiple valuation methods are necessary (because for some products there are methods that do not work). The reader develops a familiarity with various valuation methods through analysis of more complicated financial derivatives. After that, the effect of dividends from the underlying asset dividend is investigated using all valuation methods. Forward and futures prices are discussed at the end of the chapter. Both the tree and finite difference methods are applied with a spreadsheet and readers are encouraged to do hands-on activities. Sample codes of simulation in R are offered so that readers can mimic and practice. Finally, Chapter 4 contains a project on valuing options using the finite difference method in a spreadsheet. The authors' experience is that some readers will skip the spreadsheet and choose software such as R or Python to finish the project.

Chapter 4 Extensions and Further Reading:

Various derivatives are introduced in this chapter along with basic computational pricing tools. With regard to further reading on various products, readers may find *Options, Futures and Other Derivatives* by John Hull to be very helpful.

From the computational side, *Mathematical Modeling and Computation in Finance: With Exercises and Python and MATLAB Computer Codes* by Cornelis W Oosterlee and Lech A Grzelak does an excellent job both in illustrating stochastic calculus and numerical analysis in the field of quantitative finance. Their book covers the most popular models in the field for researchers and quants in the financial industry. There is one popular method in their books that is not covered in this book: the COS method through a discounted characteristic function for European option valuations. Although there is no code in the textbook itself, Matlab and Python codes used for most tables and figures are available to readers online. Readers will find that this book lays a nice foundation for Oosterlee and Grzelak's book which includes stochastic calculus as well as further information on computational tools and financial products.

End-of-Chapter Project

Finite Difference Methods for Option Pricing

This project extends the project from the previous chapter on valuing a put option. Specifically, the project demonstrates that the finite difference approaches converge to the Black-Scholes value in the case of a simple put option. For the explicit difference method, the project examines the convergence pattern. For the implicit finite difference method, the project practices programming skills with matrix algebra. The Analytic Model (in this case the Black-Scholes formula) for a European put is used as the benchmark. The project compares how fast the explicit finite difference method converges and compares it with the Binomial Tree method from the previous chapter's project. Finally, the project compares American put values from the explicit finite difference method with a benchmark. These exercises are important because finite difference methods are very popular valuation methods in the financial industry and these exercises help develop familiarity with their fundamental machineries.

A stock price is currently $60 and is expected to follow geometric Brownian motion. The stock's annual volatility is expected to be 15%. The risk-free rate is 5% per annum with continuous compounding. What is the value of a six-month European put option with a strike price of $62?

a1. Use a spreadsheet to build a 30x30 explicit finite difference method grid to approximate the put value.

a2. Report the put values with different n's in grids by increasing the number of steps with n = 30, 31, 32,…, 39. Place the put values on a graph with the X-axis as the number of grids and the Y-axis as the values of the puts. Analyze the pattern for convergence.

a3. Perform a convergence analysis as in a2, but with n = 40, 50, 60,…, 100. Graph the results as in a2. Compare the graph in a2. with the new graph in a3. What is similar about the two graphs and what is different? What does this comparison suggest regarding convergence?

a4. Show how to adjust the model if the put option is an American put.

b1. Use the implicit finite difference method to estimate the option value by choosing n = 5, 6, 7, 8, 9, and 10.

b2. Compare the results of the implicit difference method with those of the explicit finite difference method.

End of Chapter Problems (Fundamental Problems)

1. A stock price is currently \$30. Over the next 3-month period, assume that the stock price follows geometric Brownian motion with a volatility of 15%. The risk-free interest rate is 5% per annum with continuous compounding. What is the value of the following derivatives?
   a. Use a two period alternative binomial tree to find the value of a derivative, where the payoff is $[\mathrm{Max}(30 - S_T,\ 0)]^{1/2}$. S_T is the stock price in 3 months. Assume that early exercise is not allowed.
   b. Repeat the calculation in (a) assuming that early exercise is allowed and the corresponding payoff is $[\mathrm{Max}(30 - S_t, 0)]^{1/2}$ at any time t before (and including) T.
   c. Compare these results with the results of Chapter 3 Problem 5 which used a CRR binomial Tree.

2. A stock price is currently \$30. Over the next 3 months it is expected to have a volatility of 15%. The risk-free interest rate is 5% per annum with continuous compounding.
   a. Use a two-period alternative binomial tree to find the value of a derivative that pays off $\sqrt{S_T}$, where S_T is the stock price in 3 months. Assume that early exercise is not allowed.
   b. Repeat the previous calculation assuming that early exercises pays off $\sqrt{S_t}$ at any time t before (and including) T.

3. State which of the following four methods (analytic formula, CRR binomial tree, Monte-Carlo simulation, and finite difference method) is good for valuing the derivatives provided below. If several methods work well, state them all. Also, identify the feature(s) of the derivatives that led to your conclusion.
   a. The option payoff is Max $(S_T - S_{\text{ave}})$ (i.e., it is an average or Asian call option).
   b. An American-style call option.
   c. A derivative with payoff equal to $\sqrt[3]{S_T}$.
   d. A derivative with payoff equal to $\mathrm{Cos}(S_T)$.

4. Assume that the Singapore Dollar price is currently \$0.76 (US \$/SGD) and the exchange rate has geometric Brownian motion. Consider a 3-month call with a strike price of 0.755 (US\$/SGD). Assuming that the exchange rate volatility is 30%, the US risk-free interest rate is 2.5%, and the Singapore risk-free interest rate is 1.5% (both with continuous compounding), find the value for the European call using each of the following methods given:

   $S_0 = 0.76, \; T = 0.25, \; K = 0.755, \; \sigma = 0.3, \; r = 0.025, \; r_f = 0.015$

   a. Analytic method
   b. 30-step CRR Binomial Tree
   c. Explicit finite difference of size 30x30
   d. Simulation

5. A stock price is \$50 on January 1. The stock will pay a dividend of \$1 in one month (Feb. 1) and \$1 in four months (May 1). Assume that the risk-free rate of interest is 5% per annum with continuous compounding. On January 1 an investor establishes a short position in a 6-month forward contract on the stock with settlement on July 1.
   a. What is the January 1 value of the forward contract?
   b. On April 1 (three months later), the price of the stock is \$58 and the risk-free rate of interest is still 5% per annum. On April 1 what would be the forward price on a new forward contract that settles on July 1 (three months later)? What would be the April 1 value of the original short position in the forward contract (i.e., established January 1)?

6. In June, the S&P 500 index price is 2,430 and the Sep. S&P 500 futures price is 2,426. Assuming that the risk-free rate is 2% (continuously compounded), what is the implied dividend yield?
7. Consider a European call option on a futures contract on crude oil, with future price \$60, the time to the option's maturity is 4 months and the time to the futures' maturity is 5 months. The exercise price of the option is \$60, the risk-free interest rate is 3% per annum and we assume the volatility of the futures price is 20%.
   a. Based on the Black model, find the call value.
   b. Use a 30-step Binomial tree to value this call.

8. The current value of the British pound is \$1.20 and the volatility of the dollar/pound exchange rate is 15% per annum. A European call option to purchase a British pound has an exercise price of

$1.22 and a time to expiration of 6 months. Assume the U.S. risk-free rate is 2% per annum and the UK risk-free rate is 2.5%, both with continuous compounding.

a. Based on the Black-Scholes model, find the call value.

b. Use a 30-step Binomial tree to value this call.

9. The current value of the British pound is $1.20 and the volatility of the dollar/pound exchange rate is 15% per annum. An American put option to purchase a British pound has an exercise price of $1.22 and a time to expiration of 6 months. Assume the U.S. risk-free rate is 2% per annum and the UK risk-free rate is 2.5%, both with continuous compounding. Use $S_{max} = 2.40$, and make grids $\Delta S = 0.4$ and $\Delta t = 0.5/6$ to value the option. (Hint: If you use a spreadsheet, it is not hard to calculate 6x6 grids. If you don't use a spreadsheet, you can use a calculator; note that you don't need to know all the grid values, since only $f_{3,0}$ is needed. So only calculate the relevant grids.)

10. An American-style call option on the S&P 500 future Index is observed in June and is expiring in exactly three months. Assume that the call option's underlying asset is the December S&P 500 Index futures contract. The June value of the S&P 500 index futures is currently 2,426, the exercise price is 2,400, the risk-free interest rate is 2% per annum (continuously compounded), and the volatility of the index is 20% per annum. Find the call value using each of the following methods:

a. A 30 step CRR Binomial Tree

b. Explicit finite difference of size 30 x 30

End of Chapter Problems (Challenging Problems)

1. Show that $u = e^{\left(r-\frac{1}{2}\sigma^2\right)\Delta t+\sigma\sqrt{\Delta t}}$, $d = e^{\left(r-\frac{1}{2}\sigma^2\right)\Delta t-\sigma\sqrt{\Delta t}}$, and $Q = 1/2$ satisfy the equations:

$$EX_i = Q \times Lnu + (1 - Q) \times Lnd = r \times \Delta t - \frac{1}{2}\sigma^2 \times \Delta t$$

$$\text{Var}X_i = Q \times (Lnu)^2 + (1 - Q) \times (Lnd)^2 - (EX_i)^2 = \sigma^2\Delta t.$$

2. Show that if an asset price follows geometric Brownian motion, the geometric average of the price $S_{\text{ave}} = e^{\frac{1}{T}\int_0^T Ln(S(t))dt}$ is lognormal and the arithmetic average $S_{\text{ave}} = \frac{1}{T}\int_0^T S(t)dt$ is approximately lognormal.

   Now consider two more exotic options. This exercise uses the simulation approach to value the following options. In each exercise assume continuous compounding for interest rates.

   A ***floating lookback call*** is a financial derivative with a payoff $S_T - S_{\min}$.
   A ***cash-or nothing call*** is a binary option with payoff Q if $S_T > K$ and 0 otherwise.

3. Consider a floating lookback call with the following parameters: S = 50, r = 0.06, q = 0.04, σ = 0.2, and T = 0.5. Use simulation to value this call.
4. Consider a cash-or-nothing call with parameters: Q = 10, K = 100, S = 96.5, r = 0.05, q = 0.03, σ = 0.2, and T = 0.5. Use simulation to value this call.
5. Consider a 3-month down-and-out call option on cotton futures. The strike price is 54 cents per pound, the barrier is 48 cents, the current futures price is 51 cents, the risk-free interest rate is 3%, and the volatility of cotton futures is 40% per year. Use simulation to value this call option. (H = 48, K = 54, S = 51, r = 0.03, σ = 0.4, and T = 0.25).

Note

1 See Hull, John C. *Options, Futures, and other Derivatives, Ninth Edition", 2015, Pearson Education.*

5

Risk Management and Hedging Strategies

Risk is the possibility that something bad such as an injury or a loss will happen. In life, nobody likes risk. Economic risk is uncertainty regarding future utility or profitability. Hedging against risk means strategically using positions in assets or financial derivatives to offset undesirable risks such as adverse price movements. Various hedging strategies are divided into separate chapters in most textbooks. For example, hedging systematic risk (beta), duration hedging, and delta hedging are in different chapters in Hull's book. But all these hedging strategies have much in common. They belong to two main types of strategies: return-based hedging and unit-value based hedging. Their formulas are similar, but not identical. This chapter combines the hedging applications into a single chapter and discusses the mathematics, beginning with the simplest cases and models.

The chapter's discussion is organized around the type of financial derivative used to hedge as well as the complexity involved. The next section begins with one-to-one hedging using forward/futures contracts.

5.1 Simple Hedging Using Forward and Futures Contracts

Risk-bearing is a key financial activity in the process of striving for higher risk-adjusted returns. Market participants often are exposed to two types of risks: ***desirable risks***, risks that they are willing to bear in the search for higher expected returns, and ***undesirable risks*** that they would prefer to hedge away if they could be cost-effectively reduced or eliminated. Thus, the key goals of risk management are to reduce unwanted risks and manage those risks perceived as being necessary to bear or desirable to bear because they offer attractive risk premiums.

Hedging is the purposeful reduction or elimination of one or more existing risks by establishing new contracts or positions with variation that offsets some or all existing risks. Consumers often hedge their risks of losses from major storms and car accidents by purchasing insurance on their homes and cars. Individuals purchasing this insurance may be viewed as buying put options to hedge risks of property ownership. Similarly, firms manage their business risks by hedging with derivative contracts.

5.1.1 One-to-One Hedging of Commodity Risks Using Forward Contracts

Consider a farmer who grows, harvests, and sells corn. A major source of risk to the farmer is variation in the market price of corn, S_T, when it is ready for harvesting and sale at time T. The farmer is concerned that the corn price will decline before harvest. At the same time it is likely there are one or more corporations such as a grain mills and cereal producers that are concerned that they will face paying a high price to buy corn in the market at time T. If the corn price, S_T, is high at time T when the food processor needs corn, the processor will have to pay a high price to buy corn in the market. The processor's future cash flow is $- S_T$, where the negative sign indicates a payment.

Commodity forward contracts (and futures contracts) nicely meet the concerns of both the farmers growing the corn and the processors purchasing the corn. In effect, they agree prior to harvest (at time 0) to transact at time harvest time (T) using the contract forward price, F_0, rather than the market price of the commodity S_T. For convenience, at time T they typically settle the forward contract in cash rather than through physical delivery.

Chapter 1 also discusses physical and cash settlement of forward contracts. For simplicity, this chapter assumes that all forward contracts settle in cash. For example, consider a forward commodity contract (with no counterparty risk) and in which both parties agree at time 0 to exchange a specified commodity at time T at the current forward price, F_0. Table 5.1 depicts the per-bushel cash flows for the farmer with a short position in a forward contract on the commodity (corn) and a food producer with a long position.

Table 5.1 shows that a long position in a forward contract on corn generates the cash flow $S_T - F_0$ from the forward contract when the contract settles at time T (where F_0 is the forward price for the corn established at the inception of the contract). The long side (i.e., the mill or cereal producer) receives the payment $S_T - F_0$, when the difference is positive and makes the payment when it is negative. The long side (e.g., a processor such as a mill) then purchases the corn in the open market at S_T. Each processor's

TABLE 5.1

Example of Cash Flows per Bushel of Corn with and without Hedges

	Cash Flow at Time T without Hedge	Forward Settlement	Hedged Cash Flows
Farmer	$+S_T$	$F_0 - S_T$	$+F_0$
Corn Processor	$-S_T$	$S_T - F_0$	$-F_0$

transactions net to a payment of F_0 (the net price is locked-in by paying F_0 since $-F_0 = -S_T + S_T - F_0$ at time T). On the other side of the forward contract, the farmer's income is locked in using a short position in the forward contract. The farmer's short position in a forward contract on corn generates the cash flow $F_0 - S_T$ in which the farmer receives money at settlement when $F_0 - S_T > 0$ and pays $F_0 - S_T$ when it is negative. The farmer sells the corn in the open market at S_T which nets the cash flow F_0 at maturity T when combined with the payoff or payment from the forward contract.

In summary, the use of derivatives transforms each party from exposure to an uncertain cash flow at time T (S_T) to a fixed cash flow known at time T (F_0). Using forward contracts to hedge, both parties reduce unnecessary risk and can focus on their main businesses, which for the farmer is producing corn efficiently and for the food processors is using corn efficiently.

5.1.2 Natural Hedgers, Speculators, and Keynes' Theory of Commodity Returns

Both the farmer and the food processor in the previous section faced uncertainty in their income based on the future price of corn. Since market prices respond to random changes in supply and demand, in some years the corn price unexpectedly falls before the harvest – perhaps because of higher-than-projected harvests – helping the mills and other corporations, while hurting the farmer. In other years, the price of corn might rise before the harvest due to droughts. These high prices hurt the corn users and benefit the corn producers. Forward prices help both sides hedge their risk while retaining their incentives to produce and use corn efficiently.

The key to this example is that both participants can enter forward contracts in the marketplace with a size that matches the size of the position to be hedged. The farmer may enter a short position in corn forward contracts for 100,000 bushels – matching the farmer's anticipated harvest of 100,000 bushels. Similarly, in order to be fully hedged, the food processor would establish long positions in corn forwards with a size that matched its anticipated need for corn.

As illustrated in the above example, market participants use derivatives to reduce some or all of their ordinary business risks – especially those over which they have little or no control such as market prices of commodities and financial securities. If the demand of corn users to obtain long positions in corn forward contracts matches the supply of short positions desired by corn producers, the supply and demand will be nicely balanced. But what if the demand and supply from corn users and corn producers do not match at various points in time? Should market participants desiring a hedge wait patiently for the possibility that their counterparts may decide to enter the market? No. There is no reason to believe that the supply and demand for hedging vehicles will be balanced. The solution to these imbalances comes from *speculators* who enter the market as "reluctant" sellers to "anxious" buyers, and as reluctant buyers to anxious sellers. In other words, speculators (for a price) provide additional supply of forward contracts when there is excess demand and additional demand when there is excess supply at current market prices.

Keynes termed *market participants* using derivatives to hedge ordinary business risks as ***natural hedgers***. He theorized that imbalances in the demand and supply for forward contracts would be met by speculators as the imbalance caused forward prices to reach levels that offered risk premiums to the speculators. In other words, speculators enter markets when the market prices of the derivative contracts offer (or appear to offer) attractive risk premiums. If corn users seek long contracts in excess of the total short positions offered by corn producers, forward prices rise and speculators take on short positions in anticipation of eventual commodity price declines. In summary, the speculators strive to find imbalances among natural hedgers that lead to price distortions that, in turn, provide the speculators with an incentive (i.e., attractive risk premiums) to bear the risk.

5.1.3 Hedging Multiple Commodity Risks

Consider an example of a firm with reasons to hedge multiple commodities risks simultaneously. An oil refinery's business involves purchasing crude oil and refining that commodity into consumer products such as gasoline, heating oil, diesel fuel, jet fuel, and a host of other products. Consider a simplified representation of the process as converting three barrels of crude oil into two barrels of gasoline and one barrel of heating oil (a mixture often expressed simply as 3:2:1). The refinery seeks to earn a profit through generating revenues from selling the refined products that exceed the costs of purchasing and refining the crude oil. But the refinery faces the risk of increases in the price of crude oil and decreases in the prices of refined products.

The refinery seeks to reduce the risks of commodity price changes over which it has no control and which it perceives causes business risk that is both unpredictable and undesirable. However, the refinery is willing to bear other risks that it perceives as tolerable because the refinery has control over them, such as avoiding accidents, being efficient, and keeping labor costs low.

In summary, the refiner faces two major types of risk: fluctuations in the market prices of the energy products, and deviations in the efficiency with which the refinery operates. The refinery decides to *hedge* the risks of commodity price fluctuations by locking in the purchase price of crude oil and the sales prices of gasoline and heating oil using the energy derivatives market. In fact, energy derivative traders term the above relationship between the three products as the 3:2:1 ***crack spread***.

This refiner can use the 3:2:1 crack spread to hedge out the risks of energy price fluctuations. The result is that the refiner can focus on managing the risks over which the refiner has control: the degree of operational efficiency. The key is that the refiner chooses to bear the risks relating to operational inefficiency (avoiding waste, shutdowns, labor disputes, accidents, etc.) while hedging against the risks of increases in crude oil costs and decreases in the revenues from refined products. A key to the refiner's success is to construct effective hedge ratios between the refiner's commodity flows and the derivative contracts (e.g., energy futures contracts) on those commodities.

The energy refiner can benefit from well-designed hedging models. The models for hedging are simplified to the extent that the risk exposures are easily measured and that there are derivatives that trade at prices that are very highly correlated with the prices of the positions needing to be hedged. The latter sections of this chapter demonstrate the mathematics and economics of situations in which the hedging models consider various complications.

5.1.4 Hedging of Financial Risks Using Forward Contracts

Consider an endowment fund that inherits a $20 million well-diversified portfolio of U.S. stocks to fund construction of a new building with a projected cost of $20 million. We assume that our well-diversified portfolio has the same volatility as U.S. equity market index and is perfectly correlated with U.S. equity market index in this section. We will relax this condition in Section 5.2.5 where cross hedging is discussed. Suppose that due to legal issues the portfolio cannot be traded by the endowment for six months. How can the endowment fund utilize financial forward contracts to hedge the risk of fluctuations in the value of the stock portfolio during the next six months to lock in the current $20 million value? The solution using financial forward contracts is to establish a short position in a major U.S. equity market index such as the S&P 500 with a *notional* value of $20 million.

The ***notional value*** of a forward or futures contract is its quoted futures or forward price for the quantity of the underlying asset of each contract. The term *notional value* is used because participants in futures and forward contracts do not pay or receive the notional value when they enter and exit the contracts. Rather, they simply settle profits or losses in cash (although they may be required to post collateral). In other words, the value of a forward or futures contract is typically zero, at least initially. For example, if a person agrees to buy an ounce of gold at the current market price of $1,500 in three months, the value of the agreement is zero, but its notional value is $1,500. Of course, if the market price of the gold changes after the agreement is made, the agreement can begin to have positive or negative market values to the participants equal to the difference between the forward price and the market price as detailed in Chapter 4. Notional values of forward contracts (and futures contracts) work much better than market values when discussing hedges involving derivatives.

In the above example about hedging an endowment, the endowment locks in a "sales price" for its portfolio typically by establishing a short position in a forward contract or futures contract. In practice, the forward contract will be settled for cash. The hedge ratio must be set such that fluctuations in the value of the forward contracts will be very close to fluctuations in the value of the endowment (and in opposite directions). The effectiveness of the hedge depends on the extent to which the returns of the portfolio to be received match the returns of the equity index underlying the forward contract.

An investor with a financial asset who temporarily wishes to avoid bearing an asset's risk can use short positions in forward contracts to lay off the asset's risk. An example is an investor wishing to temporarily remove the risk of a portfolio, perhaps during a period in which the investor anticipates high volatility or a decline in prices. The investor is striving to avoid bearing the transactions costs of liquidating and re-establishing the portfolio.

Hedging with forward contracts on commodities shares many similarities with hedging with financial forward contracts. Some differences exist. Commodities can have *storage costs* and *convenience yields* that are not equal across market participants. Forward contracts on financial securities tend to have interest or dividend payments that *are* equal across market participants. Additionally, commodities are typically much more difficult to short sell than financial assets. This means that forward prices of commodity contracts, unlike forward contracts on financial assets, may not be driven toward a single no-arbitrage price.

5.1.5 Hedging of Currency Risks Using Forward Contracts

Entities facing currency risk (i.e., foreign exchange risk) can hedge that risk using forward contracts in much the same way that derivatives can be used to hedge commodity risk or the risks of financial assets. However, currency hedging involves two currencies – and so the concept can be a little more confusing since there are two currencies in which the transaction may be analyzed. The problem becomes easier when one of the currencies is viewed as the ***home currency***. Then, the ***foreign currency*** can simply be treated the same as any other asset or commodity.

Consider a U.S. exporter that has delivered U.S. goods to a European distributor and in return will receive 50,000 Euros in three months. The company's payoff in the spot market will be 50,000 × S_T where S_T is the Euro price in three months. Because the company will receive the Euros in three months, if the Euro depreciates in that time, the company will lose money (measured in U.S. dollars). The company faces ***foreign currency risk*** (i.e., ***exchange rate risk***).

The company wishes to lock-in its profits in terms of U.S. dollars. The company decides to enter into a short position in a three-month forward contract calling for the delivery of 50,000 Euros and receipt of U.S. dollars.

Through entering the three-month forward to deliver the Euros, the company will receive a fixed number of U.S. dollars per Euro as a result of its contract in three months. So, the company effectively locks in the exchange rate as the forward rate $F_0 = (S_T + F_0 - S_T)$. On the other hand, a U.S. importer from Europe may want to establish positions in forward contracts on exchange rates that contain long positions on the Euro and short positions on the U.S. dollar. Let's look at an example.

Example 5.1: A U.S.-based import firm must pay 10 million British pounds (GBP) in three months. Note that the risk to the company (measured in U.S. dollars) is that the price of the GBP will rise in terms of U.S. dollars causing an increased expense measured in U.S. dollars. So, the company establishes a 10 million GBP long position in a three-month forward contract at the U.S. dollar price of $1.31 per GBP. Determine what happens in the following scenarios:

a. The exchange rate rises to $1.33 per GBP in three months
b. The exchange rate declines to $1.29 per GBP in three months.

Solution of a. The payoff of the long 10 million GBP position at a forward price of $1.31 is $(S_T - F_0) \times 10 \text{ million GBP} = (1.33 - 1.31)\frac{\$}{\text{GBP}} \times 10 \text{ million GBP} = 0.2$ million dollars. The company gains from the forward market at the end of three months. However, in three months, the company needs to use 1.33 × 10 = $13.3 million to exchange to 10 million British pounds to pay its British bill. The company has locked in its cost (13.3 – 0.2 = $13.1 million) in terms of U.S. dollars.

Solution of b. The payoff of the long 10 million GBP forward contract at $1.31 is $(S_T - F_0) \times 10 \text{ million GBP} = (1.29 - 1.31)\frac{\$}{\text{GBP}} \times 10 \text{ million GBP} = -0.2$ million dollars. In this case, the company loses 0.2 million dollars at the given forward rate at the end of three months when measured in terms of U.S. dollars. The company may regret that it entered this long forward position. However, this is a consequence of hedging the risk: foregone potential profits and foregone potential losses. In three months, the company only needs to use 1.29 × 10 = $12.9 million to exchange to 10 million British pounds. The company locked in its net cost at 12.9 + 0.2 = $13.1 million.

The downside of using hedging is that the outcome is worse than not hedging some of the times. When a company hedges its risks, it is eliminating upside potential (as well as the downside risk). In the above example, when the market exchange rate goes down to $1.29, without hedging the company would have made a profit both in its main business and through a favorable movement in the foreign exchange market. Since the forward payoff is symmetric, the gain/loss is also symmetric. If we need to protect against a large loss, a forward contract involves the trade-off of losing the potential for a large gain. Options can provide downside protection while retaining upside potential. But this option protection requires an upfront premium (purchase price). An advantage of using positions in forward contracts rather than long positions in options is that forward contracts do not require an upfront premium (although they may require posting of collateral).

On an advanced note, it may be reasonable to question the rationality of being concerned about profits and losses as measured by a particular currency. Currencies fluctuate in value relative to each other. Hedging to lock in a profit based in U.S. dollars does not remove all risk to a U.S. investor – rather it concentrates the risk on fluctuations in the value of the U.S. dollar. The U.S. dollar, like all currencies, fluctuates in value. Because investors in each country tend to view their financial situation as being measured in terms of their home country's currency, there is an illusion that hedging against foreign exchange risk reduces total risk exposure. Rather, it simply focuses the exposure to currency risk on the hedger's home currency – which may or may not be a wise economic decision.

5.1.6 Hedging Using Futures Instead of Forwards

Forward and futures serve similar purposes. In the previous examples, the forward contracts can be replaced with futures contracts if they are available. Forward contracts can have customized features (such as exact size and settlement dates since it is a negotiated contract between two parties) while futures contracts are standardized in size and maturity (by the exchange on which it is traded) to facilitate widespread trading of identical contracts. In addition, futures have daily settlement (marking-to-market) which means that cash is transferred between the accounts of the long and short positions based on daily price changes. This daily cash settlement has an important implication: it accelerates the cash flows from profits and losses between each side of the contract – which changes the risks of the contracts (relative to forward contracts) due to the time value of money.

Also, if futures are used, there may not be an exact match for the size and timing of the hedge. Participants typically choose the futures contract that has a maturity closest to the time at which the risk exposure being hedged is expected to resolve. Consider the following example.

Example 5.2: A U.S.-based export firm will receive 10 million GBP in three months. The risk to the company is that the GBP's value relative to the U.S. dollar will go down. Assume that only four-month GBP futures contracts (as opposed to a three-month maturity match) are available and are trading at a futures price of \$1.31. The export company shorts contracts totaling 10 million four-month GBP futures and then closes its position after three months. Determine what happens in the following separate scenarios:

a. Three months later, the spot exchange rate is \$1.33 and the futures price is \$1.34.
b. Three months later, the spot exchange rate is \$1.29 and the futures price is \$1.30.

Solution of a. In three months, the payoff of the short 10 million four-month GBP futures at \$1.31 is: $(F_0 - F_1) \times 10$ million GBP $= (1.31 - 1.34)\frac{\$}{\text{GBP}} \times 10$ million GBP $= -0.3$ million dollars.

In three months, the firm loses \$0.3 million in the futures market. However, also in three months, the firm is able to convert the 10 million GBP in the spot market to:

$$1.33 \times 10 \text{ million} = 13.3 \text{ million U.S dollars.}$$

The total result is that the firm locked in its revenue at 13.3 – 0.3 = \$13.0 million.

Solution of b. In three months, the payoff of short 10 million four-month GBP futures at \$1.31 is: $(F_0 - F_1) \times 10$ million GBP $= (1.31 - 1.30)$ $\frac{\$}{\text{GBP}} \times 10$ million GBP $= 0.1$ million dollars.: In three months, the firm gains \$0.1 million in the at futures market. However, in three months, the firm is also able to convert the 10 million GBP in the spot market to:

$$1.29 \times 10 \text{ million} = 12.9 \text{ million U.S. dollars.}$$

The total result is that the firm locked in its revenue at 12.9 + 0.1 = \$13.0 million.

In both cases, the firm receives \$13.0 million. Note that both of the scenarios assumed that the difference between the spot price and futures price at the end of the hedge was –0.01. This quantity, $S_1 - F_1$, is called the ***basis*** (b_1). The price at which the firm exited its short position was $F_0 + b_1$. So in both scenarios, the basis was the same ($S_1 - F_1 = -0.01$) and therefore the result was the same. If the two scenarios had different bases, the results would have differed between the scenarios. This demonstrates the ***basis risk*** of hedging with futures or forward contracts in which the contracts are exited prior to the settlement date of the contracts (or the contract is settled prior to the resolution of the risk being hedged). This problem disappears when hedgers are able to use contracts that expire (settle) on the date at which the hedge is designed to end. The reason is because the futures price (or forward price) and the spot price of an asset converge (at expiration the basis of both contracts, $S_1 - F_1$, goes to zero).

5.2 One-to-h Hedging Using Forward and Futures Contracts

This section discusses several extensions of the simple case of one-to-one hedging in the previous section. In this section, the underlying asset of the forward contract is not the same as the asset to be hedged. All hedges in this section belong within two main types of strategies: return-based hedging and unit-value based hedging. ***Return-based hedging*** reduces the sensitivity of the portfolio's return due to some risk factor such as market risk or a yield change. ***Unit value-based hedging*** reduces the sensitivity of the portfolio's value due to some risk factor such as an asset price change or an asset volatility change. ***Dynamic hedging*** refers to another perspective of the hedging: the process of rebalancing through time to maintain target risk exposures. ***Cross-hedging*** is a special type of hedging in which the returns/unit value to the asset to be hedged and the hedging instrument are imperfectly correlated. In the previous section, since the underlying asset of the forward contract is the same as the asset to be hedged, the volatility and quantity of the asset being hedged and the volatility and quantity of the futures/forward contract's underlying asset are the same and the correlation between the asset and hedging instrument is perfect (one-to-one). This section first investigates various hedging strategies when the correlation between two positions is close to perfect but the volatilities of the two positions differ (Sections 5.2.1–5.2.4). Finally, we discuss when there is imperfect correlation between two positions in Section 5.2.5.

5.2.1 Return-Based Hedging Using Forward Contracts

The main focus of this section is to find a hedge ratio between the asset to be hedged and a hedging vehicle using return-based hedging. The goal is to find a ratio between the weights of the two assets in a hedged portfolio that reduces the sensitivity of the hedged portfolio's return to some risk factor. The ratio is called the ***hedge ratio***. The two assets here refer to the ***asset to be hedged***, which is the asset with risks which the hedger wishes to eliminate or reduce, and the ***hedging vehicle***, which is the asset used to hedge the risk.

A complication of modeling with forward contracts is that the contract itself typically requires no investment outlay and has little or no market value. Thus, unlike cash assets (assets with positive cash values), its *percentage return* cannot be computed. A solution is to model *fully collateralized forward contracts*. A fully collateralized forward contract is a combination of the forward contract and sufficient funds for the long side to pay for delivery of the underlying asset at the initial forward price of the contract (F_0). The quantity of money (i.e., the collateral) is placed in an asset with a riskless return, r, to fund settlement of the contract at time t and is: $-F_0e^{-rT}$. This permits computation of the return of a forward contract. For short time periods and moderate interest rates, the return of the forward contract from time t to time T is often approximated as: $(F_T - F_t)/F_t$ where F_t is the forward price for a contract initiated at time t. For futures contracts, F_t is the futures price.

A fundamental technique for modeling the optimal hedge ratio is to use coefficients from regressions of returns if the return of the hedging vehicle and the return of the asset to be hedged have a correlation close to 1 or –1. Consider the following linear regression model:

$$r_{asset} = \mathrm{b}_* r_{hedging\ \ vehicle} + \text{error}$$

The coefficient b of the regression can be interpreted as follows: If the return of the hedging vehicle is 1%, the return of the asset tends to be b%.

From elementary statistics, it is easy to see:

$$\mathrm{b} = \frac{\rho_{\text{asset, hedging vehicle}} \times \sigma_{\text{asset}}}{\sigma_{\text{hedging vehicle}}}.$$

where $\rho_{\text{asset, hedging vehicle}}$ is the correlation of the returns of the hedging asset and the asset to be hedged, σ_{asset} is the volatility of the asset (i.e., standard deviation of returns) being hedged and $\sigma_{\text{hedging vehicle}}$ is the volatility of the hedging vehicle.

Assume in a hedged portfolio containing an asset and a hedging vehicle, the weight allocations are w_{asset} and $w_{\text{hedging vehicle}}$.

In order to reduce the sensitivity of returns of a hedged portfolio containing an asset to be hedged and a hedging vehicle, make: $w_{\text{hedging vehicle}}|r_{\text{hedging vehicle}}|$ and $w_{\text{asset}}|r_{\text{asset}}|$ equal in value but with an opposite direction (sign):

$$w_{\text{hedging vehicle}}|r_{\text{hedging vehicle}}| = w_{\text{asset}}|\ r_{\text{asset}}|$$

$$w_{\text{hedging vehicle}} = w_{\text{asset}} \times |b|.$$

Inserting the above formula of b,

$$w_{\text{hedging vehicle}} = w_{\text{asset}} \left| \left(\frac{\rho_{\text{asset,hedging vehicle}}\sigma_{\text{asset}}}{\sigma_{\text{hedging vehicle}}} \right) \right|$$

Define h as a *hedge ratio*:

$$h = -\rho_{\text{asset, hedging vehicle}}\sigma_{\text{asset}}/\sigma_{\text{hedging vehicle}} \tag{5.1}$$

The ***weight allocation*** to reduce the sensitivity of the return of the portfolio is:

$$w_{\text{hedging vehicle}} = w_{\text{asset}} \times |h| \tag{5.2}$$

An absolute value is used in Equation 5.2 to make the w_{asset} and $w_{\text{hedging vehicle}}$ both positive. The hedge ratio could have a negative sign. If the correlation is positive, $h < 0$, a long position hedges a short position. If the correlation is negative, $h > 0$, both positions are held with the same direction (both long or both short).

Let's consider the weight w carefully. Assume that an investor short sells \$100 in an asset with a return of 1% with respect to a movement of a risk factor and would like to eliminate the risk with a hedging vehicle offering a return of 4% with respect to the same movement of the above risk factor. It is very clear that the investor would need to go long \$25 in the hedging vehicle. What are the $w_{\text{hedging vehicle}}$ and w_{asset} in Equation 5.2?

The answer depends on where the negative sign is placed.

Case 1:

If the negative sign is placed on one of the returns, then the weights as well as the positions in the assets must have the same sign. This can be viewed as being long \$100 in an asset with a return of –1%.

$$w_{\text{hedging vehicle}} = \frac{25}{125} = 20\%$$

$$w_{\text{asset}} = \frac{100}{125} = 80\%$$

The hedging equation is: $0.8 \times (-0.01) + 0.2 \times (0.04) = 0$.

Case 2:

If the negative sign is on the weight of one asset:

$$w_{\text{hedging vehicle}} = \frac{25}{-100 + 25} = -33.333\%$$

$$w_{\text{asset}} = \frac{-100}{-75} = 133.333\%$$

The hedging equation is: 1.33333 × (0.01) − 0.33333 × (0.04) = 0 and the returns of both assets must share the same sign.

Note that what matters is that the relative weights in both cases are 4:1. One is long and one is short (hedge ratio h = −1/4).

Consider the advantage and disadvantage of using negative weights. The disadvantage of using a negative weight is that long $100 in asset one, long $50 in asset two, and short $150 in asset three could make the total assets 0 (100 + 50 − 150 = 0), which is problematic because the denominator cannot be 0. The advantage of using a negative weight is that when multiple assets are considered and it is not known which one is supposed to be long and which one is supposed to be short, then negative signs cannot be assigned to returns. This is particularly true in Chapter 6 where negative weights are allowed. The optimization algorithm will output the sign of weights automatically. So, the convention in Chapter 6 is as in case 2: total investment weights are always 100% and the weights can be positive or negative. However, in this chapter, for convenience, there is one portfolio to be hedged and one hedging vehicle, and we assume all weights are positive. So, it is necessary to put a short position with an opposite return as in case 1 when weights and investment dollars are positive. In section 5.6, there is a short position with a negative duration while weights and investment dollars are positive. In that way, the denominator will not be 0.

Instead of hedging by weighted returns (as above), underlying assets are often counted in units. The next develops the hedge ratio based on the unit values of the assets.

5.2.2 Unit Value-Based Hedging Using Forward Contracts

The analytics of determining a hedge ratio can be performed with a focus on returns or on unit values. The decision depends on the nature of the asset to be hedged. If it is a stock portfolio, then there are multiple units of different assets and returns should be the focus. If the asset to be hedged is a particular stock or stock index, the unit value approach will tend to be easier. Sometimes, both approaches work.

This section is focused on unit values with the goal of matching the total dollar size of a long position with the total dollar size of a short position. In the previous section, position size was expressed in weight percentages. The position being hedged had a weight of w_{asset} and the position used to hedge had a weight of $w_{\text{hedging vehicle}}$. Weight percentages tend to work well when modeling with returns since returns are proportional numbers in which the total return of a portfolio is a sum of percentage-weighted individual returns. The following demonstrates how to find hedge ratio in unit value-based hedging (with a focus on unit numbers) where the total investment of a portfolio is a sum of unit-weighted individual investments. We then compare the two approaches using an example.

Suppose that we own n_A units of financial asset A at a value of A per unit. We would like to use n_B units of a forward or futures contract (asset B) with an initial value of B per contract to hedge the risk of asset A assuming that ΔA and ΔB have a correlation close to 1 or −1. The goal is to create an optimal hedge, so that a dollar change in the value of asset A will be offset as closely as possible by a dollar change in the notional value of asset B. Let ΔA and ΔB denote the per unit price change of each asset. The hedge ratio for a perfect hedge in this case is expressed as a ratio based on the dollar price changes per unit and number of units. Specifically, multiply the *per unit price change of each asset*, ΔA, by *number of units*, n_A, do the same with asset B and equate the two changes:

$$n_A \, \Delta \mathrm{A} = -n_B \, \Delta B$$

Define h as the *hedge ratio*, expressed using the unit value approach as:

$$h = -\Delta A / \Delta B$$

In reality, h cannot be estimated perfectly and most likely h is not a constant. The easiest model to use is to estimate h based on this simple linear model (assuming ΔA and ΔB have a correlation close to 1 or −1):

$$\Delta A = (-h) \times \Delta B + \text{error}$$

The ratio h is estimated based on many data points of $(\Delta B, \Delta A)$ and running a linear regression of ΔA with respect to ΔB. If the hedge ratio is constant, the regression line would be expected to pass through (0,0) and have a slope of $-h$. The hedge ratio h is the negative of the slope coefficient of this regression line, which can be calculated as:

$$h = -\rho_{\Delta A, \Delta B} \frac{\sigma_{\Delta A}}{\sigma_{\Delta B}}$$

where ΔA and ΔB are the price changes of assets A and B for the hedge period, $\rho_{\Delta A, \Delta B}$ is the correlation between price changes, ΔA and ΔB, and $\sigma_{\Delta A}$, $\sigma_{\Delta B}$ are the standard deviations of the price changes of assets A and B for the same period.

Note that this is only a perfect hedge if our assumption, that the hedge ratio is constant, holds. The goodness of fit of the regression line can be used to test the assumption. From statistics, the hedge will work well if the correlation $\rho_{\Delta A, \Delta B}$ is close to −1 or 1. The regression's R^2 tells us how good a regression line is. Note that $R^2 = (\rho_{\Delta A, \Delta B})^2$ in this single factor model.

Recall that A is the asset to be hedged and B is the hedging vehicle. The hedge ratio formula for the unit value approach is:

$$h = -\frac{\Delta\textbf{asset}}{\Delta\textbf{hedging vehicle}} = -\rho_{\Delta\text{asset},\ \Delta\text{hedging vehicle}} \frac{\sigma_{\Delta\text{asset}}}{\sigma_{\Delta\text{hedging vehicle}}} \quad (5.3)$$

Let n_{asset} and $n_{\text{hedging vehicle}}$ be the units of assets in the asset and its hedging vehicle, respectively. The formula $n_{\text{asset}}\ \Delta\text{asset} = n_{\text{hedging vehicle}}\ \Delta\text{hedging vehicle}$ can be used to solve for $n_{\text{hedging vehicle}}$. The number of ***units*** for optimal hedging is:

$$n_{\text{hedging vehicle}} = n_{\text{asset}} \times |h| \quad (5.4)$$

If $h < 0$, the hedge should be long $n_{\text{hedging vehicle}}$ units of the hedging vehicle to hedge a short position. If $h > 0$, the hedge should be short $n_{\text{hedging vehicle}}$ units of the hedging vehicle to hedge a short position. In particular, if the riskless rate is fixed, if the forward contract used to hedge risk has an underlying asset that is the same as the asset to be hedged, and if the hedge's time period matches the forward's maturity, then

$$\sigma_{\Delta\text{hedging vehicle}} = \sigma_{\Delta\text{asset}},\ \rho_{\Delta\text{asset},\Delta\text{hedging vehicle}} = 1, \quad \text{then}$$

$$h = -1$$
$$n_{\text{hedging vehicle}} = n_{\text{asset}}$$

This is one-to-one hedging. One-to-one hedging ($h = -1$) indicates that one unit of a fully collateralized short position in a forward contract perfectly hedges one unit of a long position in the underlying asset. This is consistent with the results depicted in Table 5.1 that indicate a perfect hedge. The solution is appropriate for *financial* forward contracts in a perfect market. The relationship is approximate for *commodities* due to the potential storage costs and convenience yields, which can vary between economic participants.

Note that the unit value approach formula above is very similar to the formula for the returns approach. However, there are significant differences in theory:

1. The hedge ratio formula looks similar. However, one deals with the variance and correlation of the returns and the other deals with the variance and correlation of the price changes, so they are different numerically. This is demonstrated in the next example.
2. If the correlation of the two assets (price changes or returns) is close to 1 or −1, both methods can achieve a nearly perfect hedge and then the two methods are similar.

The following example uses two approaches, unit value and return, to compare the results.

Example 5.3: Assume that that the standard deviation, $\sigma_{\Delta A}$, of monthly changes in the prices of asset A (ΔA) is \$0.60 and the standard deviation, $\sigma_{\Delta B}$, of monthly change in the prices of futures contracts on asset B (ΔB) is \$8 with perfect positive correlation between the two changes. An investor has 2 million units of asset A worth \$10 each. Using asset B, a futures contract with a notional value of \$100, as the hedging vehicle, how many futures contracts form the optimal hedge?

- Unit value-based hedging approach:

Because $\sigma_{\Delta A} = 0.6$, $\sigma_{\Delta B} = 8$, $\rho_{\Delta A,\Delta B} = 1$:

$$h = -\rho_{\Delta A,\Delta B}\frac{\sigma_{\Delta A}}{\sigma_{\Delta B}} = -1 * \frac{0.6}{8}$$

$$n_B = n_A \times |h| = 2{,}000{,}000 \times \left(1 \times \frac{0.6}{8}\right) = 150{,}000 \text{ units of } B$$

The negative sign of h in the equation for h implies a short position of 150,000 units of B.

The \$20 million of asset A should be hedged with \$150,000 units of asset B (which at \$100 each is \$15 million).

- Return-based hedging approach:

Assume that there is perfect correlation of returns and $\sigma_A = \frac{\sigma_{\Delta A}}{A}$, $\sigma_B = \frac{\sigma_{\Delta B}}{B}$.

In Section 5.2.1 the hedge between an asset to be hedged and hedging vehicle was analyzed with weights and returns. Given perfect correlation of the returns (and denoting the asset to be hedged as A and the hedging vehicle as B):

$$h = -\rho_{A,B}\sigma_A/\sigma_B$$

where h expresses the ratio between the standard deviation of returns.

The weight allocation is:

$$w_B = w_A|h| = w_A\rho_{A,B}\frac{\sigma_A}{\sigma_B}$$

Now, $\sigma_A = \frac{\sigma_{\Delta A}}{A}$ and $\sigma_B = \frac{\sigma_{\Delta B}}{B}$, where A and B are unit prices for assets A and B, respectively:

$$h = -\rho_{A,B}(\sigma_{\Delta A}/A)/(\sigma_{\Delta B}/B) = -1 \times (\$0.60/\$10)/(\$8.00/\$100) = -0.75.$$

$$w_B = 0.75w_A$$

The \$20 million of asset A should be hedged with a weight of 75% in asset B, which is \$15 million (or 150,000 units).

Note that in the unit value method and return method, the hedge ratio formulas (5.1) and (5.3) place the hedging vehicle in the denominator and the asset being hedged in the numerator. It is important to remember this fact.

Unit value-based hedging:

$$h = -\frac{\boldsymbol{\Delta}(\textbf{asset price})}{\boldsymbol{\Delta}(\textbf{hedging vehicle})} = -\rho_{\boldsymbol{\Delta}\textbf{asset},\boldsymbol{\Delta}\textbf{hedging vehicle}}\frac{\sigma_{\boldsymbol{\Delta}\textbf{asset price}}}{\sigma_{\boldsymbol{\Delta}\textbf{hedging vehicle}}}$$

$$n_{\text{hedging vehicle}} = n_{\text{asset}} \times |\boldsymbol{h}|$$

Return-based hedging:

$$\boldsymbol{h} = -\rho_{\textbf{asset return,hedging vehicle return}}\sigma_{\textbf{asset return}}/\sigma_{\textbf{hedging vehicle return}}$$

$$w_{\text{hedging vehicle}} = w_{\text{asset}}|\mathbf{h}|$$

The hedge is not static unless the assets are perfectly correlated and the hedging positions are of equal size and volatility. The next section considers dynamic hedges.

5.2.3 Dynamic Hedging

Dynamic hedging is a strategy that involves rebalancing hedge positions as market conditions change. If market changes occur that cause the parameters in calculating the hedge ratio, h, to change, h generally needs to be recalculated with the new parameters and the hedging position needs to be rebalanced. Section 5.3 shows various examples of dynamic hedges with options involved and where the hedge ratio is not constant.

There are three common dynamic hedging strategies in practice:

1. The portfolio weights can be *continuously* rebalanced to a ratio of h in which h is not constant. Continuous rebalancing is used extensively in subsequent sections.
2. The portfolio weights can be rebalanced according to a time schedule (e.g., daily), or

3. The portfolio weights can be rebalanced when the ratio of weights departs from h by a preset criterion.

In most applications, the decision involves a trade-off between transactions costs and portfolio risk with more rapid rebalancing minimizing risk but increasing transactions costs. The first strategy of continuous rebalancing is not possible, although computerized trading allows extremely rapid rebalancing. The second strategy is conducive to scheduled rebalancing such as daily rebalancing, while the third strategy is designed to transact on an "as-needed" basis. For theoretical modeling purposes, the rebalancing strategy is typically assumed to be continuous and transactions costs are ignored.

Why is rebalancing required even if the hedge ratio itself is constant? Recall that the hedging Equation 5.2 can be rewritten into:

$$\frac{w_{\text{hedging vehicle}}}{w_{\text{asset}}} = |h| = |\rho_{\text{asset, hedging vehicle}}|\,\sigma_{\text{asset}}/\sigma_{\text{hedging vehicle}}.$$

When the two assets involved in the hedge have equal variances (or volatilities) and are perfectly correlated, the hedging ratio, h, remains constant at 1. In such a one-to-one hedge, the returns of two assets are essentially the same, so there is no need to adjust the relative weights during the hedge process since the hedge ratio keeps unchanged during hedge period. When the two assets involved in the hedge have unequal variances and are imperfectly correlated, the positions to be rebalanced constantly to keep the hedging ratio, h, unchanged through time. The reason is because the return (and therefore the changes in the market values) of the two assets differ. The absolute value of the return of the more volatile asset will be higher than the absolute value of the lower volatility asset. Even though the optimal hedge ratio will remain at h, (assuming the expected volatilities are unchanged) the actual ratio between the assets will change. Therefore to keep the ratio of the actual weights of the two assets fixed requires constant rebalancing. In other words, the size of one or both positions must be altered to keep the hedge ratio at $\sigma_{\text{asset}}/\sigma_{\text{hedging vehicle}}$. The next section provides a numerical example that illustrates why constant rebalancing is needed when the hedge ratio is not one. In particular, an example of rebalancing is discussed with differing leverage that causes differing volatility. We first examine rebalance is needed even hedge ratio is a constant, we then examine rebalance is needed when hedge ratio is not a constant in Section 5.3.

5.2.4 Hedging with Perfectly Correlated Assets and Differing Leverage

A common reason for investments with similar assets to exhibit different return volatilities is leverage. Let's examine the case of well-diversified equity mutual funds (i.e., mutual funds with common stock as their primary holding). Most equity mutual funds are unlevered and therefore investors can generally expect the returns of the well-diversified funds to have roughly the same return volatility as a *direct* investment in a well-diversified portfolio of equities. Some ***exchange traded funds (ETFs)*** offer two-to-one leverage by using equity derivatives such as futures contracts to double the exposure of the fund to the volatility of the underlying equity market. These ETFs therefore tend to exhibit returns with twice the volatility of its underlying assets. A key issue is the frequency with which the mutual fund adjusts its derivatives or debt level to retain a 2–1 level of return volatility. Most levered ETFs rebalance their portfolio leverage daily.

Let's consider the implications of using a $50,000 short position in a 2–1 levered ETF to hedge the risk of a $100,000 portfolio of equities assuming that both assets are invested in the same type of equities (e.g., U.S. large cap equities). Because the daily returns of the assets are perfectly correlated (by assumption) and the volatility of the ETF is double that of the equity portfolio, the optimal hedge ratio is −0.50 dollars of the ETF per each 1 dollar long of unlevered equities. Thus the investor establishes a short position in the ETF of $1 for every $2 of asset in their equity portfolio (or a long position in an ETF designed to have a short exposure to equities).

The hedge should work fine for the first day. Let's consider the result of a 2% increase in equity prices. The 2% increase causes the long equity position to rise from $100,000 to $102,000. That gain should be offset by the short position in the ETF, which generates a loss of 4% (2–1 leverage times a 2% equity increase times −1 to reflect the short position). The 4% decline in the $50,000 short ETF position is a $2,000 loss from the ETF increasing from $50,000 to $52,000. The $2,000 gains and losses offset each other as designed.

However, note that the investor's portfolio now includes a $102,000 long position in direct equities and a $52,000 short position in the levered ETF. Since the volatilities and correlation have not changed, the portfolio is no longer hedged to its optimal hedge ratio. To return the portfolio to being hedged the positions must be returns to a hedge ratio of −0.50. The solution that maintains a $100,000 long position at the end of the trading day the investor should sell $2,000 of stock and reduce the short position in the ETF by $2,000 to restore the −0.50 hedge ratio.

The key result is this: when leverage drives volatility differences, a hedge needs to be rebalanced whenever the underlying leveraged positions change to bring back the exposures to target levels even if hedge ratio keeps a constant.

Note that the hedge ratio formulas in Section 5.2.1 involve variances and correlation. Changes in variances and correlation require rebalancing the position to maintain the hedge since the hedge ratio has changed. The next section discusses cross hedging of imperfectly correlated assets.

5.2.5 Hedging with Imperfectly Correlated Assets

The previous sections focused on hedging when assets have a perfect correlation. Here we discuss hedging between assets that are imperfectly correlated. A ***cross hedge*** is a hedge in which the asset being hedged and the underlying asset of the hedging vehicle (usually futures or forward contracts) are not the same but are somewhat highly correlated. This section explores hedging imperfectly correlated assets based on both the unit values approach and the weighted returns approach.

Examples include: (1) hedging a portfolio of equities with a financial derivative based on an equity index that is only similar to the portfolio being hedged, (2) hedging a currency exposure on a country with a currency for which futures contracts are not traded using the futures contracts on a nearby country's currency that does have futures contracts available, and (3) hedging a commodity exposure using futures contracts that specify delivery of a commodity that has a different location, grade, or other quality such as the differences found in types of crude oil.

A cross hedge has the following assumptions: *the correlation of the asset to be hedged and the asset to be used as a hedging vehicle (e.g., the underlying asset of the futures contracts when futures are used to perform the hedge) is reasonably high, yet not perfect.* If the correlation is low (i.e., the fit of the regression model, R^2, is low), the hedge may introduce more risk than it removes. Understanding the mathematics and statistics behind the hedging techniques is essential in understanding appropriate applications and potential limitations.

In order to apply the hedge ratio formulas from the cross hedging examples, begin by reviewing the formulas and then discuss the applications.

- **Hedging a stock based on unit values**
  Consider the use of a unit value approach when correlation is imperfect. Historical data can be used to estimate the slope coefficient using a linear regression. From statistics, the most common estimate of the slope coefficient is:

$$h = -\frac{\Delta(\text{asset price})}{\Delta(\text{hedging vehicle})} = -\rho_{\Delta\text{asset},\Delta\text{hedging vehicle}} \frac{\sigma_{\Delta\text{asset price}}}{\sigma_{\Delta\text{hedging vehicle}}}$$

where $\rho_{\Delta\text{asset},\Delta\text{hedging vehicle}}$ is the correlation between the price changes of the asset and hedging vehicle and $\sigma_{\Delta\text{asset}}$ and $\sigma_{\Delta\text{hedging vehicle}}$ are standard deviations for price changes for the asset and hedging vehicle. In this derivation, h is derived from dollar price changes rather than percentage returns. The results are therefore related to per unit price changes, and we can calculate that $n_{\text{hedging vehicle}}$ units of hedging vehicle to hedge against n_{asset}units of asset:

$$n_{\text{hedging vehicle}} = n_{\text{asset}} \,|\, h \,|$$

$$n_{\text{hedging vehicle}} = n_{\text{asset}} * \,|\, \rho_{\Delta\text{asset},\Delta\text{hedging vehicle}} \,|\, \frac{\sigma_{\Delta\text{asset price}}}{\sigma_{\Delta\text{hedging vehicle}}}$$

For example, consider an investor wishing to hedge a position in stock *i* using a derivative position on the overall stock market such as the S&P 500. Assume that the correlation between the price changes of the two assets is 0.83 and, for simplicity, that they both have the same unit size. Assume further that the annualized volatility of the price changes of stock *i* is 0.60 while the annualized volatility of the price changes of the overall market is 0.15. Inserting these values into the previous equation generates the hedge ratio $h = -0.83 \times 0.60/0.15 = -3.32$. This result means that for every one unit of stock *i* being hedged the manager should short 3.32 units of the market asset. Note that with a volatility of 0.60, stock *i* had four times the volatility of the market, but the hedge ratio is only 3.32 to 1. The point is that when assets have modest correlation, the optimal hedge ratio tends to use a relatively small number of units of the hedging vehicle relative to when the correlation would be high. For example, a correlation of 1 would generate a hedge ratio of –4.00 in this example. The relatively low optimal hedge ratio when correlation is imperfect makes sense because a large position in the hedging vehicle would introduce large risks from the performance of the hedging asset that is not correlated with the asset being hedged.

- **Hedging a stock portfolio with index futures based on returns**

In the returns approach, recall that in Section 5.2.1, the derived hedge ratio is:

$$h = -\rho_{\text{asset return},\text{hedging vehicle return}}\sigma_{\text{asset return}}/\sigma_{\text{hedging vehicle return}}$$

$$w_{\text{hedging vehicle}} = w_{\text{asset}} \,|\, h \,|$$

In this example, the asset to be hedged is the stock portfolio *P* and the hedging vehicle is the S&P 500 futures contract, *F*.

$$w_F = w_P \times \,|\, h \,|$$

$$h = -\rho_{r_P,r_F}\sigma_{r_P}/\sigma_{r_F}$$

Assuming that the S&P 500 futures index represents the market, then:

$$h = -\frac{\rho_{r_P,r_F}\sigma_{r_P}}{\sigma_{r_F}} = -\rho_{r_P,r_m}\sigma_{r_P}/\sigma_{r_m} = -\beta_p$$

$$w_F = w_P \times |\beta_p|.$$

Weights may not sound as intuitive as values or even the number of contracts in an investment. In the following material, we derive the formula for the number of contracts.

Let V_P denote the stock portfolio value and V_F denote the futures value for a single contract and let N be the number of contracts we need for hedging.

Since $w_F = \frac{N \times V_F}{N \times V_F + V_P}$ and $w_P = \frac{V_P}{N \times V_F + V_P}$, then:

$$\frac{N \times V_F}{N \times V_F + V_P} = \frac{V_P}{N \times V_F + V_P} \times |\beta_p|$$

$$N = \frac{V_P}{V_F} \times |\beta_p|.$$

The formula for the number of contracts using S&P 500 index futures is:

$$\boldsymbol{N} = \frac{\boldsymbol{V_P}}{\boldsymbol{V_F}} \times |\boldsymbol{\beta_p}| \tag{5.5}$$

To obtain a target level of systematic risk, β_T for the hedged portfolio (rather than eliminating systematic risk), the number of contracts needed is:

$$\boldsymbol{N} = \frac{\boldsymbol{V_P}}{\boldsymbol{V_F}} \times |\boldsymbol{\beta_T} - \boldsymbol{\beta_p}| \tag{5.6}$$

In addition, if our hedging vehicle has systematic risk not equal to the market, i.e., $\beta_{\text{hedging vehicle}} \neq 1$, the number of contracts needed is:

$$\boldsymbol{N} = \frac{\boldsymbol{V_P}}{\boldsymbol{V_F}} \times \left|\frac{\boldsymbol{\beta_T} - \boldsymbol{\beta_p}}{\boldsymbol{\beta_{\text{hedging vehicle}}}}\right| \tag{5.7}$$

Consider when the goal of a hedge is to completely or partially remove or "layoff" the systematic or market risk of portfolio P through short positions in the S&P 500 index futures contract. The remaining risk will be idiosyncratic risk that is unique to portfolio P. The manager's hope is that through diligent analysis she can construct portfolio P such that it will perform well relative to the market index in both good and bad times. By taking away the market risk, the net hedged performance will be driven by skill and luck rather than the direction of the overall market. Another reason that a manager might wish to hedge a diversified equity portfolio using S&P 500 index futures is to temporarily remove the market risk in anticipation of a market downturn without having to sell the entire portfolio, which can take time, cause substantial transactions costs, and in some cases tax consequences.

Next, consider an example of hedging a stock portfolio.

Example 5.4: Assume that a portfolio has a value of $10,000,000 and $\beta = 2$ while the S&P 500 index futures price is 2,426 (each futures contract is 250 times the futures price). How many contracts provides a hedge to eliminate systematic risk?

$$\text{Since } \beta = 2,$$
$$N = \times \frac{V_P}{V_F} |\beta| = \frac{10,000,000}{2426 \times 250} \times 2 \approx 33.$$

Note that if our hedge period is daily, even though beta (the best estimate of the value of beta) may not change daily, the values of the portfolio V_P relative to the value of the futures contracts will change, and so the optimal number of contracts, N, will change. Therefore, a dynamic hedge should be used that rebalances the number of contract through time.

The next section discusses hedging an option portfolio where h itself may need to change instantaneously. This introduces substantial differences from hedging portfolios that only include stocks, currencies, and commodities.

5.3 Delta-Hedging of Options Portfolio Using the Underlying Asset

This section discusses the hedging of an options portfolio. Unlike, forward and futures hedging, option hedging is typically a dynamic hedging strategy due to continuous changes of the hedge ratio, h. We first look at delta-hedging: hedging the underlying asset price risk in an options portfolio.

5.3.1 Introduction to Continuous-Time Delta-Hedging Involving Options

A major area in which hedging is used is in option portfolios. For example, the risk of an option is often hedged using positions in the option's underlying asset. Therefore, in this case the asset being hedged is an option and the hedging vehicle is the underlying asset of the option. An option is a nonlinear function of its underlying asset. Therefore the optimal hedge ratio will tend to change when the value of the option's underlying asset changes.

When there is an analytic formula to value a financial derivative, we can find the analytic formula for hedge ratios through partial derivatives. In the case of options, the premier formula is the Black-Scholes option pricing model: option price = $f(S,K,T,t, r,\sigma)$. The main use of the Black-Scholes formula for European options is for hedging, not for valuing the options.

The unit value approach derived in forward/future context in section 5.2 can be used when hedging options, since unlike in the case of a portfolio of stocks, it is easy to count options on the same underlying asset by units. Assume that the underlying asset, S, is used can be to hedge one unit of an option, and assume that the goal is to hedge fully the option price change due to underlying asset price change. Denote f as the option price, S as the price of the option's underlying asset, and Δf (or ΔS) as the change of the price of the option (or the underlying asset):

In the unit value approach: $h = -\frac{\Delta A}{\Delta B}$, in which A is the asset to be hedged (which is an option f) and B in this case is the hedging vehicle (the underlying asset). The underlying asset is used to hedge option risk, so $A = f$, $B = S$, and the hedge ratio formula is:

$$h = -\frac{\Delta f}{\Delta S} \tag{5.8}$$

Note that the hedging vehicle is always in the denominator and the asset being hedged is in the numerator. Sometimes, the notation is abused by ignoring the negative sign. The negative sign is ignored under the assumption that it is understood that the hedge of correlated assets requires an opposite position.

In the Black-Scholes world, h is not a constant because S continuously changes. Therefore, hedging requires an instantaneous rate of change:

$$h = -\frac{\partial f}{\partial S}$$

= the number of units of S that instantaneously hedge one unit of f.

The partial derivative of f with respect to S is so important to practitioners that it has a universally recognized term (Δ_f):

$$h = -\Delta_f = -\frac{\partial f}{\partial S} \approx -\frac{\Delta f}{\Delta S} \tag{5.9}$$

Here we distinguish the notation Δf, which is the price change of asset f, while Δ_f is the delta of f, which is the sensitivity of the price change of asset f with respect to the price change of the underlying asset. The relation between Δf and Δ_f is:

$$\Delta_f \approx \frac{\Delta f}{\Delta S}.$$

Note that the Black-Scholes option pricing model offers an analytical formula for European options that can be used to calculate ***option deltas,*** Δ_c and Δ_p, the sensitivity of an call option price and put option price, respectively, with respect to the price of the underlying asset. The hedge between one unit of option f and its underlying asset S is typically modeled as follows:

$$n_s = n_f \times |h|$$

When $n_f = 1$, the hedge is:

$$n_s = \Delta_f.$$

We use Δ_f units of the underlying asset, S, to hedge the risk of one unit of the option. This is discussed in the context of Black-Scholes formulas in Section 5.3.3.

Partial derivatives commute with a constant: $\frac{\partial x f_1}{\partial S} = x\frac{\partial f_1}{\partial S}$. So, x units of options can be hedged by x Δ units of the underlying asset. Define the ***portfolio delta*** for a portfolio of options as a linear combination of option deltas weighted by the units of options where all options must have the same underlying asset (which can be used as a hedging vehicle). The underlying asset has a delta of 1 $\left(\frac{\partial S}{\partial S} = 1\right)$, so a single-asset portfolio containing Δ units of the underlying asset has a portfolio delta of Δ.

Note that we have not discussed buying vs. shorting. It is obvious that $\frac{\partial(-f)}{\partial S} = -\frac{\partial f}{\partial S}$. Assume a portfolio f with a delta of Δ_f. What is the portfolio delta for shorting f? Since shorting f is equivalent to owning –f, shorting f has a delta of $-\Delta_f$. Therefore, a portfolio containing Δ units of the underlying asset has a positive exposure of Δ.

If a portfolio has a positive Δ, to hedge that exposure requires establishing a hedging position that has a negative Δ. A first-order hedge with respect to changes in the underlying asset is termed as a ***delta-hedge.*** Common approaches are to short an asset such as a call option (that has a positive Δ) to hedge a share of its underlying stock, or short the underlying asset to hedge a long position in a call. Corresponding arguments work for the case of a put option that has a negative Δ. For example, consider a call option with a delta of 0.25. Short positions in four of those call options would form a hedge with owning one share of the stock underlying the options. Long positions in three put options with a delta of –0.33 would hedge owning one share of the underlying stock. The details of delta-hedging a portfolio are in Section 5.3.4.

The next section explores the risk of option portfolios without delta-hedging as a prelude to further discussions of delta-hedging.

5.3.2 An Example without Delta Hedging

Before demonstrating delta-hedging formally, consider an example that illustrates why a delta-hedge is better than two other alternatives. We adopt Hull's book's idea here. The two other strategies are a ***naked position*** and a ***covered call*** position, both of which have substantial risk – which provides motivation for delta-hedging.

> **Example 5.5**: Assume a stock price follows Geometric Brownian motion with $S_0 = 57.45$ and volatility equal to 15%. An investment bank has written (sold) a six-month European call on 100,000 shares of a non-dividend-paying stock generating proceeds of $150,000. The strike price of the call is 62 and the risk-free rate is 5%. How does the bank hedge the risk of its short option exposure with respect to the price of the call's underlying stock? (Note that the bank's estimate of the Black-Scholes call value is 1.269 on one share of stock, so at $1.50 the call is over-priced relative to the Black-Scholes call value, indicating an advantage to shorting the call if the bank is right.)

Before discussing the delta-hedging, consider two strategies: a naked position in the option and a covered call.

1. Naked position in short call options: Take no action.
   What if the **stock goes up** to $70 by the time the option expires? In this case, the bank will have a large loss:
   Option proceeds minus payout: 150,000 – (70 – 62) × 100,000 = –650,000
   Net loss = 650,000
   What if the stock goes down an equal amount or at least lower than $62 by the time of the option's expiration? In this case, the call is not exercised and the bank collects $150,000:
   Option proceeds minus payout: 150,000 – 0
   Net gain = $150,000
2. Covered call position: Buy 100,000 shares today to hedge the option's risk (note that this is one-to-one unit hedging).
   What if the **stock goes down** to $50 by the time the option expires? In this case, the bank will have a net loss. The call is not exercised and the option position has zero value at expiration for a gain of $150,000.
   Stock position loss is (57.45 – 50) × 100,000 = 745,000
   Net loss = $745,000 – $150,000 = $595,000

What if the **stock goes up** to \$70 by the time the option expires? In this case, the bank will have a net gain:
Option position loss: 150,000 – (70 – 62) × 100,000 = –650,000
Stock position gain = (70 – 57.45) × 100,000 = 1,255,000
Net gain = \$1,255,000 – \$650,000 = \$605,000
The covered call position did not hedge well because the total stock position was more sensitive to changes in S than the option position. In order to make the hedge work, intuitively the sensitivities of the long stock position and the short option position must be set equal (through delta-hedging rather than one-to-one unit hedging), which is discussed in detail in the next section.

5.3.3 Delta-Hedging an Option Position Using Its Underlying Asset

This section discusses how to delta-hedge a short position in a call option using the underlying asset. Recall that the delta of an option is the sensitivity of the option value to changes in the underlying asset price.

$$\Delta_f = \frac{\partial f}{\partial S}$$

If the Black-Scholes formula is used to calculate the delta of a long position in a European call option on a non-dividend-paying stock then delta is the partial derivative of the option price with respect to the price of the underlying asset:

$$\Delta_c = \frac{\partial C}{\partial S} = N(d_1) \tag{5.10}$$

where $N(\cdot)$ is the cumulative distribution function of standard normal distribution.

The delta of a short call potion is $-\Delta_c$, which is $-N(d_1)$. To hedge one unit of the call option requires holding $N(d_1)$ units of the underlying asset with the opposite sign.

The delta of long a European put on the stock is: $N(d_1) - 1$, as shown below.

$$\Delta_p = \frac{\partial p}{\partial S} = -N(-d_1) = N(d_1) - 1 \tag{5.11}$$

The delta of short position in a European put on the stock is $1 - N(d_1)$. In order to hedge a short position in a European put, this requires maintaining a short position of $1 - N(d_1)$ units of the underlying assets at each point in time. $N(\cdot)$ is always between 0 and 1, so, the delta of a long call is between 0 and 1. Long positions in a put have a delta between −1 and 0.

Note that while the underlying asset price, S, changes, Δ_f also changes. So, delta-hedging is a first-order approximation to hedge the asset price risk and requires continuous rebalancing.

Return to Example 5.5 as an example of a simplified first-order approximation to use a long position in the underlying stock to delta-hedge a $150,000 (100,000 call options) short position in the call option. The delta of the call option is 1/3 – calculated using $N(d_1) = 0.3336$, which was rounded to 1/3 for our convenience. The 100,000 short calls (totaling $150,000) multiplied by their delta creates an exposure equivalent to that of 33,333.33 shares of stock. The hedging position is therefore a $1,915,000 long position in the stock, established at $57.45 per share for a total of 33,333.33 shares.

Consider a $6 shift in the stock price from $57.45 per share to either $51.45 or $63.45. The first-order approximations of the option values that result from an instantaneous up or down shift in the stock price are as follows:

What if the stock goes **down** to $51.45 in the next instant? In this case, the bank will break even: Each call option falls in value by approximately $2 (i.e., −$6 × 1/3) based on the delta of 1/3. The short call option position has a gain of $200,000. The stock position loss is (57.45 – 51.45) × 33,333.33 shares = $200,000.

Net result = $200,000 – $200,000 = $0 (using a first-order approximation of the option price change)

What if the **stock goes up** to $63.45 in the next instant? In this case, the bank will break even: Each call option rises in value by approximately $2 (i.e., $6 × 1/3) based on the delta of 1/3. The short call option position loses $200,000. The stock position gain is (63.45 – 57.45) × 33,333.33 shares = $200,000.

Net result = $200,000 – $200,000 = $0 (using a first-order approximation of the option price change)

The above example demonstrates a very important principle of delta-neutral hedging ($\Delta = \mathbf{0}$), which is defined formally in the next section: delta-neutral hedging reduces or even eliminates the sensitivity of a portfolio to the *direction* of the movement in the price of the underlying asset. However, it should not be concluded from this first-order approximation that delta-neutral hedging, even if perfectly applied, eliminates the impacts of price changes in the underlying asset on the hedged portfolio. As is discussed in later sections, the performance of perfectly delta-neutral portfolios can be very substantially affected by changes in the anticipated volatility in the underlying asset even when the delta-neutrality is continuously maintained.

Nevertheless, if the goal is to hedge risk, a trader should seek a portfolio with a portfolio Δ_f as small as possible so that the portfolio's sensitivity to directional changes in S is reduced. The ideal (delta neutral) case is $\Delta_f = 0$.

Since Δ_f is continuously changing, hedging requires rebalancing to keep the position delta neutral.

5.3.4 Delta-Hedging a Portfolio Containing Options by Matching the Deltas

This section discusses the delta of portfolio containing multiple options with the same underlying asset. Since partial derivatives commute with summation and a constant multiple, one can calculate the Δ_f of a portfolio that contains a linear combination (by units n) of calls and puts as the linear combinations (by units n) of the deltas for individual calls and puts:

$$\boldsymbol{\Delta}_f = \sum_{i=1}^{M} n_i \times \boldsymbol{\Delta}_{asset\ i} \tag{5.12}$$

where asset i may be a call or a put on the same underlying asset and n_i are the units of the corresponding calls or puts.

In Section 5.3.1, Δ_c shares of the underlying asset were used to hedge one call. To generalize the hedging idea further, consider a portfolio of options on some underlying asset as a hedging vehicle and the asset to be hedged as another portfolio of options on the same underlying asset. Assume that the asset to be hedged has a delta of Δ_{asset} and that the hedging vehicle has a delta $\Delta_{hedging\ vehicle}$.

By Section 5.2.2,

$$h = -\frac{\Delta\text{asset}}{\Delta\text{hedging vehicle}}$$

$$n_{\text{hedging vehicle}} = n_{\text{asset}} \times |h|,$$

where Δasset refers to the change in the value of the underlying asset and Δ hedging vehicle refers to the change in the value of the hedging vehicle. Note that Δ_{asset} is the delta of the asset and $\Delta_{hedging\ vehicle}$ is the delta of the hedging vehicle, so

$$\Delta_{asset} \approx \frac{\Delta\text{ assets}}{\Delta S},$$

$$\Delta_{hedging\ vehicle} \approx \frac{\Delta\text{ hedging vehicle}}{\Delta S},$$

$$h \approx -\frac{\Delta_{asset}\Delta S}{\Delta_{hedging\ vehicle}\Delta S} = -\frac{\Delta_{asset}}{\Delta_{hedging\ vehicle}}$$

If the units n_f are included in the portfolio Δ_f, we have $h = 1$ and:

$$\Delta_{hedging\ vehicle} = -\Delta_{asset} \tag{5.13}$$

Equivalently, the hedged portfolio (which includes the asset to be hedged and the hedging vehicle) has a delta of 0, and it is an example of ***delta-neutral hedging***.

Consider the following example regarding a portfolio delta, its corresponding meaning, and how to delta-neutral hedge a portfolio of options and their underlying asset.

Example 5.6: A call has $\Delta_c = 0.4$ and its corresponding put has $\Delta_p = 0.4\text{–}1 = -0.6$.

Abbi shorts calls on 1,000 shares of a stock and shorts puts on 500 shares of the same stock. What is the delta of the portfolio Δ_f?

$$\Delta_f = n_c \times \Delta_{-c} + n_p \times \Delta_{-p} = n_c \times (-\Delta_c) + n_p \times (-\Delta_p)$$
$$= 1{,}000 \times (-0.4) + 500 \times (+0.6) = -400 + 300 = -100$$

(Note that $\Delta_{-c} = -\Delta_c,\ \Delta_{-p} = -\Delta_p$.)

What is the meaning of the portfolio delta of –100? Since $\Delta_f = \frac{\partial f}{\partial S} \approx \frac{\Delta f}{\Delta S}$, a \$1 increase in the price of the underlying asset will cause the above portfolio to lose approximately \$100.

How can the delta of this portfolio be hedged? In order to be delta-neutral, the portfolio must have a hedging vehicle with the opposite delta as the portfolio. In Example 5.6, $\Delta_f = -100$. Any one of the three following actions (or a combination of those actions) could be used to hedge the risk:

- Buy 100 shares of stock since one share of stock has a delta of 1, or
- Buy an option portfolio on the same stock with a delta of 100, or
- Short an option portfolio on the same stock with a delta of –100.

Sometimes, using futures contracts on the underlying assets is better than using cash positions in the underlying asset because futures contracts have greater liquidity. The next section demonstrates using futures contracts for delta-hedging.

At this stage, it seems that we have not connected to our hands on approach. In fact, for a derivative with no analytic solution and we have no closed form formula for Δ_f. We could use an EXCEL spreadsheet, R simulation to calculate $\frac{\Delta f}{\Delta S}$ as an approximation of Δ_f. Hedging strategies with EXCEL are used in Chapter 7.

5.4 Delta-Hedging Using Futures

The previous sections provided examples of how to use cash securities such as underlying assets and options to delta-hedge the delta risk of a portfolio. This section discusses the use of forward contracts, futures contracts, or options on futures to delta-hedge. The general principle remains the same, namely the delta-neutral strategy: seeking to hedge a portfolio by constructing a hedging vehicle with the opposite delta of the portfolio being hedged. This section begins by discussing the calculation of deltas for forwards, futures, and options on futures.

5.4.1 The Delta of a Forward Contract

A forward contract fixes the delivery price at which delivery or settlement will take place, but subjects the parties to the risk of changes in the value of the underlying asset relative to the forward value. There is therefore risk to the *value* of the forward contract that is driven by the market price of the asset underlying the forward contract just as in the case of an option. The value of a long position in a forward contract (with cash distributions and carrying costs set to zero) at time t differs from its value at time 0 (when the forward price was established):

$$V_t = S_t - F_0 \times e^{-r(T-t)}$$

Calculate the sensitivity of this value, V, with respect to the underlying asset price, S, as the forward contract's delta:

$$\Delta_V = \frac{\partial V}{\partial S} = 1 \tag{5.14}$$

So, a long position in a forward contract has a delta of one unit, indicating that it rises and falls on a one-to-one dollar basis with its underlying asset price. This is exactly what was found in Section 5.1.

However, if there is a dividend yield q,

$$V_t = S_t e^{-q(T-t)} - F_0 \times e^{-r(T-t)}.$$

So a forward contract with a continuous dividend yield has the following delta:

$$\Delta_V = \frac{\partial V}{\partial S} = e^{-q(T-t)} < 1 \tag{5.15}$$

If there is cash dividend D,

$$V_t = S_t - PV(D) - F_0 \times e^{-r(T-t)}.$$

So the forward contract with a discrete cash dividend has a delta:

$$\Delta_V = \frac{\partial V}{\partial S} = 1.$$

Since cash dividends are highly predictable in the short run, they do not affect the forward contract's delta.

The reason that the continuous dividend yield matters in forward delta is because, by assumption, the dollar value of the dividend changes with the spot price of the asset. A fixed dividend does not change. In reality, short-term equity dividends do not fluctuate with the price of the underlying asset, so a fixed-dividend assumption is more useful in applications. For long-term contracts, especially in the context of carrying costs (negative dividends), a continuous dividend model can be more useful.

5.4.2 The Delta of a Futures Contract

Futures contracts are marked to market daily, meaning that the economic consequences of any changes in the market price of the futures contract are immediately settled between the long and short position holders. In effect, at the close of trading each day, the previous day's futures contracts are settled to the price of new contracts, and the old contracts are rolled over into positions in the new contracts. A future contract's risk is therefore directly related to changes in its daily price f_t. Assume that the futures contract has the following pricing formula: (This is true if the underlying asset has a zero correlation with the interest rate, r)

$$f_t = S_t \times e^{r(T-t)}$$

The delta is the sensitivity of the futures price with respect to the underlying asset price. The futures contract has a delta greater than one to the extent that the time value of money is positive:

$$\Delta_f = \frac{\partial f}{\partial S} = e^{r(T-t)} > 1 \tag{5.16}$$

Note that the "acceleration" of the cash payments between the long and short sides in a futures contract relative to a forward contract causes the delta of a futures contract to have a larger absolute value by a factor that reflects the time value of money between times t and T.

If there is a dividend yield q,

$$f_t = S_t e^{(r-q)(T-t)}.$$

The delta of a futures contract on an asset with a dividend yield has a delta that is diminished by the dividend yield:

$$\Delta_f = \frac{\partial f}{\partial S} = e^{(r-q)(T-t)} \tag{5.17}$$

If there is a fixed cash dividend D,

$$f_t = (S_t - PV(D))e^{r(T-t)}.$$

So the futures contract with a cash dividend has the same delta as the non-dividend stock because the present value of the dividend has already been deducted from the stock price in the formula for the futures price (i.e., $S_t - PV(D)$):

$$\Delta_f = \frac{\partial f}{\partial S} = e^{r(T-t)} > 1.$$

In summary, futures contracts have a larger delta than the underlying asset and its corresponding forward contract (as long as $r > q$). Also, periodic (fixed) cash dividends do not appear in the formula for delta. However, a continuous dividend yield that is proportional to the price of the underlying asset makes the contract's delta smaller than it would be if the underlying asset did not have a dividend.

5.4.3 Delta of an Option on a Futures Contract

Consider a European option with maturity T at time t that allows the holder of the option to enter a futures contract at a prespecified strike price K. (Note that the underlying futures contract must have a longer maturity, T_1, than the maturity, T, of the option.)

Recall the Black formula for a European call, c_t, on a futures contract from Chapter 4:

$$\widehat{c_t} = f_t e^{-r(T-t)} N(d_1) - K e^{-r(T-t)} N(d_2)$$

where $d_1 = \dfrac{Ln(f_t / K) + \left(\frac{\sigma^2}{2}\right)(T-t)}{\sigma\sqrt{T-t}}$ and $d_2 = \dfrac{Ln(f_t / K) + \left(-\frac{\sigma^2}{2}\right)(T-t)}{\sigma\sqrt{T-t}} = d_1 - \sigma\sqrt{T-t}$.

There are two different kinds of delta for this option: ***delta*** Δ_c is the sensitivity of the price of the option on the futures contract with respect to the price of the cash asset underlying the futures, while the ***futures delta*** of option on futures $\tilde{\Delta}_c$, is the sensitivity of the price of the option with respect to the futures price. First, calculate the futures delta $\tilde{\Delta}_c$:

$$\tilde{\boldsymbol{\Delta}}_c = \frac{\partial \hat{c}_t}{\partial f_t} = e^{-r(T-t)}\mathbf{N}(d_1) \tag{5.18}$$

where $d_1 = \frac{Ln(f_t / K) + \left(\frac{\sigma^2}{2}\right)(T-t)}{\sigma\sqrt{T-t}}$.

The details of the proof are left as exercises. Note that d_1 and d_2 also contain f_t and that these partial derivatives are not straightforward.

If *futures* contracts are used to hedge this option on a futures contract, then delta is the number of futures needed because the numerator of $\tilde{\boldsymbol{\Delta}}_c$ is related to the asset to be hedged which is a call option on the futures contract, while the denominator is related to the hedging vehicle (i.e., the futures). The hedge ratio is the futures' delta:

$$h = -\tilde{\Delta}_c = -\frac{\partial \hat{c}}{\partial f}.$$

Next, calculate the underlying asset's sensitivity (delta) Δ_c:

$$\Delta_c = \frac{\partial \hat{c}}{\partial S}.$$

By the chain rule, $\Delta_c = \frac{\partial \hat{c}}{\partial f}\frac{\partial f}{\partial S} = \tilde{\Delta}_c \times \Delta_f$, where Δ_f is the delta of the futures and $\widetilde{\Delta}_c$ is the futures delta discussed above.

Inserting into Equations 5.17 and 5.18:

$$\Delta_c = (e^{-r(T-t)}\mathbf{N}(d_1)) \times \left(e^{r(T_1-t)}\right) = e^{r(T_1-T)}\mathbf{N}(d_1) \tag{5.19}$$

where $d_1 = \frac{Ln(f_t / K) + \left(\frac{\sigma^2}{2}\right)(T-t)}{\sigma\sqrt{T-t}}$.

Make sure not to confuse T, the option's expiration time, with T_1, the futures settlement time. Note that T_1, the maturity for the futures contract does not appear in d_1

If the *underlying cash asset* is used to hedge this option on a futures contract, delta Δ_c is the number of underlying assets needed. This is because the numerator of Δ is related to the asset to be hedged (which is a call on the futures), while the denominator is related to the hedging vehicle (which is the underlying asset). The hedge ratio is therefore the underlying asset delta which we simply refer to as the call delta:

$$h = -\Delta_c = -\frac{\partial \hat{c}}{\partial S}.$$

With both the deltas calculated, hedges can be formed between the cash asset underlying the futures contracts, the futures contracts on the underlying assets, and various options on the futures contract. Hedges can even be formed between two or more futures contracts that differ by their maturities, which simply require use of the deltas from each position to make the final portfolio delta neutral. Several examples are provided in the following section.

5.4.4 Examples of Delta-Hedging of Options on Futures

This section examines several examples of delta-hedging options on futures contracts. Some of the examples are similar to exercises in Hull's book.

Example 5.7: Consider a European call on a futures contract that a trader believes is relatively underpriced. The trader can establish a long position in that call option and use the option's underlying futures contract to form a hedge.

Since our hedging vehicle is a futures contract, the hedge ratio is the futures' delta. The Black formula is used to calculate the delta of the futures contract, $\tilde{\Delta}_c$. Assume that the option contract's maturity time is T and the futures contract maturity time is T_1. By the Black formula, the call price is:

$$\hat{C} = f_0 e^{-rT} N(d_1) - K e^{-rT} N(d_2)$$

where $d_1 = \frac{Ln(f_0 / K) + \left(\frac{\sigma^2}{2}\right)T}{\sigma\sqrt{T}}$ and $d_2 = \frac{Ln(f_0 / K) + \left(-\frac{\sigma^2}{2}\right)T}{\sigma\sqrt{T}} = d_1 - \sigma\sqrt{T}$.
The hedging ratio is related to the futures delta:

$$h = -\tilde{\Delta}_c = -\frac{\partial \hat{C}}{\partial f} = -e^{-rT} N(d_1)$$

where $d_1 = \frac{Ln(f_0 / K) + \left(\frac{\sigma^2}{2}\right)T}{\sigma\sqrt{T}}$
Note that T is the maturity for the option and that the above example used $t = 0$. So, $e^{-rT}N(d_1)$ units of futures are needed to hedge this call on the futures contract.

Example 5.8: Consider a European call on a futures contract that a trader believes is relatively underpriced. The trader establishes a long position in that call option and uses the underlying cash asset to hedge. Assume that the option contract's maturity time is T and the futures contract maturity time is T_1.

Since the hedging vehicle is the underlying cash asset, the hedge ratio is the option's delta with respect to underlying asset. Using the chain rule to calculate the delta Δ_c:

$$\frac{\partial \hat{C}}{\partial S} = \frac{\partial \hat{C}}{\partial f}\frac{\partial f}{\partial S} = e^{-rT}N(d_1)\ e^{rT_1} = N(d_1)e^{r(T_1-T)}$$

where $d_1 = \frac{Ln(f_0 / K) + \left(\frac{\sigma^2}{2}\right)T}{\sigma\sqrt{T}}$.

So, $e^{r(T_1-T)}N(d_1)$ units of the underlying asset are needed to hedge this call on the futures contract.

Example 5.9: Consider a European call on a futures contract that a trader believes is relatively underpriced. The trader establishes a long position in that call option and uses a *different* futures contract with expiration T_2 to hedge. The futures contract is denoted as $\hat{f}$.

Since the hedging vehicle is another futures, $\hat{f}$, the hedge ratio is the option's delta with respect to the futures contract with expiration T_2. Using the chain rule to calculate the delta:

$$\frac{\partial \hat{C}}{\partial \hat{f}} = \frac{\partial \hat{C}}{\partial S}\frac{\partial S}{\partial \hat{f}} = N(d_1)e^{r(T_1-T)}e^{-rT_2}.$$

Note that d_1 depends on the option's underlying asset, which is the future contract f, not the asset used for hedging, which is $\hat{f}$. The option's underlying never changes, it is still the futures contract f. Note that d_1 also depends on the option's maturity T only.

Therefore, $d_1 = \frac{Ln(f_0 / K) + \left(\frac{\sigma^2}{2}\right)T}{\sigma\sqrt{T}}$ for all of the above examples.

The trader needs $N(d_1)e^{r(T_1-T)}e^{-rT_2}$ units of the futures contract $\hat{f}$ with maturity T_2 to hedge this call on the futures contract f with maturity T.

The previous examples have demonstrated the use of a variety of pairs of assets to be delta-hedged. Consider the goals of a speculator. The speculator scans the market looking for assets that the trader perceives as being underpriced or overpriced – perhaps due to disruptions caused by large buy or sell orders in a particular security. The speculator strives to take a long position in an asset perceived as temporarily underpriced (or a short position in an asset perceived as temporarily overpriced) and to lay off (hedge) the delta risk in assets that are fairly priced (or better yet favorably priced). The speculator hopes to unwind the trade at a profit when prices return to equilibrium levels and therefore to collect a reward for providing liquidity to the market.

The previous examples can also be viewed from the perspective of a market maker such as an investment bank that is meeting the needs of its clients by being the counterparty to their desired trades. Similarly to the role of the speculator the investment bank takes positions offsetting the client's needs and may lay off the risk with other assets.

In both cases, the more alternatives that are available for hedging (e.g., options on the cash asset, options on futures, futures and forwards on similar assets, etc.), the better the set of opportunities to identify mispriced securities and to form hedges.

The next section discusses sources of risk in options other than changes in the underlying asset.

5.5 Greek Letters Hedging of Other Option Risk Factors

In a Black-Scholes world, an option value has a formula $f(S,K,t,T,r,\sigma)$ (it is an analytic formula for European options), where S is the underlying asset price, K is the strike price, T-t is the option's time to expiration, r is the riskless interest rate, and σ is the standard deviation of the returns of the underlying asset. Previous sections have focused on delta, the partial derivative with respect to the first of those six parameters, the underlying asset price S. The partial derivatives of an option price with respect to its parameters are denoted with letters from the Greek alphabet and are collectively known as the ***Greeks*** (even though one of them, Vega, is not a Greek letter): $\Delta_f = \frac{\partial f}{\partial S}, \theta = \frac{\partial f}{\partial t}, \rho = \frac{\partial f}{\partial r}$, and $\nu = \frac{\partial f}{\partial \sigma}$. The partial derivatives represent risk sensitivity with respect to the denominators. Some of these partial derivatives can be used to form hedge ratios to hedge the risk of the corresponding factors in much the same as was demonstrated in the case of delta in the previous sections. The next section discusses Vega, which measures sensitivity to volatility.

5.5.1 Vega and Vega Neutral Hedging

Option price sensitivity to the return volatility of the underlying asset is represented by $\frac{\partial f}{\partial \sigma}$, and is entitled ***Vega*** (ν). The sensitivity of option prices to changes in volatility is extremely important in risk management. Note that while other parameters such as the price of the underlying asset are observable, the volatility parameter reflects anticipated volatility in the returns of the underlying asset over the option's life and is not directly observable.

According to the Black-Scholes formula, the Vega of a European call option on a non-dividend-paying stock is (assuming t = 0 for ease):

$$\nu_c = \frac{\partial c}{\partial \sigma} = S_0\sqrt{T}N'(d_1) \tag{5.20}$$

Note that the Vega of a long position in a regular European or American option is always positive. Further, the Vega for a long position in a European *call* and a long position in a European *put* on a non-dividend-paying stock are the same if the underlying assets, strike prices, and maturities are the same. This is a very easy corollary of put call parity:

$$p + S = c + Ke^{-rT}$$

$$\frac{\partial c}{\partial \sigma} = \frac{\partial p}{\partial \sigma} \tag{5.21}$$

Vega is largest for options that are close to the money. This can be seen from the separation of an option price into its intrinsic value and time value as the two vertical line segments (the bottom is the intrinsic value and the top is the time value) depicted in Figure 5.1.

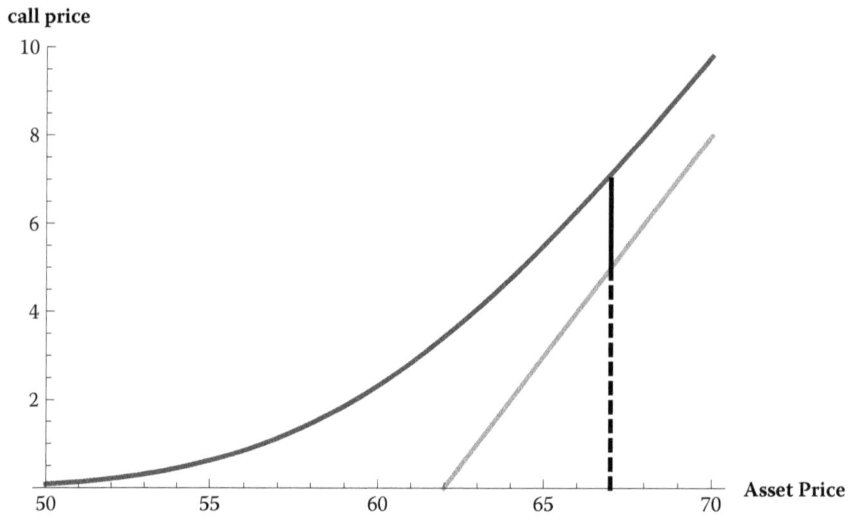

FIGURE 5.1
Value of a Call Option.

The ***intrinsic value of an option*** is the greater of its value if exercised immediately and zero:

$$\text{Intrinsic Value of Call Option at Time } t = \max(S_t - K, 0)$$

The ***time value of an option*** is the excess, if any, of the current option price above its intrinsic value:

$$\text{Time Value of Option} = \text{Option Price} - \text{Intrinsic Value}$$

The interpretation of the time value of an option is that it represents the added value that the option holder obtains from having the right to choose to exercise or not exercise an option during the option's remaining life. Nearness to the money, therefore, makes the optionality more valuable, driving the values of both Vega and gamma higher. (Gamma is another Greek letter and is discussed later.) It is time and volatility, working together (as can be seen in the Black-Scholes formula) that create the option's time value – the value of the option that would not exist with zero time to expiration remaining. When an option is near-the-money its time value is largest and therefore its Vega is largest. An option extremely far into the money or far out of the money has little time value (especially when interest rates are low or the expiration is near) and therefore has a relatively small Vega.

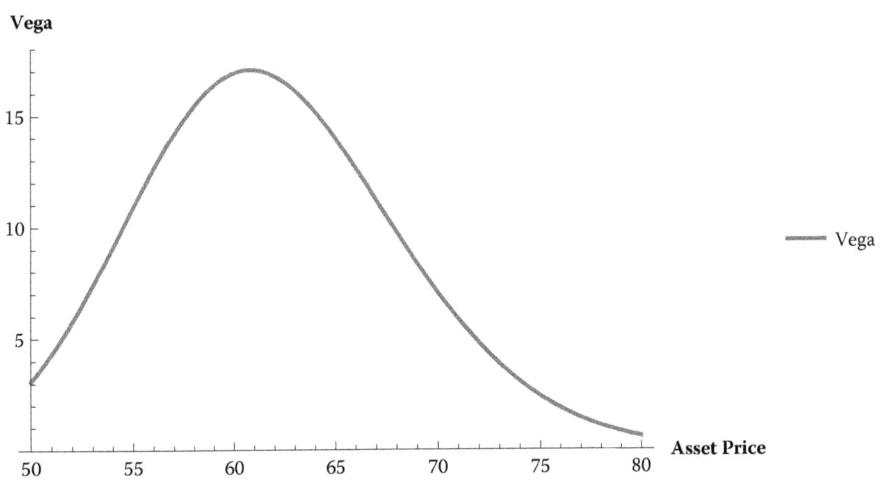

FIGURE 5.2
Vega of an Option.

The above graph of Vega against asset price uses input values from the following: $K = 62$, $\sigma = 15\%$, $r = 5\%$, $T = 0.5$ (Figure 5.2).

Note that the Vega peaks above the strike price ($K = 62$), indicating the importance of volatility on an option that is on the brink of moving into and out of the money.

Since partial derivatives commute with summation and a constant multiple, the ***portfolio Vega*** can be calculated where the portfolio contains a linear combination of calls and puts weighted by *the units of options* and all options have the same underlying asset. The portfolio Vega is the linear combination of the Vegas for the individual calls and puts.

Vega captures option price changes due to changes in the anticipated volatility of the returns of the underlying asset. So, even when other parameters do not change, a change of volatility will affect option values due to the asymmetric payoffs of options (with deltas approaching one in one direction and approaching zero in the other direction). *Vega is positive for long option positions* and *negative for short option positions* (and tends to be assumed to be zero for assets without optionality). Vega is assumed to be zero for the underlying asset. Therefore, if a portfolio is net long positions in options, if anticipated volatility declines, the portfolio value will tend to decline. Conversely, a portfolio that is net short option positions will tend to have a negative Vega. A ***Vega-neutral hedging*** strategy strives to have a portfolio Vega equal to 0.

Because the partial derivative of the underlying asset (S) with respect to volatility is zero, options must be used to manage or perfectly hedge Vega risk. In order for a portfolio with options to be hedged to Vega-neutrality, the goal is to set the Vega of the hedging position equal to the Vega of the position to be hedged (times –1). Assume that the asset to be hedged is f_2 and the hedging vehicle is f_1.

The Vega hedge ratio is $h = -\dfrac{\Delta f_2}{\Delta f_1} \approx -\left(\dfrac{\partial f_2}{\partial \sigma}\Delta\sigma\right)\Big/\left(\dfrac{\partial f_1}{\partial \sigma}\Delta\sigma\right) = -\dfrac{\nu_2}{\nu_1}$

Note that the denominator is the hedging vehicle and the numerator is the asset to be hedged.

Since long positions in options have positive ν, short positions in options are necessary to hedge the Vega of a portfolio with long options to zero.

As before, in order to eliminate volatility risk, make the hedge ratio $h = 1$. Therefore, Vega neutral hedging requires:

$$\nu_{\text{hedging vehicle}} = -\nu_{\text{asset}} \tag{5.22}$$

Note that so far in this section S and σ refer to the same underlying. If a portfolio contains various options on different underlying assets, the hedging method must use multiple hedge vehicles and multiple deltas, Vegas, and other partial derivatives on different underlying assets.

The following section introduces the formulas and properties for two other Greeks (theta and rho): $\theta = \frac{\partial f}{\partial t}$ and $\rho = \frac{\partial f}{\partial r}$. These two Greeks are not used for hedging because t is deterministic, which means it does not contain risk, and r is assumed riskless and in the case of options on equities with relatively short times to expiration, the interest rate risk is small. However, if the underlying asset is related to interest rates such as bonds or derivatives on bonds so that interest rate risk is a substantial driver of risk, then duration and the advanced hedges in Chapter 7 are the main tools for hedging (duration hedging is discussed in Section 5.6). The same is true for $\frac{\partial f}{\partial K}$, which is not defined as a Greek. Later sections relate the Greeks through equalities.

5.5.2 Rho and Theta

Rho represents the interest rate risk of an option and is defined as the rate of change of the value of a derivative with respect to the interest rate:

$$\rho = \frac{\partial f}{\partial r}$$

Assuming that the underlying asset's value is not related to the interest rate, according to Black-Scholes formula, the rho of a European call on a non-dividend-paying stock is:

$$\rho_c = \frac{\partial c}{\partial r} = KTe^{-rT}N(d_2) \tag{5.23}$$

The option price's sensitivity to *the passage of timet*, ***theta***, is given by $\frac{\partial f}{\partial t}$ and is known as the option's time decay. The theta (θ) of a derivative (or portfolio of derivatives) is the rate of change of the value with respect to the passage of time:

$$\theta = \frac{\partial f}{\partial t}$$

According to Black-Scholes formula, the theta of a European call on a non-dividend-paying stock is:

$$\theta_c = \frac{\partial c}{\partial t} = -\frac{S_0 N'(d_1)\sigma}{2\sqrt{T}} - rKe^{-rT}N(d_2) \tag{5.24}$$

Note that holding all other variables constant an option's price will decline steadily to $\max(S_T - K, 0)$ (the option's intrinsic value) at expiration. The variable t denotes the passage of time. As time passes and t grows, there is less time left, so options tend to lose values everything else equal. Therefore, the theta of long a call or put is usually *negative*. This means that, as time passes with the price of the underlying asset and its volatility held constant, the value of a long call or put option declines. There are exceptions: a deep in-the-money European put option on a stock (with little or no dividend) can have a positive theta (due to the time value of money). Also, an in-the-money call on a currency with a very high interest rate can have a positive theta (due to the opportunity cost of delaying receipt of the currency).

It is not necessary to hedge theta since theta (and the passage of time itself) is not a random variable. However, theta has some important relationships with delta and gamma, so, it will be discussed in subsequent material.

Delta-hedging has major limitations. The next two sections discusses gamma, another Greek letter, to address the major limitation of delta.

5.5.3 Advantages and Limitations of Delta-Hedging

The previous sections discussed delta-neutral portfolios in which it seemed that delta-hedging solved any concerns about sensitivity of the portfolio to changes in the price of the underlying asset price. However, as a first-order derivative, delta is only able to capture risk due to very small changes in the price of the underlying asset. If a large shift in prices occurs, the change of Δ with respect to a change of S matters. Before showing that, we first show that if a portfolio contains long options (calls or puts) together with a delta-neutral hedge, the portfolio will always gain money from an instantaneous price change in the underlying asset up or down. When the underlying asset's price changes, the delta changes and a delta-neutral portfolio needs to be rebalanced to remain delta neutral.

However, if the portfolio contains short options (calls or puts) together with a delta-neutral hedge, when the underlying asset's price changes instantaneously in either direction, the portfolio always loses money. In particular, the loss will be big if the change in delta in response to the change in S is large. Note that if there are little or no changes in the price of the underlying stock through time, long positions in options will tend to generate losses and short positions in options will tend to generate profits. This is shown in the following discussions.

Consider Figures 5.3 and 5.4. The two graphs show the price of a long position in a call and a long position in a put, respectively, as a function of the price of the underlying asset, S. The option values are calculated using the Black-Scholes formula. Figures 5.3 and 5.4 depict the relationship between the price of an option and the value of its underlying asset. The delta of each option is equal to the slope of the curved line (option price-to-underlying asset price relationship). A straight line is illustrated at an arbitrary point to illustrate the delta at a particular point and show that the delta is equal to the slope of the tangent line.

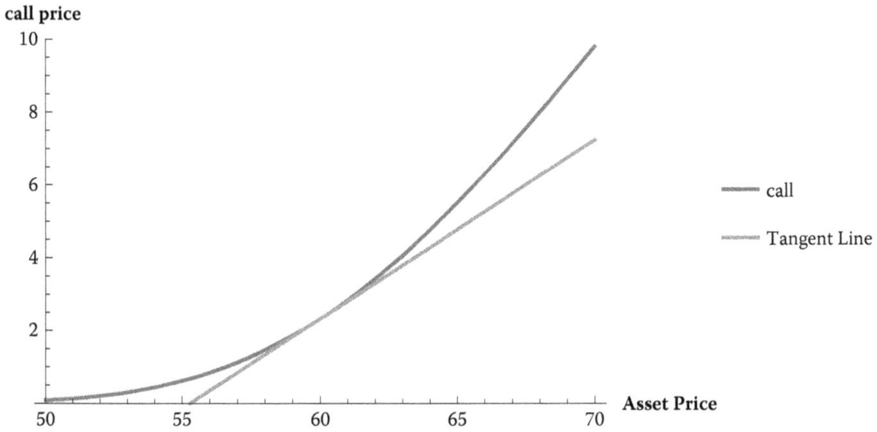

FIGURE 5.3
Call Option Prices and the Slope (to Highlight delta) at Asset Price = 60.

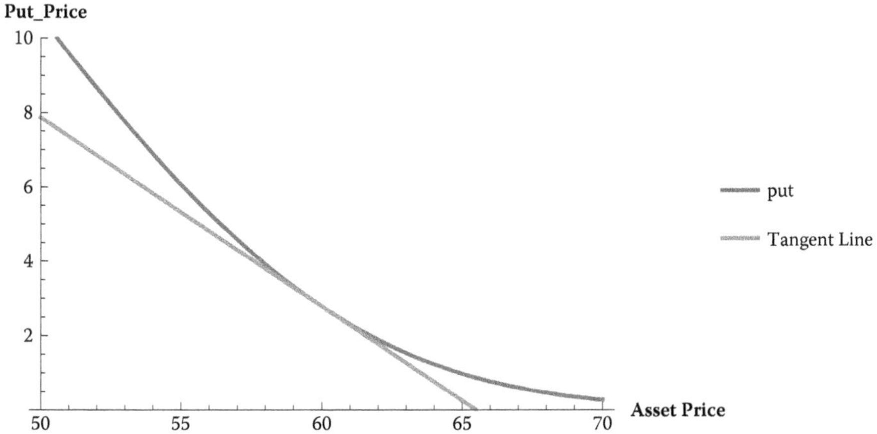

FIGURE 5.4
Put Option Prices and the Slope (to Highlight delta) at Asset Price = 60.

Note that the delta moves from near zero to near one from left (out-of-the-money) to right (in-the-money) in the case of a call. The diagrams vary the value of the underlying asset (along the horizontal axis) but use the same other parameters as following: $K = 62$, $\sigma = 15\%$, $r = 5\%$, $T = 0.5$.

Option deltas change as the price of the underlying asset changes as captured by the curve's changing slope. From the graph, delta is positive for long a call. Also, as S increases, Δ increases. Analogously, Δ is negative for long a put. So as S increases, Δ is less negative and is therefore also increasing. Therefore, the value of long an option (call or put) is a concave up function of S. The price curves stay above the tangent lines. This has a very important economic meaning for delta-neutral hedging.

Both graphs provide an important insight into the valuation of a delta-hedged portfolio. Figure 5.3 provides insight into the valuation of a delta-hedged portfolio that is long a call option and short Δ shares of the underlying asset (e.g., a stock), while Figure 5.4 illustrates a delta-hedged portfolio that is long a put option and long Δ shares of the stock.

Consider Figure 5.3. The tangent line indicates changes in the value of the short stock position (a liability) while the curved line indicates changes in the value of the long call position. When the underlying asset price rises, the liability (i.e., the value of the short position in the stock) rises, and it causes the portfolio to lose value. However, the long call rises more in value. Note that for this hedged portfolio (long a call option and short Δ shares of the underlying stock), any shift in the stock price (up or down) will cause the portfolio's net value to rise; as the underlying stock rises, the call position gains more than the short stock position loses, and when the underlying stock declines, the call position loses less than the short position gains. Therefore, ignoring the passage of time, any volatility in the underlying asset works to the benefit of the delta-hedged portfolio. This result should make intuitive sense because the portfolio is long an option and option prices benefit from higher-than-anticipated volatility. The delta neutrality makes the direction of the stock move inconsequential, but the portfolio retains the ability to benefit from volatility through its long option position.

Both figures illustrate that the delta-hedging is less effective as the change in the price of the underlying asset becomes larger (assuming no rebalancing). However, the key observation in both figures is that in both cases the imperfect hedging caused by an immediate and large change in the price of the underlying asset inures to the benefit of the investor with a long position in the call or put.

In summary, a delta-neutral portfolio that is long an option and hedged with the underlying asset will increase in value whenever an up or down shift occurs in the price of the underlying asset (ignoring the effects of the passage of time on the option price). The increase in the portfolio's value is depicted in the graphs by the difference between the curve and its tangent line. The magnitude of the difference is determined by two factors: the change of S and the change of Δ with respect to a change of S (the concavity). If there is a large change in S, the gain is large. Also, if there is a large change of Δ with respect to the change in S, even for a small change in S, the gain could be large. If the portfolio is rebalanced back to delta-neutrality after the shift, then any gains are locked in, the portfolio returns to being delta hedged, and the process starts over.

The above discussion has several important points – some of which are revisited in later sections. In the case of a long call or put being delta-hedged with a position in the underlying asset:

1. Low realized levels of volatility in the price of the underlying stock generate modest hedging profits when the portfolio is regularly brought back to delta-neutrality (ignoring the effects of the passage of time),
2. High levels of realized volatility in the price of the underlying stock generate relatively large hedging profits (ignoring the effects of the passage of time),
3. The hedging profits may or may not cover the potential losses in the option's time value as time passes, and
4. The frequency (i.e., speed) with which the hedge is rebalanced as price shifts begin to take place has an effect on the frequency and size of profits from volatility. In practice, the frequency of rebalancing can be an important issue.

Next, consider what happens in the case of a delta-neutral portfolio that is short an option and is hedged with the option's underlying asset. The portfolio will decrease in value whenever an up or down shift occurs in the price of the underlying asset (ignoring the effects of the passage of time on the option price). The value decrease is the difference between the curve and its tangent line. If the change in S and the change of Δ with respect to a change in S (concavity) are both small, there may be a small loss from delta-neutral hedging. If either the change in S or the change in Δ with respect to a change in S (concavity) is big, the portfolio will lose substantial value even though it is delta-hedged. In particular, if the change of Δ with respect to a change of S is big, even a small change in S could cause a large loss.

This discussion has emphasized the importance of changes in delta due to changes in S. The next section discusses the last Greek, gamma, which is the change of delta with respect to a change in the price of the underlying asset.

5.5.4 Gamma as a Second-Order Greek and Gamma-Neutral Hedging

As a second-order partial derivative, ***gamma*** (Γ) is the rate of change of delta (Δ) with respect to the price of the underlying asset.

$$\Gamma = \frac{\partial \Delta}{\partial S} = \frac{\partial^2 f}{\partial S^2}$$

According to the Black-Scholes formula, the gamma of a European call on a non-dividend-paying stock is:

$$\boldsymbol{\Gamma} = \frac{\partial^2 c}{\partial S^2} = \frac{\partial^2 p}{\partial S^2} = \frac{N'(d_1)}{S_0 \sigma \sqrt{T}} \tag{5.25}$$

Note that gamma for a call and a put are equal if the underlying assets, strike price, and expiration are all the same. This is a very easy corollary of put call parity:

$$p + S = c + Ke^{-rT}.$$

$$\frac{\partial^2 c}{\partial S^2} = \frac{\partial^2 p}{\partial S^2} \tag{5.26}$$

Figure 5.5 depicts gamma as a function of S. It uses the same example as before: $K = 62$, $\sigma = 15\%$, $r = 5\%$, $T = 0.5$. Gamma is greatest when options are: (1) near the money and (2) near expiration. Nearness to the money and nearness to expiration increase gamma because that is when the option is nearest to the moneyness and point in time where the likelihood that the option will be exercise is most variable.

Recall the time value of an option (i.e., the excess of the option's total value above its intrinsic value – where the intrinsic value is the greater of the immediate exercise value and zero). The time value of the option is positively related to: (1) time to expiration and (2) the anticipated volatility

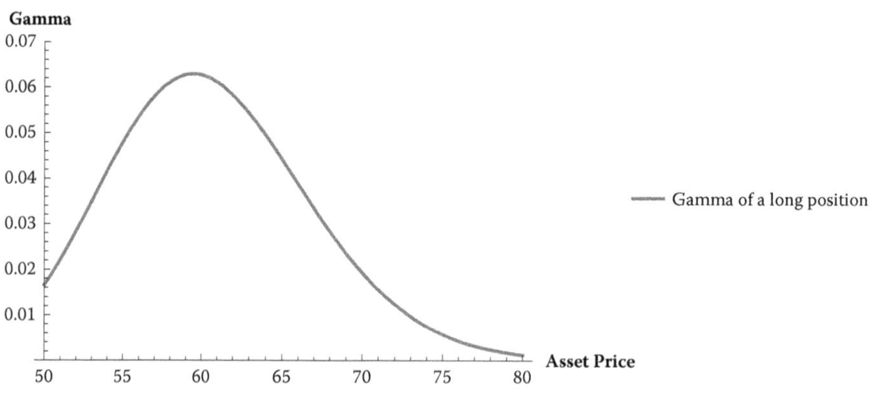

FIGURE 5.5
Gamma of a Long Position in Options.

of the underlying asset. When both of these factors are low, the time value will be small. When both of these factors are large the time value will be large. When the time to expiration is low, the option will tend to have a very large gamma if near-the-money and a small gamma elsewhere. When the time to expiration is large, the option will tend to have small values of gamma.

Figure 5.6 depicts three degrees of time value and three different degrees of curvature which is related to gamma. The graph of call prices against asset price uses input values: $K = 12$, $r = 5\%$. The values S, σ, and t vary. When the time value is high (the left curve with the longest time to expiration and highest total volatility), the curvature is small (i.e., the gamma is small) and it is spread over a wide range of the x-axis. When the time value is low (the right curve), the curvature is big (i.e., the gamma is large) near the strike price (near the money). Accordingly, gamma is highest where the curvature is sharp (near the money options near to expiration) and is lowest for options away from the money or with long times to expiration.

Analysis of a portfolio's gamma is very important. The gamma of a **long** option position is always **positive** (it is a concave up function and has a positive second derivative), while the gamma of a **short** option position is always **negative** (it is a concave down function and has a negative second derivative). The sign of gamma indicates option *ownership*. Being net long options offers larger profits from a favorable move in the option than losses from an unfavorable move in the underlying asset. This is illustrated by the convexity in diagrams of calls and puts (Figures 5.3 and 5.4) since delta is an increasing function of S for a long option position whether it is a call or a put. Being net short options offers large losses from an unfavorable move in the underlying asset and small profits from a favorable move in the option.

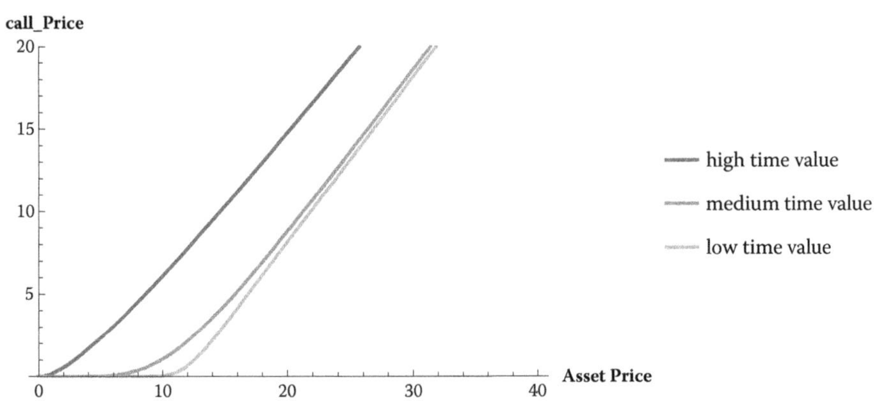

FIGURE 5.6
Call Pricing and the Time Value of Options.

Since gamma is a second partial derivative, it commutes with summation and a constant multiple. The ***portfolio gamma*** can be calculated when the portfolio contains a linear combination of calls and puts, weighted by units, as the linear combinations of the gammas of the individual calls and puts weighted by *the units* (where all options must have the same underlying asset).

Gamma indicates the degree of departure of a portfolio's value due to a change in S from that of the tangent line with a slope of delta. When gamma is small, for a moderate change of ΔS, the tangent line is still a good approximation of value and a delta-neutral hedge will experience small changes in value. When gamma is large, the curve bends sharply and even a moderate change in S can cause large changes in the delta-neutral portfolio's value. If the gamma of a delta-neutral portfolio is positive, quick changes in either direction will increase the value of the portfolio. Everything else equal, an investor should prefer positive gamma. But positive gamma comes with negative theta. Option prices adjust such that in an efficient market investors must pay for positive gamma in the form of option prices that contain time value (and likely time decay) commensurate with the gamma.

In the case of a delta-neutral portfolio, if the gamma is large then its delta changes dramatically when the underlying asset price changes and the portfolio will exhibit a sharper curve in the diagram of its price when graphed against S. This means that a delta-neutral portfolio with a high positive gamma will gain more than a portfolio with a low gamma when the underlying asset quickly rises or falls in value. Highly positive gamma also means that the portfolio needs to have more underlying shares purchased or sold to rebalance the portfolio back to delta-neutrality. So, rebalancing infrequently could cause the portfolio to give back gains due to lack of delta neutrality if the price of the underlying asset reverts back to previous values. Matching the gammas of the portfolio to be hedged and the hedging vehicle is known as ***gamma-neutral hedging***:

$$\mathbf{\Gamma}_{\text{hedging vehicle}} = -\mathbf{\Gamma}_{\text{asset}} \tag{5.27}$$

Note that the underlying asset, S, has gamma equal to zero. So unlike the case of forming a delta-neutral portfolio (when the position in the underlying asset can be changed to control the total delta), only options can be used to change the gamma of a portfolio. The management of portfolios containing options using the concepts of delta-, gamma-, and Vega-hedging are discussed in subsequent sections. Briefly, delta-neutrality protects a portfolio from changes in the direction of S. Positive gamma in portfolios with long options offer attractive gains if S exhibits greater volatility than implied by the prices of the portfolio's options when they were purchased. However, theta measures the time decay of the long option positions, which threatens the portfolio with net losses if S exhibits lower volatility than anticipated when the options were purchased. The exposure of options to profits or losses based on changes in the level of anticipated volatility is captured by the Vega of the portfolio.

5.5.5 An Example of Delta-Hedging, Vega-Hedging, and Gamma-Hedging

Note that while options contain gamma and Vega, the underlying asset S has Vega and gamma equal to zero. So, unlike in the case of delta-neutral hedging, options must be used rather than the underlying asset to manage a Vega-/gamma-neutral hedge. This section discusses management of the delta, gamma, and Vega of a portfolio or strategy. It begins with a demonstration of hedging each risk. We will use an example similar to Hull's.

Example 5.10: Assume the following three Greeks for a portfolio and two options. Note that the underlying asset (also available for trading) has delta = 1, and both gamma and Vega equal to zero. The portfolio is delta-neutral with gamma = –1,800 and Vega = –5,600. Option 1 has delta = 0.6, gamma = 1.5, and Vega = 5.0. Option 2 has delta = –0.5, gamma = 1.2, and Vega = 3.6.

a. What positions in Option 1 and the underlying asset when added to the portfolio will make the portfolio delta- and gamma-neutral?

   To achieve gamma neutrality using Option 1, go long 1,800/1.5 = 1,200 units of Option 1. However, that will change the delta to 0 + 0.6 × 1,200 = 720. Therefore, short 720 units of the underlying asset to return the portfolio to delta-neutrality. Adding a position in the underlying asset will not change the gamma or Vega.

b. Alternatively, what position in Option 1 and the underlying asset will make the portfolio delta and Vega neutral?

   To achieve Vega-neutrality using Option 1, go long 5,600/5 = 1,120 units of Option 1. However, that will change the delta to 0 + 0.6 × 1,120 = 672. Therefore, short 672 units of the underlying asset to return the portfolio to delta-neutrality. Adding a position in the underlying asset will not change gamma or Vega.

c. What position in Option 1, Option 2, and the underlying asset will make the portfolio delta-, gamma-, and Vega-neutral?

   The problem involves solving three simultaneous equations (the delta-neutral equation, the gamma-neutral equation, and the Vega-neutral equation) using three variables (units of Option 1, units of Option 2, and units of the underlying asset). The problem can be simplified because the underlying asset has exposure only to delta. Begin by ignoring delta as well as the underlying asset and solving for the portfolio that sets gamma and Vega to zero with variables w_1, units of Option 1 and w_2, units of Option 2:

$$-1,800 + 1.5w_1 + 1.2w_2 = 0 \text{ (Gamma neutral)}$$

$$-5,600 + 5.0w_1 + 3.6w_2 = 0 \text{ (Vega neutral)}$$

The solution is $w_1 = 400$ and $w_2 = 1{,}000$ indicating that the portfolio requires long positions of 400 and 1,000 in Options 1 and 2, respectively. The new portfolio delta = 400 × 0.6 + 1,000 × (−0.5) = −260. A long position of 260 units in the underlying asset brings the portfolio to delta-neutral without affecting the Vega and gamma hedge.

The next section discusses the relationship between the Greeks.

5.5.6 Relationships between Delta, Gamma, and Theta

Positive gamma measures the advantage of long positions in options illustrated by the graph of option prices against S. The cost of achieving this desirable exposure is the time value of the option. If S experiences low volatility, the negative theta of long options positions will tend to drive the option prices down through time (assuming that the market anticipates continued modest volatility). This section explores these issues.

For a delta neutral portfolio, Π, an approximation of the portfolio's value change can be formed:

$$\Delta\Pi \approx \theta\Delta t + \tfrac{1}{2}\Gamma(\Delta S)^2 \tag{5.28}$$

Note that in Equation 5.28, Δ is the symbol for a finite change, not the symbol for the partial derivative with respect to S (i.e., delta). If one is familiar with Ito's lemma, this is not a new equation. It is a discrete time version of Ito's lemma:

$$df = \frac{\partial f}{\partial S}dS + \frac{\partial f}{\partial t}dt + \frac{1}{2}\frac{\partial^2 f}{\partial S^2}(dS)^2.$$

Regard the change of the portfolio's value, $\Delta\Pi$, as a quadratic equation with respect to change of underlying asset's price, ΔS.

The above equation demonstrates the positive effect of positive gamma and the negative effect of theta for a long position in an option. Note that when a positive gamma is multiplied by ΔS^2, the impact is non-negative. However, Δt is always positive, and a long position in an option has negative theta. Therefore, the net change in Π is the combined effect of the gains from changes in S against the losses from the negative theta and the inevitable passage of time. An option's value is a function of *anticipated* volatility in the returns of the underlying asset. When *realized* volatility exceeds anticipated volatility, the effect of gamma tends to exceed the effect of theta on a long option position leading to a profit, and vice versa. This relationship assumes that the option has been delta-hedged by longing/ shorting its underlying asset. If the long position in the option is not delta-hedged, then the option also contains exposure to the direction of the price changes in the underlying asset.

The previous paragraph indicated that when *realized* volatility exceeds anticipated volatility, the effect of gamma ***tends to*** exceed the effect of theta on a long option position leading to a profit, and vice versa. It is not always true. The performance of a delta-neural, positive gamma portfolio depends not just on the *level* of realized volatility but also on details regarding that realized volatility. Profits depend on whether the realized volatility occurred when gamma is highest or lowest (e.g., near-the-money and near expiration vs. far from-the-money and far from expiration). Also, profitability relates to whether the hedge is being rebalanced frequently or infrequently and how that decision relates to any tendency in the underlying asset to experience volatility that had an *ex post* period of trending or mean-reversion. Practitioners using such strategies often succeed or fail depending on their luck or skill in selecting appropriate rebalancing intervals and timing the level and nature of the realized volatility.

Consider another relationship. For a portfolio of derivatives on a non-dividend-paying stock:

$$\theta + rS_0\Delta + \frac{1}{2}\sigma^2 S_0^2\Gamma = r\Pi \tag{5.29}$$

This formula can be viewed as the discrete version of the Black-Scholes PDE.

Untill now, we have discussed analytic formulas (closed form) for hedging with the Greeks. Note that all Greeks are not constants, therefore, all hedges by Greeks are dynamic which can be put in the context of a binomial tree, finite difference and simulation. This is because the Greeks do not need to be in a closed form formula in arbitrary derivatives other than simple options, they can be calculated through binomial tree and finite difference methods in an EXCEL spreadsheet or through simulation in R. Since we illustrate it in Chapters 3 and 4, we will not go into the details again. In Chapter 7, advanced hedges for fixed income instruments are demonstrated using an EXCEL spreadsheet. The next section discusses how to hedge interest rate risk for portfolios containing fixed-income assets using a duration hedge in closed form formulas.

5.6 Duration-Hedging of Fixed-Income Instruments and Interest Rate Risk Using Bonds and Bond Futures

Duration and convexity hedging of fixed-income securities is a fundamental tool to manage interest rate risks that is an important part of most undergraduate investment courses as well as the CFA curriculum. However, because the term structure of interest rates tends to have non-parallel shifts in the real world, advanced tools have been developed to address all types of shifts. Chapter 7 discusses hedging strategies using the HJM model framework. The HJM model is an important approach that addresses non-parallel interest rates shifts.

This section first defines duration and then examines duration-hedging of the interest rate risk in a bond portfolio.

5.6.1 Duration and Duration-Hedging

Consider the following formula for the price of a single zero coupon riskless bond with a face value of $1:

$$P = e^{-rT}$$

The partial derivative of P with respect to r is $\frac{\partial P}{\partial r} = -TP$. Dividing the partial derivative by the bond price and reversing its sign $-\frac{\partial P / \partial r}{P}$ generates a measure known as duration:

$$D = -\frac{\partial P/\partial r}{P}$$

which in the case of a pure discount bond equals its maturity, T.

In the case of a pure discount bond and when r is viewed as the bond's yield-to-maturity, the bond's duration is a measure of instantaneous interest rate risk. Duration measures the sensitivity of bond returns $\left(\frac{\Delta P}{P}\right)$ with respect to a change in its yield Δr. In the case of coupon bonds or when the zero coupon rates across different maturities are allowed to be imperfectly correlated, T is not a perfect measure of interest rate risk. In the previous sections, the hedging of equities and derivatives on equities was made tractable by assuming that all of the derivatives shared the same underlying asset (S). In fixed-income analysis, precise hedging can be modeled when interest rates of all maturities are assumed to experience instantaneous, identical, and infinitesimal shifts. *The assumption of parallel, additive, and infinitesimal term structure shifts* permits perfect modeling of hedging using duration (i.e., T), although the assumption that only parallel term structure shifts can occur is not realistic (short rates tend to fluctuate more than long rates).

The ***duration of a coupon bond*** can be described as its sensitivity of the bond return with respect to a yield shift:

$$D = -\frac{\partial P/\partial y}{P} \tag{5.30}$$

Duration is useful because:

- a bond's return has an approximately linear relation with respect to a change in its yield:

$$r_p = \frac{\Delta P}{P} \approx -D \times \Delta y \tag{5.31}$$

- a bond's value change can be approximated by:

$$\Delta P \approx -D \times \Delta y \times P \tag{5.32}$$

The return of a bond is $r_p = \frac{\Delta P}{P}$, and in the case of a zero coupon bond, its yield y is equal to the market interest rate r. A first-order approximation to the return of a T-year zero coupon bond, r_T, in response to a shift in its yield to maturity, Δy, is:

$$r_T \approx -\Delta y \;\; T$$

For a zero coupon bond, the duration, D, is just maturity T. A zero coupon bond's return is approximately equal to the negative product of its yield change and its duration. With the assumption of *parallel, additive, and infinitesimal term structure shifts*, this relationship can be extended to the duration of a coupon bond or a bond portfolio.

How do we calculate duration for a coupon bond? We first consider a simple bond portfolio containing long positions in three zero coupon bonds with maturities T_1, T_2, T_3 and portfolio weights w_1, w_2,w_3. The first-order approximated return of the portfolio, r_p (assuming identical yield shifts of Δy) is the following weighted sum of individual bond's return:

$$r_p \approx w_1 \times (-\Delta y T_1) + w_2 \times (-\Delta y T_2) + w_3 \times (-\Delta y T_3)$$

which is equivalent to:

$$r_p \approx -\Delta y\,(w_1 T_1, +w_2 T_2 + w_3 T_3).$$

So, the duration for this bond portfolio can be defined as: $D = w_1T_1, +w_2T_2+w_3T_3$

It is easy to see that coupon bonds can be viewed as a portfolio of zero coupon bonds. The duration of a coupon bond can be defined as the weighted sum of the times to each cash flow and the weight of each cash flow as a percentage of the bond's total value.

$$D = \sum_{i=1}^{n} w_i T_i$$

The ***duration of the bond portfolio***, D, can be found as the weighted sum of the durations of each bond where the weight is the percentage value of each bond in the bond's total value:

$$\boldsymbol{D} = \sum_{i=1}^{n} w_i \boldsymbol{D}_i \tag{5.33}$$

Consider the case of a bond portfolio containing long positions in two bonds – not necessarily zero coupon bonds. Assume that a portfolio has P dollars in two bonds: one with a duration D_1 and the other with a duration D_2. The first bond has a portfolio weight of w_1 and the second bond has a weight of w_2. For a yield change Δy, the first bond's dollar value change is approximately: $D_1 \times \Delta y \times w_1 \times P$, and the second bond's dollar value change is approximately $D_2 \times \Delta y \times w_2 \times P$. The total bond value change is approximately: $D \times \Delta y \times P$.

Summing the two value changes: $D \times \Delta y \times P = D_1 \times \Delta y \times w_1 \times P + D_2 \times \Delta y \times w_2 \times P$, implies:

$$D = w_1 \times D_1 + w_2 \times D_2$$

Note that the weight is the percent of the portfolio's value attributable to each bond, namely, $w_1 = \frac{P_1}{P}$ and $w_2 = \frac{P_2}{P}$, where P is the value of the portfolio, P_1 is the value for bond 1 and P_2 is the value for bond 2. In our convention, the duration of a short position in a bond is negative.

Now apply this idea to hedging. If there is a parallel yield shift, Δy, and it is small, we can adjust the weights of the underlying asset and the hedging vehicle to have the same return with respect to Δy, but the opposite sign. In doing so, the portfolio will have no sensitivity toward a small yield change. In the following discussions, weights are all positive and duration itself carries the negative sign.

Consider three zero coupon bonds with maturities of one, two, and three years, respectively, and with initial continuously compounded yields as depicted in Table 5.2. Assume that an investor with a two-year bond wishes to hedge that bond's interest rate risk with a position in a one-year bond, a three-year bond, or a mix of the two bonds.

TABLE 5.2

Bonds Returns from a Small Yield Change

Bond #	Maturity	Initial Yield	Initial Price	Price after 10 bp Shift Up	Return
1	1 Year	5.00%	95.12294	95.02787	−0.09995%
2	2 Years	6.00%	88.69204	88.51484	−0.19980%
3	3 Years	7.00%	81. 05842	80.81561	−0.29955%

- One way to find a hedged portfolio is to consider only two bonds such as hedging the two-year bond with the one-year bond. Assume that the hedged portfolio includes that the one-year bond has a weight of w_1 and the two-year bond has a weight of w_2.

The two-bond portfolio will be hedged and have a zero return when there is a yield change Δy:

$$w_2 r_2 = -w_1 r_1$$

$$w_2(-\Delta y T_2) = -w_1(-(-\Delta y T_1))$$

$$w_1 = w_2\ T_2/T_1\ = w_2 \times 2/1 = \ 2w_2$$

To create a hedge against the two-year bond requires a short position twice the size (i.e., $w_1 = 2w_2$) in the one-year bond. Assume that there is a 10 bp shift up. Inserting the returns for the bond from Table 5.2 generates the following hedged return, r_h:

$$r_h \ = \ -0.\ 19980\% + (2 \times 0.\ 09995\%) = 0.0001\%$$

The hedge is slightly off (1/100 of a basis point) because it is a first-order approximation.

- Another hedge of the two-year bond could be formed by placing weights of 50% on bonds 1 and 3 in the hedging portfolio if the two-year bond has a weight of 100%. (If the total hedged portfolio has a weight of 100%, then bond 2 has a weight of 50% and a duration of 2, and bonds 1 and 3 have a weight of 25% and durations of −1 and −3.) Note that larger interest rate shifts and non-parallel term structure shifts (i.e., when yields of different maturities shift by different amounts) will tend to create hedging errors.

The point of the exercise is that under the given assumptions of parallel, additive, and infinitesimal term structure shifts, the duration is a first-order derivative of functions much like the delta of options in the previous sections. Similar to the case of a delta hedge, the duration approach is easily expanded to coupon bonds or a bond portfolio because partial derivatives commute with summation and constant multiples. However, there is one big difference regarding weights of a delta hedge and duration hedge. In the case of delta, the weight is the number of units of the options, while in the case of duration the weight is the percentage of the value of each of the bonds comprising the total value of the bond portfolio.

Two weighting-approaches have been discussed in various places in this chapter.

- In the return approach to hedging, the weights are the percentage weights of the asset being hedged as well as the hedging vehicle in the total portfolio.
- In the unit value approach to hedging, the weights are the numbers of units.

In particular, in a discussion of duration, the weight is the percentage of the values of each of the bonds in the total value of the bond portfolio while in a discussion of Greek-letter option hedging the weight are units of options.

Next, derive the hedge ratio, h, for a bond portfolio based on duration.

Assume that the asset being hedged has a duration, D_{asset}, and the hedging vehicle has a duration of $D_{\text{hedging vehicle}}$. Assume that there is a small parallel yield shift, Δy. A hedged portfolio will have a zero return when:

$$w_{asset} r_{asset} = -w_{hedging\ vehecle} r_{hedging\ vehecle}$$

$$w_{asset}(-\Delta y D_{asset}) = -w_{hedging\ vehicle}(-\Delta y D_{hedging\ vehicle})$$

$$w_{hedging\ vehicle} = -w_{asset} \times \frac{D_{asset}}{D_{hedging\ vehicle}} = w_{asset} \times |h|$$

where the hedging ratio for a duration hedge is:

$$\boldsymbol{h} = \frac{\boldsymbol{D}_{asset}}{\boldsymbol{D}_{hedging\ vehicle}} \tag{5.34}$$

The negative duration in the hedge vehicle represents opposite positions in the asset to be hedged and the hedging vehicle. The weight allocation is:

$$w_{\textbf{hedging vehicle}} = w_{\textbf{asset}} \times |\boldsymbol{h}| \tag{5.35}$$

Sometimes, the number of contracts, N, is useful:

If the asset being hedged has one unit with a value of V_{asset} and the hedging vehicle has a value $V_{\text{hedging vehicle}}$ per contract, and the number of contracts in the hedging vehicle needed is N:

$$\frac{N \times V_{\text{hedging vehicle}}}{N \times V_{\text{hedging vehicle}} + V_{\text{asset}}} = \frac{V_{\text{asset}}}{N \times V_{\text{hedging vehicle}} + V_{\text{asset}}} \times \frac{|D_{\text{asset}}|}{|D_{\text{hedging vehicle}}|}$$

$$N = \frac{V_{\text{asset}}}{V_{\text{hedging vehicle}}} \times \left| \frac{D_{\text{asset}}}{D_{\text{hedging vehicle}}} \right| \tag{5.36}$$

Note that Equation 5.36 retains the consistency of placing the asset being hedged in the numerator and the hedging vehicle in the denominator.

5.6.2 Duration-Hedging Using Bond Futures

Bond futures can be used to hedge bond portfolios and bank loan and deposit interest risks. We will illustrate it in the followings:

5.6.2.1 Duration-Hedging a Bond Portfolio Using Bond Futures

This section discusses using bond futures or interest rate futures contracts to hedge a bond portfolio. Futures are frequently used to hedge interest rate risk because they:

- are easier to short sell
- tend to have lower transaction costs
- tend to have lower margin requirement
- tend to be more liquid

Assume that a portfolio manager is concerned about interest rate risk in the next six months. The manager can use bond/interest rate futures with six months to settlement to hedge. The portfolio manager can always roll over the hedge if needed as part of a continued hedge. Since the manager cares about the interest rate risk at the end of the maturity of the futures contracts, it is easier to model all durations as of the end of the futures maturity.

Consider two portfolios: Portfolio P is a portfolio of bonds with interest rate risk that the portfolio manager wishes to eliminate through hedging for a short period of time. Portfolio F contains short-maturity futures contracts on bonds and will be used to hedge the interest rate risk of Portfolio P for a short period of time. Consider the returns of the bond portfolios represented by durations:

$\frac{\Delta V_p}{V_p} \approx -D_P \times \Delta y$ where D_P is the duration of the bond portfolio to be hedged at the future's maturity and V_p is the bond value anticipated at the maturity of the futures contract that is to be used as the hedging vehicle. In practice, investors will typically use the bond value today as an approximation.

$\frac{\Delta V_F}{V_F} \approx -D_F \times \Delta y$, where D_F is the duration of the futures' underlying asset at the end of futures maturity and V_F is contract price for the bond/interest rate futures contract today.

From the math described in a previous section, the number of futures contracts needed is:

$$N = \frac{V_p \times |D_P|}{V_F \times |D_F|} \tag{5.37}$$

5.6.2.2 Duration-Hedging Bank Loan/deposit Interest Rate Risk Using Bond Futures

Many commercial banks have short-term liabilities (e.g., short-term CDs and deposit accounts) and long-term assets (mortgages and other loans), which leads to a duration mismatch. For a parallel yield shift, Δy, the change in the value of the loan is $\Delta L \approx -V_L \times D_L \times \Delta y$. The change in the value of deposits is $\Delta D \approx -V_D \times D_D \times \Delta y$. In order to hedge the sensitivity of the portfolio to yield changes requires $V_L \times |D_L| = V_D \times |D_D|$.

In order to lessen the interest rate risk of the maturity mismatch, the bank may target loan amounts that are only a fraction of the size of the deposits. So the bank needs to keep loan values, V_L, much smaller than the short-term deposit values, V_D. The ratio $\frac{V_L}{V_D} = |\frac{D_D}{D_L}|$ hedges the bank's interest rate risk under the assumption that the interest rates of both loans and deposits shift by small and equal amounts. However, this ratio of assets to deposits may be inconsistent with other objectives that the bank wishes to pursue. Futures contracts on bonds and other fixed income derivatives can provide solutions with greater flexibility, efficiency, and profitability.

Assume the bank's target loan value is V_L, the bank's target deposit value is V_D, the bond futures value per contract is V_F, and the corresponding durations are D_L, D_D, and D_F, respectively. Assume that N is the number of futures contracts needed for a duration hedge. Further assume that there is a parallel yield shift, Δy. The change of the total loan amount is:

$$\Delta L \approx -V_L D_L \Delta y.$$

The change in the value of the deposits is:

$$\Delta D \approx -V_D D_D \Delta y$$

Since in the bank industry $|D_L| > |D_D|$ and $V_L \times |D_L| > V_D \times |D_D|$, the bank needs to short bond futures. The change of the bond futures price is:

$$\Delta F \approx -NV_F D_F \Delta y$$

$$V_L D_L \Delta y + V_D D_D \Delta y + NV_F D_F \Delta y = 0.$$

In this case, $D_D < 0$, $D_F < 0$, and $D_L > 0$, since deposits and futures are short positions and the loan is a long position. To avoid confusion, we use absolute values in the following formula.

The number of bond futures needed to hedge the duration mismatch between loan and deposits is:

$$\boldsymbol{N} = \frac{\boldsymbol{V_L} \times |\boldsymbol{D_L}| - \boldsymbol{V_D} \times |\boldsymbol{D_D}|}{\boldsymbol{V_F} \times |\boldsymbol{D_F}|} \tag{5.38}$$

5.6.2.3 Various Examples

Next, consider several examples.

- Example of Hedging a Bond Portfolio Using Bond Futures Contracts

Example 5.11: Assume that a bond portfolio has a market value of \$10 million and an anticipated duration of seven years in six months. A six-month bond futures contract price is quoted as \$99.725 per \$100 and each contract has a face value of \$100,000. The futures contract's underlying bond is anticipated to have a duration of 10 years at settlement. The goal is to fully hedge the interest risk using the above six-month bond futures. How many contracts are needed to hedge the \$10 million portfolio at the settlement date of the futures contract? Note that $V_F = \$99,725$. So, the number of contracts needed for the full hedge is:

$$\frac{V_p \times |D_P|}{V_F \times |D_F|} = \frac{10{,}000{,}000 \times 7}{99725 \times 10} \approx 70 \text{ contracts.}$$

- Example of Hedging a Bond Portfolio Using Eurodollar Futures Contracts

A ***Eurodollar futures*** contract is a contract on short-term LIBOR rates. At the contract's expiration, the underlier is the three-month LIBOR interest rate. So, the duration of the bond at the settlement of the futures contract is 0.25. So, a one-basis-point (0.01%) move in the futures quote corresponds to a gain or loss of 0.01% × 1/4 = 0.0025%. The contract size is $1 million dollars, so the dollar change is $1,000,000 × 0.0025% = $25 per contract per basis point.

Example 5.12: Assume that a bond portfolio has a value of $10 million and its duration is assumed to be seven years in six months. The goal is to fully hedge the interest risk using six-month Eurodollar futures. Six-month Eurodollars are quoted at 97.75. The face value is $1 million per contract. Find the number of contracts needed to hedge the interest rate risk.

Using the previously formula for N: $\dfrac{V_p \times |D_P|}{V_F \times |D_F|} = \dfrac{10{,}000{,}000 \times 7}{977,500 \times \frac{1}{4}}$

$$\approx 286.$$

- Example of Hedging Interest Rate Risk Caused by a Loan and Deposit Duration Mismatch Using Bond Futures

Example 5.13: A bank manager has a $100 million loan portfolio with a duration of 10 years, and $120 million of short-term deposits with a duration of one year. Suppose the manager is concerned about possible interest rates fluctuations and wants to hedge the bank's interest rate exposure. Assume that the six-month futures contract on Treasury bonds is projected to have a duration of eight years in six months and a futures price of $132. How can the manager hedge the loan portfolio risk six months hence to have the same duration as its deposits? (Note that the futures contract on the Treasury bond has a face value of $100,000 and that its price is quoted through reference to a $100 face value.)
In this example,

$$N = \frac{100{,}000{,}000 \times 10 - 120{,}000{,}000 \times 1}{8 \times 132{,}000} = 833$$

The number of T bond futures contracts to achieve a duration hedge is 833.

5.6.3 Dollar Duration

There is a concept called *dollar duration* which we notate as DD and illustrate with continuously compounded interest rates. ***Dollar duration*** is the *dollar* change in a bond or futures contract's value due to a specified interest rate change, Δy:

$$\boldsymbol{DD = -D \times \Delta y \times V} \tag{5.39}$$

where D is duration (sometimes it is called effective duration to distinguish it from dollar duration), Δy is the yield change, and V is the portfolio value. Dollar duration is an alternative measure of interest rate risk and can be used to determine hedges. So, instead of matching duration, we can choose to match "dollar duration."

The dollar durations for the bond portfolio and bond futures are:

$$DD_p = -D_P \times \Delta y \times V_p$$

$$DD_F = -D_F \times \Delta y \times V_F$$

In terms of risk, matching dollar durations is equivalent to matching effective durations. Recall that in a duration hedge, N, the number of contracts in the hedging asset, is specified as:

$$N = \frac{V_p \times |D_P|}{V_F \times |D_F|} = \left| \frac{-D_P \times V_p \times \Delta y}{-D_F \times V_F \times \Delta y} \right|$$

So, Equation 5.39 using the above formula:

$$\boldsymbol{N = \frac{|DD_p|}{|DD_F|}} \tag{5.40}$$

Assuming that the target dollar duration is DD_T, the number of futures contracts needed is:

$$\boldsymbol{N = \frac{|DD_T| - |DD_P|}{|DD_F|}} \tag{5.41}$$

Equation 5.40 is analogous to Equation 5.37, and Equation 5.41 is analogous to Equation 5.38 since dollar duration includes the information of portfolio value V.

Consider the previous example with three changes in the following example: (1) dollar duration is used rather than effective duration, (2) there are no short-term deposits, and (3) instead of seeking a complete hedge against interest rate risk (DD = 0), the manager seeks to limit the interest rate risk to a target exposure (specifically that DD = \$0.25 million). This means that the manager is willing to tolerate a portfolio fluctuation of \$0.25 million if the interest rate shift occurs that was specified in the computation of the dollar duration.

Example 5.14: A manager has a \$100 million portfolio with a duration 10 years and wants to limit the bank's dollar duration to \$0.25 million. Assume that the futures contract on Treasury notes has a duration of eight years and a futures price of \$132. The manager selects a 10 basis point interest rate shift up as the basis for calculating DDs. How can the manager hedge the portfolio's risk down to the \$0.25 million limit?

First, calculate the dollar duration of the current portfolio: $DD_p = -D \times \Delta y \times V_p = -10 \times 0.001 \times 100 = -1$ million. Next, the dollar duration of the futures contract needs to be calculated. The notional value of this futures contract is $V_F = 132{,}000$, so $DD_F = -D \times \Delta y \times V_F = 8 \times 0.001 \times 132,000 =\$1{,}056$ per contract. To bring the $DD_p = -1$ million down to $DD_p = -0.25$ million requires that the hedging portfolio of futures contracts has $DD_F = 0.75$ million. Therefore:

$$N = \frac{750{,}000}{1056} = 710.23.$$

The manager needs to short 710 contracts of the Treasury futures to hedge the portfolio's risk down to a *DD* of \$0.25 million.

Finally, consider a practical issue: Due to the Treasury bond futures contract's ***cheapest to deliver*** (CTD for short) convention in the financial industry, some adjustments need to be made in a real world problem. If the duration for the cheapest-to-delivery bond is known instead of the duration of the hypothetical bond specified in the futures contract, the price of the cheapest-to-deliver bond can be used in the computation of the DD_{CTD}. Then divide DD_{CTD} by the bond's conversion factor to get DD_F. Here we need to give more background on the concept of a cheapest-to-deliver bond.

Derivative markets tend to function better when there are larger supplies of the underlying asset available in the market to be delivered. Many futures and forward contracts therefore allow delivery by the short side of the contract to be fulfilled by delivering assets that differ in "grades" or quality subject to a price adjustment. For example, a futures contract on wheat might specify that the deliverable grain can differ by protein content with delivery of higher protein levels credited at higher prices.

Some financial futures contracts also allow for a spectrum of quality or grades. For example, Treasury bond futures contracts allow for delivery of various bonds with differing maturities and coupons, although there is a hypothetical asset specified in the contract with a specific maturity and coupon (that does not even exist). Each actual bond that can be delivered to satisfy the contract is subject to a price adjustment factor that tends to reduce the dispersion in the attractiveness of the various bond issues for delivery. Nonetheless, there is a cheapest-to-deliver bond at all times and it tends to be relative to that bond that the futures price trades (when there is adequate liquidity available). However, when interest rates change, the bond that is cheapest-to-deliver bond may change, which can be viewed as an option of the short side of the contract to deliver the bond with the lowest factor-adjusted value.

5.6.4 Duration Hedging with Second-Order Partial Derivatives

Using a first-order partial derivative-based fixed-income model may generate a hedge that is not effective for large changes in interest rates or yields. In other words, if Δy is large, using the instantaneous rate of change D as a hedge ratio may cause a large approximation error as can be seen in a Taylor expansion. This is very similar to the error in delta-hedging an options portfolio when there are large shifts in the price of the underlying asset.

Recall the earlier example of three zero coupon bonds with maturities ranging from one to three years. Assuming a uniform yield shift of 10 basis points for each bond, the risk of the two-year bond could be effectively hedged using a short position in the one-year bond with twice the value of the two-year bond. Let's modify that example by examining a 100 basis point shift instead.

To hedge the two-year bond with the one-year bond still requires: $w_1/w_2 = T_2/T_1 = 2/1 = 2$. Inserting the returns for the bond from Table 5.3 generates the following hedged return, r_h:

$$r_h = -1.9801\% + (-2 \times -0.9950\%) = 0.0099\%.$$

The previous hedging error of 1/100th of a basis point for a 10 basis point shift in yields has grown to 99/100ths of a basis point for a 100 basis point shift in yield. So, it is clear that the duration hedge may not perform well if yields change by a large amount.

This section introduces hedging of second derivative risk for fixed-income assets and is much like hedging gamma in the previous sections on options. The second-order risk measure for fixed-income risk analysis is referred to as ***convexity***:

$$\textbf{Convexity} = \mathbf{C} = \frac{1}{V}\frac{\partial^2 V}{\partial y^2} \tag{5.42}$$

TABLE 5.3

Bonds Returns and a Large Yield Change

Bond #	Maturity	Initial Yield	Initial Price	Price After 100 bp Shift Up	Return
1	1 Year	5.00%	95.12294	94.17645	−0.99502%
2	2 Years	6.00%	88.69204	86.93582	−1.98013%
3	3 Years	7.00%	81. 05842	78.66279	−2.95545%

By a second-order Taylor expansion $\Delta V \approx \frac{\partial V}{\partial y}\Delta y + \frac{1}{2}\frac{\partial^2 V}{\partial y^2}(\Delta y)^2$:

$$\boldsymbol{\Delta V / V = -D\Delta y + \frac{1}{2}C\,(\Delta y)^2} \tag{5.43}$$

For a zero coupon continuously compounded bond, the pricing and convexity formulas are as follows:

$$V = e^{-rT}$$

$$C = T^2$$

The second-order approximation provided in option hedging by gamma hedging is analogous to the application of convexity as a second-order approximation in hedging against large yield shifts. Returning to the previous example, using duration and convexity to approximate the return for one-year zero coupon bond (and including the second-order terms for a long position):

$$\frac{\Delta V}{V} = -D\Delta y + \frac{1}{2}C\,(\Delta y)^2 = -1 \times 0.01 + \frac{1}{2} \times 1^2 \times 0.01^2 = -0.00995.$$

The return for the two-year zero coupon bond:

$$\frac{\Delta V}{V} = -D\Delta y + \frac{1}{2}C\,(\Delta y)^2 = -2 \times 0.01 + \frac{1}{2} \times 2^2 \times 0.01^2 = -0.0198$$

The above return approximations using both the first- and second-order derivatives match the table numbers well. Note that the convexity measure adds a larger change in the long-term bond, so the optimal hedge requires a little less than two contracts.

Next, the formulas for the hedge ratio, weights, and number of contracts are derived.

Use both duration and convexity to hedge a large interest rate change. Using "asset" to denote the asset to be hedged and "hedging vehicle" as the position used for hedging purposes, form the hedged portfolio P. In order for the return of portfolio P to be zero:

$$w_{\text{asset}}\left|r_{\text{asset}}\right| = w_{\text{hedging vehicle}}\left|r_{\text{hedging vehicle}}\right|$$

To make the portfolio interest rate neutral (so that $r_p = 0$ when there is a change in the yield) find the optimal number of contracts in the hedge:

$$\frac{w_{\text{hedging vehicle}}}{w_{\text{asset}}} = \frac{|r_{\text{asset}}|}{|r_{\text{hedging vehicle}}|} = \frac{|-D_{\text{asset}}\Delta y + \frac{1}{2}C_{\text{asset}}(\Delta y)^2|}{|-D_{\text{hedging vehicle}}\Delta y + \frac{1}{2}C_{\text{hedging vehicle}}(\Delta y)^2|}$$

So, we have

$$\boldsymbol{w}_{\textbf{hedging vehicle}} = \boldsymbol{w}_{\textbf{asset}}\frac{|-D_{\text{asset}}\Delta y + \frac{1}{2}C_{\text{asset}}(\Delta y)^2|}{|-D_{\text{hedging vehicle}}\Delta y + \frac{1}{2}C_{\text{hedging vehicle}}(\Delta y)^2|} \tag{5.44}$$

Inserting the weights:

$$w_{\text{hedging vehicle}} = \frac{V_{\text{hedging vehicle}}}{V_{\text{hedging vehicle}} + V_{\text{asset}}}, \quad \text{and } w_{\text{asset}} = \frac{V_{\text{asset}}}{V_{\text{hedging vehicle}} + V_{\text{asset}}}$$

produces the following formula related to values:

$$\boldsymbol{V}_{\textbf{hedging vehicle}} = \boldsymbol{V}_{\textbf{asset}} \times \frac{|-\boldsymbol{D}_{\textbf{asset}}\boldsymbol{\Delta y} + \frac{1}{2}\boldsymbol{C}_{\textbf{asset}}(\boldsymbol{\Delta y})^2|}{|-\boldsymbol{D}_{\textbf{hedging vehicle}}\boldsymbol{\Delta y} + \frac{1}{2}\boldsymbol{C}_{\textbf{hedging vehicle}}(\boldsymbol{\Delta y})^2|} \tag{5.45}$$

If Δy is small, then the convexity is inconsequential,

$$\boldsymbol{V}_{\textbf{hedging vehicle}} = \boldsymbol{V}_{\textbf{asset}}\frac{|\boldsymbol{D}_{\textbf{asset}}|}{|\boldsymbol{D}_{\textbf{hedging vehicle}}|}.$$

When hedging using future contracts, the number of contracts is important.

Assume a bond portfolio with a value of V_P and a futures contract with a contract price of V_F. Use N futures contracts to hedge the bond portfolio. This equation is:

$$V_F \times N = V_P \times \frac{|D_P|}{|D_F|}$$

$$N = \frac{|D_P|\ V_P}{|D_F|\ V_F}$$

Adding convexity produces the following equation for the number of contracts:

$$N = \frac{\left|\left(-D_P \Delta y + \frac{1}{2} C_P (\Delta y)^2\right)\right| V_P}{\left|\left(-D_F \Delta y + \frac{1}{2} C_F (\Delta y)^2\right)\right| V_F}$$

Simplifying:

$$\boldsymbol{N} = \frac{V_P \times \left| -D_P + \frac{1}{2} C_P \Delta y \right|}{V_F \times \left| -D_F + \frac{1}{2} C_F \Delta y \right|} \tag{5.46}$$

Note that the formula for N in Equation 5.46 depends on the value of Δy. If Δy is very small, the terms containing Δy can be ignored and the formula is the same as Equation 5.37. Using an interest rate futures contract to hedge a bond is fundamentally the same position as using an equivalent hedge with a physical bond.

Similar as duration, the **convexity of a coupon bond or a bond portfolio** is simply a weighted average of the squared-longevities of the various cash flows weighted by the percentage of their present values expressed relative to the value of the entire bond portfolio. The reason is as follows: zero coupon bonds have a convexity of T^2. Since partial derivatives commute with summation and constant multipliers, the convexity of a coupon bond must be the weighted average of the convexities of the coupon bond's cash flows. Similarly, the convexity of a bond portfolio is the weighted average of the convexities of the portfolio's assets. We also keep the convention that long and short positions have convexity with opposite signs and we keep absolute values in the formulas.

Like gamma, convexity is a measure of curvature. In the case of convexity, it is the curvature of the price-yield relationship. The risks caused by large interest rate shifts rather than small shifts can be managed by hedging the convexity of the total portfolio.

From a mathematical perspective, a convexity hedge in bond portfolios is parallel to a gamma-hedge in option portfolios. In both cases, attention must be paid to two issues: controlling second order effects (convexity and gamma) in addition to controlling first-order effects (duration and delta).

Like duration, an important assumption implicit in using convexity for hedging is that it assumes that the term structure experiences parallel shifts. Thus the use of convexity tends to provide improved management of large parallel interest rate shifts. It should be noted that convexity uses T^2 which happens to be an excellent measure of the risk of a bond to *slope* shifts in the term structure. In other words, convexity can be used to measure the response of fixed-income products to shifts in the term structure that are linearly related to longevity. Thus, managing both duration and convexity can substantially improve interest rate risk management (although hedging with both risk measures often requires more complex and larger positions). Hedging for non-parallel interest rate shifts is discussed in Chapter 7.

Having discussed various hedging vehicles including forward/futures contracts and options, this chapter concludes with a discussion of hedging using one more derivative: a swap.

5.7 Hedging Various Risks Using Swap Contracts

This section discusses three different types of swaps. A ***swap*** involves two parties exchanging cash flows based on some specified risk factor. We use diagrams to illustrate the cash flows exchanged.

Consider an easy type of swap: a plain vanilla interest rate swap.

- Plain vanilla interest rate swap

A ***plain vanilla interest rate swap*** is called plain vanilla based on the idea that it is relatively simple and popular. It is designed to allow market participants to convert a fixed-interest rate exposure (cash flows based on a fixed interest rate) to a variable interest rate exposure (cash flows that vary with short-term rates), or vice versa.

Assume that several years ago a company viewed the economy as being in a persistent low-interest environment. The company therefore issued a floating rate loan. However, more recently, the company has become concerned that LIBOR rates may rise due to tightening by the Fed. The company would like to switch from floating rate financing to fixed rate financing. The company can use a swap to hedge the existing floating interest rate payments and lock in a fixed rate (i.e., lock in fixed interest payments) by a swap as the ***pay-fixed side***.

Swaps are often described using diagrams such as the diagram in Figure 5.7 for a plain vanilla interest rate swap.

The period of the swap is usually quarterly, but it could be any other time period. The ***floating rate*** is usually based on LIBOR. The ***fixed rate*** is negotiated between the parties at the inception of the swap. Chapter 7

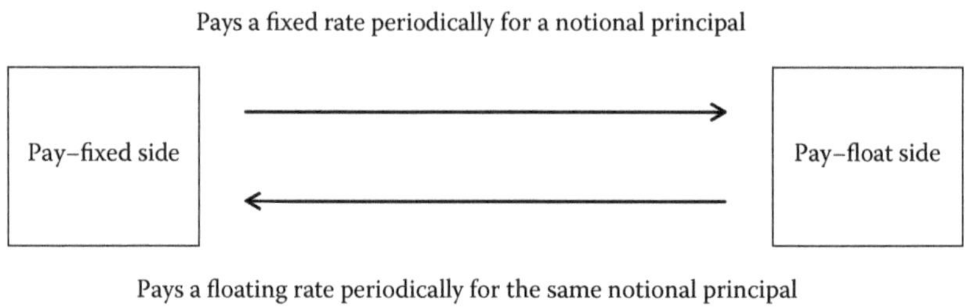

FIGURE 5.7
Plain Vanilla Interest Rate Swap.

discusses how the fixed rate is derived mathematically. The floating rate is observed at the beginning of each period, but the payment based on that rate is usually made at the end of each period, which is termed the ***paid in arrears*** method. The cash flows exchanged are found by multiplying the fixed and floating rates by the notional value (i.e., principal or amount) of the swap and then netting them so that only the netted cash flow is exchanged. Also, the notional principal of each side is the same, so at the end of the swap, there is no exchange of principal – only the net cash flow based on the differential between the fixed and floating rate.

Let's look more closely at the cash flow payoff in each period. For a two-year swap there would be eight quarterly payments between the parties based on the difference between the floating and fixed rates. Assume t represents the end of the period of the swap.

$$\begin{aligned}\textbf{Pay} - \textbf{fixed payment at } t = {} & \textbf{Notional principal} \\ & \times (\textbf{swap fixed rate} - \textbf{LIBOR at } t - \textbf{1}) \\ & \times \textbf{days/360} \end{aligned} \tag{5.47}$$

Note above that the "pay-fixed payment at t" is a netting of the cash flows: it adds in the swap fixed rate and subtracts out the variable rate (LIBOR) so that when the value is positive it represents the net cash flow due from the pay-fixed to the counterparty and when it is negative it represents the net cash flow due to the pay-fixed side.

The party receiving the floating rate is the ***swap buyer*** and the party paying the floating rate is the ***swap seller***. The fixed rate determined at the inception of the swap is called the ***price of the swap*** and it is negotiated and calculated at the initiation of the swap which is demonstrated in Chapter 7. The focus in this chapter is *why* each side of the swap wants to enter the swap contract.

From the perspective of the party receiving the fixed payment (the swap seller), a plain vanilla swap has the same exposure as owning a portfolio that is long a fixed coupon bond and short a floating rate coupon bond.

Conversely, to the party paying the fixed payment (and receiving the floating rate as the swap buyer), a plain vanilla swap has the same exposure as owning a portfolio that is long a floating-rate bond and short a fixed coupon bond. In both cases the bonds have equal principal values and no possibility of default.

It is therefore easy to see that in the case of the previously introduced company that had borrowed by issuing a floating rate loan (i.e., is effectively short a floating rate bond), the company can effectively switch to being financed by a fixed rate note by simply entering a swap (of equal size and longevity as the bond) as a floating-rate receiver (the swap buyer).

- Total return swap

Another popular swap is a total return swap. Simply put, a ***total return swap*** allows two parties to exchange cash flows based on the difference between the realized total returns of two assets (e.g., the returns of one riskless bond and one risky security or portfolio).

For example, assume that a company invests in a corporate bond. After one year, the company becomes concerned that the credit of the bond may deteriorate. The company could sell the bond. However, financial derivatives such as swaps can often provide more liquid, less disruptive, or tax-preferred methods of managing the risks. Perhaps the company is optimistic about the bond's long-term future and may wish to lay off the credit risk on only a temporary basis. The company may therefore choose to enter a total return swap as "Party A".

First, consider the cash flows being exchanged between the two parties to the swap in Figure 5.8.

Let's assume that the parties negotiated that the payment from B to A would be a fixed rate. Note that Party A passes the total return of the risky bond to Party B and receives a fixed payment. If the bond's credit is downgraded, the price of the bond likely falls, reducing the payment from

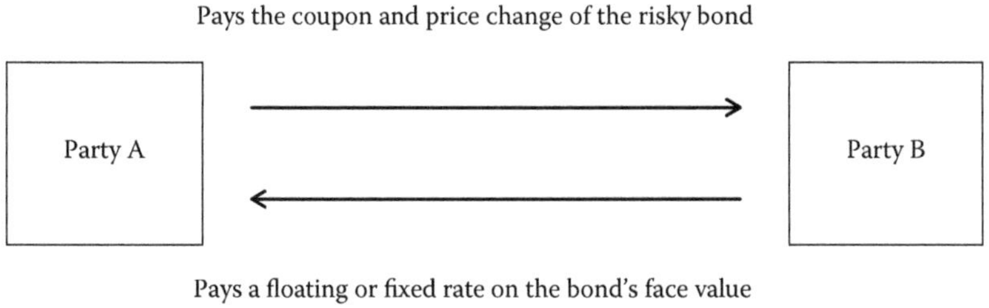

FIGURE 5.8
Total Return Swap.

A to B. Party A has hedged the risk of its risky bond and now is exposed to a fixed rate bond free of credit/default risk (assuming that the counterparty to the swap is free of default risk).

A total return swap is different from an asset swap (to be discussed in detail in Chapter 8). An asset swap delivers the coupon of a bond (if any) but not the change in the bond's price. So, the party in an asset swap that delivers the risky coupon retains the price risk. In this sense, an asset swap does not fully transfer the credit risk of the risky bond between the parties.

A total return swap is also different from a credit default swap (CDS for short), which is also detailed in Chapter 8. A CDS hedges default risk, but not all credit risk. For example, a major credit downgrade of a bond will cause market value losses to its owner, but unless a specified *credit event* takes place the CDS will not trigger a payment to the CDS protection buyer.

Swaps always involve two parties. The needs of the parties differ. That is why swaps take place. Sometimes both parties enter the swap to better hedge or manage their risks. Sometimes one party seeks the swap to hedge its risk while the other party (e.g., a major bank) serves as a counterparty to the swap as part of the bank's ongoing business of providing swaps at prices that offer the bank an attractive risk-adjusted return. There can be ***counterparty risk to financial derivatives*** (including swaps): the risk to one party that the other party will not be able to fulfill their obligations from a financial derivative contract.

- Foreign currency swap

Finally, this section discusses foreign currency swaps. The distinguishing characteristic of a ***foreign currency swap*** is that both sides of the contract experience cash flows in two different currencies. Consider the following diagram (Figure 5.9):

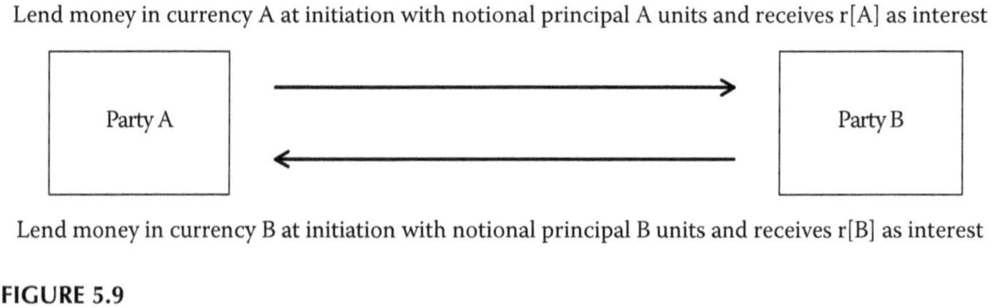

FIGURE 5.9
Foreign Currency Swap.

Unlike most other swaps, a foreign exchange swap has three distinct payment types: the inception payments, the interest payments, and the final principal payments. The diagram indicates that Party A to the contract lends money to Party B in one currency (usually Party A's currency), while Party B lends money to Party A in a different currency (usually Party B's currency). Party A then receives from Party B interest in the currency that it lent and Party B receives interest in the currency that it lent. Finally, at the swap's termination, each Party receives back the same currency and same notional amount that it lent.

Why would two parties from different countries engage in this transaction? The most common reason is that each party is able to obtain more favorable borrowing terms in its own country than in the markets of other countries – presumably due to the familiarity and comfort that lenders possess through lending within their own borders. A currency swap allows two firms to each borrow in their home currency and then exchange the cash flows involved in that transaction with each other with the net result being that both firms can indirectly and *effectively* borrow in foreign markets at terms that are better than as if they tried to borrow directly in the foreign markets.

For example, assume that Company A is a China-based firm that wishes to borrow U.S. dollars because it is doing business in the United States and perhaps because Company A views borrowing in U.S. dollars as providing a currency hedge. Conversely, Company B is a U.S.-based firm that wishes to borrow Chinese yuan because it is doing business with China and perhaps because Company B views borrowing in yuan as providing a currency hedge. However, both companies observe that the terms that they would receive for borrowing directly in foreign markets are not attractive. So Company A (the Chinese firm) borrows yuan in China, Company B borrows dollars in the United States and they both exchange the funds with each other at the prevailing exchange rate. So each firm receives the funding that they sought. Over the life of the swap (as well as the life of the loans), the Chinese Company (A) pays the U.S. company interest in U.S. dollars while the U.S. Company (B) pays the Chinese company interest in Chinese yuan. At the termination of the swap, the original principal payments are reversed. (Note: This occurs even though the exchange rate is likely to have changed.)

Let's assume that Company A (the Chinese company) has borrowing costs (i.e., interest charges) in U.S. dollars of 10% and in Chinese yuan of 8%. Company B's borrowing cost in the United States is 8% and in China is 9%.

The foreign currency swap hedges the foreign currency risk and does so with cost savings to each firm. First, note the cash flows in the swap. Usually, the notional principal satisfies the following relation:

$$\text{Notional principal A} = \text{Notional principal B} \times \text{exchange rate at initiation}$$

Of course, Company A (the Chinese company) could directly borrow in the United States but it would pay 10%, which involves high interest charges and a foreign loan. Instead, Companies A and B enter the currency swap. Company A gets a loan in China in yuan at 8% interest, and lends the Chinese yuan to Company B at 8.5%. Company B gets a loan in the United States in dollars at 8% interest, and lends the U.S. dollars to Company A at 9%.

Let's look at the cash flows in detail. Assume Company A needs $1 million to do business in the U.S. for a three-year period, and it is equivalent to 6.65 million yuan now. Company A borrows 6.65 million yuan with an annual interest rate of 8% in the Chinese market for three years, and delivers the yuan to Company B as a three-year loan at 8.5% annual interest (in yuan). Company A requires that Company B pay the principal back at the end of three years in the form of the 6.65 million yuan. Note that Company A locks in three years of net income of 0.5% interest in yuan relative to having borrowed directly in China. Simultaneously, Company A gets $1 million using the swap and pays an interest rate charge of 9%, which is 1% cheaper than if Company A borrowed directly in the United States. At the end of the third year, Company A returns $1 million to company B. The key point is this: a currency swap allows both parties to lower their risks and lower their interest costs. Also, lenders can focus on lending within geographic areas within which they are most familiar and can most efficiently gather relevant information.

Hedging Formulas Summary

Hedging is important in finance. There are various hedging methods because there are different assets being hedged and there are different hedging vehicles. However, they share many things in common.

This chapter demonstrated three types of hedging:

1. Return-based hedging

Determine a weight allocation for the hedge:

$$w_{\text{hedging vehcle}} = w_{\text{asset}} \times |h|$$

where w is the percentage of the asset (hedging vehicle) in the total value of the hedged portfolio.

- In equity-based hedging use regression to select the hedge ratio:

$$h = -\rho_{\text{asset return,hedging vehicle return}}\sigma_{\text{asset return}}/\sigma_{\text{hedging vehicle return}}$$

In particular, in a stock portfolio hedge, if the hedging vehicle is a market index and the asset is stock (or a stock portfolio), then the hedge ratio is the stock's (or stock portfolio's) systematic risk, β. Note that in regression based hedging, the correlation of the hedging vehicle and the asset being hedged must be high. Otherwise, the hedge is likely to be ineffective.

$$h = -\rho_{\text{stock return,market}}\sigma_{\text{stock return}}/\sigma_{\text{market return}} = -\beta_P$$

The hedge formula is:

$$w_{\text{hedging vehicle}} = w_{\text{asset}} \times |h|$$

$$h = -\beta_P$$

If the hedging vehicle has a beta other than one: $h = -\frac{\beta_p}{\beta_{\text{hedging vehicle}}}$.

- In duration hedging, the goal is to eliminate the sensitivity of the portfolio's return due to a small yield change

In a duration hedge, the sensitivity of portfolio's return due to a small yield change is eliminated (this is known as interest rate immunization). The hedge ratio is:

$$w_{\text{hedging vehicle}} = w_{\text{asset}} \times |h|$$

where $h = \frac{D_{asset}}{D_{hedging\ vehicle}}$

Instead of considering the portfolio weights, sometimes the portfolio's dollar value is considered.

The weight of the hedging vehicle and the asset to be hedged can be easily represented by portfolio values:

$$w_{\text{hedging vehicle}} = \frac{V_{\text{hedging vehicle}}}{V_{\text{hedging vehicle}} + V_{\text{asset}}}$$

$$w_{\text{asset}} = \frac{V_{\text{asset}}}{V_{\text{hedging vehicle}} + V_{\text{asset}}}$$

where V is the total value of asset/hedging vehicle. If weights are inserted into the equation $w_{\text{hedging vehicle}} = w_{\text{asset}} \times |h|$, the resulting relation between the value and hedge ratio is:

$$V_{\text{hedging vehicle}} = V_{\text{asset}} \times |h|$$

If the notation is altered to use $V_{\text{hedging vehicle}}$ to represent the per contract value of the hedging vehicle, the number of contracts needed, N, is:

$$N = \frac{V_{\text{asset}}}{V_{\text{per contract of hedging vehicle}}} \times |h|$$

where h is the hedge ratio.

In summary,

$$\boldsymbol{w_{\text{hedging vehicle}} = w_{\text{asset}} \times |h|}$$

$$\boldsymbol{V_{\text{hedging vehicle}} = V_{\text{asset}} \times |h|}$$

$$\boldsymbol{N = \frac{V_{\text{asset}}}{V_{\text{per contract of hedging vehicle}}} \times |h|}$$

where

$$\boldsymbol{h = -\rho_{\text{asset return,hedging vehicle return}}\,\sigma_{\text{asset return}}/\sigma_{\text{hedging vehicle return}}}$$

$$\boldsymbol{h = -\beta_P \text{ for a stock portfolio}}$$

$$\boldsymbol{h = \frac{D_{\text{asset}}}{D_{\text{hedging vehicle}}} \text{ for a bond portfolio}}$$

If there is a target systematic risk, β, or a target duration, D, the hedge ratios are:

$$\boldsymbol{h = \frac{\beta_{\text{Target}} - \beta_P}{\beta_{\text{hedging vehicle}}} \text{ for a stock portfolio}}$$

$$\boldsymbol{h = \frac{D_{\text{Target}} - D_{\text{asset}}}{D_{\text{hedging vehicle}}} \text{ for a bond portfolio}}$$

Note that futures contracts on a stock index (with a beta assumed to be equal to one) or bond futures are often used as a hedging vehicle.

2. Unit-value-based hedging

We can find the number of units needed to hedge:

$$n_{\text{hedging vehicle}} = n_{\text{asset}} \times |h|$$

where h is the hedge ratio:

$$h = -\frac{\text{change (asset price)}}{\text{change (hedging vehicle)}}$$

- In regression-based hedging:

$$h = -\rho_{\Delta\text{asset}, \ \Delta\text{hedging vehicle}} \frac{\sigma_{\Delta\text{asset price}}}{\sigma_{\Delta\text{hedging vehicle}}}$$

- In Greek letter hedging, the goal is to eliminate the price sensitivity with respect to a risk factor.

$$h = -\frac{\partial f_{\text{asset}}}{\partial B} / \frac{\partial f_{\text{hedging vehicle}}}{\partial B}$$

where f represents the unit value (price), and B is a risk factor.
In particular, if risk factor B is asset price S, it is delta-hedging (where Δ is delta):

$$\Delta_{\text{asset}} = \frac{\partial f_{\text{asset}}}{\partial S}$$

$$\Delta_{\text{hedging vehicle}} = \frac{\partial f_{\text{hedging vehicle}}}{\partial S}$$

$$h = -\Delta_{\text{asset}} / \Delta_{\text{hedging vehicle}}$$

$$n_{\text{hedging vehicle}} = n_{\text{asset}} \times (\Delta_{\text{asset}} / \Delta_{\text{hedging vehicle}})$$

$$n_{\text{hedging vehicle}} = n_{\text{asset}} \times (\nu_{\text{asset}} / \nu_{\text{hedging vehicle}})$$

$$n_{\text{hedging vehicle}} = n_{\text{asset}} \times (\Gamma_{\text{asset}} / \Gamma_{\text{hedging vehicle}})$$

In summary:

$$n_{\text{hedging vehicle}} = n_{\text{asset}} \times |h|$$

where

$$h = -\rho_{\Delta\text{asset},\ \Delta\text{hedging vehicle}} \frac{\sigma_{\Delta\text{asset price}}}{\sigma_{\Delta\text{hedging vehicle}}} \text{ for regression based hedge,}$$

$$h = -\Delta_{\text{asset}}/\Delta_{\text{hedging vehicle}} \text{ for delta neutral hedging,}$$

$$h = -\nu_{\text{asset}}/\nu_{\text{hedging vehicle}} \text{ for vega neutral hedging,}$$

$$h = -\Gamma_{\text{asset}}/\Gamma_{\text{hedging vehicle}} \text{ for gamma neutral hedging.}$$

Since a portfolio's delta is weighted by units, consider the portfolio Δ. So the delta-hedging strategy is to equate the two deltas:

$$\Delta_{\text{hedging vehicle}} = -\Delta_{\text{asset}},$$

producing delta-neutral hedging.

Similarly:

$$\nu_{\text{hedging vehicle}} = -\nu_{\text{asset}},$$

$$\Gamma_{\text{hedging vehicle}} = -\Gamma_{\text{asset}},$$

which are called Vega-neutral and gamma-neutral hedges.

One-to-one hedging is a special case of unit-value-based hedging where the hedge ratio is –1. Forward contracts or futures contracts may be used to hedge the underlying asset to avoid commodity risk or foreign currency risk.

3. Swap hedging

Swaps are contracts to exchange a stream of cash flows. Hedging was discussed for the following types of swaps: interest rate swaps, total return swaps, and currency swaps. Credit default swaps and asset swaps are detailed in Chapter 8.

Finally, note that hedging is usually not static, it is dynamic. Sometimes, the hedge needs to be adjusted because the hedge ratio changes. Sometimes the hedge needs to be adjusted even though hedge ratio is constant.

Chapter Summary

This chapter is about risk management and hedging. Risk management and hedging are two key reasons why financial mathematics is crucial to the financial industry and financial markets. From a Mathematics point of view, this book views hedging strategies as consisting of two approaches: unit-value hedging and return-based hedging. Commonly used hedging strategies such as systematic risk (beta) hedging and duration hedging are return-based hedging. However, Greek letter hedging of options is based on unit-value hedging. Due to the underlying mathematics, the hedge ratio formulas are very similar: the portfolio's target beta (or duration or Greek letter) minus the portfolio's existing beta (or duration or Greek letter) is in the numerator, while the hedging vehicle beta (or duration or Greek letter) is in the denominator. Another similarity between the hedging approaches is that the calculation of duration and Greek letters both use partial derivatives. However, due to the different approaches mentioned above, Greek letter hedging weights are expressed as units of options, while duration hedging weights are expressed as percentages of the dollar value relative to the total dollar value. The chapter includes a project to practice constructing a hedge and to practice additional simulation.

Chapter 5 Extensions and Further Reading

Financial risk management is a very large topic. This book has explored financial management in limited depth. One fundamental risk management tool not covered here is VaR (value at risk). Miller's book, listed below, provides various risk management tools that this book was not able to include. *Risk Management and Financial Institutions* by John Hull is especially good at explaining models intuitively.

1. *Risk Management and Financial Institutions* by John C. Hull
2. *Quantitative Risk Management* by Michael B. Miller

End-of-Chapter Project

How to Perform Arbitrage in the Real World—A Naïve Example

The current price of a stock is \$60. Over the next six months, the stock price is expected to follow Geometric Brownian motion with an annual volatility equal to 15% and to pay no dividends. The risk-free rate is 5% per annum, continuously compounded. Now consider that you have sold (written) a six-month European put option with a strike price \$62 for \$4.00. It is known that the Black-Scholes value of this put option is \$2.79 based on the 15% volatility. You believe that you have made a great trade by selling the put for more than its value according to the Black-Scholes model. But is it really a good decision? Suppose that after selling the put for \$4 you use the money to book a six-month vacation to China. But the stock price falls to \$50 by the time you return. You will need to pay the put buyer Max(K–S_T) = \$12 to settle the option and you will be bankrupted.

This project shows how to safely lock in a profit on selling an overpriced put and using delta-hedging under the assumption that the underlying stock experiences no more volatility than as was expected.

1. Fill in the blanks in the following table given the stock trajectory that is described in the first three columns. (You may use a spreadsheet to calculate the Δ of your position – namely short a put.) Note that $T = 1/2$. Assume that you are not required to post cash to maintain your position. Once you short sell shares, you immediately lend the proceeds to the bank to earn interest. Of course, if you buy shares, you need to borrow money. Note:
   For column #4: Short a put with $\Delta = N(-d_1) = -N(d_1) + 1$ where:

$$d_1 = \frac{Ln\left(\frac{S(t)}{K}\right) + \left(r + \frac{\sigma^2}{2}\right)(T - t)}{\sigma\sqrt{(\mathrm{T} - \mathrm{t})}}$$

and $T = 0.5$, $r = 0.05$, $\sigma = 0.15$, $K = 62$. T-t is changing.
For column #5: Action you need to take: Sell (Buy) number of shares
For column #7: Cumulative money lent (borrowed) = cumulative lent

(borrowed) from last period±interest generated from last period + this period's lent (borrowed)

No.	t	S_T	Δ of Your Short Put Position. Note that $T-t$ Is Changing in the Following Rows	Sell (Buy) # of Shares	Amount of Money Lent (Borrowed)	Cumulative Money Lent (Borrowed)	Interest Due in $[t,t+\Delta t]$
0	0	60					
1	0.05	63.44					
2	0.10	62.38					
3	0.15	62.25					
4	0.20	68.05					
5	0.25	65.66					
6	0.30	68.99					
7	0.35	71.04					
8	0.40	73.34					
9	0.45	69.62					
10	0.50	71.00					

2. Generate an alternative stock trajectory such that $S_{0.5} < 62$, by modifying the R commands below:

```
S0=20
a=0.14
b=0.2
T=1
n=100
deltaT=T/n
S=rep(S0,n+1)
W=rnorm(n,0,1)
for(i in 1:n){
S[i+1]=S[i]+a*S[i]*deltaT+ b*S[i]*sqrt(deltaT)*W[i]
}
plot(S, main="Generalized Wiener Process",type="l")
S
```

Fill the information from the program into a table and a spreadsheet.

3. Do a formal report on how much money you gained or lost from the trajectory originally provided and the simulation, as well as the trajectory you simulated with $S_{0.5} < 62$. Note that there are three positions: the put option, the shares of the underlying stock used to hedge, and a bank account with interest for borrowing and lending. The profits and losses from all three positions must be considered and summed.
4. If $S_{0.5} > 62$, what is your Δ at the expiration? If $S_{0.5} < 62$, what is your Δ at the expiration? Draw Δ as a function of time, mark Δ in red and in green for the two cases.

Bonus (Choose one, not both)

1. Program the following: Generate 1,000 tables and do 10,000 time steps for each table. Calculate the total payoff based on delta-hedging. The gain is $1.21 on average. Calculate the standard error.
2. Write a story about hedging (delta/gamma, etc.) by reading several papers online. Quote at least five papers or other resources.

End-of-Chapter Problems (Fundamental Problems)

1. A U.S.-based export firm will receive 10–12 million GBP in three months. The risk to the company is that the GBP's value relative to the U.S. dollar will go down. Assume that only four-month GBP futures contracts (as opposed to a three-month maturity match) are available and are trading at a futures price of $1.32.
   a. The export company shorts futures contracts totaling 10 million GBP in four-month GBP futures and then closes its position after three months. Determine what happens in the following separate scenarios:
   Three months later, the company receives 12 million GBP, the spot exchange rate is $1.29, and the futures price is $1.30.
   Three months later, the company receives 12 million GBP, the spot exchange rate is $1.33 and the futures price is $1.34.
   b. The export company shorts futures contracts totaling 12 million four-month GBP futures and then closes its position after three months.
   Determine what happens in the following separate scenarios:
   Three months later, the company receives 12 million GBP, the spot exchange rate is $1.29, and the futures price is $1.30.
   Three months later, the company receives 12 million GBP, the spot exchange rate is $1.33, and the futures price is $1.34.
   c. The export company shorts futures contracts totaling 10 million four-month GBP futures and then closes its position after three months. In addition (and at the same time the futures contracts were shorted), the company goes long three-month puts on 2 million GBP at the strike price $1.31. (Assume that the put price is 0.01 dollars.)

Determine what happens in the following separate scenarios:
Three months later, the company receives 12 million GBP, the spot exchange rate is \$1.29, and the futures price is \$1.30.
Three months later, the company receives 12 million GBP, the spot exchange rate is \$1.33, and the futures price is \$1.34.

d. The export company shorts contracts totaling 10 million four-month GBP futures and then closes its position after three months. In addition (and at the same time the futures contracts were shorted), the company goes long three-month puts on 2 million GBP at the strike price \$1.31. (Assume that the put price is 0.01 dollars.)
Determine what happens in the following separate scenarios:
Three months later, the company receives 10 million GBP, the spot exchange rate is \$1.29, and the futures price is \$1.30.
Three months later, the company receives 10 million GBP, the spot exchange rate is \$1.33, and the futures price is \$1.34.

2. Assume that the standard deviation, $\sigma_{\Delta A}$, of monthly changes in the price of asset A (ΔA) is \$0.80 and the standard deviation, $\sigma_{\Delta B}$, of monthly changes in the prices of futures contracts on asset B (ΔB) is \$10, and that there is perfect positive correlation between the two changes. A trader has 2 million units of asset A worth \$100 each and use asset B, a futures contract with a notional value of \$1,000, as the hedging vehicle. How many futures contracts does the trader need?
   a. Using the unit value approach
   b. Using the return approach.

3. Assume that a trader's portfolio has a value of \$10 million and $\beta = 2$. The trader needs to reduce the systematic risk to $\beta = 1$. The S&P 500 index futures price is 2,400 (each futures contract is 250 times the futures price). How many contracts provides an optimal hedge?

4. A manager has a \$5 million portfolio with a beta of 1.1. The manager wants to reduce the beta exposure to 0.5. The manager decides to use equity index futures contracts to hedge. Assume the beta on the futures contract is 1.05 and the futures price is 2,000 (each futures contract is 250 times the futures price).
   a. How can the manager's goal be achieved?
   b. If the index increases in value by 2%, calculate the error of the hedge due to rounding in the futures position

c. Explain why there is often hedging error from the fact that the portfolio and the futures contract are not perfectly correlated with the index.

5. Assume a stock price follows geometric Brownian motion with $S_0 = 60$ and volatility equal to 15%. A six-month European call has a strike price of 62 and the continuously compounded risk-free rate is 5%. Find the call price as well as the requested Greeks based on the Black-Scholes model.
   a. Delta
   b. Gamma
   c. Vega
   d. Theta
   e. Check that the Greeks satisfy $\theta + rS_0\Delta + \frac{1}{2}\sigma^2 S_0^2 \Gamma = rc$

6. Assume a stock price follows geometric Brownian motion with $S_0 = 60$ and volatility equal to 15%. A bank has written (sold) a six-month European call on 100,000 shares of a non-dividend-paying stock generating proceeds of $300,000. The strike price of the call is 62 and the risk-free rate is 5%.
   a. If the bank wants to be delta neutral, how many shares of the stock should the bank buy?
   b. Assume that the next day, the stock price rises to $61, how should the bank adjust its position to keep delta neutral?
   c. Assume that one month later, the stock price is $60, how should bank adjust its position to keep delta neutral?

7. The initial price of a stock is $60/share and the initial value of a put on that stock is $2.79 with a delta of –0.51 and one-half year to maturity. A bank sells 1,000 contracts (i.e., 100,000 puts). To delta-hedge the position, the bank short sells $100{,}000 \times 0.51 = 51{,}000$ shares of stock at $60/share. The bank deposits all of its cash at 3% interest annum. Two weeks later, the price of the stock rises to $63.44 and the delta of the put becomes –0.31. Calculate how much money was received by the bank initially and how much money the bank has two weeks later after the portfolio adjustments. (Assume simple interest.)

8. Consider a futures contract on crude oil. The time to the futures' settlement is five months. The risk-free interest rate is 3% per annum. What would be the initial position in a six-month oil futures contract to delta hedge the five-month oil futures contract?

9. Consider a European put option on a futures contract on crude oil. The time to the option's maturity is four months and the time to the futures' maturity is five months. The current futures price is \$60, the exercise price of the option is \$60, the risk-free interest rate is 3% per annum, and we assume the volatility of the futures price is 20%. Based on the Black-Scholes model, find:
   a. The put price
   b. The initial position in the five-month oil futures necessary to delta-hedge on put option
   c. What would be the initial position in a six-month oil futures contract to delta-hedge the put option?

10. Assume that a company delta-hedged its portfolio which contained long positions in out-of-the-money put and call options on a stock index. Analyze what happens in the following two scenarios:
    a. The stock market becomes very volatile.
    b. The stock market becomes very stable.

11. Assume that a company delta-hedged its options portfolio that contains long positions in many at-the-money put and call options on a stock using shares of the underlying stock. Analyze what happens in the following scenarios:
    a. The stock market becomes very volatile.
    b. The stock market becomes very stable.

12. A current portfolio combines various options (calls and puts with different strike prices and maturities) on the same underlying asset:

1,000 long calls, each with: delta = 0.5, gamma = 2.0, and Vega = 1.0
500 short calls, each with: delta = 0.3, gamma = 0.6, and Vega = 0.2
1,000 long puts, each with: delta = –0.5, gamma = 2.0, and Vega = 1.0
Two other options on the same underlying asset are available in the market to be used for hedging:
Option #1 with: delta = 0.6, gamma = 1.0, and Vega = 0.8
Option #2 with: delta = 0.8, gamma = 3.0, and Vega = 0.6

a. What are the values of delta, gamma, and Vega for the current portfolio?
b. What positions in Option #1 and the underlying asset can be established to make the portfolio delta and Vega neutral?
c. What positions in Option #2 and the underlying asset can be established to make the portfolio delta and gamma neutral?

d. What positions in Options #1 and #2, and the underlying asset can be established to make the portfolio gamma, Vega, and delta neutral?

13. Consider a delta-neutral portfolio that includes a long option. The book mentioned that if the gamma of a delta-neutral portfolio is positive, instantaneous changes in the underlying asset in either direction will increase the value of the portfolio.
    a. Illustrate from the concavity of a graph why both increases and decreases in the underlying asset price are good.
    b. Assume that a decrease of $5 in an underlying asset price occurs that increases the value of a particular portfolio that contains positions in the asset and one or more options. If the portfolio is not rebalanced to change its delta and if the asset price increases $5, use a graph to explain why the benefit to the portfolio of the original decline in the asset price is lost.

14. Assume that a bond portfolio has a market value of $100 million and will have a duration of seven years in six months. A six-month bond futures contract's price is quoted as $101.5 per $100 and each contract has a face value of $100,000. The futures contract's underlying bond is projected to have a duration of nine years in six months. How many contracts duration-hedge the $100 million portfolio at the settlement date of the futures contract?

15. Assume that a bond portfolio has a value of $100 million and its duration is assumed to be seven years in six months. How many Eurodollar futures contracts ($1 million face value per contract and quoted $97) will duration-hedge the portfolio's interest rate risk?

16. A manager has a $100 million fixed-income portfolio with an effective duration of 8.0. Suppose the manager is concerned that interest rates will suddenly rise 40 basis points.
    a. What is the portfolio's dollar duration?
    b. If the manager is interested in limiting the portfolio's exposure of loss to $1 million, should the manager take a long or short position in bond futures contracts?
    c. Assume that the manager's goal is to limit the portfolio's exposure of loss to $1 million for a 40 bp interest rate increase. The manager chooses a futures contract with a dollar duration, for a 40 bp change, equal to $1,100. What size of futures position allows the manager to achieve the goal?

d. Continue assuming that the manager is interested in limiting the portfolio's exposure of loss to $1 million. The manager chooses a futures contract on a Treasury bond as a hedge. Assume the cheapest-to-deliver bond has a conversion factor 1.1 and a dollar duration, for a 40 bp change, equal to $1,210. What size and sign futures position will allow the manager to achieve the goal?

e. Continue assuming that the manager is interested in limiting the portfolio's exposure of loss to $1 million. The manager chooses a future contract on a Treasury bond as a hedge. Assume that the futures contract has a face value of $100,000 and is currently trading at $110. The effective duration of the futures contract is 4.0. What size of futures position will allow the manager to achieve the goal?

End-of-Chapter Problems (Challenging Problems)

Sometimes the ideal financial instruments for hedging are not available or they are unattractive for some reason such as taxes, transactions costs or regulatory constraints. In that case financial managers may be able to create them synthetically. For example, there can be some income tax advantages in the United States to receiving income from derivative securities. For example, *Section 1256 securities* are derivatives such as futures contracts that allow investors to receive a substantial portion of income in the form of long-term capital gains. Synthetic securities also help traders set up arbitrage transactions found by comparing the expected returns of comparable cash securities and synthetic securities and then arbitraging any substantial differences.

Here are two examples:

Synthetic risk-free asset = long a stock portfolio and short a futures contract on a stock index

Synthetic equity = long risk-free asset + long a futures contract on the equity

1. Creating synthetic equity: An investor holds $100 million in risk free asset yielding 3%. Over the next six months, the investor wishes to have a synthetic equity position of $100 million. The investor decides to establish a long position in six-month S&P 500 index futures contracts (the index has a dividend yield of 2%). The futures contract has a price of $2,700 and its multiplier is 250. Determine the proper strategy for achieving the goal of a synthetic equity exposure to the S&P 500. Use continuously compounded interest rates and continuous dividend payments.

a. What will be the value of the $100 million position in risk-free in six months?

b. What number of futures contracts should the investor enter long for the synthetic equity position? Note: Please round the answer to an integer. What is the exact amount of money equitized today (i.e., the amount of formerly money-like assets transformed into having a risk exposure similar to a direct investment in equities)?

c. In order to match today's equitized amount, the investor may need to buy or sell a small amount of risk-free asset. What is that amount?

d. How many units of the index will the investor hold by being long the number of futures contracts you have chosen?

e. What is the payoff of this synthetic strategy? (Assume that the ending S&P 500 index price is S_T)

2. Creating synthetic cash: An investor has $100 million invested in a well-diversified portfolio mimicking S&P 500 index. The investor wants to lock in recent gains for the next three months by creating a synthetic risk-free asset position out of the equity. Assume that the risk-free rate is 3% compounded continuously. The index underlying a futures contract has a continuous dividend yield of 2%. The futures contract has a price of $2,700 and a multiplier of 250. Determine the proper strategy for achieving the goal.

   a. What will the value of the cash account be in three months?

   b. What is the number of futures that the investor should short? Please round your answer to an integer. What is the amount of money in the synthetic cash position?

   c. How many units of the index has the investor effectively "sold"?

   d. How much will the synthetic cash position be worth at the end of three months?

Sometimes a portfolio manager uses a hedging strategy to create the position he/she wants. Here is an important application of hedging.

3. A manager has a $100 million portfolio that consists of 50% stocks and 50% bonds. The beta of the stock portfolio is 0.8 and the duration of the bond portfolio is 5. Assume that the manager wishes to achieve an effective mix of 60% stocks and 40% bonds. Since the move is only temporary, the manager decides to use equity futures contracts with a unit value of $300,000 and a beta of 1.05 and bond

futures contracts with a price of $100,000 and duration of 8.00 to achieve the goal.

a. What futures positions should the manager establish?

b. What are the potential problems and hedge errors with this hedging strategy?

4. A manager has a $100 million portfolio that consists of 70% stocks and 30% bonds. The manager wants to achieve an exposure equivalent to a mix of 50% stock and 50% bond. Assume the beta of the current stock portfolio is 1.3 and duration of the current portfolio is five years. The price of the equity index futures contracts is $900 and each contract has a multiplier of 250. Assume that the equity index futures contract has a beta of 1.00, that the bond futures contract has a duration of eight years, the price of the bond futures contract is $99, and each contract has a face value of $100,000. Determine the futures positions for a hedge strategy.
5. A manager of $100 million of large-cap equities would like to shift 50% of its position to mid-cap equities. The beta of the large cap position and middle cap position are 0.8 and 1.2, respectively. The beta of the large-cap futures contract is 0.75 and the futures price is $9,800 per contract. The beta of the mid-cap futures contract is 1.25 and the futures price is $240,000 per contract. Determine the futures position for a hedge strategy.

6

Portfolio Management

This chapter begins with an examination of the Markowitz Frontier when short sales are allowed. In particular, the chapter discusses the optimized portfolios both without a risk-free asset and with a risk-free asset. The capital asset pricing model (CAPM) is derived using the frontier in the Markowitz mean-variance framework under the key assumption that a risk-free asset exists. The chapter then introduces linear programming and quadratic programming to derive Markowitz's Frontier in a general setting. Next, the Markowitz Frontier is analyzed under the restriction that short sales are not allowed. Finally, the chapter discusses the optimization problem under the assumption of a Geometric Brownian motion framework. Throughout this chapter, portfolio management is studied under the assumption that returns of the portfolio's constituent assets have distributions close to normal. However the returns of most financial assets do not closely approximate a normal distribution, owing especially to their leptokurtosis (i.e., "fat tails").

6.1 Markowitz Frontier

In the 1950s, Harry Markowitz introduced a quantitative tool to manage a portfolio of stocks. Markowitz modeled the ability of an investor to decrease the risk of her investments by holding additional risky assets, a phenomenon known as diversification. Markowitz postulated that risk-averse investors construct portfolios to minimize risk for a given level of expected return (or, equivalently, maximize expected return based on a given level of risk). Markowitz assumes that investors base decisions solely on expected return and return standard deviation, a two-moment approach to risk assuming either that returns are normally distributed or that investors have quadratic utility functions. This section derives the Markowitz Frontier from an analytic perspective allowing short-selling. The CAPM model is derived in a mathematical framework at the end of the section.

6.1.1 Analytical Solution for the Markowitz Frontier without a Risk-Free Asset

Chapter 2 provides an introduction to Markowitz portfolio theory. A ***portfolio*** is a mix of securities such as stocks, bonds etc. The ***weight***, w_i, of a security in a portfolio is the percentage or portion of the portfolio's total value invested in ith security. The ***return*** of a security is the investment's percentage gain or loss over a specified period of time. ***Expected return*** is the long-run average return. Recall that the ***expected return of a portfolio*** is simply the weighted average of the expected returns of the portfolio's constituent assets. The ***variance of a portfolio's return*** depends on the variances of each asset and the *correlation coefficients* between the returns of each pair of assets.

Chapter 2 previously studied the special case of a two-asset portfolio. In that case, all different combinations of weights in the two assets form a hyperbola in the plane with standard deviation σ as the x-axis and expected return r as the y-axis as illustrated in Figure 6.1.

$$\begin{cases} r_p = w_X \times r_X + w_Y \times r_Y \\ \sigma_P = \sqrt{w_X^2 \sigma_X^2 + w_Y^2 \sigma_Y^2 + 2 w_X w_Y \sigma_X \sigma_Y \rho_{XY}} \end{cases}$$

where $w_X + w_Y = 1$ and $0 \le w_X \le 1$.

Figure 6.1 imposes $0 \le w_X \le 1$. Thus, short selling of assets is not allowed.

Consider (σ_p, r_p) with respect to changes in weight allocations. First insert $w_Y = 1 - w_X$ into both equations. Note that the above equations change to $\begin{cases} y = r_p = f(w_X) \\ x = \sigma_p = g(w_X) \end{cases}$.

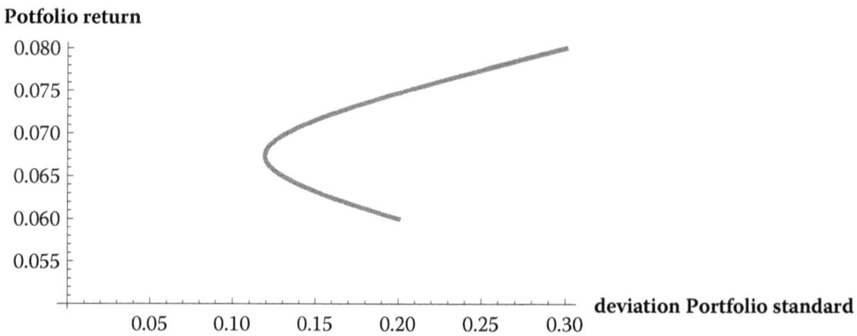

FIGURE 6.1
A Portfolio Consisting of Two Assets with Short Selling Disallowed.

The above equations are now parametric equations with one parameter w_X. The pair of equations represents a curve. Solving the parameter w_X as a function of y in the first equation and substituting this for w_X in the second equation yields a hyperbolic x–y equation.

This chapter generalizes the two-asset case and in addition allows for negative weights (short selling). For any required expected return, the weights that minimize the variance of the portfolio are considered *optimal*. In other words, the investor's goal is to minimize the level of risk for a given level of expected return.

Let's look at an example with three assets.

$$\begin{aligned} \textit{Find} \quad & \underset{w_X, w_Y, w_Z}{\text{Min}} \ \sigma_p^2 = \underset{w_X, w_Y, w_Z}{\text{Min}} \ \{w_X^2\sigma_X^2 + w_Y^2\sigma_Y^2 + w_Z^2\sigma_Z^2 + 2w_X w_Y \sigma_X \sigma_Y \rho_{XY} \\ & \qquad + 2w_X w_Z \sigma_X \sigma_Z \rho_{XZ} + 2w_Y w_Z \sigma_Y \sigma_Z \rho_{YZ}\} \\ \text{where:} \quad & E(p) = w_X \times r_X + w_Y \times r_Y + w_Z \times r_Z = \text{a known constant } r_p \\ & w_X + w_Y + w_Z = 1 \end{aligned}$$

Given the expected return vector $\{r_X, r_Y, r_Z\}$, standard deviations, and covariances for three assets, there are three unknowns $\{w_X, w_Y, w_Z\}$, and two constraints. This optimization problem is a Lagrange multiplier problem. Note that the conditions $0 \leq w_X \leq 1$ or $0 \leq w_Y \leq 1$ or $0 \leq w_Z \leq 1$ are not imposed.

In general, there are n assets, which correspond to n variables of weight allocations $\{w_1, w_2, \ldots, w_n\}$ and only two constraints. The first constraint is on the expected return for the portfolio. The other constraint is that the weights must sum to one. Note two conventions for this chapter: (1) the return of an asset is the return for a long/buy position of the asset, and (2) that the total weight sums to one, which means that the initial net investment cannot be 0. Here is an example, asset A grows from \$100 to \$120, asset B grows from \$40 to \$60, asset C grows from \$50 to \$70. The returns are {20%, 50%, 40%}. If we establish a long position in Asset A, and short positions in Assest B and C, our initial outlay is: 100 – 40 – 50 = \$10. Later, we receive: 120 – 60 – 70 = –\$10, which is a return of on the portfolio of –200%. In terms of weights: w = {100/10, –40/10, –50/10} = {1000%, –400%, –500%}. The weight sums to one and the weighted sum of returns is : 1000% ∗ 20% + (–400%) ∗ 50% + (–500%) ∗ 40% = –200%.

Note that for any fixed return for the portfolio, r_p, there is one optimal solution for the variance, σ_p^2. Consider a graph of all optimal solutions of σ_p^2 while r_p varies. This graph forms a hyperbola and it is called the ***Markowitz Frontier*** or ***the minimum variance set*** (see Figure 6.2). All possible portfolios containing these n assets must lie on or inside this hyperbola.

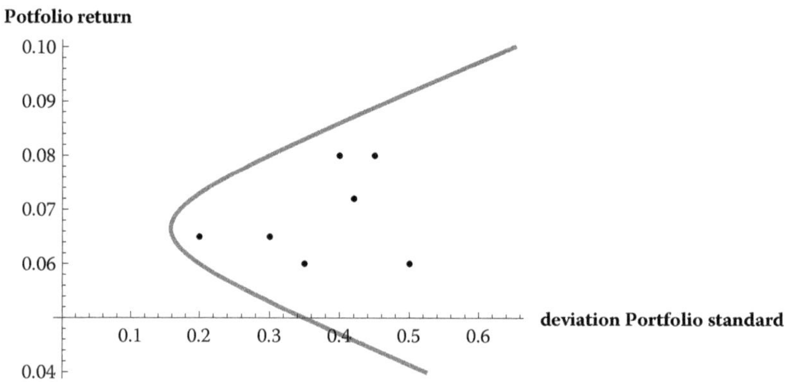

FIGURE 6.2
A general illustration of Markowitz Frontier.

Next, matrix notation is introduced to form this optimization problem.

Given $\overrightarrow{r} = \begin{pmatrix} r_1 \\ \cdots \\ r_n \end{pmatrix}$, a column vector of expected returns of n assets and Σ, a covariance of the returns of n assets, assume $\overrightarrow{w} = \begin{pmatrix} w_1 \\ \cdots \\ w_n \end{pmatrix}$ is the column vector of weight allocations of portfolio in n assets. Let $r_p = E(p)$ be the expected return for the portfolio and σ_p^2 the variance of the returns of the portfolio. The variance of the portfolio can be solved in terms of the covariance matrix and the weights vector, $\sigma_p^2 = \overrightarrow{w}^T \Sigma \overrightarrow{w}$. Furthermore, denote $\overrightarrow{1}$ as a column vector of ones: $\begin{pmatrix} 1 \\ \cdots \\ 1 \end{pmatrix}$.

It is necessary to assume that no securities have the same expected returns, and the covariance matrix is non-singular. Consequently, no security is a linear combination of the other securities. Since a weight allocation, $\overrightarrow{w}$, uniquely determines a portfolio, we sometimes describe a portfolio as portfolio $\overrightarrow{w}$, which represents the portfolio invested in the assets with the corresponding weight allocation $\overrightarrow{w}$.

The optimization problem in matrix format is:

$$\text{Find} \quad \underset{\overrightarrow{w}}{\text{Min}} \tfrac{1}{2} \overrightarrow{w}^T \Sigma \overrightarrow{w} \text{ subject to:}$$

$$\overrightarrow{w}^T \overrightarrow{r} = r_p$$

$$\overrightarrow{w}^T \overrightarrow{1} = 1.$$

Note that $\frac{1}{2}\sigma_p^2$ is minimized instead of σ_p^2 just for a calculation convenience in the proof.

Consider the following theorem.

6.1 MARKOWITZ FRONTIER THEOREM – WITHOUT A RISK-FREE ASSET AND WITH SHORT-SELLING ALLOWED

There are n risky assets available in the market, with expected return vector $\vec{r} = \begin{pmatrix} r_1 \\ \cdots \\ r_n \end{pmatrix}$ and a covariance of return matrix Σ (assumed as positive definite) given. The vector $\vec{w} = \begin{pmatrix} w_1 \\ \cdots \\ w_n \end{pmatrix}$ is denoted as the weight allocation for n assets and $\vec{1}$ as a column vector of $\begin{pmatrix} 1 \\ \cdots \\ 1 \end{pmatrix}$. We denote r_p as the expected return of the portfolio and σ_p^2 as the variance of the return of the portfolio.

A Markowitz Frontier portfolio is a portfolio, p, with a weight allocation $\vec{w}$ that minimizes the risk (variance or standard deviation of the portfolio returns) while keeping the expected return of the portfolio constant as r_p:

Find weight allocation $\vec{w}$ to

$$\min_{\vec{w}} \tfrac{1}{2}\, \vec{w}^T \Sigma \vec{w} \tag{6.1}$$

where $\vec{w}^T\vec{r} = r_p$, $\vec{w}^T\vec{1} = 1$.

Note that r_p is the expected return for the portfolio and $\sigma_p^2 = \vec{w}^T \Sigma \vec{w}$ is the portfolio variance.

For a given r_p, the optimal weight allocation $\vec{w}$ is given by:

$$\vec{w} = \left(\frac{c}{d}\Sigma^{-1}\vec{r} - \frac{b}{d}\Sigma^{-1}\vec{1}\right) r_p + \left(\frac{a}{d}\Sigma^{-1}\vec{1} - \frac{b}{d}\Sigma^{-1}\vec{r}\right) \tag{6.2}$$

where, $a = \vec{r}^T\Sigma^{-1}\vec{r}$, $b = \vec{r}^T\Sigma^{-1}\vec{1} = \vec{1}^T\Sigma^{-1}\vec{r}$, $c = \vec{1}^T\Sigma^{-1}\vec{1}$ and $d = ac - b^2$.

In addition, the expected return of the optimal portfolio, r_p, and its corresponding standard deviation, σ_p, satisfy the following equation:

$$\frac{\sigma_p^2}{1/c} - \frac{\left(r_p - \frac{b}{c}\right)^2}{d/c^2} = 1 \tag{6.3}$$

Remarks

The last equation, Equation 6.3, is similar to the hyperbola in Figure 6.2, which is called the ***Markowitz Frontier*** or ***Minimum Variance Set***. In particular, the top half of the hyperbola is called the ***Markowitz Efficient Frontier***, which is more useful than the whole hyperbola.

Before proving the theorem, first review some mathematics.

> **MATHEMATICS REVIEW 6.1**
>
> The calculus of the gradient vector in matrix form:
>
> If a multi-variable function $F(w_1, \ldots, w_n) = \frac{1}{2}\vec{w}^T\Sigma\vec{w}$, where $\vec{w} = \begin{pmatrix} w_1 \\ \cdots \\ w_n \end{pmatrix}$, and Σ is an $n \times n$ matrix, then the gradient vector is $\frac{\partial F}{\partial \vec{w}} = \Sigma\vec{w}$.
>
> If a multi-variable function $G(w_1, \ldots, w_n) = \vec{w}^T\vec{r}$, where $\vec{r}$ is a constant vector, then the gradient vector is $\frac{\partial G}{\partial \vec{w}} = \vec{r}$.

> **MATHEMATICS REVIEW 6.2**
>
> Method of Lagrange Multipliers:
>
> To find the minimum value for the function $F(w_1, \ldots, w_n)$ subject to constraints $G_1(w_1, \ldots, w_n) = c_1$ and $G_2(w_1, \ldots, w_n) = c_2$ (and assuming that the minimum value exists):
>
> a. Find the solutions of the system of equations:
>   Gradient vector of $F = \lambda \times$ Gradient vector of $G_1 + \mu \times$ Gradient vector of G_2
>
> $$G_1(w_1, \ldots, w_n) = c_1$$
> $$G_2(w_1, \ldots, w_n) = c_2$$
>
> where $(w_1, \ldots, w_n, \lambda, \mu)$ are variables. There are $(n + 2)$ equations, n of them involving gradient vectors and two equations are constraints. The variables λ and μ are called ***Lagrange multipliers.***
>
> b. Evaluate F at all points $\vec{w} = (w_1, \ldots, w_n)$ that result from step a; the smallest is the minimum value of F.

Next, the theorem is proved.

Proof:

Based on the mathematics review, the following equation is to be solved:

$$\Sigma\vec{w} = \lambda\vec{r} + \mu\vec{1}$$
$$\vec{w}^T\vec{r} = r_p$$
$$\vec{w}^T\vec{1} = 1$$

Since Σ is a positive definite matrix, it follows that the solutions of the above equations are necessary and sufficient for a global optimum. From equation $\Sigma\vec{w} = \lambda\vec{r} + \mu\vec{1}$, we can write the equation as (Σ is positive definite, so Σ^{-1} exists):

$$\vec{w} = \lambda\Sigma^{-1}\vec{r} + \mu\Sigma^{-1}\vec{1}.$$

Multiplying by $\vec{r}^T$:

$$\vec{r}^T\vec{w} = \lambda\vec{r}^T\Sigma^{-1}\vec{r} + \mu\vec{r}^T\Sigma^{-1}\vec{1}.$$

Multiplying by $\vec{1}^T$:

$$\vec{1}^T\vec{w} = \lambda\vec{1}^T\Sigma^{-1}\vec{r} + \mu\vec{1}^T\Sigma^{-1}\vec{1}.$$

Note that $\vec{r}^T\vec{w} = \vec{w}^T\vec{r} = r_p$, $\vec{1}^T\vec{w} = \vec{w}^T\vec{1} = 1$. Recall for vector/matrix A, B, that $(AB)^T = B^TA^T$. For a scalar k, $k^T = k$. Substitute r_p and 1 into the above equations of $\vec{r}^T\vec{w}$ and $\vec{1}^T\vec{w}$:

$$r_p = \lambda\vec{r}^T\Sigma^{-1}\vec{r} + \mu\vec{r}^T\Sigma^{-1}\vec{1}$$
$$1 = \lambda\vec{1}^T\Sigma^{-1}\vec{r} + \mu\vec{1}^T\Sigma^{-1}\vec{1}.$$

Denote a $= \vec{r}^T\Sigma^{-1}\vec{r}$, b $= \vec{r}^T\Sigma^{-1}\vec{1} = \vec{1}^T\Sigma^{-1}\vec{r}$, and c $= \vec{1}^T\Sigma^{-1}\vec{1}$. Since Σ is symmetric and positive definite, $(\Sigma^{-1})^T = \Sigma^{-1}$. Rewriting these two equations in matrix form:

$$\begin{pmatrix} r_p \\ 1 \end{pmatrix} = \begin{pmatrix} a & b \\ b & c \end{pmatrix}\begin{pmatrix} \lambda \\ \mu \end{pmatrix}.$$

Since Σ is a positive definite matrix, $a > 0$ and $c > 0$. It is not hard to prove $d > 0$, which is left as an exercise. Solve the system of equations:

$$\begin{pmatrix} \lambda \\ \mu \end{pmatrix} = \begin{pmatrix} a & b \\ b & c \end{pmatrix}^{-1}\begin{pmatrix} r_p \\ 1 \end{pmatrix}.$$

The determinant of $\begin{pmatrix} a & b \\ b & c \end{pmatrix}$ is $d = ac - b^2$.

Using the determinant, invert the matrix:

$$\begin{pmatrix} \lambda \\ \mu \end{pmatrix} = \frac{1}{d}\begin{pmatrix} c & -b \\ -b & a \end{pmatrix}\begin{pmatrix} r_p \\ 1 \end{pmatrix}.$$

Therefore, $\lambda = \frac{cr_p - b}{d}$, and $\mu = \frac{-br_p + a}{d}$, where $a = \vec{r}^T\Sigma^{-1}\vec{r}$, $b = \vec{r}^T\Sigma^{-1}\vec{1} = \vec{1}^T\Sigma^{-1}\vec{r}$, $c = \vec{1}^T\Sigma^{-1}\vec{1}$, and $d = ac - b^2$.

The solutions for λ and μ are substituted into $\vec{w} = \lambda\Sigma^{-1}\vec{r} + \mu\Sigma^{-1}\vec{1}$. This results in $\vec{w} = \left(\frac{cr_p - b}{d}\right)\Sigma^{-1}\vec{r} + \left(\frac{-br_p + a}{d}\right)\Sigma^{-1}\vec{1}$.

Simplifying the expression:

$$\vec{w} = \left(\frac{c}{d}\Sigma^{-1}\vec{r} - \frac{b}{d}\Sigma^{-1}\vec{1}\right)r_p + \left(\frac{a}{d}\Sigma^{-1}\vec{1} - \frac{b}{d}\Sigma^{-1}\vec{r}\right),$$

where $a = \vec{r}^T\Sigma^{-1}\vec{r}$, $b = \vec{r}^T\Sigma^{-1}\vec{1} = \vec{1}^T\Sigma^{-1}\vec{r}$, $c = \vec{1}^T\Sigma^{-1}\vec{1}$, and $d = ac - b^2$.

Note that any portfolio on the Markowitz Frontier has a weight allocation given by the above equation. Similarly, any portfolio with this weight allocation is on the Markowitz Frontier.

Note that $\vec{w}$ is a function of r_p. In order to get σ_p^2 as a function of r_p, insert $\vec{w}$ into $\sigma_p^2 = \vec{w}^T\Sigma\vec{w}$. This results in a relation between σ_p^2 and r_p (the details are left for the exercises):

$$\frac{\sigma_p^2}{1/c} - \frac{\left(r_p - \frac{b}{c}\right)^2}{\frac{d}{c^2}} = 1.$$

The relation between the variance and expected return of an optimal portfolio is hyperbolic.■

Remarks

1. Equation $r_p = \frac{b}{c} + \sqrt{\frac{c\sigma_p^2 - 1}{c^2/d}}$ represents the top half of the hyperbola and equation $r_p = \frac{b}{c} - \sqrt{\frac{c\sigma_p^2 - 1}{c^2/d}}$ represents the bottom half of the hyperbola. The top half of the hyperbola is called the ***Markowitz Efficient Frontier***.
2. When $r_p = \frac{b}{c}$, the minimum variance portfolio is $\sigma_p^2 = \frac{1}{c}$. This portfolio corresponds to the vertex of the hyperbola.

3. In the case of two assets, all portfolios are on the Markowitz Frontier. In the case of more than two assets, the optimal portfolios form a hyperbola. Conversely, any portfolio on the hyperbola represents an optimal portfolio. In fact, given a point on the hyperbola with a fixed r_p, the equation in Theorem 6.1 gives a unique weight allocation so that the corresponding portfolio has a minimum variance. In addition, we have $\vec{w}^* = \left(\frac{a}{d}\Sigma^{-1}\vec{1} - \frac{b}{d}\Sigma^{-1}\vec{r}\right)$, which is a frontier portfolio with $r_p = 0$ and $\vec{w}^{**} = \left(\frac{c}{d}\Sigma^{-1}\vec{r} - \frac{b}{d}\Sigma^{-1}\vec{1}\right) + \left(\frac{a}{d}\Sigma^{-1}\vec{1} - \frac{b}{d}\Sigma^{-1}\vec{r}\right)$, which is a frontier portfolio with $r_p = 100\%$.

It can be shown that any frontier portfolio weight is a linear combination of $\vec{w}^*$ and $\vec{w}^{**}$.

$$\vec{w} = (\vec{w}^{**} - \vec{w}^*)r_p + \vec{w}^* = (1 - r_p)\vec{w}^* + r_p\vec{w}^{**}.$$

This is called the ***two fund theorem,*** namely the Markowitz Frontier can be represented by a linear combination of the two efficient portfolios.

6.2 THE TWO FUND THEOREM

Two efficient portfolios with weight allocations $\vec{w}^*$ and $\vec{w}^{**}$ can be established so that all other efficient portfolios can be expressed as a linear combination of the two efficient portfolios (with weight allocations $\vec{w}^*$ and $\vec{w}^{**}$). The weight allocations $\vec{w}^*$ and $\vec{w}^{**}$ are:

$$\vec{w}^* = \left(\frac{a}{d}\Sigma^{-1}\vec{1} - \frac{b}{d}\Sigma^{-1}\vec{r}\right) \tag{6.4}$$

$$\vec{w}^{**} = \left(\frac{c}{d}\Sigma^{-1}\vec{r} - \frac{b}{d}\Sigma^{-1}\vec{1}\right) + \left(\frac{a}{d}\Sigma^{-1}\vec{1} - \frac{b}{d}\Sigma^{-1}\vec{r}\right) \tag{6.5}$$

In order to find the expected return of r_p, invest $(1 - r_p)$ in fund $\vec{w}^*$ and r_p in $\vec{w}^{**}$.

The two fund theorem has an important implication for practical investment management. It states that two efficient investment products (such as passive mutual funds) can be used to meet the preferences of all investors (who seek an efficient portfolio).

Consider two important examples.

Example 6.1: Given three uncorrelated assets with $\vec{r} = \begin{pmatrix} 0.1 \\ 0.2 \\ 0.3 \end{pmatrix}$ and $\Sigma = I$ (i.e., the identity matrix, namely $\sigma_1^2 = \sigma_2^2 = \sigma_3^2 = 1$), find the equation of the Markowitz Frontier as well as the expected return and standard deviation of the minimum variance portfolio.

Given:

$$
\begin{aligned}
a &= \vec{r}^T \Sigma^{-1} \vec{r} = \vec{r}^T \vec{r} = 0.14, \\
b &= \vec{r}^T \Sigma^{-1} \vec{1} = \vec{r}^T \vec{1} = 0.6, \\
c &= \vec{1}^T \Sigma^{-1} \vec{1} = \vec{1}^T \vec{1} = 3, \text{ and} \\
d &= ac - b^2 = 0.14 \times 3 - 0.6^2 = 0.06
\end{aligned}
$$

Insert the above values into the optimal allocation formula:

$$
\vec{w} = \left(\frac{c}{d}\Sigma^{-1}\vec{r} - \frac{b}{d}\Sigma^{-1}\vec{1}\right) r_p + \left(\frac{a}{d}\Sigma^{-1}\vec{1} - \frac{b}{d}\Sigma^{-1}\vec{r}\right) = \begin{pmatrix} -5 \\ 0 \\ 5 \end{pmatrix} r_p + \begin{pmatrix} 4/3 \\ 1/3 \\ -2/3 \end{pmatrix}
$$

$$
\frac{\sigma_p^2}{1/3} - \frac{(r_p - 0.2)^2}{0.06/9} = 1
$$

Using $r_p = \frac{b}{c}$, $\sigma_p = \sqrt{\frac{1}{c}}$, solve for the minimum variance portfolio: $r_p = 0.2$ and $\sigma_p = \sqrt{1/3}$.

Example 6.2: An inferior security improves the Markowitz Frontier.

Consider two securities: A and B. Security A has expected return 0.08 and variance 0.15, Security B has expected return 0.12 and variance 0.20, and the covariance between the returns of Security A and Security B is 0.08.

a. Find the equation of the Markowitz Frontier for two assets.
b. Consider adding one more security, C, with expected return 0.04 and variance 0.25. The covariance between Security C and Security A is 0.12 and the covariance between Security C and Security B is 0.20. Find the equation of the Markowitz Frontier for the three assets.

Solution:

a. In the first case: $\vec{r} = \begin{pmatrix} 0.08 \\ 0.12 \end{pmatrix}$, $\Sigma = \begin{pmatrix} 0.15 & 0.08 \\ 0.08 & 0.20 \end{pmatrix}$

Find that $\Sigma^{-1} = \frac{1}{0.0236}\begin{pmatrix} 0.20 & -0.08 \\ -0.08 & 0.15 \end{pmatrix}$

$a = \vec{r}^T \Sigma^{-1} \vec{r} = 0.08068$
$b = \vec{r}^T \Sigma^{-1} \vec{1} = \vec{1}^T \Sigma^{-1} \vec{r} = 0.76271$
$c = \vec{1}^T \Sigma^{-1} \vec{1} = 8.05085$
$d = ac - b^2 = 0.067797.$

The Markowitz Frontier is:

$$\frac{\sigma_p^2}{1/c} - \frac{\left(r_p - \frac{b}{c}\right)^2}{d/c^2} = 1$$

which is equivalent to: $8.05085\sigma_p^2 - 956.0381356(r_p - 0.094737)^2 = 1$.
The Markowitz Frontier is plotted in Figure 6.3.

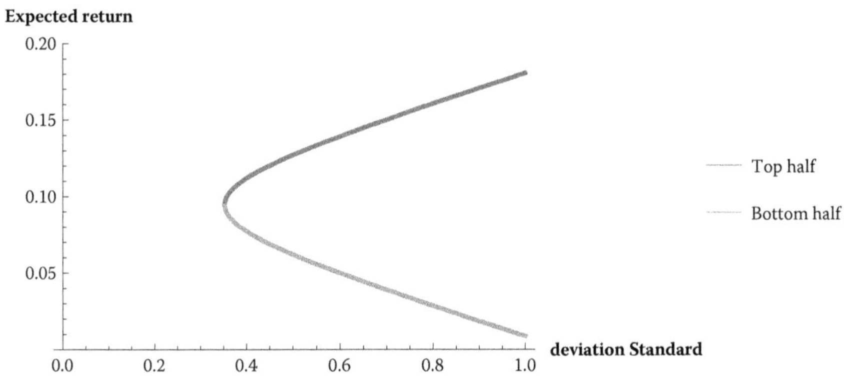

FIGURE 6.3
Markowitz Frontier for two assets.

b. In the second case, $\vec{r} = \begin{pmatrix} 0.08 \\ 0.12 \\ 0.04 \end{pmatrix}$, $\Sigma = \begin{pmatrix} 0.15 & 0.08 & 0.12 \\ 0.08 & 0.20 & 0.20 \\ 0.12 & 0.20 & 0.25 \end{pmatrix}$

The Markowitz Frontier can be solved using Mathematica, which generates Figure 6.4.

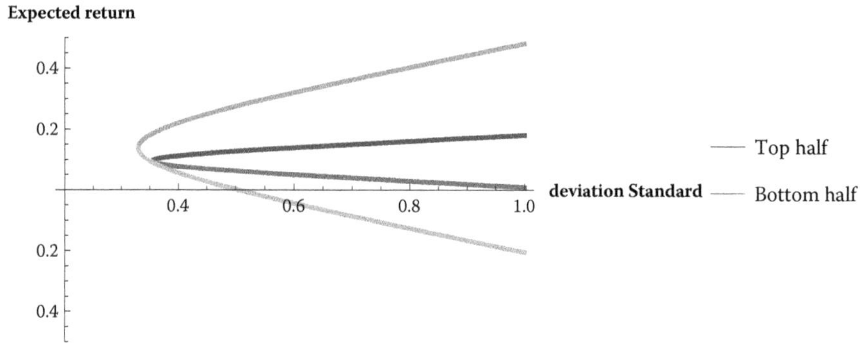

FIGURE 6.4
Two Markowitz Frontiers for two assets and three assets.

The horizontal axis of Figure 6.4 represents σ_p and the vertical axis represents r_p. The inside hyperbola (Case a) is the frontier for portfolios consisting of Securities A and B. The outside hyperbola (Case b) in Figure 6.4 is the Markowitz Frontier for portfolios consisting of Securities A, B, and C. Consider the change of the Markowitz Frontier from Case a to Case b when Security C is added. Security C seems inferior to Securities A and B because it has a lower expected return and higher variance than A and B. But adding the inferior security (Security C) to the portfolio actually improves the Markowitz Frontier.

Mathematica Code 6.1
Generate Markowitz Frontier with two and three assets, respectively

```
SigI=Inverse[{{0.15,0.08},{0.08,0.2}}];
r=Transpose[{{0.08,0.12}}];
a=Transpose[r].SigI.r;
l=Transpose[{{1,1}}];
b=Transpose[r].SigI.l;
c=Transpose[l].SigI.l;
d=a c-b^2;
f1[x_]:=b/c+Sqrt[(c x^2-1)/(c^2/d)]
f2[x_]:=b/c-Sqrt[(c x^2-1)/(c^2/d)]
Plot[{f1[x],f2[x]},{x,0,1},PlotStyle→{Thickness
[0.01],Thickness[0.01]}]
Sig1={{0.15,0.08,0.12},{0.08,0.2,0.2},
{0.12,0.2,0.25}};
Sig1I=Inverse[Sig1];
r1=Transpose[{{0.08,0.12,0.04}}];
```

```
a1=Transpose[r1].Sig1I.r1;
l1=Transpose[{{1,1,1}}];
b1=Transpose[r1].Sig1I.l1;
c1=Transpose[l1].Sig1I.l1;
d1=a1 c1-b1^2;
f3[x_]:=b1/c1+Sqrt[(c1 x^2-1)/(c1^2/d1)]
f4[x_]:=b1/c1-Sqrt[(c1 x^2-1)/(c1^2/d1)]
Plot[{f1[x],f2[x],f3[x],f4[x]},{x,0.2,1},PlotStyle→
{Thickness[0.01],Thickness[0.01],   Thickness[0.01],
Thickness [0.01]}, PlotRange→{-1,1}]
```

The above solution for the Markowitz Frontier assumes that there is no risk-free asset. Next, the analysis is extended to the case where a risk-free asset exists.

6.1.2 Analytical Solution for the Markowitz Frontier When There Is a Risk-Free Asset

Markowitz's seminal work in the 1950s (for which he was awarded the Nobel Prize in Economics) permitted all risky assets but did not include a riskless asset. In the 1960s, Sharpe, Lintner, and Mossin independently extended Markowitz's framework to include a risk-free asset. This section extends Markowitz's model by including a risk-free asset with an expected (and certain) return r_f. The new optimization problem becomes:

$$\text{Find weight allocation } \vec{w} \text{ to } \min_{\vec{w}} \tfrac{1}{2}\vec{w}^T\Sigma\vec{w}$$

$$\vec{w}^T\vec{r} + (1 - \vec{w}^T\vec{1})r_f = r_p.$$

The optimal allocation can be found using the method from the previous section:

$$\vec{w} = \Sigma^{-1}(\vec{r} - r_f\vec{1})\frac{r_p - r_f}{E}$$

where $E = (\vec{r} - r_f\vec{1})^T\Sigma^{-1}(\vec{r} - r_f\vec{1}) = a - 2br_f + cr_f^2$.

6.3 MARKOWITZ FRONTIER THEOREM – EXTENDED TO INCLUDE A RISK-FREE ASSET AND WITH SHORT SALES ALLOWED

There are n risky assets with expected return vector $\vec{r} = \begin{pmatrix} r_1 \\ \cdots \\ r_n \end{pmatrix}$ and a positive definite covariance of return matrix Σ. In addition, there is a risk-free asset that offers a return of r_f. The vector $\vec{w} = \begin{pmatrix} w_1 \\ \cdots \\ w_n \end{pmatrix}$ is the weight allocation for n risky assets, $\vec{1}$ is a column vector of $\begin{pmatrix} 1 \\ \cdots \\ 1 \end{pmatrix}$, r_p is the expected return of the portfolio, and σ_p^2 is the variance of the return of the portfolio.

The Markowitz Frontier portfolio is a portfolio that minimizes the standard deviation of the portfolio return holding the expected return of the portfolio constant. It is equivalent to the following optimization problem:

Find weight allocation $\vec{w}$ to

$$\underset{\vec{w}}{\mathbf{Min}} \tfrac{1}{2} \vec{w}^T \Sigma \vec{w} \tag{6.6}$$

where

$$\vec{w}^T \vec{r} + (1 - \vec{w}^T \vec{1}) r_f = r_p$$

For a given r_p, the optimal weight allocation $\vec{w}$ is given by:

$$\vec{w} = \Sigma^{-1}(\vec{r} - r_f \vec{1}) \frac{r_p - r_f}{E} \tag{6.7}$$

where $E = (\vec{r} - r_f \vec{1})^T \Sigma^{-1} (\vec{r} - r_f \vec{1}) = a - 2br_f + cr_f^2$, $a = \vec{r}^T \Sigma^{-1} \vec{r}$, $b = \vec{r}^T \Sigma^{-1} \vec{1} = \vec{1}^T \Sigma^{-1} \vec{r}$, $c = \vec{1}^T \Sigma^{-1} \vec{1}$.

In addition, the expected return, r_p, of the Markowitz Frontier portfolio (the optimal portfolio) and its corresponding standard deviation, σ_p, satisfy the following equation:

$$r_p = \begin{cases} r_f + \sqrt{E}\,\sigma_p & \text{if} r_p > r_f \\ r_f - \sqrt{E}\,\sigma_p & \text{if} r_p < r_f \end{cases} \tag{6.8}$$

Proof
Rewrite the optimization constraint to:

$$\vec{w}^T(\vec{r} - r_f\vec{1}) = r_p - r_f.$$

Using the Lagrange multipliers method, solve the equation related to the gradient vectors together with the constraint equation:

$$\Sigma\vec{w} = \lambda(\vec{r} - r_f\vec{1})$$
$$\vec{w}^T(\vec{r} - r_f\vec{1}) = r_p - r_f.$$

The first equation is equivalent to:

$$\vec{w} = \lambda\Sigma^{-1}(\vec{r} - r_f\vec{1})$$
$$\vec{w}^T = \lambda(\vec{r} - r_f\vec{1})^T\Sigma^{-1}.$$

Insert $\vec{w}^T$ into the constraint equation:

$$\vec{w}^T(\vec{r} - r_f\vec{1}) = r_p - r_f,$$

to obtain:

$$\lambda(\vec{r} - r_f\vec{1})^T\Sigma^{-1}(\vec{r} - r_f\vec{1}) = r_p - r_f.$$

Solve for λ,

$$\lambda = \frac{r_p - r_f}{(\vec{r} - r_f\vec{1})^T\Sigma^{-1}(\vec{r} - r_f\vec{1})} = \frac{r_p - r_f}{E}.$$

Insert λ into the equation for $\vec{w}$,

$$\vec{w} = \Sigma^{-1}(\vec{r} - r_f\vec{1}) \times \lambda$$
$$= \Sigma^{-1}(\vec{r} - r_f\vec{1})\frac{r_p - r_f}{E}$$

where $E = (\vec{r} - r_f\vec{1})^T\Sigma^{-1}(\vec{r} - r_f\vec{1}) = a - 2br_f + cr_f^2$,
where $a = \vec{r}^T\Sigma^{-1}\vec{r}$, $b = \vec{r}^T\Sigma^{-1}\vec{1} = \vec{1}^T\Sigma^{-1}\vec{r}$, $c = \vec{1}^T\Sigma^{-1}\vec{1}$.
Note that $\vec{w}$ is a function of r_p. Insert $\vec{w}$ into $\sigma_p^2 = \vec{w}^T\Sigma\vec{w}$ and simplify to derive the Markowitz Frontier:

$$\sigma_p^2 = \frac{(r_p - r_f)^2}{E}.$$

The details are saved for the exercises.

Since Σ is positive definite, $E > 0$ and the equation can be simplified.

$$\sigma_p = \begin{cases} \frac{r_p - r_f}{\sqrt{E}} & \text{if } r_p > r_f \\ -\frac{r_p - r_f}{\sqrt{E}} & \text{if } r_p < r_f \end{cases}$$

Equivalently,

$$r_p = \begin{cases} r_f + \sqrt{E}\,\sigma_p & \text{if } r_p > r_f \\ r_f - \sqrt{E}\,\sigma_p & \text{if } r_p < r_f \end{cases}$$

The Markowitz Frontier is composed of two half lines starting from the point $(0, r_f)$. The slope of the lines are $\sqrt{E}$ and $-\sqrt{E}$. ∎

The Markowitz Frontier is composed of two half lines when the risk-free asset is included. These two lines are related to the hyperbola of the Markowitz Frontier without the risk-free asset in the following theorem. (In this section, short sales are allowed. In other words, negative weights are permitted for any asset.)

6.4 TANGENT PORTFOLIO THEOREM

If $r_f < \frac{b}{c}$, then the upper half line $r_p = r_f + \sqrt{E}\,\sigma_p$ is *tangent* to the Markowitz Efficient Frontier containing only risky assets:

$$\frac{\sigma_p^2}{1/c} - \frac{(r_p - \frac{b}{c})^2}{d/c^2} = 1.$$

The ***tangent portfolio e*** has the following mean and standard deviation:

$$(\sigma_p, r_p) = \left(\frac{\sqrt{E}}{b - cr_f}, \frac{a - br_f}{b - cr_f}\right) \tag{6.9}$$

If $r_f > \frac{b}{c}$, then the bottom half line, $r_p = r_f - \sqrt{E}\,\sigma_p$, is tangent to the bottom half of the Markowitz Frontier containing only n risky assets and which excludes the risk-free asset:

$$\frac{\sigma_p^2}{1/c} - \frac{(r_p - \frac{b}{c})^2}{d/c^2} = 1.$$

The ***tangent portfolio e'*** has the following expected return and standard deviation:

$$(\sigma_p, r_p) = \left(\frac{-\sqrt{E}}{b - cr_f}, \frac{a - br_f}{b - cr_f}\right) \tag{6.10}$$

If $r_f = \frac{b}{c}$, then neither line intersects the Markowitz Frontier containing only risky assets:

$$\frac{\sigma_p^2}{1/c} - \frac{\left(r_p - \frac{b}{c}\right)^2}{d/c^2} = 1$$

Proof

Consider the line passing through $(0, r_f)$ tangent to:

$$\frac{\sigma_p^2}{1/c} - \frac{\left(r_p - \frac{b}{c}\right)^2}{d/c^2} = 1$$

Rewrite the Markowitz Frontier hyperbola for risky assets in x-y form since the standard deviation is on the x-axis and the expected return is on the y-axis.

$$\frac{x^2}{1/c} - \frac{\left(y - \frac{b}{c}\right)^2}{\frac{d}{c^2}} = 1.$$

This is equivalent to:

$$cx^2 - \frac{c^2}{d}\left(y - \frac{b}{c}\right)^2 = 1.$$

With a risk-free asset, the set of efficient portfolios lies on the line tangent to this hyperbola. Thus, we need to find the line passing through $(0, r_f)$ tangent to the risky frontier. Assume that the tangent line intersects this hyperbola at the point (x_0, y_0). Then take the derivative of the hyperbola with respect to x to find the slope of the line.

$$2cx - 2\frac{c^2}{d}\left(y - \frac{b}{c}\right)\frac{dy}{dx} = 0$$

$$\frac{dy}{dx} = \frac{(d)(x)}{c\left(y - \frac{b}{c}\right)}$$

Knowing that the tangent point is (x_0, y_0), use the assumption to find the tangent line equation:

$$y - y_0 = \frac{dx_0}{c\left(y_0 - \frac{b}{c}\right)}(x - x_0).$$

The efficient frontier with a risk-free asset has the intercept $(0, r_f)$. Insert this point for x and y in the above equation.

$$r_f - y_0 = \frac{dx_0}{c\left(y_0 - \frac{b}{c}\right)}(-x_0)$$

The point (x_0, y_0) lies on the hyperbola and must satisfy:

$$cx_0^2 - \frac{c^2}{d}\left(y_0 - \frac{b}{c}\right)^2 = 1.$$

Using the two equations with respect to (x_0, y_0), solve as $y_0 = \frac{d + b^2 - bcr_f}{bc - c^2 r_f}$ $= \frac{ac - bcr_f}{bc - c^2 r_f} = \frac{a - br_f}{b - cr_f}$. This equation makes sense if $b - cr_f \neq 0$. Substituting the curve equation and solving for x_0.

$$x_0 = \sqrt{\frac{a - 2br_f + cr_f^2}{(b - cr_f)^2}} = \frac{\sqrt{E}}{|b - cr_f|}$$

The slope of this tangent line passing through (x_0, y_0) and $(0, r_f)$ is given by:

$$\frac{y_0 - r_f}{x_0} = \frac{\frac{a - br_f}{b - cr_f} - r_f}{\frac{\sqrt{E}}{|b - cr_f|}} = \frac{\frac{E}{b - cr_f}}{\frac{\sqrt{E}}{|b - cr_f|}} = \sqrt{E}\,\frac{|b - cr_f|}{b - cr_f} = \begin{cases} \sqrt{E} & \text{if } r_f < \frac{b}{c} \\ -\sqrt{E} & \text{if } r_f > \frac{b}{c} \end{cases}$$

So, if $r_f < \frac{b}{c}$, then the upper half line (with slope $\sqrt{E}$), $r_p = r_f + \sqrt{E}\,\sigma_p$, is tangent to the Markowitz Efficient Frontier,

$$\frac{\sigma_p^2}{1/c} - \frac{\left(r_p - \frac{b}{c}\right)^2}{d/c^2} = 1$$

and the tangent point is $\left(\frac{\sqrt{E}}{(b - cr_f)}, \frac{(a - br_f)}{(b - cr_f)}\right)$.

If $r_f > \frac{b}{c}$, then the bottom half line (with slope $-\sqrt{E}$), $r_p = r_f - \sqrt{E}\,\sigma_p$, is tangent to the Markowitz Frontier,

$$\frac{\sigma_p^2}{1/c} - \frac{\left(r_p - \frac{b}{c}\right)^2}{dc^2} = 1$$

and the tangent point is $\left(\frac{-\sqrt{E}}{b - cr_f}, \frac{a - br_f}{b - cr_f}\right)$.

If $r_f = \frac{b}{c}$, then there is no tangent line. ■

Remarks

1. For an investor with a specific required or target expected rate of return, the optimal portfolio is the portfolio with the lowest risk. We first considered the case where no risk-free asset exists in which the set of optimal (efficient) portfolios lie on a hyperbola. The hyperbola is called the ***Markowitz Frontier***. Introducing a risk-free asset to the optimal portfolio problem moves the set of optimal portfolios onto two lines tangent to the hyperbola. Non-optimal portfolios lie inside the hyperbola or lines. The optimal portfolio problem can be viewed from another perspective: for a given or target level of risk σ_p, the optimal portfolio is in fact the one with the highest expected return r_p. According to this specification only the upper half of the hyperbola and the top line are optimal. The upper half of the hyperbola is called the ***Markowitz Efficient Frontier*** while the upward-sloping line is the set of efficient portfolios including the riskless asset. In practice, the bottom half of the hyperbola and the downward sloping line are not optimal investment allocations.
2. We derived the Markowitz Efficient Frontier with no restrictions on the weight allocations. This allows for a negative weight, which implies that short selling is permissible. The weights can be restricted so that short sales are not allowed with the constraints: $w_i \geq 0$. The resulting optimal solutions will change. This is detailed in the next section.
3. In the presence of a risk-free asset and where $r_f < \frac{b}{c}$, all optimal portfolios can be formed with a linear combination of the risk-free asset and the ***tangent portfolio*** e. This is called the ***one fund theorem***.
4. In the presence of a risk-free asset and where $r_f > \frac{b}{c}$, the one fund theorem exists. In other words, all optimal portfolios will lie on the top line and can be formed as a combination of the risk-free asset and a short position in the tangent portfolio e'.
5. In the presence of a risk-free asset, where $r_f = \frac{b}{c}$, optimal weight $\overrightarrow{w} = \Sigma^{-1}(\overrightarrow{r} - r_f\overrightarrow{1})\frac{r_p - r_f}{E}$, we have:

$$1^T\overrightarrow{w} = (1^T\Sigma^{-1}\overrightarrow{r} - r_f 1^T\Sigma^{-1}1)\frac{r_p - r_f}{E} = \left(b - \frac{b}{c}c\right)\frac{r_p - r_f}{E} = 0.$$

The optimal portfolio in this case has a weight sum of 0 as a whole though it has non-zero weights in individual assets in risky assets and invests everything in the risk-free asset.

Consider the following:

Example 6.3: Adding a risk-free asset to Example 6.2.

Consider three assets (A, B, and C) with the expected return vector and covariance matrix being same as in Example 6.2. In addition, there is a risk-free asset. Find solutions to the following:

a. There is a risk-free asset that offers r_f = 5%. What is the line equation of efficient portfolios?
b. There is a risk-free asset that offers r_f = 13.82%. What is the line equation of efficient portfolios?
c. There is a risk-free asset that offers r_f = 15%. What is the line equation of efficient portfolios?
d. Use the risk-free asset in part (a). What is the tangent portfolio's return and standard deviation? If an expected return of 20% for the portfolio is required, what is the corresponding weight allocation?
e. Use the risk-free asset in part (b). What is the tangent portfolio's return and standard deviation? If an expected return of 20% for the portfolio is required, what is the corresponding weight allocation?
f. Use the risk-free asset in part (c). What is the tangent portfolio's return and standard deviation? If an expected return of 20% for the portfolio is required, what is the corresponding weight allocation?

To solve these problems, first calculate E = $a - 2br_f + cr_f^2$, $a = \vec{r}^T\Sigma^{-1}\vec{r}$, $b = \vec{r}^T\Sigma^{-1}\vec{1} = \vec{1}^T\Sigma^{-1}\vec{r}$, $c = \vec{1}^T\Sigma^{-1}\vec{1}$. (You may use Mathematica for this calculation. The codes are very similar to code 6.1.)

a. Find $E = 0.20315$. First, find the lines by

$$r_p = \begin{cases} 0.05 + \sqrt{0.20315}\,\sigma_p & \text{if } r_p > r_f \\ 0.05 - \sqrt{0.20315}\,\sigma_p & \text{if } r_p < r_f \end{cases}$$

Figure 6.5 uses Mathematica to plot the figure.

b. Using $E = 0.13165$, the result is shown in Figure 6.6:

c. Using $E = 0.13292$, the result is shown in Figure 6.7.

d. First, calculate $b/c = 0.1382$ and r_f= 5%. Therefore, $r_f < \frac{b}{c}$. The tangent portfolio is on the top half of the curve. The tangent portfolio coordinates are found using $e = \left(\frac{\sqrt{E}}{b - cr_f}, \frac{a - br_f}{b - cr_f}\right)$ = (0.5561, 0.3001). The tangent portfolio has σ_e = 55.61% and r_e = 30.01%.

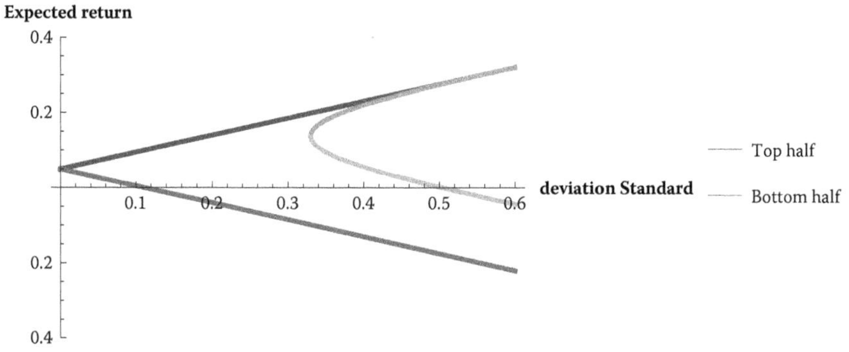

FIGURE 6.5
Markowitz Frontier with Risk-Free Asset having a rate of 5%.

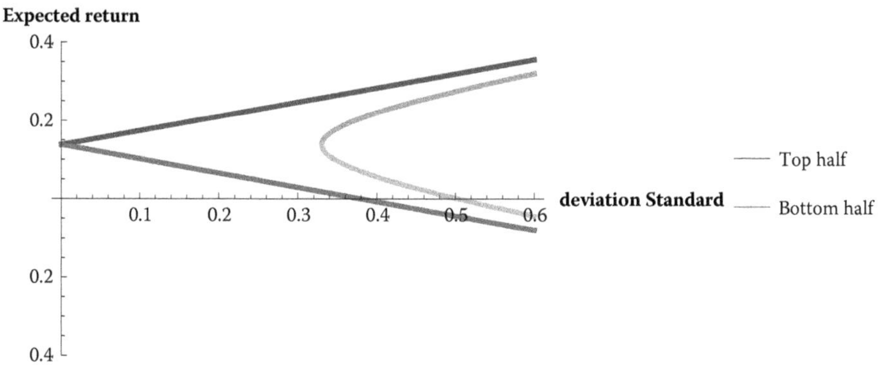

FIGURE 6.6
Markowitz Frontier with the Risk-Free Asset having a rate of 13.82%.

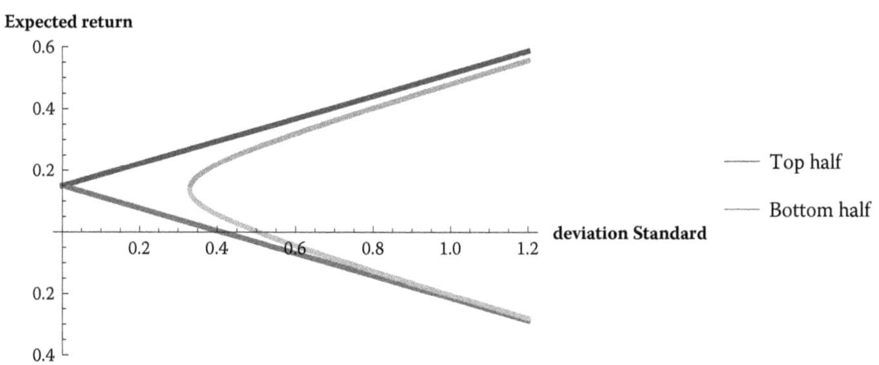

FIGURE 6.7
Markowitz Frontier with the Risk-Free Asset having a rate of 15%.

The weight allocation for (σ_r, r_p) is:

$$\begin{aligned}\vec{w} &= \Sigma^{-1}(\vec{r} - r_f\vec{1})\frac{r_p - r_f}{E}\\ &= (0.5667, \quad 1.6665, \quad -1.6347)\end{aligned}$$

The weight of the risk-free asset is 1 – (0.5667 + 1.6665 – 1.6347) = 0.4016. The desired optimal portfolio invests 40.16% in the risk-free asset and 59.84% in the tangent portfolio.
The calculation of the expected return of the portfolio can be verified by investing in Assets A, B, and C and the risk-free asset:

$$0.4016 \times 0.05 + 0.5667 \times 0.08 + 1.6665 \times 0.12 - 1.6347 \times 0.04 = 20\%.$$

Invest 40.16% in risk-free asset, 56.67% in Asset A, 166.65% in Asset B, and short 163.47% in Asset C.
Alternatively, check the expected return of the portfolio by investing in the tangent portfolio.

$$0.4016 \times 0.05 + 0.5984 \times 0.3001 = 19.966\%.$$

The result of the previous calculation is slightly off from 20% due to rounding error.
The portfolio standard deviation, σ_p, can be found two ways.
First, use the portfolio weight allocation:

$$\begin{aligned}\sigma_p^2 &= \vec{w}^T\Sigma\vec{w}\\ \sigma_p &= 0.3328\end{aligned}$$

Also, calculate the standard deviation from the tangent portfolio:

$$\sigma_p = 0.5984 \times \sigma_e = 0.5984 \times 0.5561 = 0.3328$$

So, in order to get a 20% return, use 33.28% as the standard deviation.

e. Calculate b/c = 0.1382 as before, r_f= 13.82% (given), and $r_f = \frac{b}{c}$.
   The weight allocation is

$$\begin{aligned}\vec{w} &= \Sigma^{-1}(\vec{r} - r_f\vec{1})\frac{r_p - r_f}{E}\\ &= (0.0714, 0.7369, -0.8082)\end{aligned}$$

The weight invested in the risk-free asset is 1 – (0.0714 + 0.7369 – 0.8082) = 1. The optimal portfolio is 100% in the risk-free asset and a net zero weight on risky assets.

The expected return of the portfolio of Assets A, B, and C can be checked:

$$100\% \times 0.1382 + 0.0714 \times 0.08 + 0.7369 \times 0.12 - 0.8082 \times 0.04 = 20\%.$$

Therefore, invest 100% in risk-free asset, 7.14% in Asset A, 73.69% in Asset B, and short 80.82% Asset C.

The value σ_p can also be obtained by the portfolio weight allocation:

$$\sigma_p^2 = \vec{w}^T \Sigma \vec{w}$$
$$\sigma_p = 0.1703$$

In order to obtain a 20% return, the portfolio must have a 17.03% standard deviation. Note that our demanded return is higher than the risk free rate, thus there must be investment in risky assets although the total weight is 0% in risky assets.

f. $b/c = 0.1382$, $r_f = 15\%$, and $r_f > \frac{b}{c}$. The tangent portfolio is on the bottom half of the curve. The coordinates are $e' = \left(\frac{-\sqrt{E}}{b - cr_f}, \frac{a - br_f}{b - cr_f}\right) = (3.3714, -1.0791)$. The tangent portfolio has standard deviation $\sigma_e = 337.14\%$ and expected return $r_e = -107.91\%$.

The weight allocation is

$$\vec{w} = \Sigma^{-1}(\vec{r} - r_f \vec{1})\frac{r_p - r_f}{E}$$
$$= (0.0262, 0.5559, -0.6229)$$

The weight of the risk-free asset is 1 – (0.0262 + 0.5559 – 0.6229) = 1.0407.

The portfolio is invested 104.07% in the risk-free asset and short-sells 4.07% of the tangent portfolio.

The expected return of the portfolio can be checked by investing in Assets A, B, and C:

$$1.0407 \times 0.15 + 0.0262 \times 0.08 + 0.5549 \times 0.12 - 0.6229 \times 0.04 = 20\%.$$

Therefore, invest 104.07% in risk-free asset, 2.62% in Asset A, 55.49% in Asset B, and short 62.29% of Asset C.

Alternatively, check the expected return of the portfolio by investing in the tangent portfolio.

$$1.0407 \times 0.15 - 0.0407 \times (-1.0791) = 20\%.$$

The value of σ_p can be found in two ways. One uses the portfolio weight allocation:

$$\sigma_p^2 = \overrightarrow{w}^T \Sigma \overrightarrow{w}$$
$$\sigma_p = 0.1371$$

It can also be calculated from the tangent portfolio:

$$\sigma_p = 0.0407 \times \sigma_e = 0.0407 \times 3.3714 = 0.1372$$

In order to obtain a 20% return, the portfolio will have a 13.7% standard deviation.

Remark

Note that in a, b, and c, Theorem 6.3 was used to find the Markowitz Frontier. We have not checked the relation between r_f and $\frac{b}{c}$ as stated in Theorem 6.4. In d, e, and f, we checked this relation and found that the results are consistent.

The previous sections set up Markowitz mean-variance portfolio theory in Theorems 6.1–6.4. The next subsection derives the CAPM model, which is a direct corollary of the Markowitz mean-variance portfolio theory.

6.1.3 The Capital Market Line, CAPM, and Security Market Line

This section extends the Markowitz mean-variance portfolio theory with the assumptions: (1) that the risk-free asset is available to all investors for both lending and borrowing, (2) that there are no market imperfections, (3) that asset returns are either normally distributed or that investors have quadratic utility functions, and (4) that all investors share identical expectations of the expected return, variance, and covariances for all assets (i.e., investors have homogeneous expectations).

Given these assumptions, the optimal portfolio for every investor is a linear combination of the risk-free asset and the tangent portfolio e, which is also called the ***market portfolio, m***, in equilibrium. Everyone's decision is homogenous with regard to the tangent portfolio. The only difference is that the investors' weight allocations may differ between the tangent portfolio and risk-free asset. The tangent portfolio must be the market portfolio. The market portfolio includes all risky assets available in the market with asset weights proportional to their total values.

Optimal portfolios must stay on a line in the σ–r plane. The line passes through $(0, r_f)$ and (σ_m, r_m). This line is called the ***capital market line***. The slope of the line is:

$$\frac{r_m - r_f}{\sigma_m}.$$

The equation is:

$$r = r_f + \frac{r_m - r_f}{\sigma_m}\sigma$$

The ***market portfolio***, r_m, is the portfolio of all investable assets with the weight of each asset equal to the proportion of that asset's total value relative the sum of the total values of all assets.

Having introduced the market portfolio, the systematic risk of an individual asset is introduced. The ***systematic risk*** of an individual asset is not the total risk, σ_i, rather it is the portion of the total risk that is correlated to the risk of the market portfolio.

Consider how the equilibrium expected return of an individual asset relates to its systematic risk, which is specified by the CAPM.

6.5 THE CAPM

Assume that long and short positions in the risk-free asset (i.e., borrowing and lending at the riskless rate) are allowed and that everyone has access to all the assets in the market (the market portfolio). Further assume that the estimates of expected return and covariance (including variance) are the same for all assets and for all investors (homogeneous expectations) and that investors use a mean and variance framework to make investment decisions (owing either to asset returns being normally distributed or investor utility functions being quadratic). In the absence of trading imperfections such as transactions costs, the optimal portfolio for all investors is a linear combination of the risk-free asset and the market portfolio. Further, the CAPM specifies the equilibrium expected return of each asset.

Denote the expected return of the market portfolio as r_m, the variance of the market portfolio as σ_m^2, the expected return of individual asset i as r_i, the covariance of the individual asset and the market as $\sigma_{i,m}$, and the return of risk-free asset as r_f. In equilibrium, the expected return of an individual asset must satisfy the following equation:

$$r_i = r_f + \frac{\sigma_{i,m}}{\sigma_m^2}(r_m - r_f) = r_f + \beta_i(r_m - r_f) \tag{6.11}$$

where $\beta_i = \frac{\sigma_{i,m}}{\sigma_m^2}$. Note that β_i is a measure of systematic risk; it measures the risk of the asset in terms of the responsiveness of its returns to the returns of the market portfolio.

Proof
Assume that a person invests x portion in asset i, and $1 - x$ in the market portfolio (which includes asset i). Clearly, this is not an optimal choice unless $x = 0$.
Note that:

$$r_p = xr_i + (1 - x)r_m, \sigma_p^2 = x^2\sigma_i^2 + (1 - x)^2\sigma_m^2 + 2x(1 - x)\sigma_{i,m}$$

As x is varied, a curve, Γ, forms that intersects with the Markowitz Frontier involving all risky assets in the market at $x = 0$. The curve Γ cannot cross the capital market line, otherwise, there would be a portfolio on the curve that is optimal but not on the capital market line. Since there is one point on this curve Γ that intersects with the capital market line, the capital market line must be tangent to the curve Γ at the market portfolio point.
The capital market line has the slope: $\frac{r_m - r_f}{\sigma_m}$. Calculate the slope of this curve, Γ, at the point of the market portfolio with $x = 0$:

$$\frac{dr_p}{d\sigma_p}\Big|_{x=0} = \frac{dr_p/dx}{d\sigma_p/dx}\Big|_{x=0} = \frac{r_i - r_m}{(\sigma_{i,m} - \sigma_m^2)/\sigma_m}$$

The derivation of $d\sigma_p/dx\,|_{x=0} = (\sigma_{i,m} - \sigma_m^2)/\sigma_m$ is left for an exercise.
Set the two slopes equal to each other:

$$\frac{r_i - r_m}{(\sigma_{i,m} - \sigma_m^2)/\sigma_m} = \frac{r_m - r_f}{\sigma_m}$$

Simplify the previous expression:

$$(r_i - r_m)\sigma_m^2 = (r_m - r_f)(\sigma_{i,m} - \sigma_m^2)$$

$$r_i\sigma_m^2 - r_m\sigma_m^2 = (r_m - r_f)\sigma_{i,m} - (r_m - r_f)\sigma_m^2$$

$$r_i\sigma_m^2 = (r_m - r_f)\sigma_{i,m} + r_f\sigma_m^2$$

$$r_i = r_f + (r_m - r_f)\sigma_{i,m}/\sigma_m^2$$

$$r_i = r_f + (r_m - r_f)\beta_i \ \blacksquare$$

The ***total risk***, σ_i, of each risky asset may be viewed as being comprised of variation that is *uncorrelated* with the performance of the market portfolio (called ***unique, diversifiable, nonsystematic***, or ***idiosyncratic risk***), and risk measured by β_i that is *correlated* with the performance of the market portfolio (called ***non-diversifiable, systematic,*** or ***beta risk***).

β_i is a measure of an asset's systematic risk. The key insight of the CAPM is that an asset's additional expected return above the risk-free asset, $r_i - r_f$, is proportional to the extra expected return of market, $r_m - r_f$ (termed the ***market risk premium***). Therefore, investors require and receive a risk premium for systematic risk but are not consistently rewarded for bearing diversifiable risk.

If the equation

$$r_i = r_f + (r_m - r_f)\beta_i$$

is drawn in the $\beta - r$ plane, it is called the ***security market line***. It is the relation between the systematic risk, β, of an individual asset and the expected return of an individual asset. It is different from the capital market line in two ways. First, the capital market line is in $\sigma - r$ plane and the security market line is in the $\beta - r$ plane. The second difference is that all optimal portfolios are on the capital market line while all individual assets with different systematic risk betas are on the security market line. The intercept for both of these two lines is the risk-free asset. The market portfolio lies on the capital market line at $\sigma = \sigma_m$ and on the security market line at $\beta = 1$.

In investment books, readers can find numerous applications of the CAPM model. In a famous paper by Eugene F. Fama and Kenneth R. French, "The CAPM is Wanted, Dead or Alive," the authors discuss perspectives on the model. It should be noted that the CAPM is a general equilibrium model, meaning that it describes all asset prices endogenously (within the model). In the CAPM model, all assets lie on the security market line.

In practice, single factor market models are often used that are very similar to the CAPM but differ in two regards: (1) they allow some assets to lie off of the security market line and (2) they are not general equilibrium models. Single factor market models simply note that each asset's return can be broken into diversifiable and systematic components. The CAPM requires all assets to have expected returns determined exclusively by their systematic risk (β), the riskless rate, and the expected return of the market.

Previous sections discussed Markowitz portfolio theory and the CAPM model. Next we explore a question: How are assets weighted if short sales are not allowed? In other words, the next section, examines the case of disallowing short sales. Note that the derivations of the previous theorems were based on the Lagrange multiplier method, which does not allow inequality constraints such as $w_i \geq 0$. The tool used in the next section is called quadratic programming and is often used in the investment industry.

6.2 Portfolio Optimization: Linear and Quadratic Programming

The previous section investigated the Markowitz mean-variance portfolio theory model and its extension into the CAPM through analytic solutions. That section used the method of Lagrange multipliers, which requires equalities as constraints. What if short sales are not allowed? Disallowing short sales imposes an inequality constraint on portfolio weights: $w_i \geq 0$. This section investigates numerical solutions of this problem using R software. Linear programming and quadratic programming are very useful tools in quantitative finance.

6.2.1 Finding the Markowitz Frontier Using Quadratic Programming

The package ***quadprog*** can be installed from R to solve optimum points on a quadratic target function under linear constraints. Both equalities and inequalities are accepted. In this first subsection we continue to deal with the case of short sales being allowed using the problem from the previous section, first without risk-free asset, then with risk-free asset. Quadratic programming is introduced using the quadratic programming package "quadprog." The new results are compared to results in the previous section. After that, the problem is solved again with short sales disallowed – first without risk-free asset and then with risk-free asset.

Recall that the Markowitz Frontier theorem minimizes the risk while keeping the expected return of the portfolio constant. The problem is formulated as:

$$\text{Find weight allocation } \vec{w} \text{ to } \underset{\vec{w}}{\text{Min}}\, ½\vec{w}^T \Sigma\, \vec{w} \text{ where}$$

$$\vec{w}^T \vec{r} = r_p$$

$$\vec{w}^T \vec{1} = 1$$

Note that r_p is the given expected return for the portfolio and $\sigma_p^2 = \vec{w}^T \Sigma \vec{w}$ is the portfolio variance.

The R package *quadprog* has a built-in function useful to this exercise:

qp = solve. QP(Dmat, dvec, Amat, bvec, meq = x)

The code is used to minimize the function $-\vec{w}^T\vec{d} + 1/2w^T D w$ under the constraint $\vec{w}^T A \geq \vec{b}$.

Define $\vec{w} = \begin{pmatrix} w_1 \\ \cdots \\ w_n \end{pmatrix}$ as the independent variable, $\vec{d}$ is a column vector which we refer to as dvec. D is an $n \times n$ matrix referred to as *Dmat*, A is an $n \times m$ matrix referred to as *Amat*, m is the number of constraints, $\vec{b}$ is a given vector referred to as bvec, and x is the number of equality constraints in the optimization problem.

In this Markowitz Frontier problem, as in Section 6.1.1:

$w = \begin{pmatrix} w_1 \\ \cdots \\ w_n \end{pmatrix}$, dvec $= \begin{pmatrix} 0 \\ \cdots \\ 0 \end{pmatrix}$, $Dmat = \Sigma$, $Amat = \begin{pmatrix} 1 & r_1 \\ \cdots & \cdots \\ 1 & r_n \end{pmatrix}$, bvec $= (1, r_p)$ because the constraints

$$\vec{w}^T \vec{r} = r_p$$
$$\vec{w}^T \vec{1} = 1$$

in matrix form are: $\vec{w}^T \begin{pmatrix} 1 & r_1 \\ \cdots & \cdots \\ 1 & r_n \end{pmatrix} = (1, r_p)$.

Here, the parameter is "x = 2", since there are two constraints, both equalities.

Consider the following example for illustration.

Example 6.4: This example repeats Example 6.2 using the R package quadprog.

Consider:

$$\vec{r} = \begin{pmatrix} 0.08 \\ 0.12 \\ 0.04 \end{pmatrix}, \ \Sigma = \begin{pmatrix} 0.15 & 0.08 & 0.12 \\ 0.08 & 0.20 & 0.20 \\ 0.12 & 0.20 & 0.25 \end{pmatrix}$$

Given $r_p = 0.1$, find the Markowitz Frontier portfolio (σ_p, r_p) and its weight allocation under the assumption that there is no risk-free asset. First, install the package "quadprog" and load the package "quadprog". Then, run the following R-code:

R-Code 6.1 FIND OPTIMAL PORTFOLIO WITH NO RISK-FREE ASSET, SHORT SALES ALLOWED

#Given r_p, find the optimal portfolio σ_p

```
library(quadprog)
rp = 0.10
Dmat = matrix(c
(0.15,0.08,0.12,0.08,0.2,0.2,0.12,0.2,0.25),3,3)
dvec = c(0,0,0)
Amat = matrix(c(1,1,1,0.08,0.12,0.04),3,2)
bvec = c(1,rp)
qp = solve.QP(Dmat, dvec, Amat, bvec, meq=2)
rp
sigmap = sqrt(2 * qp$value)
sigmap
weight = qp$solution
weight
```

Note that in the above codes, *qp* saves quadratic-programming-related results: *qp*\$value saves the minimal value of $1/2\overrightarrow{w}^T \Sigma \overrightarrow{w}$ and *qp*\$soluion saves the optimal weight allocation. The resulting Markowitz Frontier portfolio $(\sigma_p, r_p) = (0.34635, 0.1)$ and the weight allocation is {0.7153846 0.3923077 −0.1076923}.

Note that this portfolio generates the expected return 10%:

$$0.08 \times 0.7153846 + 0.12 \times 0.3923077 + 0.04 \times (-0.1076923) = 10\%.$$

Next, generate the whole Markowitz Frontier and illustrate it in a graph. First, select some range of r_p, say [−0.4,0.6]. Then, cut the interval 10,000 times. For each r_p, solve the frontier point (σ_p, r_p). Finally, draw the graph. The code utilizes a key tool in programming known as a "*for* loop" that performs a task multiple times over a range of values of r_p for an index variable.

R-Code 6.2 Find Markowitz Frontier with no risk-free asset, short sales allowed

```
#Markowitz frontier codes
library(quadprog)
n = 10000
r = seq(-0.4, 0.6, by=(0.6+0.4)/n)
sigma = rep(0, n+1)
for(i in 1:n+1)
{
Dmat = matrix(c
(0.15,0.08,0.12,0.08,0.20,0.20,0.12,0.20,0.25),3,3)
dvec = c(0,0,0)
Amat = matrix(c(1,1,1,0.08,0.12,0.04),3,2)
bvec = c(1, r[i])
qp = solve.QP(Dmat, dvec, Amat, bvec, meq=2)
sigma[i] = sqrt(2 * qp$value)
}
plot(sigma, r, col="green")
```

Figure 6.8 depicts the Markowitz Frontier generated by R.

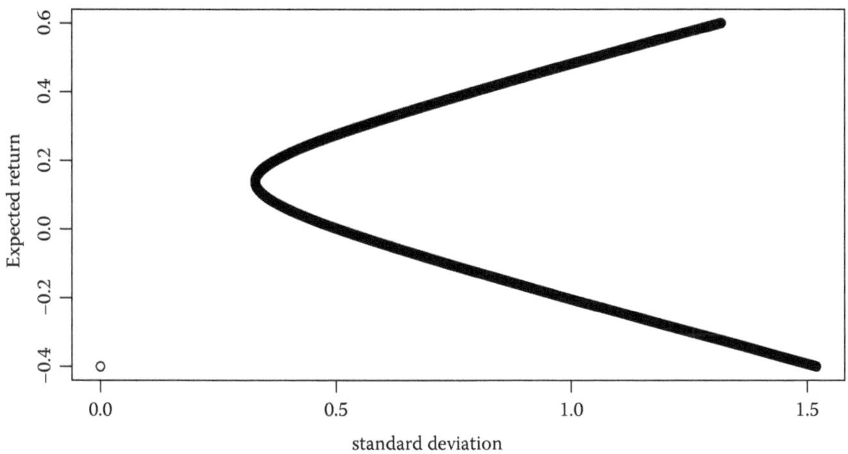

FIGURE 6.8
Markowitz Frontier for three assets by quadratic programing.

Note that the curve matches the analytic solution of the outside curve in Figure 6.4.

Now consider the case when a risk-free asset exists.

Example 6.5: Consider Example 6.3d where $r_f = 0.05 < \frac{b}{c}$.

$$\vec{r} = \begin{pmatrix} 0.08 \\ 0.12 \\ 0.04 \end{pmatrix}, \Sigma = \begin{pmatrix} 0.15 & 0.08 & 0.12 \\ 0.08 & 0.20 & 0.20 \\ 0.12 & 0.20 & 0.25 \end{pmatrix}, r_f = 0.05.$$

a. Find the Markowitz frontier portfolio with $r_p = 20\%$.
b. Find the tangent portfolio and corresponding weight allocation.

Solution for (a): The program can solve the following optimization problem by modifying code 6.1:

Find the weight allocation $\vec{w}$ to $\underset{\vec{w}}{\mathbf{Min}} \tfrac{1}{2}\vec{w}^T \boldsymbol{\Sigma} \vec{w}$ where

$$\vec{w}^T \vec{r} + (\mathbf{1} - \vec{w}^T \vec{\mathbf{1}}) r_f = r_p$$

The constraint is equivalent to $\vec{w}^T (\vec{r} - \vec{\mathbf{1}} r_f) = r_p - r_f$.

Change code 6.1 to: $Amat = \begin{pmatrix} r_1 - r_f \\ \cdots \\ r_n - r_f \end{pmatrix}$

bvec = $(r_p - r_f)$, $r_p = 20\%$ and meq = 1. The result matches the results in 6.3 (d): weight: {0.5666533, 1.6664759, −1.6347089}, where $r_p = 0.2$ and $\sigma_p = 0.3327987$.

Since $r_f = 0.05 < \frac{b}{c}$, the tangent portfolio is on the top half of the curve.

Solution for (b): Here we use a different approach to find the tangent portfolio. Construct lines by connecting all the points on the top half of the Markowitz Frontier hyperbola without the risk-free asset and the risk-free portfolio $(0, r_f)$. Then find the tangent line with the maximum slope,

$$\frac{r_e - r_f}{\sigma_e} = \max_p \left\{ \frac{r_p - r_f}{\sigma_p} \right\}.$$

Use the following R-code to find the tangent portfolio (σ_e, r_e) and weight allocation for tangent portfolio.

Sample R-Code 6.3 Find tangent portfolio when risk-free asset exists, short sales allowed

```
#Find the tangent portfolio and weight allocation of the tangent portfolio
library(quadprog)
n =10000
r = seq(-0.4, 0.6, by = (0.6 + 0.4) / n)
sigma = rep(0, n+1)
rf = rep(0.05, n+1)
sharp = rep(0, n+1)
maxsharp = 0
weight = rep(0, 3)
for(i in 1:n+1)
{
Dmat = matrix(c
(0.15,0.08,0.12,0.08,0.20,0.20,0.12,0.20,0.25), 3, 3)
dvec = c(0,0,0)
Amat = matrix(c(1,1,1,0.08,0.12,0.04), 3, 2)
bvec = c(1, r[i])
qp = solve.QP(Dmat, dvec, Amat, bvec, meq=2)
sigma[i] = sqrt(2 * qp$value)
sharp[i] = (r[i] - rf[i]) / sigma[i]
if(sharp[i] > maxsharp)
{
maxsharp = sharp[i]
weighttangent = qp$solution
rtangent = r[i]
sigmatangent = sigma[i]
}
}
rtangent
sigmatangent
weighttangent
```

The tangent portfolio has coordinates (σ_e, r_e) = (0.5562175, 0.3007). The weight allocation for r_e is {0.9469615 2.7852692 − 2.7322308}.
The results prove that the efficient frontier with a risk-free asset is the line tangent to the top half of the Markowitz Frontier hyperbola excluding the risk-free asset. As long as the tangent portfolio (σ_e, r_e) is identified, the efficient frontier including a risk-free asset can be formed by connecting $(0, r_f)$ and (σ_e, r_e).

Next, redo problem (a) and identify the portfolio with r_p = 20% that lies on the tangent line using the tangent portfolio.

The portfolio is a linear combination of the risk-free asset and the tangent portfolio. Assume that the weight in the risk-free asset is x.
Solve the following equation for x:

$$0.05 \times x + (1 - x)r_e = r_p,$$

which is equivalent to:

$$x = \frac{r_p - r_e}{r_f - r_e}$$

In order to get the weight allocation for the risky assets, simply adjust the weight in the tangent portfolio via multiplication by (1−x). We use the following codes to do this task.

Sample R-Code 6.4 Find optimal portfolio when risk-free asset exists, short sales allowed (This code needs the output of code 6.3 for rtangent, weighttangent, sigmatangent)

```
#Find optimal portfolio weight
rp = 0.20
rf1 = 0.05
rfweight = (rp - rtangent) / (rf1 - rtangent)
weight = (1 - rfweight) * weighttangent
sigmap = (1 - rfweight) * sigmatangent
weight
rp
sigmap
```

The above approach identifies the following weight allocations for (σ_p, r_p): {0.5665905, 1.6664954, −1.6347611}, where $r_p = 0.2$ and $\sigma_p = 0.3327987$. The efficient frontier is the line connecting $(0, r_f)$ and (σ_e, r_e). These results are very close to the analytic solution in Example 6.3d.

When an analytic solution to the optimization problem can be found, quadratic programming may not be needed. However, when the constraint is not an equality, such as disallowed short sales, it is necessary to use quadratic programming, detailed in the next subsection.

6.2.2 Finding the Markowitz Frontier Using Quadratic Programming When Short Sales Are Disallowed

While we can't find an analytic solution when short selling is disallowed, we can use R-code to solve the quadratic programming problem and find the Markowitz Frontier.

Example 6.6 is similar to Example 6.4 except that the short sales are disallowed.

Example 6.6: No risk-free asset is allowed or involved.

$$\vec{r} = \begin{pmatrix} 0.08 \\ 0.12 \\ 0.04 \end{pmatrix}, \Sigma = \begin{pmatrix} 0.15 & 0.08 & 0.12 \\ 0.08 & 0.20 & 0.20 \\ 0.12 & 0.20 & 0.25 \end{pmatrix}$$

Find the Markowitz Frontier portfolio with $r_p = 10\%$ without short selling. In other words, impose $w_i \geq 0$ for $i = 1, 2, 3$. Note that in the solution to Example 6.4, short-selling was allowed.

The code is very similar. The constraint matrix, *Amat*, needs to be expanded to five constraints, with three inequalities: $w_i \geq 0$ for $i = 1, 2, 3$:

$$\text{Amat} = \begin{pmatrix} 1 & r_1 & 1 & 0 & 0 \\ 1 & r_2 & 0 & 1 & 0 \\ 1 & r_3 & 0 & 0 & 1 \end{pmatrix}, \text{bvec} = (1, r_p, 0, 0, 0).$$

$$(w_1 \quad w_2 \quad w_3) \begin{pmatrix} 1 & r_1 & 1 & 0 & 0 \\ 1 & r_2 & 0 & 1 & 0 \\ 1 & r_3 & 0 & 0 & 1 \end{pmatrix} \begin{pmatrix} = \\ = \\ \geq \\ \geq \\ \geq \end{pmatrix} (1, r_p, 0, 0, 0).$$

Still specify meq=2. The first two are equality constraints while the rest default as inequalities. Here is the code to identify the efficiency frontier portfolio.

R-Code 6.5 FIND OPTIMAL PORTFOLIO WITH NO RISK-FREE ASSET, SHORT SALES NOT ALLOWED

```
#Given rp, find the optimal portfolio σp
library(quadprog)
rp = 0.10
Dmat = matrix(c(0.15, 0.08, 0.12, 0.08, 0.2, 0.2, 0.12,
0.2, 0.25), 3, 3)
dvec = c(0,0,0)
Amat = matrix(c
(1,1,1,0.08,0.12,0.04,1,0,0,0,1,0,0,0,1),3,5)
bvec = c(1,rp,0,0,0)
qp = solve.QP(Dmat, dvec, Amat, bvec, meq=2)
rp
sigmap = sqrt(2 * qp$value)
sigmap
weight=qp$solution
weight
```

The weight allocation (σ_p, r_p) is {0.5, 0.5, 0} and:

$$r_p = 0.1$$
$$\sigma_p = 0.357071$$

Similarly, the whole Markowitz Frontier if found by varying r_p:

R-Code 6.6 FIND MARKOWITZ FRONTIER WITH NO RISK-FREE ASSET, SHORT SALES NOT ALLOWED

```
#Markowitz frontier codes
library(quadprog)
n =10000
r = seq(0.04, 0.12, by = (0.12 - 0.04)/ n)
sigma = rep(0, n+1)
for(i in 1:n+1)
{
Dmat =matrix(c(0.15, 0.08, 0.12, 0.08, 0.2, 0.2, 0.12,
0.2, 0.25), 3, 3)
dvec = c(0, 0, 0)
Amat = matrix(c(1, 1, 1, 0.08, 0.12, 0.04, 1, 0, 0, 0, 1,
0, 0, 0, 1), 3, 5)
bvec = c(1, r[i], 0, 0, 0)
qp = solve.QP(Dmat, dvec, Amat, bvec, meq=2)
sigma[i] = sqrt(2 * qp$value)
}
plot(sigma, r, col = "blue")
```

Figure 6.9 depicts the Markowitz Frontier as a graph. The Markowitz Frontier is composed of several pieces of hyperbolas with some degenerated to lines.

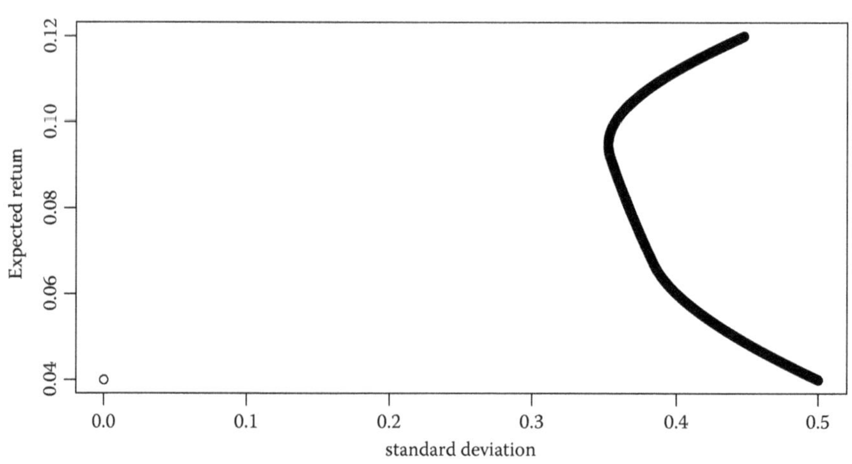

FIGURE 6.9
Markowitz Frontier without Short Selling.

Next, the case involving the risk-free asset, similar to Example 6.5, is examined. The only restriction is that short sales are not allowed.

Example 6.7: The risk-free asset has $r_f = 0.05$.

$$\vec{r} = \begin{pmatrix} 0.08 \\ 0.12 \\ 0.04 \end{pmatrix}, \quad \Sigma = \begin{pmatrix} 0.15 & 0.08 & 0.12 \\ 0.08 & 0.20 & 0.20 \\ 0.12 & 0.20 & 0.25 \end{pmatrix},$$

Find the efficient frontier portfolio with $r_p = 10\%$ and without short sales. In other words, $w_i \geq 0$ for $i = 1, 2, 3$. Also, borrowing (i.e., short selling the riskless asset) is not permitted: $\sum_{i=1}^{3} w_i \leq 1$.

The previous sections used the tangent portfolio argument to identify the Markowitz Frontier under the assumption that short selling is permitted (i.e., possible portfolio weights are not constrained to be non-negative). With short sales disallowed, the problem can be approached more directly.

Find the weight allocation $\vec{w}$ to $\underset{\vec{w}}{\text{Min}}\, ½\vec{w}^T\Sigma\vec{w}$ where

$$\vec{w}^T\vec{r} + (1 - \vec{w}^T\vec{1})r_f = r_p$$
$$w_i \geq 0 \text{ for } i = 1, 2, 3.$$
$$\textstyle\sum_{i=1}^{3} w_i \leq 1$$

Rewrite the constraints as $\vec{w}^T(\vec{r} - r_f\vec{1}) = r_p - r_f$, $w_i \geq 0$ for $i = 1, 2, 3$, $\sum_{i=1}^{3} w_i \leq 1$.

Here are the resulting constraints:

$$Amat = \begin{pmatrix} r_1 - r_f & 1 & 0 & 0 & -1 \\ r_2 - r_f & 0 & 1 & 0 & -1 \\ r_3 - r_f & 0 & 0 & 1 & -1 \end{pmatrix}, \quad \text{bvec} = (r_p - r_f, 0, 0, 0, -1).$$

$$(w_1 \quad w_2 \quad w_3)\begin{pmatrix} r_1 - r_f & 1 & 0 & 0 & -1 \\ r_2 - r_f & 0 & 1 & 0 & -1 \\ r_3 - r_f & 0 & 0 & 1 & -1 \end{pmatrix}\begin{pmatrix} = \\ \geq \\ \geq \\ \geq \\ \geq \end{pmatrix}(r_p - r_f, 0, 0, 0, -1).$$

Here is the code:

R-Code 6.7 FIND OPTIMAL PORTFOLIO WHEN RISK-FREE ASSET EXISTS, SHORT SALES NOT ALLOWED

```
library(quadprog)
rp = 0.10
rf = 0.05
Dmat = matrix(c(0.15, 0.08, 0.12, 0.08, 0.2, 0.2, 0.12,
0.2, 0.25), 3, 3)
dvec = c(0,0,0)
Amat = matrix(c(0.08-rf, 0.12 - rf, 0.04 - rf, 1, 0, 0, 0,
1, 0, 0, 0, 1, -1, -1, -1),3,5)
bvec = c(rp - rf, 0, 0, 0, -1)
qp = solve.QP(Dmat, dvec, Amat, bvec, meq=1)
rp
sigmap = sqrt(2 * qp$value)
sigmap
weight = qp$solution
weight
```

The resulting portfolio weights are {0.03454231, 0.69948187, 0} and:

$$r_p = 0.1$$
$$\sigma_p = 0.3192175$$

In Examples 6.6 and 6.7 with short sales disallowed, the weight allocation on Asset C is 0. This is not a surprising result. Asset C has a lower expected return and higher standard deviation relative to the other assets. When short sales are not allowed, it is optimal not to invest in these assets. When short sales are allowed, investors can use short positions in inferior assets to improve the risk-adjusted return.

Next, we identify the Markowitz frontier by varying r_p. Include the risk-free asset and impose that short-selling and borrowing are disallowed.

R-Code 6.8 FIND MARKOWITZ FRONTIER WHEN RISK-FREE ASSET EXISTS, SHORT SALES NOT ALLOWED

```
#Markowitz frontier codes without short sells
library(quadprog)
n = 10000
r = seq(0.04, 0.12, by = (0.12 - 0.04)/ n)
sigma = rep(0, n+1)
for(i in 1:n+1)
{
Dmat = matrix(c(0.15, 0.08, 0.12, 0.08, 0.20, 0.20,
0.12, 0.20, 0.25), 3, 3)
dvec = c(0, 0, 0)
Amat = matrix(c(0.08 - rf, 0.12 - rf, 0.04 - rf, 1, 0, 0, 0,
1, 0, 0, 0, 1, -1, -1, -1), 3, 5)
bvec = c(r[i] - rf, 0, 0, 0, -1)
qp = solve.QP(Dmat, dvec, Amat, bvec, meq=1)
sigma[i] = sqrt(2 * qp$value)
}
plot(sigma, r, col="red")
```

Figure 6.10 graphs the results without short-selling. It is not surprising that the frontier is composed of two lines with each line connecting a risky asset with the riskless asset.

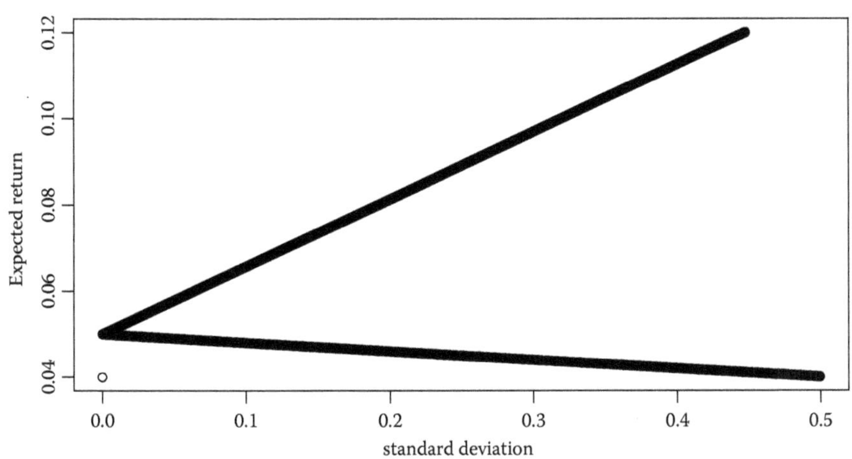

FIGURE 6.10
Markowitz Frontier with the Risk-Free Asset and without Short Selling.

6.2.3 Testing for Arbitrage Using Linear Programming

Previous sections demonstrated the application of quadratic programming to portfolio theory where the objective function is quadratic and the constraints are linear. In finance, linear programming is as important as quadratic programming when the objective function is linear and the constraints are linear. This section demonstrates an application of linear programming to identify an optimal portfolio when arbitrage opportunities are assumed to be possible.

The overview of the process is as follows: Attempt to construct a portfolio with a minimum initial cash investment (preferably negative) and a range of possible payoffs that are non-negative (preferably positive). If a portfolio can be constructed that generates immediate cash (i.e., the short sales exceed the long purchases) and if that portfolio generates nothing except non-negative potential cash flows, then the portfolio would be an arbitrage opportunity.

This application is adapted from the book *Introduction to Modern Portfolio Optimization with NuOPT, S-PLUS and S+Bayes* by Scherer and Martin. Unlike Scherer and Martin, we demonstrate programming using R. Specifically, we use the package ***"lpSolveAPI"*** in R.

Consider the following example in which both the objective function and the constraints are linear (permitting the use of linear programming). The linear programming model identifies constrained optima.

Minimize: $c_1x_1 + c_2x_2 + c_3x_3$ under the constraints:

$$a_{11}x_1 + a_{12}x_2 + a_{13}x_3 \geq b_1$$
$$a_{21}x_1 + a_{22}x_2 + a_{23}x_3 \geq b_2$$
$$a_{31}x_1 + a_{32}x_2 + a_{33}x_3 \geq b_3$$
$$a_{41}x_1 + a_{42}x_2 + a_{43}x_3 \geq b_4$$
$$a_{51}x_1 + a_{52}x_2 + a_{53}x_3 \geq b_5$$
$$a_{61}x_1 + a_{62}x_2 + a_{63}x_3 \geq b_6$$

The objective function captures the initial cash flow of establishing a portfolio with the c parameters indicating the price for each asset. If $\{c_1, c_2, c_3\} = \{1, 1, 1\}$, then price for each asset is \$1. The x's are the number of units invested in each asset. Each row of the constraints represents a different economic state (i.e., a different set of growth factors for the securities) with each a representing a growth factor on an asset and b representing the payoff. The program searches for the optimal number of portfolio units (denoted by the x's).

To use the lpSolveAPI package, first specify the number of rows and columns in the constraints. In this case there are six rows and three columns, with the six rows denoting economic states and the three columns denoting the three assets. Then input the left-side coefficients of the constraint inequalities (possible security returns) as a matrix of growth rates. The vector $\{b_1, b_2, b_3, \ldots b_6\}$ is input as the right side of the constraints, providing payoff boundaries for the six economic states.

In the following example, select $\{c_1, c_2, c_3\} = \{1, 1, 1\}$ indicating that each asset has an initial price of \$1. Therefore, $\{x_1, x_2, x_3\}$ indicates the initial allocation to the three assets in units (with negative values indicating short sales that generate immediate cash and positive values indicating long positions that require immediate cash expenditures). Therefore, $c_1x_1 + c_2x_2 + c_3x_3$ represents the total initial payment if positive or proceeds if negative. Let r_{ij} be the return of jth asset in ith economic state. So, $1 \leq i \leq 6$ and $1 \leq j \leq 3$ in the above example. Assume that the asset grows by a factor of $a_{ij} = e^{r_{ij}}$. The matrix is: $A = (a_{ij})$ in the above inequalities.

Using the package "MASS" in R, the returns r_{ij} are generated using a multivariate normal distribution with expected return and covariance matrix as follows: $\vec{r} = \begin{pmatrix} 0.08 \\ 0.12 \\ 0.04 \end{pmatrix}$, $\Sigma = \begin{pmatrix} 0.15 & 0.08 & 0.12 \\ 0.08 & 0.20 & 0.20 \\ 0.12 & 0.20 & 0.25 \end{pmatrix}$.

The economic meaning is as follows: invest $c_1x_1 + c_2x_2 + c_3x_3$ in the three assets. For example, x_i dollars is invested in asset i because the price of each asset was assumed to be \$1. Each row of the left-hand side of the constraints is constructed as three random growth factors for the three assets in each of the six economic states. Finally, $\{b_1, b_2, b_3, b_4, b_5, b_6\} = \{0, 0, 0, 0, 0, 0\}$ forces each economic state to have a non-negative end-of-period value.

The objective is to minimize the initial payment subject to the ending cash flows in all six states being non-negative. If the minimum initial payment is negative and all possible subsequent cash flows are non-negative, then there is an arbitrage opportunity. While the above framework is limited to three securities and six possible economic states, the states (rows) serve the role of providing simulations, so in practice the number of states simulated would be large.

R-Code 6.9 TEST ARBITRAGE FOR SIX RANDOM STATES

```
library(MASS)
mean = c(0.08, 0.12, 0.04)
cov = matrix(c(0.15, 0.08, 0.12, 0.08, 0.2, 0.2, 0.12,
0.2, 0.25), 3, 3)
s = exp(mvrnorm(6, mean, cov))
A = matrix(s, ncol = 3)
A
library(lpSolveAPI)
my.lp = make.lp(6, 3)
set.column(my.lp, 1, A[,1])
set.column(my.lp, 2, A[,2])
set.column(my.lp, 3, A[,3])
set.objfn(my.lp, c(1, 1,1))
set.constr.type(my.lp,  c(">=",  ">=",  ">=",  ">=",
">=", ">="))
set.rhs(my.lp, c(0, 0, 0,0,0,0))
set.bounds(my.lp, lower=c(-100, -100, -100), upper=c
(100, 100, 100),columns=c(1, 2, 3))
solve(my.lp)
get.objective(my.lp)
get.variables(my.lp)
get.constraints(my.lp)
```

A sample run generated:

A = 0.8279407 1.2542184 1.0374495
1.7014322 1.7359021 1.6447726
0.5699035 0.6337180 0.6442901
1.1728972 1.2072638 1.0663865
0.6038509 0.5654845 0.4897162
1.3402780 2.1019694 2.0037428
$X = (-47.90719, 100.00000, -55.98305)$

The above results indicate that there are six economic states. In the first state, asset one grows by a factor of 0.8279407 (losing money since the factor is less than one), asset two grows by a factor of 1.2542184 (gaining over 25%), and asset three grows by a factor of 1.0374495 (also gaining money). The solution is to short \$47.91 of asset one, purchase \$100 of asset two, and short \$55.98 of asset three. The portfolio generates an initial positive cash flow of \$3.89. (The minimum initial payment is negative: −47.90719 + 100.00000 + (−55.98305) = −3.89).

The short position in asset one is worth −\$47.91 × 0.8279407 = −\$39.77 at the end of the period, so the position can be terminated by paying \$39.77 to offset the short position. Asset two is a long position that grows to 100 × 1.2542184 = \$125.42. The short position in asset three is worth −55.98 × 1.0374495 = −\$58.08 at the end of the period, so the position can be terminated by paying \$58.08 to offset the short position. The net cash flows at the end of the periods is −39.77 + 125.42 − 58.08 = 27.68. There is an arbitrage opportunity because the strategy generates proceeds of \$3.89 at the start of the period and \$27.68 at the end of the period in state one (and no negative outcomes in other states).

Since the program randomly simulates returns, the result of the optimization will depend on the random seed in R. Note that the code provides the flexibility of allowing each asset to have a different expected return (mean) and allowing the user to specify an entire covariance matrix of returns.

Another run of the program generated a minimum at 0, so there was no arbitrage in that run. Other runs will yield a negative minimum, indicating arbitrage. The "set.seed" function in base R allows for consistency among trials.

Note that the boundary is set to [−100, 100], indicating that units cannot exceed 100 in absolute value. Without a boundary the program could go unbounded. The boundary can be changed to other numbers based on the desired maximum principal.

The output of "solve(my.lp)" is important. Here are possible outputs and their interpretations.

0: "optimal solution found"
1: "the model is sub-optimal"
2: "the model is infeasible"
3: "the model is unbounded"
4: "the model is degenerate"
5: "numerical failure encountered"
6: "process aborted"
7: "timeout"
9: "the model was solved by presolve"
10: "the branch and bound routine failed"
11: "the branch and bound was stopped because of a break-at-first or break-at-value"
12: "a feasible branch and bound solution was found"
13: "no feasible branch and bound solution was found"

An output of "0" indicates that there was a valid solution to the linear programming system.

The linear programming approach detailed in this section allows an investor to identify an optimal portfolio given a view on the expected returns and covariances for a set of securities.

The next section discusses how optimization problems integrate with the stochastic model introduced in previous chapters.

6.3 The Optimal Growth Portfolio

In the Markowitz models discussed in this chapter's previous sections, optimal portfolios were defined as those that achieved maximum expected return given a standard deviation or a minimum standard deviation given an expected return. These models do not necessarily assume any particular return distributions of the assets. However, in reality, if the return distributions depart markedly from the normal distribution, the risk of a portfolio may not be well captured by its variance because kurtosis and skewness may be major concerns. This section assumes that all n assets follow Geometric Brownian motion with a pairwise constant correlation. The objective is to maximize the portfolio's expected growth rate. We first discuss the case when there is no risk-free asset, and then discuss adding the risk-free asset. The tool for this section is the quadratic programming available using the R package.

6.3.1 The Portfolio of Maximal Growth Rate without the Risk-Free Asset

This subsection begins by formatting the quadratic programming question for a stochastic model. Assume that each of n assets follows multivariate Geometric Brownian motion with a correlation as follows:

$$dS_i = \mu_i S_i dt + \sigma_i S_i dZ_i$$
$$dZ_i dZ_j = \rho_{ij} dt$$

where
$\Sigma = (\rho_{ij}\sigma_i\sigma_j)$ and $1 \le i, j \le n$.

If the matrix Σ is known, then the diagonal elements $\{\sigma_1^2, \sigma_2^2, \ldots, \sigma_n^2\}$ and ρ_{ij} are known.

Therefore, the expected growth rate vector $\vec{\mu} = \{\mu_1, \mu_2, \ldots, \mu_n\}$ and the covariance matrix Σ determine the above stochastic processes. It is still assumed that Σ is positive definite. The instantaneous rate of return of the portfolio satisfies:

$$\frac{P}{dP} = \sum_{i=1}^{n} w_i \frac{dS_i}{S_i}$$
$$= (\sum_{i=1}^{n} w_i \mu_i)dt + \sqrt{(\overrightarrow{w}^T \Sigma \overrightarrow{w})}\, dZ$$

This can be seen by:

$$\frac{dS_i}{S_i} \sim N(\mu_i dt,\ \sigma_i^2 t)$$
$$\mathrm{Cov}\left(\frac{dS_i}{S_i}, \frac{dS_j}{S_j}\right) = \rho_{ij}\sigma_i\sigma_j dt$$

A linear combination of normal distributions is a normal distribution if the joint distribution is a multivariate normal distribution:

$$\sum_{i=1}^{n} w_i \frac{dS_i}{S_i} \sim N((\sum_{i=1}^{n} w_i \mu_i)dt,\ (\overrightarrow{w}^T \Sigma \overrightarrow{w})dt)$$
$$\frac{P}{dP} = (\sum_{i=1}^{n} w_i \mu_i)dt + \sqrt{(\overrightarrow{w}^T \Sigma \overrightarrow{w})}\, dZ$$

By Ito's lemma,

$$\mathrm{dLnP} \sim N\left((\sum_{i=1}^{n} w_i \mu_i) - \frac{1}{2}\overrightarrow{w}^T \Sigma \overrightarrow{w}\right)dt,\ (\overrightarrow{w}^T \Sigma \overrightarrow{w})dt)$$
$$\mathrm{E(dLnP)} = \left((\sum_{i=1}^{n} w_i \mu_i) - \frac{1}{2}\overrightarrow{w}^T \Sigma \overrightarrow{w}\right)dt$$

In order to identify the portfolio with maximal growth, E($dLnP$):

$$\begin{aligned} &\textbf{\textit{maximize}} \quad (\sum_{i=1}^{n} w_i \mu_i) - \frac{1}{2}\overrightarrow{w}^T \boldsymbol{\Sigma} \overrightarrow{w} \\ &\textbf{\textit{under the constraint}} \quad \sum_{i=1}^{n} w_i = 1 \end{aligned} \tag{6.12}$$

Note that the quadratic program package in R finds a constrained *minimum* not *maximum*. So, the sign of the objective function can be changed (i.e., multiplied by −1) to accomplish maximization:

$$\begin{aligned} &\textbf{\textit{minimize}} \quad -(\sum_{i=1}^{n} w_i \mu_i) + \frac{1}{2}\overrightarrow{w}^T \boldsymbol{\Sigma} \overrightarrow{w} \\ &\textbf{\textit{under the constraint}} \quad \sum_{i=1}^{n} w_i = 1 \end{aligned} \tag{6.13}$$

Assume the following data: $\overrightarrow{\mu} = \begin{pmatrix} 0.08 \\ 0.12 \\ 0.04 \end{pmatrix}$, $\Sigma = \begin{pmatrix} 0.15 & 0.08 & 0.12 \\ 0.08 & 0.20 & 0.20 \\ 0.12 & 0.20 & 0.25 \end{pmatrix}$.

Here is the R-code:

R-Code 6.10 FIND PORTFOLIO OF MAXIMAL GROWTH RATE WITH NO RISK-FREE ASSET, SHORT SALES ALLOWED

```
#weight allocation for optimal growth model
library(quadprog)
Dmat = matrix(c(0.15, 0.08, 0.12, 0.08, 0.2, 0.2, 0.12,
0.2, 0.25), 3, 3)
dvec = c(0.08, 0.12, 0.04)
Amat = matrix(c(1, 1, 1), 3, 1)
bvec = c(1)
qp = solve.QP(Dmat, dvec, Amat, bvec, meq=1)
growth = -qp$value
growth
weight = qp$solution
weight
```

The solution in "qp$solution" is the weight, w, that minimizes: $-\ w^T \begin{pmatrix} \mu_1 \\ \cdots \\ \mu_n \end{pmatrix} +$ $1/2w^T\Sigma w$ under the constraint $\sum_{i=1}^{n} w_i = 1$, which is equivalent to max $(\sum_{i=1}^{n} w_i\mu_i) - \frac{1}{2}\overrightarrow{w}^T\Sigma\overrightarrow{w}$ under the constraint $\sum_{i=1}^{n} w_i = 1$.

The calculated weight allocation is {0.9113924, 2.4177215, −2.3291139}. Note that multiplying "qp$value" by (−1) identifies the maximum. The maximal growth rate for the portfolio is 0.1496203 = 14.96%. The optimal weight allocation is {0.9113924, 2.4177215, −2.3291139}.

6.3.2 The Portfolio of Maximal Growth Rate Including the Risk-Free Asset

Next, extend the previous example to the case where a risk-free asset exists with a return of r_f. Then,

$$\frac{P}{dP} = \sum_{i=1}^{n} w_i \frac{dS_i}{S_i} + (1 - \sum_{i=1}^{n} w_i) r_f dt.$$

Adding a deterministic component $(1 - \sum_{i=1}^{n} w_i)\ r_f dt$ leads to:

$$\text{dLnP} \sim N\left(\left((\sum_{i=1}^{n} w_i\mu_i) - \frac{1}{2}\overrightarrow{w}^T\Sigma\overrightarrow{w} + (1 - \sum_{i=1}^{n} w_i)r_f\right)dt,\ (\overrightarrow{w}^T\Sigma\overrightarrow{w})dt\right)$$

$$E(\text{dLnP}) = \left((\sum_{i=1}^{n} w_i\mu_i) - \frac{1}{2}\overrightarrow{w}^T\Sigma\overrightarrow{w} + (1 - \sum_{i=1}^{n} w_i)r_f\right)dt.$$

In order to find the portfolio with maximal growth, maximize the following objective function under no constraints:

$$\left(\sum_{i=1}^{n} w_i \mu_i\right) - \frac{1}{2}\vec{w}^T \Sigma \vec{w} + \left(1 - \sum_{i=1}^{n} w_i\right) r_f,$$

This can be simplified to:

$$\left\{r_f + \sum_{i=1}^{n} w_i(\mu_i - r_f) - \frac{1}{2}\vec{w}^T \Sigma \vec{w}\right\},$$

Multiply the objective function by –1 to find the minimum.

Since $-r_f$ is a constant it is equivalent to optimize:

$$\left\{-\left(\sum_{i=1}^{n} w_i(\mu_i - r_f)\right) + \frac{1}{2}\vec{w}^T \Sigma \vec{w}\right\}$$

with no constraints.

In using the R package, set dvec = $\{\mu_i - r_f\}$.

Again using $\vec{\mu} = \begin{pmatrix} 0.08 \\ 0.12 \\ 0.04 \end{pmatrix}$, $\Sigma = \begin{pmatrix} 0.15 & 0.08 & 0.12 \\ 0.08 & 0.20 & 0.20 \\ 0.12 & 0.20 & 0.25 \end{pmatrix}$ and $r_f = 0.10$ leads to the following R-code.

R-Code 6.11 FIND PORTFOLIO OF MAXIMAL GROWTH RATE WHEN RISK-FREE ASSET EXISTS, SHORT SALES ALLOWED

```
#weight allocation for optimal growth model
library(quadprog)
rf = 0.10
Dmat = matrix(c(0.15, 0.08, 0.12, 0.08, 0.2, 0.2, 0.12,
0.2, 0.25), 3, 3)
rfv = rep(rf,3)
dvec = c(0.08, 0.12, 0.04) - rfv
Amat = matrix(c(0, 0, 0), 3, 1)
bvec = c(0)
qp = solve.QP(Dmat, dvec, Amat, bvec, meq=1)
growth = rf - qp$value
growth
weight = qp$solution
weight
```

We found that the optimal weight allocation is {0.4186047 1.8674419 −1.9348837}. The maximal growth rate for the portfolio is 0.1725 = 17.25%.

Optimization is important in portfolio management. This chapter offers a fundamental framework from a theoretical viewpoint as well as tools such as linear programming and quadratic programming from a practical viewpoint. Further applications can emphasize stochastic programming, which is beyond the scope of the book.

Chapter Summary

This chapter introduces a theoretic framework of the Markowitz portfolio theory. The set of optimal portfolios include those portfolios with the minimum variance σ_p^2 for each level of expected return r_p. When short sales are allowed, the optimal portfolio weights are derived with Lagrange multipliers and matrix algebra. In Markowitz's original work, short sales are allowed (i.e., negative portfolio weights are permissible). If short sales are not allowed, quadratic programming needs to be used to derive optimal portfolios, both with and without the risk-free asset. The chapter demonstrates the use of the R package to solve this quadratic programming problem. Also, the chapter discusses linear programming using the R package since it is useful in portfolio management. Finally, Geometric Brownian motion is assumed for asset returns and solutions are demonstrated based on the approach of finding the portfolio of maximal growth rate using quadratic programming. On the theoretical side, the CAPM is proved as an extension of Markowitz portfolio theory. In addition to sample R-codes, sample codes of analytic solution through matrix algebra in Mathematica are offered so that readers can mimic and practice. The chapter includes a project to construct the Markowitz Frontier for 10-stocks portfolio.

Chapter 6 Extensions and Further Reading

Portfolio management is a large topic both in finance and in operation research. This book has devoted only one chapter on portfolio management in order to emphasize other areas. There are many books available on portfolio management. Some are highly quantitative, while others are not. Portfolio management has a connection with data science. A major goal of this book is to help readers build a foundation both in models and in a variety of basic coding skills. Both Mathematica and R were used in this chapter.

The following are good books for further exploration of portfolio management.

1. *Online Algorithms for the Portfolio Selection Problem* by Robert Dochow
2. *Introduction to Modern Portfolio Optimization with NuOPT, S-PLUS and S+Bayes* by R. Douglas Martin and Bernd Scherer
3. *Foundations for Financial Economics*, Chi-fu Huang and Robert H. Litzenberger, Prentice Hall

End-of-Chapter Project

An Optimal Portfolio Containing 10 Stocks

This project provides a hands-on experience. Readers are asked to download real historical data to estimate expected returns and the return covariance matrix. Readers then construct optimal portfolios. Specifically, one asset is added each time, so that the change of the Markowitz Frontier can be visualized. Constructing portfolios is a standard practice in the investment industry.

Pick 10 stocks, for example General Motors, Disney, Coca-Cola, General Mills, Home Depot, Microsoft, Kellogg, Facebook, Hewlett-Packard, and Wal-Mart. Collect the daily prices for these 10 securities for the past three years and convert the prices into daily returns (e.g., Daily return = [day 2 price – day 1 price]/day 1 price). Use the data labeled as adjusted prices (to adjust for dividends and splits) which is available in the historical data section of Yahoo! Finance. Using the historical data, calculate the historic mean return vector $\vec{r}$ and historic return covariance matrix Σ for these 10 stocks. Then annualize the daily mean return and covariance matrix by multiplying each by 252 (the typical number of trading days in a year) to provide for a more intuitive scale. Use the estimated covariance matrix $\mathbf{\Sigma}$ and the estimated expected return vector $\vec{r}$ to find the Markowitz Frontiers.

a. Generate nine different frontiers with the analytic method, where the horizontal axis represents σ and the vertical axis represents r. (Make sure that you pick the correct x and y scale for σ and r so that your graph makes sense) First, generate the Markowitz Frontier for two securities General Motors and Disney. Then generate the Markowitz Frontier again by adding one more security, Coca-Cola. By repeating this process, eventually nine different Markowitz Frontiers are generated. This will demonstrate the potential diversification benefits of adding securities. While more securities improve diversification, the amount of improvement from each additional security diminishes. Historical results tend to show that around 20–30 securities will provide modest diversification benefits. However, research shows that in practice owning even 100 different securities does not provide exceptional diversification (see http://www.aaii.com/journal/article/how-many-stocks-do-you-need-to-be-diversified).

The below diagram depicts the results of the authors' data (Figure 6.11). (The data and the codes are available on a website provided by Professor Qin Lu of the Math Department at Lafayette College.)

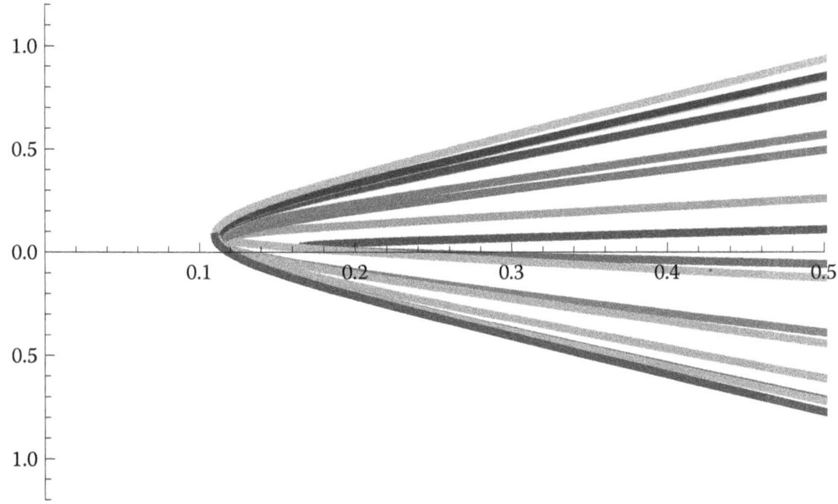

FIGURE 6.11
Evolvement of Markowitz Frontier for 10 Stocks.

b. Now consider involving all 10 stocks. Use quadratic programming to find an optimal allocation without a risk-free asset. What is the standard deviation of a portfolio to obtain a 10% expected return? What are the weight allocations? Now include a risk-free asset with a return of 2%. Find the optimal allocation with this risk-free asset. What is the portfolio's expected return and standard deviation? What are the weight allocations? Answer the above questions in the cases of short sales allowed and short sales not allowed.

c. (optional) Now assume Geometric Brownian motion growth. Use the estimated expected return and covariance matrix as μ_i and Σ to find the optimal growth of a portfolio without a risk-free asset. Then extend the results to include a risk-free asset with a return of 2%.

End-of-Chapter Problems (Fundamental Problems)

1. There are two securities: A and B. Security A has an expected return of 0.08 and a variance of 26/25; Security B has an expected return of 0.12 and a variance of 81/25; the covariance between Security A and

Security B is −41/25. (No risk-free asset is available, but short sales are allowed.)

a. Find the equation of the Markowitz Frontier for two assets.

b. Find the return and standard deviation of the minimum variance portfolio in this two-security case.

2. Consider adding one more security, C, with expected return 0.11 and variance 6/25 to the information in Problem 1. The covariance between Security C and Security A is 9/25 and the covariance between Security C and Security B is −19/25.

   a. Find the equation of the Markowitz Frontier for the three assets.

   b. Find the return and standard deviation of the minimum variance portfolio in this three-security case.

   c. Use a graph to show the two-security frontier and three-security frontier.

3. Use the data in Problem 1 and Problem 2 to answer the following questions subject to the following assumptions: all optimal portfolios must have expected return $r_p = 10\%$, there is no risk-free asset available, and short sales are allowed.

   a. Find the optimal portfolio weights with two securities using equation $r_p = w_X \times r_X + w_Y \times r_Y$. Use Theorem 6.1 to redo the optimal portfolio weights with two securities. Are the results the same? What is the standard deviation of the returns of the optimal portfolio?

   b. By Theorem 6.1, find the optimal portfolio weights using three securities. What is the standard deviation of the returns of the optimal portfolio?

   c. Using quadratic programming, find the optimal portfolio weights using three securities. What is the standard deviation of the returns of the optimal portfolio?

4. Using the data in Problems 1 and 2, consider the three-securities case. Assume that no risk-free assets are available and that short sales are allowed. Find the weight allocations in Securities A, B, C for two funds: Funds 1 and 2 as mentioned in Theorem 6.2. What are the returns and standard deviations of Funds 1 and 2? In particular, how can we represent an optimal portfolio with a return $r_p = 10\%$ as a linear combination of the two funds?

5. Return to the three securities (A, B, and C) with the given expected return vector and covariance matrix in Problems 1 and 2. Assume that there is a risk-free asset and that short sales are allowed.

a. If the risk-free asset offers $r_f = 5\%$, what is the line equation of efficient portfolios? Draw the line and Markowitz Frontier on the same graph.

b. If there is a risk-free asset that offers $r_f = 10.5778\%$, what is the line equation of efficient portfolios? Draw the line and Markowitz Frontier on the same graph.

c. If there is a risk-free asset that offers $r_f = 15\%$, what is the line equation of efficient portfolios? Draw the line and Markowitz Frontier on the same graph.

d. Use the risk-free asset $r_f = 5\%$ from part (a). What is the expected return and standard deviation of the tangent portfolio? If an expected return of 20% is required for the portfolio, what is the corresponding weight allocation? What is the tangent portfolio weight allocation? Check that the tangent portfolio weight allocation gives the corresponding return and standard deviation.

e. Use the risk-free asset $r_f = 10.5778\%$ as in part (b). What is the expected return and standard deviation of the tangent portfolio? If an expected return of 20% is required for the portfolio, what is the corresponding weight allocation?

f. Use the risk-free asset 15% as in part (c). What is the expected return and standard deviation of the tangent portfolio? If an expected return of 20% is required for the portfolio, what is the corresponding weight allocation? What is the tangent portfolio weight allocation? Check that the tangent portfolio weight allocation gives the corresponding return and standard deviation.

6. Redo problem 5d, e and f using quadratic programming.

7. Using the data in Problems 1 and 2, answer the following question assuming that a portfolio expected return $r_p = 10\%$ is required. Also assume that short sales are not allowed and that no risk-free assets are available.
Using quadratic programming, find the optimal portfolio weights using the three Securities A, B, and C. What is the standard deviation of the returns of the optimal portfolio?

8. Use the data in Problems 1 and 2. Assume that expected portfolio return, $r_p = 10\%$, is required. Assume that short sales are not allowed but that there is a risk-free asset available.

   a. Assuming $r_f = 5\%$, use quadratic programming to find the optimal portfolio weights based on the three securities: A, B, and C. What is the standard deviation of the return of the optimal portfolio?

   b. Assuming $r_f = 10.5778\%$, use quadratic programming to find the optimal portfolio weights based on the three securities: A, B, and C. What is the standard deviation of the returns of the optimal portfolio?

c. Assuming $r_f = 15\%$, use quadratic programming to find the optimal portfolio weights based on the three securities: A, B, and C. What is the standard deviation of the returns of the optimal portfolio?

9. Consider a world in which the CAPM model holds. Assume that portfolio M is the portfolio on the efficiency frontier that lies on a line that is tangent to the frontier from the riskless rate on the vertical axis. Assume that portfolio M has an expected return equal to 9% and a standard deviation of returns equal to 12% for each of the following exercises (even when the riskless rate is changed).

   a. Find the expected return of an asset with $\beta = 0.5$ assuming that the riskless rate is 4%.

   b. Find the expected return of an asset with $\beta = 0.8$ assuming that the riskless rate is 3%.

   c. Consider a portfolio comprised of two assets: Asset M and the riskless asset. Find the expected return and standard deviation of returns of the portfolio with $\beta = 0.5$ assuming that the riskless rate is 4%.

   d. Consider a portfolio comprised of two assets: Asset M and the riskless asset. Find the expected return and standard deviation of returns of the portfolio with $\beta = -0.5$ assuming that the riskless rate is 4% (allow borrowing at the riskless rate which means that the riskless asset can be short sold).

10. In the proof of Theorem 6.1, it was shown that the optimal portfolio has the following weight allocation:

$$\vec{w} = \left(\frac{c}{d}\Sigma^{-1}\vec{r} - \frac{b}{d}\Sigma^{-1}\vec{1}\right)r_p + \left(\frac{a}{d}\Sigma^{-1}\vec{1} - \frac{b}{d}\Sigma^{-1}\vec{r}\right),$$

where $a = \vec{r}^{\,T}\Sigma^{-1}\vec{r}$, $b = \vec{r}^{\,T}\Sigma^{-1}\vec{1} = \vec{1}^{\,T}\Sigma^{-1}\vec{r}$, $c = \vec{1}^{\,T}\Sigma^{-1}\vec{1}$, and $d = ac - b^2$. Insert the equation $\sigma_p^2 = \vec{w}^{\,T}\Sigma w$, and then simplify to derive the Markowitz Frontier in the case of no risk-free asset available, but short sales allowed with:

$$\frac{\sigma_p^2}{1/c} - \frac{\left(r_p - \frac{b}{c}\right)^2}{\frac{d}{c^2}} = 1.$$

11. The proof of Theorem 6.3 shows that the optimal portfolio has a weight allocation:

$$\vec{w} = \Sigma^{-1}(\vec{r} - r_f \vec{1})\lambda = \Sigma^{-1}(\vec{r} - r_f \vec{1})\frac{r_p - r_f}{E}$$

where $E = (\vec{r} - r_f \vec{1})^T \Sigma^{-1} (\vec{r} - r_f \vec{1}) = a - 2br_f + cr_f^2$.

Insert $\sigma_p^2 = \vec{w}^T \Sigma \vec{w}$ and then simplify to derive the Markowitz Frontier in the case of both the risk-free asset and short sales allowed:

$$\sigma_p^2 = \frac{(r_p - r_f)^2}{E}.$$

12. Assume Σ is a positive definite matrix, $\vec{r}$ and $\vec{1}$ are vectors as defined previously in this chapter. Let $a = \vec{r}^T \Sigma^{-1} \vec{r}$, $b = \vec{r}^T \Sigma^{-1} \vec{1} = \vec{1}^T \Sigma^{-1} \vec{r}$, $c = \vec{1}^T \Sigma^{-1} \vec{1}$, and $d = ac - b^2$. Show that $d > 0$.
13. In the proof of the CAPM model, show that $d\sigma_p / dx \,|_{x=0} = (\sigma_{i,m} - \sigma_m^2)/\sigma_m$
    where $\sigma_p^2 = x^2\sigma_i^2 + (1 - x)^2\sigma_m^2 + 2x(1 - x)\sigma_{i,m}$.
14. Assume that there are three assets. Verify that the following two statements are equivalent:
    Statement 1:

$$\text{Find } \underset{w_X, w_Y, w_Z}{\text{Min}}\ \sigma_p^2 = \underset{w_X, w_Y, w_Z}{\text{Min}}\ \{w_X^2\sigma_X^2 + w_Y^2\sigma_Y^2 + w_Z^2\sigma_Z^2 + 2w_X w_Y \sigma_X \sigma_Y \rho_{XY} + 2w_X w_Z \sigma_X \sigma_Z \rho_{XZ} + 2w_Y w_Z \sigma_Y \sigma_Z \rho_{YZ}\}$$

$$\text{where: } E(p) = w_X * r_X + w_Y * r_Y + w_Z * r_Z = \text{a known constant } r_p$$
$$w_X + w_Y + w_Z = 1$$

Statement 2:
Find $\underset{\vec{w}}{\text{Min}}\ \sigma_p^2 = \vec{w}^T \Sigma \vec{w}$
where $\vec{w}^T \vec{r} = r_p$, $\vec{w}^T \vec{1} = 1$,

$$\vec{r} = \begin{pmatrix} r_x \\ r_y \\ r_z \end{pmatrix}, \quad \Sigma = \text{covariance} = \begin{pmatrix} \sigma_X^2 & \sigma_X\sigma_Y\rho_{XY} & \sigma_X\sigma_Z\rho_{XZ} \\ \sigma_X\sigma_Y\rho_{XY} & \sigma_Y^2 & \sigma_Y\sigma_Z\rho_{YZ} \\ \sigma_X\sigma_Z\rho_{XZ} & \sigma_Y\sigma_Z\rho_{YZ} & \sigma_Z^2 \end{pmatrix}, \quad \vec{w} = \begin{pmatrix} w_x \\ w_y \\ w_z \end{pmatrix},$$

$$\vec{1} = \begin{pmatrix} 1 \\ 1 \\ 1 \end{pmatrix}.$$

15. Demonstrate that:
    a. If a multi-variable function $F(w_1, \ldots, w_n) = \frac{1}{2}\vec{w}^T\Sigma\vec{w}$, where $\vec{w} = \begin{pmatrix} w_1 \\ \cdots \\ w_n \end{pmatrix}$, and Σ is n×n matrix, then the gradient vector is $\frac{\partial F}{\partial \vec{w}} = \Sigma\vec{w}$.
    b. If a multi-variable function $G(w_1, \ldots, w_n) = \vec{w}^T\vec{r}$, where $\vec{r}$ is a constant vector, then the gradient vector is $\frac{\partial G}{\partial \vec{w}} = \vec{r}$.

16. Return to the three securities (A, B, and C) in Problems 1 and 2.
    a. Find the portfolio of maximal growth rate without a risk-free asset.
    b. Find the portfolio of maximal growth rate with a risk-free asset where $r_f = 5\%$.

End-of-Chapter Problems (Challenging Problems)

1. For any given point (σ_p, r_p) on the hyperbola (Markowitz Frontier), there is a unique portfolio with expected return r_p and the minimum standard deviation denoted as σ_p. Show that for any point (σ_p, r_p) inside the Markowitz open convex set (which excludes the frontier), there exist different weight allocations corresponding to a single point (σ_p, r_p).
2. Consider a portfolio with $n + 1$ securities.
   Denote $\vec{h} = \left(\frac{c}{d}\Sigma^{-1}\vec{r} - \frac{b}{d}\Sigma^{-1}\vec{1}\right)$ and $\vec{g} = \left(\frac{a}{d}\Sigma^{-1}\vec{1} - \frac{b}{d}\Sigma^{-1}\vec{r}\right)$. Based on Theorem 6.1, and assuming that the optimal weight allocation is $\vec{w} = \vec{h}\, r_p + \vec{g}$, and denoting $\vec{h} = \begin{pmatrix} h_1 \\ \cdots \\ h_{n+1} \end{pmatrix}$ and $\vec{g} = \begin{pmatrix} g_1 \\ \cdots \\ g_{n+1} \end{pmatrix}$:
   a. Show that if $h_{n+1} = 0$ and $g_{n+1} = 0$, then adding the n+1th security does not improve the Markowitz Frontier (including the first n securities).
   b. If $h_{n+1} = 0$ and $g_{n+1} \neq 0$, then adding the n+1th security does improve the Markowitz Frontier (including the first n securities).
   c. If $h_{n+1} \neq 0$, then adding the n+1th security improves the Markowitz Frontier, including the first n securities except one point. In other words, the Markowitz Frontier with n securities and the Markowitz Frontier with n+1 securities are tangent at one point.

3. Consider a portfolio with n+1 securities.
   Denote $\vec{h} = \left(\frac{c}{d}\Sigma^{-1}\vec{r} - \frac{b}{d}\Sigma^{-1}\vec{1}\right)$ and $\vec{g} = \left(\frac{a}{d}\Sigma^{-1}\vec{1} - \frac{b}{d}\Sigma^{-1}\vec{r}\right)$. Based

on Theorem 6.1, assuming that the optimal weight allocation is $\vec{w} = \vec{h}\, r_p + \vec{g}$, and denoting $\vec{h} = \begin{pmatrix} h_1 \\ \cdots \\ h_{n+1} \end{pmatrix}$ and $\vec{g} = \begin{pmatrix} g_1 \\ \cdots \\ g_{n+1} \end{pmatrix}$ show that $h_{n+1} = g_{n+1} = 0$ is equivalent to $(\Sigma^{-1}\vec{r})_{n+1} = (\Sigma^{-1}\vec{1})_{n+1} = 0$.

4. Consider a portfolio with n securities. Denote $\vec{h} = \left(\frac{c}{d}\Sigma^{-1}\vec{r} - \frac{b}{d}\Sigma^{-1}\vec{1}\right)$, $\vec{g} = \left(\frac{a}{d}\Sigma^{-1}\vec{1} - \frac{b}{d}\Sigma^{-1}\vec{r}\right)$, and $\vec{h} = \begin{pmatrix} h_1 \\ \cdots \\ h_n \end{pmatrix}$ and $\vec{g} = \begin{pmatrix} g_1 \\ \cdots \\ g_n \end{pmatrix}$, and assume that for all $1 \le i \le n$, and $h_i \ne 0$ we can order the list $\left\{-\frac{g_i}{h_i}\right\}$ from smallest to the largest. Using the notation $\left\{-\frac{g_{(i)}}{h_{(i)}}\right\}$ where $-\frac{g_{(i)}}{h_{(i)}}$ is the ith smallest value in the list, draw n parallel lines $y = -\frac{g_{(i)}}{h_{(i)}}$ in $\sigma - r$ plane. The lines cut the Markowitz Frontier hyperbola into n+1 segments. Show that within the ith segment of the hyperbola with $r_p \in \left(-\frac{g_{(i)}}{h_{(i)}}, -\frac{g_{(i+1)}}{h_{(i+1)}}\right)$, the sign of the weight of a particular security will never change.

5. Is there a portfolio on the Markowitz Frontier consisting of only the kth security? If so, what is the necessary and sufficient condition for this to be true?

References

[1]. Burnside, D. J., Chambers, D. R., & Zdanowicz, J. S. (2004). How Many Stocks Do You Need to Be Diversified. *American Association of Individual Investors Journal*, 6, 16–20.

[2]. Fama, E., French, K., "The CAPM is wanted, dead or alive", *The Journal of Finance* 51, 5 1996, 1947–1958.

7

Interest Rate Derivatives Modeling and Risk Management in the HJM Framework

This chapter examines modeling fixed-income securities and interest rate options in the Heath Jarrow Morton (HJM for short) model framework. More specifically, this chapter follows the one-factor HJM model as detailed in a graduate-level textbook by Jarrow (2020).

One reason to use the HJM model is that it is based on arbitrage-free pricing and uses a complete market method. Another reason is that the model provides an excellent application of the binomial tree pricing method. The third reason is that the HJM approach is popular in the financial industry even though it is not often detailed in undergraduate financial mathematics textbooks. Jarrow (2020) asserts that this "... new approach to fixed income securities has revolutionized the industry. The HJM model is employed by commercial and investment banks to price and hedge numerous types of fixed income securities and interest rate options."

The fourth reason is that many popular interest rate models can be viewed as special cases (or generalizations) of HJM models. For example, the Hull-White model and the Ho-Lee model are special cases of HJM models. Also, the LIBOR market model (LMM) dynamics can be placed in the HJM model framework. One limitation of HJM models is that they can be computationally intensive. A potential drawback is that HJM models are expressed in terms of instantaneous forward rates that are not directly observable in the market and which therefore introduce complications in calibrating parameters. It should be noted that the LMM is related to observed products in the market. Finally, note that the HJM model provides the foundation of the LMM model.

This chapter summarizes the main ideas of the HJM model in general and Jarrow (2020) in particular into a single chapter that is accessible to undergraduate students.[1] Although Monte Carlo simulation based on stochastic calculus is often used to price financial derivative products of interest in the HJM model framework due to the challenges of non-recombining tree nodes, Jarrow (2020) presents the HJM model in a binomial tree setting in order to make the model more accessible to readers. The key feature of this chapter is that it teaches the tree methods using EXCEL spreadsheets because they are so powerful in visualizing the processes and concepts and because they facilitate student learning through hands-on activities. The key benefits to learning the

HJM model through EXCEL spreadsheets are that it is easier to learn and it is easy to expand the time horizon. However, the spreadsheet approach is very awkward and space-consuming (in the context of HJM modeling) when used to form large, non-recombining trees. Thus, although the use of EXCEL spreadsheets in this chapter is valuable as a pedagogical tool, it is not useful as a valuation tool for analyses requiring 15 or more periods since non-recombining trees expand exponentially with the number of time steps.

Previous chapters in this book discuss the fundamental models to value financial derivatives. In presenting these fundamental models it was assumed that there was a single interest rate, r, that was constant. However, the assumption of a single, constant interest rate causes a fundamental problem or model error if the derivative's underlying price, S_t, is interest rate sensitive such as being a bond (or if S_t is an interest rate). Even if the underlying asset price S_t is not directly related to interest rates, if the derivative's maturity is long, its value will depend on the time value of money and so it is erroneous to assume that r is constant.

The previous chapters briefly distinguished forward contract prices and futures contract prices. If the underlying asset price's movement has no correlation with interest rate movements, then forward prices must equal futures prices. Obviously futures prices will equal forward prices when r is constant since a constant interest rate has zero correlation with any asset price. However, when considering forwards and futures on bonds it is unreasonable to assume that interest rates are constants. Since the movement of bond prices and interest rates are negatively correlated, there is a divergence between the price of a forward contract on a bond and price of an otherwise identical futures contract on the same bond.

As mentioned by Jarrow (2020), corporate bonds and municipals bonds are highly correlated with Treasury bonds; their prices and yields usually move in a similar fashion. As with Jarrow's approach, this chapter builds models based on Treasury/LIBOR rates which can be generalized to corporate rates and municipal rates.

This chapter begins by demonstrating the evolvement of zero coupon bonds, forward rates, short rates, and money market funds and introduces risk-neutral pricing in a binomial tree setting. The chapter then replicates bonds in an arbitrage-free framework. After that, the chapter values options on bonds, swaps, caps, and floors, forwards, and futures on bonds. In all derivative pricings, only two trees are necessary in the one-factor HJM model: one zero coupon bond price evolvement tree and the short-rate tree. Note that the tree here only includes the tree node values and does not include the probabilities for up and down nodes. Finally, the basic calibration of the HJM model parameters based on market data is discussed.

The next section overviews relations between zero coupon bonds, forward rates, short rates, and money markets in a deterministic world and then introduces the evolvement of zero coupon bonds, forward rates, short rates, and money markets as stochastic processes in a binomial tree setting.

7.1 Zero Coupon Bonds, Forward Rates, Short Rates, and Money Markets in a Deterministic World

To introduce notation and provide a foundation for analysis, consider fixed-income valuation in a deterministic world. Various textbooks and other materials use different notations and conventions for interest rates. This section summarizes the notations and conventions that are used later in the chapter to build up the HJM models.

In Chapter 2, the T-year risk-free bond price today ($t = 0$) was denoted as $B(0, T)$. Chapter 2 did not distinguish the notation between a zero coupon bond and a coupon bond. In this chapter, a T-year zero coupon bond price today is denoted as $P(0,T)$, a T-year coupon bond price today (where the coupon C is constant) is denoted as CB(0,T). The coupon per period expressed as a dollar value is denoted as C and is usually paid semiannually unless otherwise specified.

Denote r_{t_i} as the zero coupon rate or interest rate for the period $[0, t_i]$ where $t_i \in \{t_0 = 0, t_1, \ldots, t_{n-1}, t_n = T\}$. For example, denote $r_{3.5}$ as interest rate for the period [0, 3.5]. The formulas for the value of a zero coupon bond, $B(0, t)$, and a coupon bond, CB(0, T), both with a face value of F are:

$$B(0, t) = Fe^{-r_t \times t}$$
$$\mathbf{CB}(0, T) = C\sum_{i=1}^{n} e^{-r_{t_i} t_i} + Fe^{-r_T T}$$

In this chapter, we assume $F = 1$. Inserting $F = 1$ into the first equation above produces Equation 7.1. Equation 7.1 expresses the value of a zero coupon bond with a face of \$1 and maturity in t years as ***P(0, t)***. Note that $P(0,0) = 1$ and that $P(0, t)$ can be viewed as a discount factor that discounts \$1 due in t periods back to its value at time 0 (today). Inserting $F = 1$ and discount factor $P(0, t_i)$ into second equation above, we get Equation 7.2. Equation 7.2 expresses the value of a T-year coupon bond price at time 0 as CB(0,T), where the ***T-year coupon bond CB(0,T)*** pays a fixed coupon periodically and a final principal payment at time T. Here CB(0, T) is the value of a coupon bond with a face value \$1 which pays a coupon dollar value C every half year. In fact, CB(0, T) can be viewed as a portfolio of zero coupon bonds.

$$P(0, t) = e^{-r_t \times t} \tag{7.1}$$

$$CB(0, T) = C\sum_{i=1}^{n} P(0, t_i) + 1 \times P(0, T) \tag{7.2}$$

where $t_i \in \left\{\frac{1}{2}, 1, \ldots, T - \frac{1}{2}, T\right\}$

In Chapter 2, f_{T_1,T_2} was the forward rate over the period $[T_1, T_2]$. In other words, it is the interest rate for the period $[T_1, T_2]$, viewed at time t before time T_1.

In particular, f_{T_1,T_2} represents today's ($t = 0$) forward rate for the period $[T_1, T_2]$. In this chapter greater flexibility is achieved by defining ***forward rates*** $f(t, T_1, T_2)$ in which three time parameters are used: t, T_1, and T_2 and the first parameter, t, indicates the point in time that the forward rate is being expressed or observed and time period $[T_1, T_2]$ is when the interest rate applies. If it is clear that the time period over which the forward rate applies (T_1 to T_2) is fixed as one time period (e.g., year or Δ), the parameter T_2 can be omitted so that ***forward rate*** $f(t, T_1)$ is the interest rate for the period $[T_1, T_1 + 1]$ *or* $[T_1, T_1 + \Delta]$ viewed at time t (which occurs before time T_1).

There is another potential source of confusion about the notation for forward rates. Some books use the forward rate $f(t, T_1)$ as the ***instantaneous forward rate***: $f(t, T_1) = \lim_{T_2 \to T_1} f(t, T_1, T_2)$. Thus, it is important to remember that in this chapter, $f(t, T_1)$ is not today's forward rate over the period $[t, T_1]$ nor the instantaneous forward rate; it is the forward rate for a period of one year or Δ year starting at time T_1 and observed at time t.

In the remaining material for this chapter, there is a very important expositional change that is made in order to simplify the mathematics: interest rates and forward rates henceforth are expressed as wealth ratios rather than changes. For example, in Equation 7.1, a continuously compounded interest rate such as $r_t = 5\%$ is inserted into a function such as $e^{-r_t \times t}$ in order to form a discount factor. In this chapter's remaining material, interest rates and forward rates are often expressed as the sum of one plus the rate. So in much of the material that follows (and using continuous compounding of 5% to illustrate): $r = 1.05127$ for discrete compounding and $\tilde{r} = 0.05$ for continuous compounding. Therefore a simple one-year discrete discounting of a cash flow, say C, will be simplified to C/r rather than $C/(1 + r)$ or $Ce^{-\tilde{r}}$. This approach will be applied to both spot rates and forward rates.

Next, consider the formalization of this expositional change in Equation 7.3. In Chapter 2, the forward rate convention was continuous compounding. However, and consistently with Jarrow's textbook, this chapter uses discrete-time forward rates. As discussed in the previous paragraph, the forward rate, $f_\Delta(t, T_1)$, is defined as one plus the non-annualized discrete rate at period $[T_1, T_1 + \Delta]$ observed at t. It is set equal to the value – based on continuous compounding – as shown in Equation 7.3 and noting that $\tilde{f}(t, T_1)$ is expressed as a rate rather than as a wealth ratio.

$$f_\Delta(t, T_1) = e^{\tilde{f}(t, T_1) \times \Delta} \tag{7.3}$$

where the ***continuously compounded forward rate per year,*** $\tilde{f}(t, T_1)$, is quoted for the period $[T_1, T_1 + \Delta]$ at time t and $f_\Delta(t, T_1)$ is the discrete-time forward rate (expressed as a wealth ratio) in period $[T_1, T_1 + \Delta]$ also viewed at time t. Since $\tilde{f}(t, T_1)$ is always positive, $f_\Delta(t, T_1)$ is always greater than 1.

Note that $f_\Delta(t, T_1)$ is not an annualized rate unless Δ is 1. Therefore, the corresponding discount factor is $1/f_\Delta(t, T_1)$. It discounts $1 from time $T_1 + \Delta$ to T_1 as observed at time t.

In the case that Δ is one year, the sub-index, Δ, may be dropped from Equation 7.3. So the ***forward rate f(t, T)*** is the forward rate (wealth ratio) for time period $[T, T+1]$ observed at time t and $f(2, 3)$ is the forward rate for the time period [3, 4] based on the time value of money as observed at time $t = 2$.

In all remaining sections in this chapter except the last one, it is assumed that $\Delta = 1$. All formulas can be extended to an arbitrary period Δ easily.

First, observe the relation between $P(0, T)$ and $f(0, s)$ where $0 \le s < T$.

$$P(0, T) = \frac{1}{f(0, 0)f(0, 1)\dots f(0, T-1)} \tag{7.4}$$

Note that by defining forward rates as wealth ratios, the expression in the denominator of Equation 7.4 can be conveniently expressed as a simple product rather than having to be written as: $(1 + f(0, 0)) \times (1 + f(0, 1)) \dots (1 + f(0, T-1))$.

The ***T-year zero coupon bond price at time t, P(t, T),*** denotes the price of a bond with a face of $1 and maturity T years observed at time t. Note that $P(t, T)$ can also be viewed as a discount factor that discounts $1 at T to time t.

In addition, we have a relation between $P(t, T)$ and $f(t, s)$ where $t \le s < T$.

$$P(t, T) = \frac{1}{f(t, t)f(t, t+1)\dots f(t, T-1)} \tag{7.5}$$

It is easy to deduce that:

$$f(t, T) = P(t, T)/P(t, T+1) \tag{7.6}$$

In Equation 7.5, zero coupon interest rates (r_t) are not used or needed. The economic importance of Equations 7.5 and 7.6 is that the mathematics of the time value of money is being performed based on forward rates rather than zero coupon interest rates.

The direct modeling of values using forward rates rather than zero coupon interest rates is a key innovation within the HJM approach. In order to model the interest rates evolvement, the ***short rates r(t)*** (or spot rates, which have not been defined in prior sections) for all prospective points in time are defined as forward rates:

$$r(t) = f(t, t) \tag{7.7}$$

where $r(t)$ is the one-period interest rate (expressed as a wealth ratio) subsequently prevailing for the beginning of period t. For textbooks using continuous time rather than discretized time, $r(t)$ would be an instantaneous short rate in which Δ goes to 0.

Finally, define a ***money market fund m(T)***, as shown in (7.8).

$$m(T) = 1 \times r(0) \times r(1) \times \ldots \times r(T-1) \tag{7.8}$$

Equation (7.8) may be viewed as the accumulated value at time T of \$1 reinvested each period at the then-prevailing short rate, $r(t)$, from period 0 to period T. Note that in practice a money market fund grows period by period into the future from time $t = 0$ to time T. Thus, $m(T)$ stands for the money market fund's value after T years starting from \$1.

Note that we generally cannot directly observe $P(t, T)$, $f(t, T)$, $r(t)$, $m(t)$ today (i.e., $t = 0$) because the values are from the perspective of a future point in time (t). The values of $P(t, T)$, $f(t, T)$, $r(t)$, and $m(t)$ are generally estimated from other current market data based on sophisticated mathematical models such as HJM models. However, $P(0, s)$, $f(0, s)$ where $0 \le s < T$, $P(0, T)$, $r(0)$, and $m(1)$ can be readily observed in markets through simple mathematical models such as stripping coupon bonds.

Having defined zero coupon bonds, $P(t, T)$, forward rates, $f(t, T)$, short rates, $r(t)$, and money market fund accumulations, $m(t)$, next consider the relations in a deterministic world through the following example. All previous formulas can be generalized easily in a non-deterministic world by adding a state variable, s_t, which represents the up/down state in a binomial tree setting. For example, in the above formulas, $P(t, T)$ changes to $P(t, T, s_t)$ and $f(t, T)$ to $f(t, T, s_t)$. However, the following two formulas about liquidity premiums $L_1(t, T)$ and $L_2(t, T)$ are not zero empirically in the real non-deterministic world using the real-world probability measures.

Example 7.1: Assume that the time period is $\Delta = 1$ and that all interest rates are deterministic in each period so that there is no risk and no-risk premium.

Using a four-year model, what is the minimal set of information needed to obtain all information in the set:

$$\Omega = \{P(0,1), P(0,2), P(0,3), P(0,4), P(1,2), P(1,3), P(1,4), P(2,3), P(2,4), P(3,4), f(0,0), f(0,1), f(0,2), f(0,3), f(1,1), f(1,2), f(1,3), f(2,2), f(2,3), f(3,3), r(0), r(1), r(2), r(3), m(1), m(2), m(3), m(4)\}?$$

The answer for the minimal set is not unique. Following are just three examples:

$$A = \{r(0), r(1), r(2), r(3)\}$$

$$B = \{P(0, 1), P(0, 2), P(0, 3), P(0, 4)\}$$

$$C = \{f(0, 0), f(0, 1), f(0, 2), f(0, 3)\}$$

It is not surprising that all three sets contain an equal number of elements.
First, removing redundant information from Ω, it is clear that:

$$r(0) = f(0, 0),\ r(1) = f(1, 1),\ r(2) = f(2, 2),\ r(3) = f(3, 3).$$

So, $\{r(0), r(1), r(2), r(3)\} = \{f(0, 0), f(1, 1), f(2, 2), f(3, 3)\}$.

and it is not hard to see that $\{r(0), r(1), r(2), r(3)\} \cong \{m(1), m(2), m(3), m(4)\}$
where we define "$\cong$" as an equivalent relation which implies they can be deduced from each other.
Given $r(0)$, $r(1)$, $r(2)$, $r(3)$, it is easy to find $m(1) = r(0), \ \ldots,\ m(4) = r(0) \times r(1) \times r(2) \times r(3)$.
Conversely, given $m(1), m(2), m(3), m(4)$, it is easy to find $r(0) = m(1)$, $r(1) = m(2)/m(1), \ldots, r(3) = m(4)/m(3)$.
After redundant information is removed from Ω:

$$\Omega \cong \{P(0, 1), P(0, 2), P(0, 3), P(0, 4), P(1, 2), P(1, 3), P(1, 4), P(2, 3), P(2, 4), P(3, 4), f(0, 0), f(0, 1), f(0, 2), f(0, 3), f(1, 1), f(1, 2), f(1, 3), f(2, 2), f(2, 3), f(3, 3)\} = \Omega_1$$

Next, remove more redundant information, define

$$\Omega_P = \{P(0, 1), P(0, 2), P(0, 3), P(0, 4), P(1, 2), P(1, 3), P(1, 4), P(2, 3), P(2, 4), P(3, 4)\}$$

$$\Omega_f = \{f(0, 0), f(0, 1), f(0, 2), f(0, 3), f(1, 1), f(1, 2), f(1, 3), f(2, 2), f(2, 3), f(3, 3)\}$$

Since $P(t, T) = \frac{1}{f(t,t)f(t,t+1)\ldots f(t,T-1)}$, then $\Omega_f \to \Omega_P$.
Since $f(t, T) = P(t, T)/P(t, T+1)$, then $\Omega_P \to \Omega_f$
So, $\Omega_P \cong \Omega_f \cong \Omega_1 \cong \Omega$.
We have reduced Ω with 28 elements to Ω_P or Ω_f with 10 elements.
Next show B $\cong$ C.
Given $P(0,1)$, $P(0,2)$, $P(0,3)$, $P(0,4)$, we have $f(0,0) = P(0,0)/P(0,1)$,

$f(0, 1) = P(0, 1)/P(0, 2)$, $f(0, 2) = P(0, 2)/P(0, 3)$, $f(0, 3) = P(0, 3)/P(0, 4)$

Conversely, $P(0, 1) = 1/f(0, 0)$, $P(0, 2) = 1/(f(0, 0) \times f(0, 1))$, $P(0, 3) = 1/(f(0, 0) \times f(0, 1) \times f(0, 2))$, $P(0, 4) = 1/(f(0, 0) \times f(0, 1) \times f(0, 2) \times f(0, 3))$

Note that so far the result in a deterministic world can be generalized easily to a non-deterministic world. However, the following argument needs the assumption of a deterministic world with no risk and no-risk premium.

Show $A \cong \Omega_P$.

There are two kinds of risk premiums $L_1(t, T)$ and $L_2(t, T)$:

$$L_1(t, T) = \frac{P(t+1, T)}{P(t, T)} - r(t)$$

$$L_2(t, T) = f(t, T) - r(T)$$

The ***risk premium*** $\boldsymbol{L_1(t, T)}$ is the premium at time t of investing in zero coupon bonds with maturity T. In the real world, a longer maturity, T, would usually offer a higher-risk premium. $L_1(t, T)$ is generally not 0.

Risk premium $\boldsymbol{L_2(t, T)}$ states the relationship between the forward rates and future spot rates. In a real world, $L_2(t, T)$ is generally not 0.

Since this is a deterministic world, there is no risk and hence no-risk premium. If there is no risk premium $L_1(t, T)$,

$$\frac{P(t+1, T)}{P(t, T)} - r(t) = 0$$

The quick corollaries are:

$$P(t, T) = \frac{P(t+1, T)}{r(t)}$$

$$P(t, T) = \frac{1}{r(t)r(t+1)\ldots r(T-1)}$$

Therefore, A can be used to generate Ω_P. Note that $A \cong \Omega_P \cong \Omega_A \cong \Omega_1 \cong \Omega$.

Finally, show $A \cong C$.

Show that if A can be generated given C and the reverse is true (A generates Ω).

Here the assumption of a deterministic world is needed so that there is no risk and hence no-risk premium $L_2(t, T)$. Therefore, risk premium $L_2(t, T) = 0$.

$$f(t, T) - r(T) = 0$$

$$f(t, T) = r(T)$$

Given $f(0, 0), f(0, 1), f(0, 2), f(0, 3)$, we have $r(0) = f(0, 0)$, $r(1) = f(0, 1)$, $r(2) = f(0, 2)$, $r(3) = f(0, 3)$.

There are some properties that a deterministic world and a non-deterministic share and some properties that they do not share.

Remarks:

Similarity:

For fixed T, if all $P(t, s)$ where $0 \leq t < s \leq T$ or all $f(t, s)$ where $0 \leq t \leq s < T$ have been modeled, then all $P(t, s)$, $f(t, s)$, $r(t)$ and $m(t)$ can be modeled.

Differences:

1. In a non-deterministic world, the liquidity premiums $L_1(t, T)$ and $L_2(t, T)$ are not 0 if expressed as a real probability measure. However, in a risk-neutral world, both liquidity premiums are 0. Note that the equations for the two-risk premiums in a non-deterministic world are:

$$L_1(t, T) = E_t\left(\frac{P(t+1, T)}{P(t, T)}\right) - r(t)$$

$$L_2(t, T) = f(t, T) - E_t(r(T))$$

2. The short rates $r(t)$ can be used to determine the value of all zero coupon bond prices, $P(t, s)$, forward rates, $f(t, s)$, and money market funds, $m(t)$, in a deterministic world. This is not true in a non-deterministic world where the short rates on the nodes together with its risk-neutral probability measure can be used to find the remaining values.

The next section explores a non-deterministic world and will demonstrate that knowing the evolvement of short rates $r(t)$ alone is not generally sufficient to value of all bonds, forward rates, and money market returns.

7.2 Risk-Neutral Valuation in a One-Factor HJM Model

The one-factor HJM model is a continuous-time model in HJM's origination. Jarrow presents the model completely using a binomial tree format in his book. This chapter adopts and adapts Jarrow's method in a manner that makes it highly accessible to undergraduate students through EXCEL spreadsheets.

In order to present the risk-neutral HJM model, we first discuss the evolvement of zero coupon bonds, forward rates, short rates, and money markets in a binomial trees setting.

7.2.1 Evolvement of Zero Coupon Bonds, Forward Rates, Short Rates, and Money Market Funds

This section uses a binomial tree to model the zero coupon bond price evolvement as well as the evolvement of forward rates, short rates, and money market fund values.

Consider a four-year time period with a time interval $\Delta = 1$. Four years and $\Delta = 1$ are selected because on one hand it is not too trivial and on the other hand $\Delta = 1$ can be generalized easily to an arbitrary time interval Δ and the number of time periods can be generalized easily to any positive integer n.

Given a four-year zero coupon bond with today's value denoted as $P(0, 4)$, the value can have the following evolvement as visualized using the bond price tree, $P(t, 4)$:

$t = 0$	1	2	3
$P(0, 4)$	$P(1,4, u)$	$P(2,4,uu)$	$P(3,4,uuu)$
			$P(3,4,uud)$
		$P(2,4,ud)$	$P(3,4,udu)$
			$P(3,4,udd)$
	$P(1,4, d)$	$P(2,4,du)$	$P(3,4,duu)$
			$P(3,4,dud)$
		$P(2,4,dd)$	$P(3,4,ddu)$
			$P(3,4,ddd)$

where u is the up state and d is the down state.

Note that unlike a binomial tree in stock price evolvement, interest rate trees in HJM models do not have recombining nodes. Importantly, the tree has 2^{T-1} nodes in the final column.

When $t = 1$, the four-year zero coupon bond (with three years remaining to maturity) has two possible prices: $P(1,4, u)$ in an up state and $P(1,4, d)$ in a down state. The last column of the tree contains $2^{T-1} = 2^3$ elements. At $t = 4$, $P(4,4) = 1$. For EXCEL spreadsheet purposes, the tree occupies $2^{T-1} \times T$ cells. There are 2^{T-1} rows and T columns starting from column 0 to column $T - 1$. Note that in the tth column, the tree contains 2^t elements. As shown above, the cells with values are spread evenly among the 2^{T-1} available cells. For example, at $t = 2$, there are $2^2 = 4$ elements, which are spread evenly among the $2^3 = 8$ cells, leaving one empty space between each pair of values. This structure is used throughout the remainder of this chapter.

The following illustration is the evolvement of the bond price $P(t,3)$ tree, $P(t,2)$ tree, and $P(t,1)$ tree for three-, two-, and one-year zero coupon bonds, respectively. To be consistent, we term $P(t,1)$ a tree even though it has only one element.

$t = 0$	1	2	$t = 0$		$t = 0$
$P(0, 3)$	$P(1,3, u)$	$P(2,3,uu)$	$P(0,2)$	$P(1,2, u)$	$P(0,1)$
		$P(2,3,ud)$			
	$P(1,3, d)$	$P(2,3,du)$		$P(1,2, d)$	
		$P(2,3,dd)$			

Note that the same space structure is retained as in the case of the $P(t, 4)$ tree to make copying and pasting easy.

To provide clarity regarding notation: In $P(t, s)$, the first variable, t, is the time evolvement and the second variable, s, is the maturity. So for a particular zero coupon bond, maturity s is fixed first. Then price evolvement is discussed at time t before s. For example $P(2,3,uu)$ is the price of a three-year zero coupon bond in year 2 if the state is uu.

Similarly, the forward rate $f(t, 3)$ tree, $f(t, 2)$ tree, $f(t, 1)$ tree, and $f(0, 0)$ tree are as follows:

$t = 0$	1	2	3
$f(0, 3)$	$f(1,3, u)$	$f(2,3,uu)$	$f(3,3,uuu)$
			$f(3,3,uud)$
		$f(2,3,ud)$	$f(3,3,udu)$
			$f(3,3,udd)$
	$f(1,3, d)$	$f(2,3,du)$	$f(3,3,duu)$
			$f(3,3,dud)$
		$f(2,3,dd)$	$f(3,3,ddu)$
			$f(3,3,ddd)$

$t = 0$	1	2	$t = 0$	1	$t = 0$
$f(0,2)$	$f(1,2, u)$	$f(2,2,uu)$	$f(0,1)$	$f(1,1, u)$	$f(0,0)$
		$f(2,2,ud)$			
	$f(1,2, d)$	$f(2,2,du)$		$f(1,1, d)$	
		$f(2,2,dd)$			

From Example 7.1 or directly from Equations 7.5 and 7.6, it is easy to verify that the four-bond price evolvement trees are equivalent to four-forward-rate evolvement trees. For example, $f(1, 3, u) = \frac{P(1,3,u)}{P(1,4,u)}$, $f(1, 2, u) = \frac{P(1,2,u)}{P(1,3,u)}$, $f(1, 1, u) = \frac{P(1,1)}{P(1,2,u)}$. Reversely, $P(3, 4, uuu) = \frac{1}{f(3,3,uuu)}$. The details are left for readers to check in exercises.

Here is the short-rate $r(t)$ tree and money market fund $m(t)$ tree:

$t = 0$	1	2	3	$t = 0$	1	2	3	4
$r(0)$	$r(1, u)$	$r(2,uu)$	$r(3,uuu)$	$m(0)$	$m(1)$	$m(2,u)$	$m(3,uu)$	$m(4,uuu)$
			$r(3,uud)$					$m(4,uud)$
		$r(2,ud)$	$r(3,udu)$				$m(3,ud)$	$m(4,udu)$
			$r(3,udd)$					$m(4,udd)$
	$r(1, d)$	$r(2,du)$	$r(3,duu)$			$m(2,d)$	$m(3,du)$	$m(4,duu)$
			$r(3,dud)$					$m(4,dud)$
		$r(2,dd)$	$r(3,ddu)$				$m(3,dd)$	$m(4,ddu)$
			$r(3,ddd)$					$m(4,ddd)$

Based on (7.8), it is easy to construct the $m(t)$ tree from the $r(t)$ tree (noting that $m(0) = 1$). We demonstrated a convention here in the binomial tree set: The ***parent node*** implies the values one column backward if all trees are lined up in a vertical format by time t or column t. From the spreadsheet, any m-cell is constructed from the parent m-cell multiplied by the corresponding r-cell in the same column:

$$m(1) = m(0) \times r(0),\ m(2, u) = m(1) \times r(1, u),\ \ldots, m(4, ddd) = m(3, dd) \times r(3, ddd).$$

Note that in this chapter, when we mention a tree, it only carries the node values which do not include the probability measures for up and down states.

Defining the set of all the above 10 trees as Ω, similar as in the deterministic world, the following four equivalent sets with five trees in each set can be used to generate Ω: $A \cong B \cong C \cong D \cong \Omega$:

A = {four bond price trees, short rate tree}
B = {four bond price trees, money market fund tree}
C = {four forward rate trees, short rate tree}
D = {four forward rate trees, money market fund tree}

At first glance it may appear that the short rate trees with node values, $r(t)$, can be used to build the bond price tree, $P(t, T)$, based on the following equations related to a liquidity premium:

$$P(t, T) = \frac{E_t(P(t+1, T))}{r(t)}$$

$$P(t, T) = E_t\left(\frac{1}{r(t)r(t+1)\ldots r(T-1)}\right).$$

However, there are two fundamental problems in applying the above equations in the case of a non-deterministic world. First, what is the probability measure used to calculate the expected values? If estimated real probabilities are used, the equations fail because of non-zero risk premiums in a real world. Recall in Chapter 3 that a risk-neutral probability measure avoids the problem of specifying a non-zero risk premium. However, the second problem is that it requires extra parameters to calculate risk-neutral probabilities, presumably derived from market data. Given these problems, the short-rate tree $r(t)$ node values alone cannot be used to value all bond prices, $P(t, T)$.

How many trees are needed? Are all five trees in the above set A (or B or C or D) needed to derive Ω? Before answering these questions, consider the following simple mathematics:

To exactly fit a line, two points need to be specified since a line has two parameters. Namely, this problem requires two equations. To exactly fit a second-degree polynomial requires the input of three points since a degree two polynomial has three parameters (so three equations are needed). To exactly fit a degree three polynomial requires four points since a degree three polynomial has four parameters – requiring four equations.

This is indeed the answer for HJM models that have a number of factors selected by the choice of the researcher. Thus HJM models can be one-factor models, two-factor models, etc. *A* one-factor HJM model can use a binomial tree format and a two-factor HJM model can use a trinomial tree format. In a trinomial format each parent node has up, middle and down states. An N

factor HLM model will have $N + 1$ states. Since the number of factors determines the number of the parameters, it also determines the number of trees that need to be inputted. Since this is an undergraduate textbook, we will focus on a one-factor HJM model. The method is not hard to generalize from one factor to multifactor. For multi-factor models, readers can consult Jarrow's book for a brief discussion and his papers for details.

The next section assumes a binomial model for evolvement similar to that which was constructed in this section which is consistent with a one-factor HJM model. It discusses which trees are used as inputs and what is the minimum set of trees which can replicate other trees.

7.2.2 HJM One-Factor Model

Since this chapter's approach avoids the complexity of the martingale property, the discussion of theory in a one-factor HJM model is limited. A one-factor HJM model with a less strict theoretical model framework can be found in Jarrow's book. A one-factor HJM model with a strict theoretical model framework can be learned through a continuous-time asset pricing approach based on measure theory.

Recall the following convention in this book: When the material references one tree's node and the parent node in another tree, it simply means: (1) move one column back to find the parent node, and (2) then move to the analogous location in another tree. A similar convention applies to one tree's node and the child node in another tree.

Our approach begins with a review of the binomial tree approach of derivative pricing in Chapter 3 with a stock as the underlying asset. In the CRR tree, there are three parameters: a "u" or up factor, a "d" or down factor, and the risk-neutral probability $Q = \frac{e^{r\Delta t} - d}{u - d}$. The stock tree is formed using S_0, u, d. Then for any payoff, a risk-neutral probability Q and a constant interest rate r are used to roll back the tree one time step at a time until the value of the derivative is found in the leftmost node. So, for derivative pricing purposes, the risk neutral probability Q and the short rate r are the only inputs needed.

We use a parallel strategy in building a one-factor HJM model. A bond price evolvement, namely a $P(t, T)$ tree, is used as the underlying asset tree – where T is the largest time maturity in all bond trees. Clearly, the interest rate, r, is not a constant in the HJM model. A short rate $r(t)$ tree is used as the discount tree. Our questions are:

- Is one tree with only node values adequate?
- Are two trees with only node values adequate?
- What is the economic implication of this decision?

If there is one tree, the short rate tree of $r(t)$ with only node values, there is not enough information to determine the risk-neutral probabilities Q, which we use as probability weights to discount back. Based on the above formula in early chapters, the risk-neutral probabilities Q depend not only on r but also on u and d which are not functions of the short rates, they are functions of the underlying asset $P(t, T)$. So a second tree is needed to determine risk-neutral probabilities. Consider the following claim:

> *Two trees are needed in derivative pricing in the one-factor HJM model: the P(t, T) tree and the r(t) tree. Here, these trees only include the node values of the trees.*

Let's explore the claim that two trees with node values, the $P(t, T)$ tree and the $r(t)$ tree, are adequate to price any financial products with payoffs at time T. The following material uses a four-period tree to justify the claim.

Note that for derivative pricing purposes, the risk-neutral probability tree $Q(t)$ and short-rate $r(t)$ tree are the only inputs needed. The $Q(t)$ tree will be built through the $P(t, T)$ tree.

The $P(t,4)$ tree can be used to illustrate the process of finding the u and d evolvement tree. By taking the quotient of a child node and the parent node:

$t = 0$	1	2	3
	$u(0, 4)$	$u(1,4, u)$	$u(2,4,uu)$
			$d(2,4,uu)$
		$d(1,4, u)$	$u(2,4,ud)$
			$d(2,4,ud)$
	$d(0, 4)$	$u(1,4, d)$	$u(2,4,du)$
			$d(2,4,du)$
		$d(1,4, d)$	$u(2,4,dd)$
			$d(2,4,dd)$

Having constructed the u, d, and $r(t)$ trees, next build the $Q(t)$ tree. Here, some notations are needed for variables s_t which represent states:

$$s_0 = \Phi, \text{ empty set,}$$

$$s_1 = \{u, d\},$$

$$s_2 = \{uu, ud, du, dd\},$$

$$s_3 = \{uuu, uud, udu, udd, duu, dud, ddu, ddd\}.$$

Start by mimicking the formula $Q = \frac{e^{r\Delta t} - d}{u - d}$:

$$\mathbf{Q}(t, s_t) = \frac{r(t, s_t) - d(t, T, s_t)}{u(t, T, s_t) - d(t, T, s_t)} \tag{7.9}$$

Note that it is not necessary to use an exponential function, $e^{r\Delta t}$, if discrete time-compounded rates are being used (as is the convention in applications of the HJM model).

It is clear that the $Q(t)$ risk-neutral probability tree can be built using u and d, and the $r(t)$ tree:

$t = 0$	1	2
$Q(0)$	$Q(1, u)$	$Q(2,uu)$
		$Q(2,ud)$
	$Q(1, d)$	$Q(2,du)$
		$Q(2,dd)$

Consider why the $Q(t)$-tree doesn't have a third column. It is because u and d are needed to generate Q. Another perspective is that in order to discount the payoff backs it is only necessary to use Q for the up node and $1-Q$ for the down node.

Having formed the short-rate $r(t)$ tree and the $Q(t)$ risk-neutral probability tree, apply backward induction as detailed in Chapter 3, to get the derivative price today. The following material illustrates the roll back procedures:

Step 1. Any payoff at $t = 4$ can be rolled back to $t = 3$:

Note that for period 4 payoffs, interest rates of period 3 are used as discount factors. There are eight states of interest rates $r(3, s_3)$ where $s_3 = \{uuu, uud, udu, udd, duu, dud, ddu, ddd\}$ as follows.

$t = 3$
$r(3,uuu)$
$r(3,uud)$
$r(3,udu)$
$r(3,udd)$
$r(3,duu)$
$r(3,dud)$
$r(3,ddu)$
$r(3,ddd)$

Therefore, there are eight payoffs at $t = 4$ and 8 payoffs at $t = 3$.

The formula to derive any period 3 payoff is:

$$\text{Period 3 payoff} = \text{Final period payoff}/r(3, s_3).$$

Step 2. From $t = 3$, use the following formula to roll back to $t = 2$ and further:

Discount (by backward induction) similar as in Chapter 3 as follows:

$$\textbf{Rollback value} = (Q(t,\, s_t) \times \textbf{Payoff}_u + (1 - Q(t,\, s_t)) \times \textbf{Payoff}_d))/r(t,\, s_t) \tag{7.10}$$

The difference is that in Chapter 3 both Q and r are constants. Here both Q and r are stochastic trees.

From a pure manipulation perspective, only the zero coupon $P(t, T)$ tree and the short-rate $r(t)$ tree with node values are needed to value any derivative with payoff at T. This is the HJM one-factor model, although we have introduced it only through manipulation.

The next question is: Are all the columns of the $P(t, T)$ tree and the short-rate tree $r(t)$ needed?

In fact, notice that the last column of the $P(t, 4)$ tree and the last column of the short-rate tree are not independent (assuming that the bond has a face value of \$1):

$P(3,4,uuu) = 1/r(3,uuu)$
$P(3,4,uud) = 1/r(3,uud)$
$P(3,4,udu) = 1/r(3,udu)$
$P(3,4,udd) = 1/r(3,udd)$
$P(3,4,duu) = 1/r(3,duu)$
$P(3,4,dud) = 1/r(3,dud)$
$P(3,4,ddu) = 1/r(3,ddu)$
$P(3,4,ddd) = 1/r(3,ddd)$

So, the input in terms of trees is: (1) the full $P(t, T)$ tree, and (2) the short-rate $r(t)$ tree except the last column. Alternatively, it is sufficient to have a full short rate $r(t)$ tree and all of the $P(t, T)$ tree except its last column. Either is sufficient input to build the HJM one-factor model and to generate all the rest $P(t, s)$ trees where $0 \le s < T$. Here the trees include node values but do not include probability measures. So, in the HJM one-factor model, only two trees with node values are needed, not five trees. More factors require more parameters and more trees. HJM multifactor trees require some/all the rest of the $P(t, s)$ trees. The number of input trees depends on the number of factors.

Now consider the reverse question: Can a full short-rate tree $r(t)$ and a $Q(t)$ risk-neutral probability tree replicate the $P(t, T)$ tree? The answer is yes.

Example 7.2: Build a $P(t,4)$ tree using a full $r(t)$ tree and a $Q(t)$ risk-neutral probability tree.

First, build the last column using $P(3, 4, s_3)$ as the reciprocal of the $r(3, s_3)$ for $s_3 \in \{uuu, uud, \ldots, ddd\}$.

By Equation 7.9, $Q(t,\; s_t) = \frac{r(t, s_t) - d(t,T,s_t)}{u(t,T,s_t) - d(t,T,s_t)}$.

Specifically:

$$Q(2, uu) = \frac{r(2,uu) - d(2,4,uu)}{u(2,4,uu) - d(2,4,uu)} = \frac{r(2,uu) - P(3,4,uud) / P(2,4,uu)}{P(3,4,uuu) / P(2,4,uu) - P(3,4,uud) / P(2,4,uu)}$$

$$= \frac{P(2,4,uu) \times r(2,uu) - P(3,4,uud)}{P(3,4,uuu) - P(3,4,uud)}$$

Simplification can produce:

$$P(2, 4, uu) = (Q(2, uu) \times P(3, 4, uuu) + (1 - Q(2, uu)) \times P(3, 4, uud))/r(2, uu).$$

The above equation exactly matches backward induction. So a full short-rate tree $r(t)$ and risk neutral-probability tree $Q(t)$ can be used to replicate the $P(t, T)$ tree.

Note that if continuous time models are considered, the stochastic equation of $r(t)$ in a risk- neutral world itself carries the tree node values as well as the risk-neutral probability measure. Therefore, it is only necessary to model one process, either $r(t)$, $f(t, T)$, $P(t, T)$, or $m(t)$. Two processes are not necessary and all the remaining processes can be derived using any process.

Note that there is nothing special about the payoff $P(3, 4, s_3)$ in the previous application of Equation 7.10; any derivative tree can be found through this process of backward induction. The manipulation makes sense. Basically, there is one unique risk-neutral probability existing such that all assets values prices are consistent. This can be viewed as a concrete version of the existence of the unique risk-neutral measure under which the martingale property holds.

The above process can be used to find the tree of $P(t, 3)$, which is the three-year zero coupon bond price evolvement, the tree of $P(t, 2)$, and the tree of $P(t, 1)$. It is only necessary to use part of the $r(t)$ tree and the $Q(t)$ risk-neutral probability tree as our input. The necessary part is the tree components before maturity. This also justifies why a longest maturity tree is needed as an input.

Note that the above argument is not a proof for the existence of a risk-neutral measure. It is simply an intuitive analysis. If the $P(t, 3)$, $P(t, 2)$, and $P(t, 1)$ trees are used at the same time as inputs, the result does not guarantee a coherent risk-neutral probability measure $Q(t)$. This is similar to any case in which the number of equations exceeds the number of variables.

Since the above argument is a mathematical manipulation, we consider the economic interpretation. The argument implies that in a one-factor HJM framework (in a binomial tree setting), any interest rate derivative can be replicated by a zero coupon bond $P(0, T)$ and a money market fund $m(t)$, noting that $m(t)$ is equivalent to $r(t)$.

Once the replication strategy has been derived, we can have more confidence in the above manipulations. The following discussion and the rest of the book have not made a distinction between the terms *replication* and *hedging*. Hedging involves establishing as opposite position to the original position in order to eliminate risk. ***Replication*** means finding a position with identical risk of the original position. If a replicating position exists, it can be used to create a hedge by changing long positions to short positions and vice versa. The terms *hedge* or *hedging* will usually be used in the material that follows in place of replication.

The hedging strategy in a binomial setting is very similar to the delta-neutral hedging strategies discussed in Chapters 3 and 5. In hedging with "the Greeks," the hedge is specified by the numbers of units, not the percentage weights. In the HJM model, hedging will be described in units rather than market weights and the hedging vehicle will be bond *P*(0, *T*) and money market fund m(t) (in a one-factor HJM model). The notation $\Delta_{P(0,T)}$ is the number of units needed in *P*(0, *T*) and Δ_m is the number of units in the money market fund *m*(*t*). Hedges are established in the parent node and held until the children's node in either the up or down state (whichever has occurred). At that point, one of the children's nodes becomes the new parent node and the hedge needs to be rebalanced.

Equation 7.11 is similar to the hedge ratios in Chapters 3 and 5. *P*(0, *T*) serves as the hedging vehicle.

$$\Delta_{P(0,T)} = \frac{\textbf{Asset to be hedged in up node} - \textbf{Asset to be hedged in down node}}{\textbf{hedging vehicle } P(0,\, T)\textbf{in up node} - \textbf{hedging vehicle } P(0,\, T)\textbf{in down node}} \tag{7.11}$$

A fundamental condition must be satisfied in order to make the hedging strategy work. Specifically, holding $\Delta_{P(0,T)}$ units of *P*(0, *T*) and Δ_m unit of the money market fund *m*(*t*) in the parent node must replicate the asset to be hedged in parent node. Equation 7.12 expresses the condition:

$$\begin{aligned}&\textbf{Asset to be hedged in parent node} = \Delta_{P(0,T)} \times \textbf{hedging vehicle } P(0,\, T)\textbf{in parent node}\\&+ \Delta_m \times \textbf{money market fund in parentnode}\end{aligned} \tag{7.12}$$

Equation 7.13 rearranges Equation 7.12 to expresses the condition conducive to solving for Δ_m.

$$\Delta_m = \begin{pmatrix}\textbf{Asset to be hedged in parent node}\\ - \Delta_{P(0,T)} \times \textbf{hedging vehicle } P(0,\, T)\textbf{in parent node}\end{pmatrix} \\ /\textbf{money market fund in parent node.} \tag{7.13}$$

In this way, $\Delta_{P(0,T)}$ units of *P*(0, *T*) and Δ_m units of the money market fund (in parent node) complete the replicating condition and ensure that the asset is hedged. The second condition needed is that this strategy must also replicate the payoff in children nodes. The technical details are left to the reader as exercises.

Consider the following example that illustrates the satisfaction of the second condition.

Note that in all the examples until the end of Section 7.5, the exact same input trees are used: the $P(t,4)$ tree and the $r(t)$ tree for our convenience since these two input trees are all that are needed in the one-factor HJM model.

Example 7.3: Given a $P(t, 4)$ tree and an $r(t)$ tree both over four years:

$P(t, 4)$ tree:

0.9609803	0.9732417	0.9836759	0.9925695
			0.9905975
		0.9799735	0.9911376
			0.9887877
	0.9679386	0.9809295	0.9914787
			0.9892188
		0.9765981	0.9898377
			0.9871453

The $r(t)$ tree:

1.01	1.0088190	1.0080388	1.0074861
			1.0094918
		1.0101933	1.0089416
			1.0113394
	1.0111837	1.0096023	1.0085945
			1.0108987
		1.0121784	1.0102667
			1.0130220

a. What is the value of $P(0, 3)$?

b. If $P(0, 3)$ is 0.98 in the market, is there an arbitrage "opportunity"? How would an arbitrageur take advantage of a market price for a $P(0, 3)$ of 0.98 by trading a combination of $P(0, 3)$, $P(0, 4)$, and $m(t)$ as a self financing portfolio?

Solution of a. Note that the last column of the $r(t)$ tree is the reciprocal of the last column of the $P(t,4)$ tree.

Begin by building the u and d trees which are found as each child node divided by its parent node in the $P(t,4)$ tree. Note that the cells with numbers marked with (c) are created and other cells are copied from the the cells marked with (c). The formula for creating the cells is a straightforward quotient. For example, the cell 1.0127593 is found by dividing the up child (0.9732417) by the parent (0.9609803) and the cell 1.0072408 is found by dividing the down child by its parent.

$t = 0$	1	2	3
	$u(0, 4)$	$u(1,4, u)$	$u(2,4,uu)$
			$d(2,4,uu)$
		$d(1,4, u)$	$u(2,4,ud)$
			$d(2,4,ud)$
	$d(0, 4)$	$u(1,4, d)$	$u(2,4,du)$
			$d(2,4,du)$
		$d(1,4, d)$	$u(2,4,dd)$
			$d(2,4,dd)$
	1.0127593(c)	1.0107211(c)	1.0090412(c)
			1.0070365(c)
		1.0069169(c)	1.0113922(c)
			1.0089943(c)
	1.0072408	1.0134212	1.0107543
			1.0084505
		1.0089463	1.0135569
			1.0107999

Next, build up the $Q(t)$ risk-neutral probability tree using Equation 7.9, as shown below. The cells marked with (c) are created and the other cells are copied:

$t = 0$	1	2
$Q(0)$	$Q(1, u)$	$Q(2,uu)$
		$Q(2,ud)$
	$Q(1, d)$	$Q(2,du)$
		$Q(2,dd)$
0.4999911(c)	0.5000050(c)	0.4999867(c)
		0.5000059
	0.4999850	0.4999687
		0.5000009

Note that the upper-right cell (0.4999867) in the $Q(t)$ tree comes from $\frac{1.0080388 - 1.0070365}{1.0090412 - 1.0070365}$, where the denominator is the difference between u and d in the upper-right corner of the u and d trees. The numerator $r(t)$ is the parent r in the second to the last column in the $r(t)$ tree. The remaining cells copy the top of the column and paste the cell down through the whole column. The $Q(t)$ risk-neutral probability tree is completed column by column.

Next, in order to build the $P(t, 3)$ tree, first generate $P(2, 3, s_2)$, which is the reciprocal of the second to last column in the $r(t)$ tree:

$t = 2$
0.9920253(c)
0.9899096
0.9904890
0.9879681

The top cell with (c) is created as the reciprocal while the other cells are copied down from the top cell.

Next, roll back by probability-weighting the up and down nodes by the parent node's probability from the tree $Q(t)$ and discount back by the parent node's interest rate from the tree $r(t)$. The probability is the corresponding parent node in the probability tree $Q(t)$.

$$(0.9920253 \times 0.5000050 + 0.9899096 \times (1 - 0.5000050))/1.0088190$$
$$= 0.9823045$$

Step by step, $P(0, 3)$ can be built.

$t = 0$	1	2
$P(0, 3)$	$P(1,3, u)$	$P(2,3,uu)$
		$P(2,3,ud)$
	$P(1,3, d)$	$P(2,3,du)$
		$P(2,3,dd)$
0.9705901(c)	0.9823045(c)	0.9920253(c)
		0.9899096
	0.9782877	0.9904890
		0.9879681

The value of $P(0,3)$ is 0.9705901.

Solution of b. It is clear that the given market price of $P(0,3)$ is too high. Therefore, the arbitrage strategy should short sell $P(0,3)$ in the market. The risk of shorting $P(0,3)$ must be hedged to perform arbitrage. To do so, build a long synthetic (i.e., replicating) $P(0,3)$ portfolio using $P(0,4)$ and $m(t)$. Note that this is a self-financing strategy.

The asset to be hedged is $P(0,3)$ and the hedging vehicle is $P(0,4)$.

First, build the $m(t)$ tree, which is based on the $r(t)$ tree. The $m(t)$ tree is built up to $t = 2$. In fact, in designing the hedging strategy alone, it only requires $m(0)$ and $m(1)$. This is because at $t = 2$, the bond price $P(2,3)$ is known at every state. Therefore, the hedge strategy is only needed at periods 0 and 1. However, for checking purposes, find $m(2, u)$ and $m(2, d)$.

Note that $m(0) = 1$, $m(1) = m(0) \times r(0)$, $m(2, u) = m(0) \times r(0) \times r(1, u)$, and $m(2, d) = m(0) \times r(0) \times r(1, d)$.

$t = 0$	1	2
$m(0)$	$m(1)$	$m(2, u)$
		$m(2, d)$
1(c)	1.01(c)	1.0189072(c)
		1.0212955

Since $P(0,3)$ has three columns, the $\Delta_{P(0,4)}$ tree only has two columns. The convention of cell marked with (c) being created and other cells being copied is continued:

$\Delta_{P(0,4)}(0)$	$\Delta_{P(0,4)}(1,u)$
	$\Delta_{P(0,4)}(1,d)$
0.7574501(c)	0.5714542(c)
	0.5820054

$$\Delta_{P(0,4)}(0) = \frac{P(1, 3, u) - P(1, 3, d)}{P(1, 4, u) - P(1, 4, d)}$$

$$\Delta_{P(0,4)}(1, u) = \frac{P(2, 3, uu) - P(2, 3, ud)}{P(2, 4, uu) - P(2, 4, ud)}$$

$$\Delta_{P(0,4)}(1, d) = \frac{P(2, 3, du) - P(2, 3, dd)}{P(2, 4, du) - P(2, 4, dd)}$$

The Δ_m tree only has two columns:

$$\Delta_m(0) = \left(P(0, 3) - \Delta_{P(0,4)}(0) \times P(0, 4)\right)/m(0)$$

$$\Delta_m(1, u) = \left(P(1, 3, u) - \Delta_{P(0,4)}(1, u) \times P(1, 4, u)\right)/m(1)$$

$$\Delta_m(1, d) = \left(P(1, 3, d) - \Delta_{P(0,4)}(1, d) \times P(1, 4, d)\right)/m(1)$$

$\Delta_m(0)$	$\Delta_m(1, u)$
	$\Delta_m(1, d)$
0.2426955(c)	0.4219222(c)
	0.4108338

At time 0, the arbitrager should buy $\Delta_m(0) = 0.2426955$ units of the money market fund at \$1 and $\Delta_{p(0,4)}(0) = 0.7574501$ units of $P(0, 4)$. The cost to buy this replicating portfolio is:

$$\Delta_m(0) \times m(0) + \Delta_{p(0,4)}(0) \times P(0, 4) = 0.9705901$$

which matches the theoretic $P(0, 3)$ price found in (a). The arbitrager short sells $P(0, 3)$ in the market with proceeds of 0.98. The immediate net proceeds form the gain, which is 0.98–0.9705901.

In the period 1 up state: The arbitrager needs to keep $\Delta_{P(0,4)}(1, u) = 0.5714542$ units of $P(1,4, u)$ and $\Delta_m(1, u) = 0.4219222$ units of $m(1)$. The replicating up portfolio value is:

$$\Delta_m(1, u) \times m(1) + \Delta_{p(0,4)}(1, u) \times P(1, 4, u) = 0.9823045$$

which matches the theoretic $P(1, 3, u)$ price.

Check to be sure that the strategy is self-financed from period 0 to period 1 in the up state. Holding $\Delta_m(0)$ units of $m(0)$ and $\Delta_{p(0,4)}(0)$ units of $P(0, 4)$ grows to $P(1, 3, u)$ in the up state:

$$\begin{aligned}&\Delta_m(0) \times m(1) + \Delta_{p(0,4)}(0) \times P(1, 4, u) = 0.9823045\\ &= P(1, 3, u).\end{aligned}$$

So it is self-financed; namely the arbitrager can rebalance $\Delta_m(0)$ to $\Delta_m(1, u)$ and $\Delta_{p(0,4)}(0)$ to $\Delta_{P(0,4)}(1, u)$ without needing to expend cash.

In the period 1 down state: The arbitrager needs to keep $\Delta_{P(0,4)}(1, d) = 0.5820054$ units of $P(1, 4, d)$ and $\Delta_m(1, d) = 0.4108338$ units of $m(1)$. The replicating down portfolio value is:

$$\Delta_m(1, d) \times m(1) + \Delta_{p(0,4)}(1, d) \times P(1, 4, d) = 0.9782877$$

which matches the theoretic $P(1, 3, d)$ price.

Check to be sure that the strategy is self-financed from period 0 to period 1 in the down state. Holding $\Delta_m(0)$ units of $m(0)$ and $\Delta_{p(0,4)}(0)$ units of $P(0,4)$ does grow to $P(1,3, d)$ in the down state:

$$\Delta_m(0) \times m(1) + \Delta_{p(0,4)}(0) \times P(1, 4, d) = 0.9782877$$
$$= P(1, 3, d).$$

So it is self-financed.

Consider the following two equations that were generated above for the up node and down node, respectively:

$$\Delta_m(0) \times m(1) + \Delta_{p(0,4)}(0) \times P(1, 4, u) = 0.9823045 = P(1, 3, u)$$
$$\Delta_m(0) \times m(1) + \Delta_{p(0,4)}(0) \times P(1, 4, d) = 0.9782877 = P(1, 3, d).$$

The relation between the two equations is not surprising because the hedge ratio is $\Delta_{p(0,4)}$:

$$\Delta_{p(0,4)}(0) = \frac{P(1,3,u) - P(1,3,d)}{P(1,4,u) - P(1,4,d)}.$$

Now, consider the hedging or replicating portfolio in the uu state of period 2:

$$\Delta_m(1, u) \times m(2, u) + \Delta_{P(0,4)}(1, u) \times P(2, 4, uu) = 0.9920253$$
$$= P(2, 3, uu)$$

Note that at the end of period 2 there is no need to rebalance; the long synthetic bond matches the short bond. So it is risk free.

This example demonstrates that the $P(t, 4)$ tree and the $r(t)$ tree (without its last column), with node values, complete the market in an HJM one factor model. This has been illustrated in a binomial tree setting. It is not necessary in a one-factor HJM model to have three other trees such as $P(t, 3)$, $P(t, 2)$, and $P(t, 1)$ given (which represent the short maturity zero coupon bonds). In fact, if the market prices of short maturity zero coupon bonds, $P(0, 3)$, $P(0, 2)$, and $P(0, 1)$, do not match the theoretical prices, there would be an arbitrage opportunity. For the bond alone, a hedge is unnecessary. We can use the HJM one factor model to indentify the arbitrage opportunity and grasp it.

From a risk management perspective, it is not necessary or even advisable to use duration-based and convexity-based modeling to hedge risk because the term structure of interest rates tends to have non-parallel shifts in the real world.

Finally, compare the above hedge with the duration hedge. Consider how many units of bond $P(0,4)$ are needed to form a duration hedge.

$$N = \frac{P(0, 3)D(P(0, 3))}{P(0, 4)D(P(0, 4))} = \frac{0.98 \times 3}{0.9609803 \times 4} = 0.7648440$$

From a duration perspective, a hedger should use 0.7648440 units of $P(0, 4)$ to hedge $P(0, 3)$. However, the one-factor HJM model was shown above to use 0.7574501 units of $P(0, 4)$ to hedge $P(0, 3)$. The potential risk of the duration hedge (based on the HJM model as the evolvement) is left to the readers to check. In fact, modified duration should be used in the above calculation instead of duration since continuous compounding is not used. However, due to a flat forward rate structure which we will see in a later section, the hedge ratio is the same.

7.3 Valuation and Risk Management of Options on Bonds in One-Factor HJM Model

This section discusses two examples: one that values a European option on a zero coupon bond and another that values an American option on a coupon bond.

Example 7.4: Given a $P(t,4)$ tree and an $r(t)$ tree over a four-year period (as in Example 7.3), answer the following two questions:

a. What is the value of a two-year European call option on a four-year zero coupon bond with a strike price of $K = 0.98$.
b. What is the option's replicating strategy?

As in previous section, begin by forming the risk-neutral probability tree, $Q(t)$, using a u and d tree and the $r(t)$ tree:

$t = 0$	1	2			
0.4999911(c)	0.5000050(c)	0.4999867(c)	$Q(0)$	$Q(1, u)$	$Q(2,uu)$
		0.5000059			$Q(2,ud)$
	0.4999850	0.4999687		$Q(1, d)$	$Q(2,du)$
		0.5000009			$Q(2,dd)$

Since the call expires at the end of two years, the next step is to find the payoff in year 2 using the payoff formula: Max($P(2,4, s_2)$-K,0) where s_2 represents various states at $t = 2$. Note that $P(2,4, s_2)$ is the $t = 2$ column in $P(t,4)$ tree. The completed call tree and a part of the $P(t,4)$ tree are laid out horizontally with the call tree on the left and the $P(t,4)$ tree on the right in Spreadsheet 7.1:

Spreadsheet 7.1

European Call on four-year bond, tau = 2 and $K = 0.98$			$P(t,4)$ tree leaving last column		
$K = 0.98$					
$t = 0$	1	2			
0.0011294(c)	0.0018219(c)	0.0036759(c)	0.9609803	0.9732417	0.9836759
		0			0.9799735
	0.0004596	0.0009295		0.9679386	0.9809295
		0			0.9765981
Call $C(t,4)$ tree					
$C(0,4)$	$C(1,4,u)$	Max($P(2,4,uu)$-K,0)	$P(0,4)$	$P(1,4,u)$	$P(2,4,uu)$
		Max($P(2,4,ud)$-K,0)			$P(2,4,ud)$
	$C(1,4,d)$	Max($P(2,4,du)$-K,0)		$P(1,4,d)$	$P(2,4,du)$
		Max($P(2,4,dd)$-K,0)			$P(2,4,dd)$

There are three steps used to form the trees in Spreadsheet 7.1.

Step 1: At $t = 2$, calculate the last column of the call tree in the top-right cell with a value of Max($P(2,4,uu)$-K, 0) and copy it down (continuing with the convention that created cells are marked with (c) and copied cells are unmarked) to fill in the whole column.

Step 2: At $t = 1$, calculate the $C(1,4,u)$ cell using: (up child $\times$ $Q(1,u)$+down child $\times$ $(1-Q(1,u)))/r(1,u)$. If the $r(t)$ tree and $Q(t)$ trees are arranged vertically, copy the $C(1,4,u)$ cell to the $C(1,4,d)$ cell. The results are the values of 0.0018219 and 0.0004596.

Step 3: In Chapter 3, it was possible to copy other cells using one discounting cell. But in this chapter each column contains a different number of empty spaces. Therefore, calculate the $C(0,4)$ cell using: (up child $\times$ $Q(0)$+down child$\times$ $(1-Q(0)))/r(0)$.

The resulting call price is 0.0011294 found in the $t = 0$ column of the call tree.

Having created the call and bond price trees in Spreadsheet 7.1, the next step is to create the replicating strategy trees shown in Spreadsheet 7.2. Creation of these trees requires some conventions. In the material that follows, "current cell in different trees" refers to a "fixed location in different trees." When a current cell is selected, the parent cell is one column to the left and a child cell is one column to the right and it could refer to different trees.

Denote Δp as the number of units from bond $P(0,4)$ needed to replicate the option in the current cell (along with the money market fund). Note that for the current cell in the Δp tree, the children nodes in the call $C(t,4)$ tree and $P(t,4)$ tree are required for calculation:

Spreadsheet 7.2

Delta of $P(0,4)$		Delta of m		Money market tree	
0.2568884(c)	0.9928425(c)	−0.2457353(c)	−0.9549048(c)	1	1.01
	0.2145957		−0.2052039		
$\Delta p(0)$	$\Delta p(1,u)$	$\Delta m(0)$	$\Delta m(1,u)$	$m(0)$	$m(1)$
	$\Delta p(1,d)$		$\Delta m(1,d)$		
		0.0011294			
		$\Delta p(0) \times P(0,4) + \Delta m(0) \times m(0)$			

$$\Delta p = \frac{\text{up child in call} - \text{down child in call}}{\text{up child in } P(0,4)\text{tree} - \text{down child in } P(0,4)\text{tree}}$$

Denote Δm as the number of units needed in the money market fund, $m(t)$, to replicate the option in the current cell. The formula for Δm is:

$$\Delta m = (\text{current in call tree} - \text{current in } \Delta\, p \text{ tree} \times \text{current in } P(0,4)\text{tree})$$
$$/\text{current in money market tree}$$

Finally, the initial value of the replicating strategy is: $\Delta p(0) \times P(0,4) + \Delta m(0) \times m(0)$. Check that the replicating strategy generates the same initial call option value as the call tree:

$$\Delta p(0) \times P(0,4) + \Delta m(0) \times m(0) = \text{the initial call price} = 0.0011294$$

The second example (Example 7.5) values an American call on a coupon bond. Note that the value of an American call on a coupon bond may not equal the value of a European call on the same coupon bond.

Example 7.5: Given a $P(t,4)$ tree and an $r(t)$ tree in the four-year period (as in Example 7.3), find the answers to the following questions: (Assume that if the option is exercised at t = 2, the option holder will not receive the coupon at t = 2.)

a. What is the value of a two-year American call option on a four-year 4% coupon bond (paid every two years for simplicity) with a strike price of $K = 1.0195$.
b. What is the replicating strategy to the call option using the coupon bond as a hedge vehicle?

As before, in Example 7.4, begin by generating the $Q(t)$ tree (and use the previous $Q(t)$ tree for this example). Since the call expires at the end of two years, the final payoff is given by Max(CB(2,4, s_2)-K,0).

It is clear that the solution requires building the coupon bond tree, CB(t,4), first. Note that the bond's coupon is only paid at $t = 2$ and $t = 4$ as stated in the problem for simplicity.

Step 1: Generate the top cell in the $t = 3$ column. The current cell in the CB (t,4) tree = 1.04/current cell in $r(t)$ tree. Copy the cell down.

Step 2: Generate the top cell in the $t = 2$ column. The current cell in CB(t,4) tree = (up child in CB(t,4) tree × current node in $Q(t)$ tree + down child in CB (t,4) tree × (1 – current node in $Q(t)$ tree))/current cell in $r(t)$ tree. Then copy the cell down.

Step 3: Generate the top cell in the $t = 1$ column. The current cell in CB(t,4) tree = ((up child in CB(t,4) tree + coupon) × current node in $Q(t)$ tree + (down child in CB(t,4) tree + coupon) × (1 – current node in $Q(t)$ tree))/current cell in r (t) tree. Then copy the cell down. Note that the coupon is paid at $t = 2$.

Step 4: Generate the top cell in the $t = 0$ column. Note that there is no coupon in year 1, so it is the same as in Step 2.

Spreadsheet 7.3

Four-year coupon bond with coupon rate 4% paid on second and fourth year

t = 0	1	2	3	4	C
1.0386314(c)	1.0518217(c)	1.0230229(c)	1.0322723(c)	1.04(c)	0.04
			1.0302213	1.04	
		1.0191725	1.0307831	1.04	
			1.0283392	1.04	
	1.0462138	1.0201667	1.0311379	1.04	
			1.0287876	1.04	
		1.0156620	1.0294311	1.04	
			1.0266312	1.04	
CB(t,4) tree					
CB(0,4)	CB(1,4, u)	CB(2,4,uu)	CB(3,4,uuu)		
			CB(3,4,uud)		
		CB(2,4,ud)	CB(3,4,udu)		
			CB(3,4,udd)		
	CB(1,4, d)	CB(2,4,du)	CB(3,4,duu)		
			CB(3,4,dud)		
		CB(2,4,dd)	CB(3,4,ddu)		
			CB(3,4,ddd)		

Next, build the American call on this CB bond:

Spreadsheet 7.4

Two-year American call on coupon bond 1.0195				Coupon bond		
$t = 0$	1	2	K	0	1	2
0.0292254(c)	0.0323217(c)	0.00352292(c)		1.0386314	1.0518217	1.02302292
		0				1.01917248
	0.0267138	0.0006667			1.0462138	1.02016670
		0				1.01566204
AC(t,2) tree				CB(t,4) tree		
AC(0,2)	AC(1,2, *u*)	AC(2,2,*uu*)		CB(0,4)	CB(1,4, *u*)	CB(2,4,*uu*)
		AC(2,2,*uu*)				CB(2,4,*ud*)
	AC(1,2, *d*)	AC(2,2,*uu*)			CB(1,4, *d*)	CB(2,4,*du*)
		AC(2,2,*uu*)				CB(2,4,*dd*)

Note that in the American call, the backward induction formula needs to be changed as follows.

The current cell in any discounting column (except the last column) in the AC(t,2) tree = MAX{(up child in AC(t,2) tree × current node in $Q(t)$ tree + down child in AC(t,2) tree × (1 − current node in $Q(t)$ tree))/current cell in $r(t)$ tree, current cell in CB(t,4) tree-K}. It is necessary to compare the roll back value with the immediate exercise value. Only generate the top cell in each column and then copy it down.

For hedging/replicating, lay out this example as in Example 7.4 and then copy the cells to obtain the following trees. Note that the American call is exercised at t = 1, so no hedge is needed at t = 1.

Spreadsheet 7.5

Delta of CB(0,4)		Delta of m		Money market tree	
1	no hedge	− 1.00940594	no hedge		
	no hedge		no hedge		
$\Delta p(0)$	$\Delta p(1, u)$	$\Delta m(0)$	$\Delta m(1, u)$	$m(0)$	$m(1)$
	$\Delta p(1, d)$		$\Delta m(1, d)$		
		0.0292254			
		$\Delta p(0) \times P(0,4) + \Delta m(0) \times m(0)$			

Lastly, check that $\Delta p(0) \times P(0,4) + \Delta m(0) \times m(0)$ = the initial price = 0.0292254.

7.4 Valuation and Risk Management of Caps, Floors, and Interest Rate Swaps in a One-Factor HJM Model

This section values caps, floors, and swaps. All of these are related to floating-rate bonds.

A ***floating-rate bond*** is a coupon bond whose coupon rate varies with LIBOR. Specifically, the coupon is set equal to the LIBOR rate at the beginning of each period and is paid at the end of the period.

In theory, the market price of a floating-rate bond equals its face value at each coupon reset date. This is because owning the bond offers the same interest income as having the bond's face value inside a bank account that offers LIBOR as interest. If the coupon rate is reset continuously, then the bond's value equals its face value continuously. If it resets periodically (e.g., quarterly), then at each point in time the bond has the same value and interest rate risk as having a bond with a time-to-maturity equal to the time-to-reset.

An ***interest rate cap*** is insurance offered to a floating rate bond issuer (or other counterparty) that limits the interest rate per period to a maximum amount, K. A ***caplet*** is defined as an interest rate cap with only a single time period. An ***interest rate floor*** is insurance offered to a floating rate bond buyer (or other counterparty) that limits the interest rate per period to a minimum amount, K. A ***floorlet*** is defined as an interest rate floor with only a single time period. In a perfect market, the price of a cap or floor is the arbitrage-free value of T-maturity cap or floor. The price of a cap or floor is negotiated and calculated at the initiation of the product. A cap is a sequence of caplets and each caplet may be viewed as a call option on the short rate with a strike of K. Note that the short rate at maturity does not matter since all rates refer to the interest rate in the next period. In other words, in a tree setting, the last node payoff is related to the interest rate of its parent.

A ***plain vanilla interest rate swap*** was introduced and defined in Chapter 5. There are two parties involved that exchange cash flows with each other as follows: One party pays a fixed rate periodically based on a notional principal, L, and the other party pays a floating rate periodically. The floating rate is set equal to the LIBOR rate at the beginning of each period and is paid at the end of each period based on the same notional principal. The party receiving the floating rate is the "swap buyer" and the party paying the floating rate is the "swap seller." The fixed rate, c, is determined at the inception of the swap and is called the ***price of the swap*** and it is negotiated at the initiation of the swap. A plain vanilla interest rate swap may be viewed as a swap of a fixed rate coupon bond (in which the fixed rate is termed the swap price) and a floating rate (based on LIBOR) coupon bond with the same notional principal values L.

For example, assume the notional principal, L, is \$1. The swap can be viewed as exchanging the cash flows of a fixed-rate coupon bond for the cash flows of a floating-rate bond (with both bonds having face values equal to \$1, identical payment dates, and identical maturity dates). Note that the face values of the two bonds are equal so there is no net cash flow of principal values at the termination of the swap. Assuming the length of the time period between each cash flow exchange is $\Delta = 1$ and is equal to the time interval between coupon payments on the fixed- and floating-rate bonds, then it is easy to see that the fixed-rate bond's coupon rate, c, is exactly equal to the swap price. This can be generalized easily to non-annual payments.

In this section, the market data is the LIBOR-related market data. The section derives the prices of caps, floors, and swaps. In addition, the replicating portfolio of the product is discussed. All are done in one-factor HJM model framework.

7.4.1 Pricing and Hedging a Cap

This section only discusses a cap (e.g., a T-year cap with a strike of K) with annual cash flows because the analysis of a floor is very similar and it is easy to generalize to non-annual payments. Further, this section only prices a caplet because the price of a cap is the sum of the prices of its caplets. So, this section generalizes last section's European call option to a series of European call options. The only difference is that the underlying asset of the calls in this section is not a bond; it is the parent node's short rate.

Pricing a t-year caplet: A t-year caplet is equivalent to a t-year call on the short rate.

The payoff cell in the last column of the caplet tree = Max(parent cell in $r(t)$ tree-K,0). Since there is only one payoff occurring, the cells in all previous columns are just the discounted future expected payoff in a risk-neutral world, as rolled back in Section 7.3.

Example 7.6: Given a $P(t,4)$ tree and the $r(t)$ tree as before, consider a three-year cap at strike, $K = 1.01$: (a) Determine the price of the cap and (b) use the bond tree $P(0,4)$ to form a hedge.

The three caplet prices and their replicating portfolios are as shown below. The $P(t,4)$ tree and $r(t)$ tree are given. The column of $t = 2$ in $r(t)$ tree is used to create the column of $t = 3$ in three year caplet tree. The first step is creating the the cells marked with (c) in the three-year caplet tree and then copying and pasting to create other cells. Then, the three-year caplet tree and the $P(t,4)$ tree are listed horizontally. It is easy to calculate the hedge ratio later. The process continues through the two-year and one-year caplet trees, which are listed together.

Spreadsheet 7.6

Three-year caplet tree				Four-year zero coupon bond			
$t = 0$	1	2	3	$t = 0$	1	2	3
0.0005738(c)	9.48189E05(c)	0(c)	0.0000000(c)	0.9609803	0.9732417	0.9836759	0.9925695
			0.0000000(c)				0.9905975
		0.0001913	0.0001933			0.9799735	0.9911376
			0.0001933				0.9887877
	0.0010641	0	0.0000000		0.9679386	0.9809295	0.9914787
			0.0000000				0.9892188
		0.0021522	0.0021784			0.9765981	0.9898377
			0.0021784				0.9871453

Two-year caplet			Four-year zero coupon bond		
$t = 0$	1	2	$t = 0$	1	2
0.0005796(c)	0(c)	0(c)	0.9609803	0.9732417	0.9836759
		0(c)			0.9799735
	0.0011706	0.0011837		0.9679386	0.9809295
		0.0011837			0.9765981

One-year caplet		Four-year zero coupon bond	
$t = 0$	1	$t = 0$	1
0(c)	0(c)	0.9609803	0.9732417
	0		0.9679386

Use Equations 7.11 and 7.13 (where the asset to be hedged is the given caplet) to generate the hedging tree for the three-year, two-year, and one-year caplets:

Spreadsheet 7.7

Hedge three-year caplet

Delta of P(0,4)			Delta of m			Money market tree(c)		
t = 0	t = 1	2	t = 0	t = 1	t = 2	t = 0	t = 1	t = 2
− 0.1827815(c)	−0.0516826(c)	0	0.1762232(c)	0.0498955	0	1(c)	1.01(c)	1.0189072(c)
		0(c)			0.0001878(c)			
	−0.4968809	0		0.4772419	0			1.0212955
		0			0.0021073			

Hedge two-year caplet

Delta of P(0,4)		Delta of m		Money market tree	
t = 0	t = 1	t = 0	t = 1	t = 0	t = 1
−0.2207404(c)	0	0.2127067(c)	0(c)	1(c)	1.01(c)
	0		0.0011590		

Hedge one-year caplet

Delta of P(0,4)	Delta of m	Money market tree
t = 0	t = 0	t = 0
0	0	1

Adding the prices of the three caplets together generates the price of the cap: cap price = 0.0011533. In addition, the three hedging trees could be added together to get the hedging tree for the cap.

7.4.2 Pricing and Hedging a Swap

This section shows how to find the price of an interest rate swap and to replicate (i.e., hedge) the swap. It is interesting that in order to price a T-year swap, it is only necessary to know $P(0, s)$ where $0 < s \leq T$, the initial nodes of different bond trees and not the full $P(t, T)$ tree and the tree of $r(t)$ (which are the inputs of the one-factor HJM model). Initial nodes of the bond tree represent today's bond with different maturities and they can be calibrated directly from the market prices of coupon bonds, while the other values in the $P(t, T)$ tree (except the initial node) follow a stochastic process in HJM models.

In the previous section, it was shown that in a one-factor HJM model setting, the $P(t, T)$ tree and the $r(t)$ tree (except its last column) can be used to generate the $P(t, s)$ tree for $s < T$.

Is there is a consistency between the market-calibrated constants in $P(0, s)$ from coupon bonds and those that are model-generated given the $P(t, T)$ and $r(t)$ trees? The answer is yes. In Section 7.6 (while constructing the $P(t, T)$ tree and $r(t)$ tree in a one-factor HJM model), today's bond with different maturities, namely $P(0, s)$ (where $0 < s \leq T$) is market-calibrated from coupon bonds first. Then the initial forward rates $f(0, s)$ were found by Equation 7.6 using $P(0, s)$ (where $0 \leq s < T$) as inputs. Then market data are used to calibrate the parameters, $\eta(t)$. The forward rates $f(0, s)$, which are obtained from $P(0, s)$, together with the parameters $\eta(t)$, are used to generate the $f(t, s)$ tree (where $0 \leq s < T$). The forward-rate trees $f(t, s)$ are then used to generate the $P(t, T)$ and $r(t)$ trees. Note that values of $P(0, s)$ (where $0 < s \leq T$) are used as inputs to generate the $P(t, T)$ and $r(t)$ trees. It is not surprising that they match.

Note that only the initial node values $P(0, s)$ (where $0 < s \leq T$) are used in pricing the swap. Trees in a one-factor HJM model are not necessary to find the price of a T-year swap; the swap price is consistent with the corresponding HJM model. However, the HJM tree is usually used to replicate/hedge the swap.

To price a swap using $P(0, s)$ (where $0 < s \leq T$), assume that the cash flows are exchanged annually (i.e., for a swap Δ is set to one year). Note, however, that the formula can be generalized easily to quarterly payments and that most swaps are paid quarterly in the market. This example assumes that $L = 1$, $\Delta = 1$ for simplicity and that the swap lasts T years.

Competition will tend to drive swap prices towards values that are fair to both parties. Therefore, the value of the implied price of the swap's fixed-payment side (which resembles a fixed-rate coupon bond) tends to be the same as the value of the swap's floating-rate side (which resembles a floating-rate coupon bond).

$$\boldsymbol{C} \sum_{s=1}^{T} \boldsymbol{P}(\mathbf{0}, s) + \mathbf{1} \times \boldsymbol{P}(\mathbf{0}, \boldsymbol{T}) = \mathbf{1}$$

where $s \in \{1, 2, \ldots, T-1, T\}$ and C is the fixed coupon rate as well as the swap price.

Expressing the above equation into a formula for C:

$$\boldsymbol{C} = \frac{\mathbf{1} - \boldsymbol{P}(\mathbf{0}, \boldsymbol{T})}{\sum_{s=1}^{T} \boldsymbol{P}(\mathbf{0}, s)} \tag{7.14}$$

Therefore, the swap price (i.e., C) is in fact model-free in the sense that the HJM model is not needed to find the swap's price. It is also true that the HJM model is not needed to determine the hedge strategy. In fact, a portfolio of zero coupon bonds can be used as the replicating portfolio. However, not all the needed zero coupon bonds in the model-free replicating portfolio may trade or some bonds may have high transactions costs. This section provides a replicating portfolio containing only a single zero coupon bond $P(0, T)$ and the money market fund $m(t)$ to hedge by using a one-factor HJM model framework.

From a binomial valuation perspective, swaps are very different than options in two major ways:

1. Number of payoffs: An ordinary option has one payoff (this payoff could be a value of 0) during its life, whether European or American. The value of a European option at any time is therefore the discounted expected future value of a single payoff. Its value can be calculated by creating a tree which has potential payoffs in the last column and no payoffs in other columns. In an American option case, the immediate exercise value is compared in each tree node with the value of non-exercise to determine whether the options should be exercise early or deferred. But in either case an option will generate a single cash flow in one time period. However, a swap has multiple payoffs between the two parties of the derivative at the different time periods.
2. Option price vs. swap price: An option's price is the same as the option's value at each node. So, option prices change during the option life. Similar to forward, the swap price is viewed as a fixed term in a swap contract that describes the size of the fixed interest payments. Although the swap price does not change during the swap's life, its value to each of the counterparties does change. Presumably the initial value to each counterparty is zero. The swap value tree, VS(t), is comprised of the discounted expected future payoffs. The ***swap's value*** at each node is the difference between the values of the fixed-rate coupon bond minus the floating-rate bond (which is always L = \$1 in this chapter) at that node. In order to construct a VS(t) tree, generate a coupon bond tree and subtract the value of the floating bond (\$1). Other than future payoffs, there is the current node payoff in swap. So, it is necessary to generate a ***swap's cash flow*** tree, CF(t). Each node of the cash flow tree CF(t) is the difference of the swap rate C minus the short rate $r(t)$ of the parent node (minus 1 if expressed as a wealth ratio). Note that in calculating the CF(t) tree, there is a one-column shift since the rate is determined at the beginning of the period (based on the prevailing LIBOR) yet it is paid at the end of the period (in arrears). In order to find the replicating bond hedge ratio, Δ_P, requires the sum of the swap value and the swap cash flow value. However, in order to find the replicating money market fund hedge ratio, Δ_m, requires only the swap value because the net swap cash flow has been exchanged.

The following formulas are from a modification of Equations (7.11) and (7.13):

$$\Delta_p(t, S_t) = \frac{\mathbf{VS}(t+1, S_t u) + \mathbf{CF}((t+1, S_t u) - (\mathbf{VS}(t+1, S_t d) + \mathbf{CF}((t+1, S_t d))}{(P(t+1, T, S_t u) - P(t+1, T, S_t d))} \tag{7.15}$$

$$\Delta_m(t, S_t) = \mathbf{VS}(t, S_t) - \Delta_p(t, S_t) \times P(t, T, S_t) \tag{7.16}$$

Example 7.7 illustrates the valuation method for swaps.

Example 7.7: The followings are given: $P(0,3) = 0.9705901$, $P(0,2) = 0.9802960$, and $P(0,1) = 0.9900990$.
Note that these are consistent with the $P(t,4)$ tree and the $r(t)$ tree given in Example 7.3. Also, the $P(t,4)$ tree and the $r(t)$ tree are given as before.

a. Find the price of a three-year swap that pays annually and has $L = 1$.
b. Replicate the swap using the zero coupon bond in $P(0,3)$.

Solution of a. First, find the swap price using the prices of zero coupon bonds with different maturities today (for $P(0, s)$ where $1 \le s \le 3$) and their partial sum for $T = 3$. The spreadsheet is used as follows:

Spreadsheet 7.8

	Three-Year Swap	
	$P(0, s)$	Partial Sum
		0
1	0.9900990	0.99009901
2	0.9802960	1.97039506
3	0.9705901	2.94098521
		0.01
		C = (1 − P(0,3))/Partial Sum

The swap price: $C = (1 - P(0,3))/\text{Partial Sum} = 0.010000$.

b. First, generate the swap cash flow tree, CF(t):

The payoff at any node is equal to the opposite of the parent node in the short rate $r(t)$ tree minus 1 minus the swap rate C.
For example, at $t = 3$:

$$CF(3, uuu) = -(r(2, uu) - 1 - C)$$
$$CF(3, uud) = -(r(2, uu) - 1 - C).$$

The result is shown as follows:

Spreadsheet 7.9

Spot rates $r(t)$ tree

$t = 0$	1	2	3
1.0100000	1.0088190	1.0080388	1.0074861
			1.0094918
		1.0101933	1.0089416
			1.0113394
	1.0111837	1.0096023	1.0085945
			1.0108987
		1.0121784	1.0102667
			1.0130220

Swap cash flow CF(t) tree

$t = 0$	1	2	3
0(c)	0.0000000(c)	0.0011810(c)	0.0019612(c)
			0.0019612(c)
		0.0011810(c)	−0.0001933
			−0.0001933
	0.0000000	−0.0011837	0.0003977
			0.0003977
		−0.0011837	−0.0021784
			−0.0021784

Second, generate the swap value tree, VS(t).

Note that rolling back VS(t) itself is not enough to generate the VS(t) tree because at each node the sum of the children in the VS(t) tree and children in CF(t) are needed. As mentioned before, another way to do it is to view the future cash flows as the difference between a fixed coupon bond and a floating-rate bond. Calculate the VS(t) tree as CB(t,3) − 1 at every node.

A coupon bond tree was generated in Section 7.3. The idea is similar and so it is not repeated here. Note that the last coupon 0.01 dollar is considered in CF tree, so in couple bond tree, we put 1.00 in the column $t=3$. Here is the result:

Spreadsheet 7.10

Three-year coupon bond CB(t,3) tree

t = 0	1	2	3
1.0000000(c)	1.0020401(c)	1.0019456(c)	1.00(c)
		0.9998087	1.00
	0.9979599	1.0003939	1.00
		0.9978478	1.00

Align the following trees VS(t) and CF(t) horizontally:

Spreadsheet 7.11

Swap VS(t) tree				**Swap cash flow CF(t) tree**			
t = 0	1	2	3	t = 0	1	2	3
0.0000000	0.0020401	0.0019456	0	0	0.0000000	0.0011810	0.0019612
			0				0.0019612
		−0.0001913	0			0.0011810	−0.0001933
			0				−0.0001933
	−0.0020401	0.0003939	0		0.0000000	-0.0011837	0.0003977
			0				0.0003977
		−0.0021522	0			-0.0011837	−0.0021784
			0				−0.0021784

VS(0)	VS(1, u)	VS(2,uu)	VS(3,uuu)	CF(0)	CF(1, u)	CF(2,uu)	CF(3,uuu)
			VS(3,uud)				CF(3,uud)
		VS(2,ud)	VS(3,udu)			CF(2,ud)	CF(3,udu)
			VS(3,udd)				CF(3,udd)
	VS(1, d)	VS(2,du)	VS(3,duu)		CF(1, d)	CF(2,du)	CF(3,duu)
			VS(3,dud)				CF(3,dud)
		VS(2,dd)	VS(3,ddu)			CF(2,dd)	CF(3,ddu)
			VS(3,ddd)				CF(3,ddd)

Use Equations 7.15 and 7.16 to generate the hedging tree based on the hedging vehicle tree, P(t,3), that was generated in Section 7.2; the swap value tree, VF(t); and the swap cash flow tree, CF(t), mentioned above:

Spreadsheet 7.12

Three-year zero coupon bond			Delta of $P(0,4)$			Delta of m		
$t = 0$	1	2	$t = 0$	1	2			
0.9705901	0.9823045	0.9920253	1.0157921(c)	1.010003(c)	0	−0.9859177(c)	−0.98029(c)	0.001909(c)
		0.9899096			0			−0.00019(c)
	0.9782877	0.9904890		1.010000	0		−0.98031	0.000386
		0.9879681			0			−0.00211
$P(0,3)$	$P(1,3, u)$	$P(2,3,uu)$	$\Delta p(0)$	$\Delta p(1,u)$		$\Delta m(0)$	$\Delta m(1, u)$	
		$p(2,3,ud)$						
	$p(1,3,d)$	$P(2,3,du)$		$\Delta p(1,d)$			$\Delta m(1,d)$	
		$p(2,3,dd)$				0		
						$\Delta p(0) \times P(0,4) + \Delta m(0) \times m(0)$		

Note that in all the examples in Sections 7.4 and 7.5, the four-year bond price evolvement tree, $P(t,4)$, and the short-rate tree, $r(t)$, are the inputs to price options, caps, floors, and swaps in a one-factor HJM tree. This continues in the next section where the prices of forward and futures contracts on a three-year zero coupon bond $P(0,3)$ are found. However, the three-year zero coupon bond, $P(0,3)$, is the input generated by the $P(t,4)$ tree and the short-rate tree, $r(t)$.

7.5 Valuation and Risk Management of Forward and Futures Contracts on a Bond in a One-Factor HJM Model

A forward contract on a bond has two maturity times: the forward maturity time, T_1, and the bond maturity time, T_2, with $T_1 < T_2$. A ***forward contract on a bond***, $F(t, T_1, T_2)$, is a forward contract at time t that settles (matures) at time T_1 and has an underlying asset of bond $P(t, T_2)$. Note that a forward contract on a bond, $F(t, T_1, T_2)$, is very different from a forward rate, $f(t, T_1, T_2)$. $F(t, T_1, T_2)$ is the time t agreeable price to buy (a long position) or sell (a short position) in the underlying bond, $P(t, T_2)$, at time T_1. Alternatively, $f(t, T_1, T_2)$ is an interest rate – or the price of a contract on an interest rate – between $[T_1, T_2]$. A ***futures contract on a bond***, $\mathcal{F}(t, T_1, T_2)$, is on bond $P(t, T_2)$ and is similar to a forward contract on the same bond at time t that matures (settles) at time T_1.

In this chapter, $F(0, T_1, T_2)$ is used to denote both the forward contract and forward price today, whereas $\mathcal{F}(0, T_1, T_2)$ is used for a futures contract and the price of a futures contract on a bond today. This section begins with the derivation of $F(0, T_1, T_2)$ outside of the HJM model.

Recall long a forward contract on a zero coupon bond is an agreement today to pay $F(0, T_1, T_2)$ at time T_1 to receive a zero coupon bond that offers the cash flow \$1 at its maturity at time T_2.

Consider the following two portfolios containing zero coupon bonds initiated at time 0:

A. Long one unit of a bond: $P(0, T_2)$
B. Long $F(0, T_1, T_2)$ units of a bond, $P(0, T_1)$, and long a forward contract, $F(0, T_1, T_2)$, on a zero coupon bond.

Note that:
Portfolio A generates a cash flow of \$0 at T_1 and \$1 at T_2.

Portfolio B generates: (1) a cash flow of $F(0, T_1, T_2)$ dollars from maturity of bond $P(0, T_1)$ at time T_1, (2) rolled over at time T_1 to make a payment of $F(0, T_1, T_2)$ dollars to fulfill the obligation of taking delivery on the forward contract to receive the underlying bond, and (3) a cash flow of \$1 at T_2 from the bond's maturity. So the net cash flow is \$0 at T_1 and \$1 at T_2.

Since both portfolios deliver \$1 at time T_2, they must have equal prices today:

$$P(0, T_2) = F(0, T_1, T_2) \times P(0, T_1)$$

which generates a formula for the price of the forward contract as shown in Equation 7.17.

$$\boldsymbol{F(0, T_1, T_2) = \frac{P(0, T_2)}{P(0, T_1)}} \tag{7.17}$$

Similar to a swap price, the forward price $F(0, T_1, T_2)$ is model-free in the sense that it is not necessary to impose or use any HJM model to determine a price. Two bonds can be used to hedge the risk. In this section, a portfolio of a single zero coupon bond $P(0, T_2)$ and a money market fund $m(t)$ is used to construct a hedge in a one-factor HJM model framework.

The forward contract's payoff relates to the price of the underlying bond $P(0, T_2)$ at time T_1 and the forward price $F(0, T_1, T_2)$. Note that in one-factor HJM model tree, the price of $P(0, T_2)$ at time T_1 is denoted as $P(T_1, T_2, s_{T_1})$. So, the forward has one payoff, $P(T_1, T_2, s_{T_1}) - F(0, T_1, T_2)$, and no cash flow prior to that payoff. Accordingly, it is only necessary to hedge the forward value tree, VF(t), which is the discounted value of this single payoff $P(T_1, T_2, s_{T_2}) - F(0, T_1, T_2)$ at T_1. So rolling back the payoff $P(T_1, T_2, s_{T_2}) - F(0, T_1, T_2)$ at T_1 will build the forward value tree, VF(t). The VF(t) tree is the asset to be hedged.

The following example illustrates the idea.

Example 7.8: Consider a two-year forward on a three-year zero coupon bond given the $P(t,4)$ tree and the $r(t)$ tree as before.

a. Find the forward price $F(0,2,3)$.
b. Hedge the forward using $P(0,3)$.

The $P(t,3)$ tree can be found given the $P(t,4)$ tree and the $r(t)$ tree. Note that $P(0,2)$ can be found as 0.9802906 and $P(0,3)$ can be found as 0.9705901.

$$F(0, 2, 3) = \frac{\boldsymbol{P(0,3)}}{\boldsymbol{P(0,2)}} = \frac{0.9705901}{0.9802906} = 0.9900990$$

Next, generate node values of the forward value tree, VF(t).
The last column is $t = 2$.

$$\text{VF}(2, 3, uu) = P(2, 3, uu) - F(0, 2, 3)$$

Note that $P(2,3,uu)$ is a column in the $P(t,3)$ tree.
We continue to roll back to get the initial node.
Here are the cells in the EXCEL spreadsheets which list the $P(t,3)$ tree, the $r(t)$ tree, and the $Q(t)$ tree first and then the forward value VF(t) tree (i.e., the two-year forward contracts on a three-year zero coupon bond):

Spreadsheet 7.13

Three-year zero coupon bond $P(t,3)$ tree			Spot rates $r(t)$ tree	$Q(t)$ tree		
$t = 0$	1	2	$t = 0$	1	$t = 0$	1
0.9705901	0.982305	0.9920253	1.01	1.0088190	0.4999911	0.5000050
		0.9899096				
	0.978288	0.9904890		1.0111837		0.4999850
		0.9879681				
P(0,3)	P(1,3, u)	P(2,3,uu)	r(0)	r(1, u)	Q(0)	Q(1, u)
		P(2,3,ud)				
	P(1,3, d)	P(2,3,du)				
				r(1, d)		Q(1, d)
		P(2,3,dd)				
0.9900990		0.9802960				
F(0,2,3)		P(0,2)				

Forward value VF(t) tree (two-year forward on three-year zero coupon bond)		
$t = 0$	1	2
2.71923E − 08(c)	0.0008609(c)	0.0019263(c)
		−0.0001894
	−0.0008608	0.0003900
		−0.0021309
VF(0,3)	VF(1,3, u)	VF(2,3,uu)
		VF(2,3,ud)
	VF(1,3, d)	VF(2,3,du)
		VF(2,3,dd)

In this example, the forward value tree, VF(t), is the asset to be hedged. P(0,3) is the vehicle used to hedge. The details are left to readers in exercises.

How can the futures price, $\mathcal{F}(0, T_1, T_2)$, be obtained? Note that unlike the forward contract, the futures contract has a cash flow at each node because it is market-to-market (settled) regularly to reset its value to 0. Therefore, calculating the futures price $\mathcal{F}(0, T_1, T_2)$ is not easy. The one-factor HJM model is used next to price as well as to replicate it.

The futures price, $\mathcal{F}(0, T_1, T_2)$, is equal to the price of the underlying bond, $P(0, T_2)$, at time T_1 but is not discounted back to today (time 0). So, use backward induction to get $\mathcal{F}(t, T_1, T_2)$ from the $P(t, T_2)$ tree. The time discount factor short rate $r(t)$ is omitted.

Consider the following example.

Example 7.9: Given the $P(t,4)$ and $r(t)$ trees as before, consider a two-year futures contract on a three-year zero coupon bond $P(0,3)$.

a. Find the futures price $\mathcal{F}(0,2,3)$.
b. Hedge the futures contract using $P(0,3)$.

First, use the $P(t,4)$ and $r(t)$ trees to generate the $P(t,3)$ tree as in Example 7.8. The $P(t,3)$ tree is then used in the following:

$$\mathcal{F}(2,2,3,uu) = P(2,3,uu)$$
$$\mathcal{F}(1,2,3,u) = \mathcal{F}(2,2,3,uu) \times Q(1,u) + \mathcal{F}(2,2,3,ud) \times (1 - Q(1,u))$$
$$\mathcal{F}(1,2,3,d) = \mathcal{F}(2,2,3,du) \times Q(1,d) + \mathcal{F}(2,2,3,dd) \times (1 - Q(1,d))$$
$$\mathcal{F}(0,2,3) = \mathcal{F}(1,2,3,u) \times Q(0) + \mathcal{F}(1,2,3,d) \times (1 - Q(0))$$

The following is the spreadsheet for the futures price tree:

Spreadsheet 7.14

Futures with expiration two years on three-period zero coupon bond

$t = 0$	1	2
0.990097971(c)	0.9909675(c)	0.9920253(c)
		0.9899096
	0.9892285	0.9904890
		0.9879681
$\mathcal{F}(0,2,3)$	$\mathcal{F}(1,2,3, u)$	$\mathcal{F}(2,2,3,uu)$
		$\mathcal{F}(2,2,3,ud)$
	$\mathcal{F}(1,2,3, d)$	$\mathcal{F}(2,2,3,du)$
		$\mathcal{F}(2,2,3,dd)$

Note that with the same underlying three-year zero coupon bond, the forward price is 0.9900990 and the futures price is 0.990097971. The futures price is a little smaller than the forward price. This is consistent with results from an earlier chapter: When the underlying asset and interest rates are negatively correlated (as they are in the case of futures contracts on bonds), the futures price is lower than the price of an otherwise identical forward price due to interim cash flows from the marking to market of the futures contract. Since the interest rates are low, it does not seem to matter that much. However, in high interest rate environments, the change is large.

To create the hedge, since the futures value tree is 0 at any node, we only need to generate the futures cash flow tree (which is the children futures price minus the parent futures price):

$$\mathrm{CF}(0,2,3) = 0$$
$$\mathrm{CF}(1,2,3, u) = \mathcal{F}(1,2,3, u) - \mathcal{F}(0,2,3)$$
$$\mathrm{CF}(1,2,3, d) = \mathcal{F}(1,2,3, d) - \mathcal{F}(0,2,3)$$
$$\mathrm{CF}(2,2,3,uu) = \mathcal{F}(2,2,3,uu) - \mathcal{F}(1,2,3, u)$$
$$\mathrm{CF}(2,2,3,ud) = \mathcal{F}(2,2,3,ud) - \mathcal{F}(1,2,3, u)$$
$$\mathrm{CF}(2,2,3,du) = \mathcal{F}(2,2,3,du) - \mathcal{F}(1,2,3, d)$$
$$\mathrm{CF}(2,2,3,dd) = \mathcal{F}(2,2,3,dd) - \mathcal{F}(1,2,3, d)$$

The following spreadsheet forms the CF(0,2,3) tree:

Spreadsheet 7.15

Cash flow tree CF(t) (two-year futures on three-year zero coupon bond)

$t = 0$	1	2
0(c)	0.0008695(c)	0.0010578(c)
		−0.001058(c)
	−0.000869	0.0012605
		−0.00126
CF(0,2,3)	CF(1,2,3, u)	CF(2,2,3,uu)
		CF(2,2,3,ud)
	CF(1,2,3, d)	CF(2,2,3,du)
		CF(2,2,3,dd)

The futures cash flow tree is the asset to be hedged and $P(0,3)$ is the hedging vehicle. The details for the hedge are left as exercises. Note that futures price tree is not involved in hedge directly. Similar to swap, futures value tree and cash flow tree are used to hedge. Futures value tree is 0 since at every node, the contract value goes to 0 after cash is paid. That is why only cash flow tree is the asset to be hedged in futures contract. In forward contract, cash flow tree is 0, so forward value tree is the asset to be hedged. Also note that money market fund hedging unit Δm is only directly related to the value tree.

Having investigated the forward and futures prices, consider options on futures. Options on futures use the futures tree as the underlying. Build the option tree based on futures. The details are not specified here.

To summarize:

For purposes of finding a price:

- For an option, use the HJM model.
- For a forward, the HJM model is not necessary.
- For a futures contract, use the HJM model.
- For a cap/floor, use the HJM model.
- For a swap, the HJM model is not necessary.

For the purposes of hedging and replicating:

- For options, hedge with the option price/value tree.
- For forward contracts, hedge with the value tree VF(t). (The cash flow tree is a zero tree.)
- For futures contracts, hedge with the cash flow tree CF(t). (The value tree is a zero tree.)

- For caps/floors, hedge with the caplet/floorlet price/value tree and then sum.
- For swaps, hedge with the value tree VS(t) combined with the cash flow tree CF(t).

In summary, the two trees, with node values only, namely the zero coupon bond price tree, P(t,T), and the short-rate tree, r(t), have been assumed (where T must be longer than the derivative's underlying bond) to price and hedge various derivatives in the one-factor HJM model framework. In calculating Δp, both the value tree and cash flow tree are used, while in calculating Δm, only the value tree is used.

The next section describes how to generate the $P(t, T)$ and $r(t)$ trees and the inputs to the one-factor HJM model.

7.6 Calibrating the HJM Model Parameters from the Market Prices

Section 7.2 showed that two trees with node values only (not including the probability measures) are sufficient inputs to use the one-factor HJM model. This section compares three sets of two trees that can be used to complete the market: sets A, B, and C.

A = {bond price P(t, T) tree, short rate $r(t)$ tree except last column}
B = {bond price $P(t, T)$ tree, risk neutral probability $Q(t)$ tree}
C = {full short rate $r(t)$ tree, risk neutral probability $Q(t)$ tree}.

Next, show $B \cong A$.

From previous sections, it is clear that $P(t, T)$ and $r(t)$ can be used to find $Q(t)$. So, $A \rightarrow B$.

Next illustrate $B \rightarrow A$: By Equation 7.9, the risk-neutral probability is $\mathbf{Q} = \frac{r-d}{u-d}$. So, the $r(t)$ tree can be formed (except the last column) using the $P(t, T)$ tree and the $Q(t)$ risk-neutral probability tree. The last column of the $r(t)$ tree comes from the last column of the $P(t, T)$ tree. Therefore, $B \rightarrow A$.

Finally, show $C \cong A$. Note that the $P(t, T)$ tree can be replicated and completed by C from previous sections, so $C \rightarrow A$. It is mentioned above that $P(t, T)$ and $r(t)$ can be used to find $Q(t)$. So, $A \rightarrow C$.

Each set serves a different purpose. From the perspective of derivative valuation set C is the best. From the perspective of hedging and portfolio-replicating set A is the best. From the perspective of modeling parameters, let's consider whether A, B, or C is the best. In each of the sets A, B, and C, there are $(1 + 2 + 4 + 8 + 1 + 2 + 4) = 22$ variables and there are many degrees of freedom. As in the case of the alternative binomial tree in Chapter 4, the assumption is $Q = 1/2$. Thus, for every node of the $Q(t)$ tree, it seems that ½ is a natural choice. That leaves $22 - 7 = 15$ degrees of freedom. Set A is ruled out since it does not contain $Q(t)$. Further, $P(0,4)$, $P(0,3)$, $P(0,2)$, and $P(0,1)$ can be observed in the market (or at least they can be derived through calibration using coupon bonds). Therefore, set B is better than set C (and set A) to model parameters.

However, consider set B carefully, it has the following problem: The $P(t, T)$ and $Q(t)$ trees imply the values of $P(0, T - 1),\ldots, P(0,1)$. Therefore, it is difficult to make the model fit the observed $P(0, T - 1),\ldots, P(0,1)$ values since the calculation involves backward induction for all of the $P(t, T - 1),\ldots, P(t,1)$ trees to get the initial values of $P(0, T - 1),\ldots, P(0,1)$.

The HJM approach models forward-rate trees:

The HJM models capture the full dynamics of the entire forward rate curve. The following discussion uses the HJM one-factor model to illustrate the idea.

A single-factor HJM model sets $Q(t)$, the risk-neutral probability tree, to be constant at any node. For convenience, assume that $Q(t, s_t) = 1/2$. In addition, it is clear that $f(0,3) = P(0,3)/P(0,4)$, $f(0,2) = P(0,2)/P(0,3)$, $f(0,1) = P(0,1)/P(0,2)$, and $f(0,0) = 1/P(0,1)$. The initial nodes of the forward trees are known, which leaves $15 - 4 = 11$ degrees of freedom.

Why does modeling the $f(t, s)$ (where $0 \leq s < T$) trees avoid the problem that the $P(t, T)$ tree has?

This is illustrated next in a discrete setting, although the HJM model is originally developed in continuous time.

Use $T = 4$ to illustrate why. It is intuitively clear that using only the $f(t,3)$ tree and that $Q(t) = 1/2$ will not provide complete information regarding the $f(t,2)$ tree and the $f(t,1)$ tree. The reason is that the $f(t,3)$ tree only describes the evolvement of interest rates in the time period [3,4]. So the $f(t,3)$ tree does not provide enough information in other time periods. Note that, unlike in the case of the $P(t, T)$ tree, it is now possible to use $f(0,2)$, $f(0,1)$, and $f(0,0)$ as given inputs as well since the tree $f(t,3)$ does not provide the needed values for $f(0,2)$, $f(0,1)$, and $f(0,0)$, while $P(t,4)$ tree offers $P(0,3)$, $P(0,2)$, and $P(0,1)$ as outputs. Therefore, we can use independent Wiener processes to model forward rates, which is easier than modeling bond price processes.

HJM chooses to model the four trees $f(t,3)$, $f(t,2)$, $f(t,1)$, and $f(0,0)$ together. It is clear that the four trees can't be modeled independently. HJM models the instantaneous forward rates to follow the similar Ito process with different related parameters.

$$dX = \mu(X, t)dt + \sigma(X, t)dZ$$

In particular, HJM constructs the process as the Ito process below. Here is the model setting:

$$d\tilde{f}(t, s) = \mu(t, s)dt + \sigma(t, s)dZ_s(t)$$

where $\tilde{f}(t, s)$ is the continuously compounded *instantaneous* forward rate and $\mathbf{Z}_s(\mathbf{t})$ is the d-dimensional Wiener process where d is the number of forward rate processes being modeled and s is greater than or equal to zero and less than d. In this section, the four processes, $f(0,0)$, $f(0,1)$, $f(0,2)$, and $f(0,3)$, need to be modeled. Note that $d = 4$ and $s = 0,1,2,3$. Of course, $f(0,0)$ and $\mathbf{Z}_0$ are trivial and do not really need to be modeled. Since $f_\Delta(t, T) = e^{\tilde{f}(t,T)\times\Delta}$, then $\tilde{f}(t, T) = \frac{\log f_\Delta(t, T)}{\Delta}$.

An important result within the HJM framework is the no-arbitrage relation between $\mu(t, s)$ and $\sigma(t, s)$. The details involve a martingale property that is not covered here. We use the result that in a risk-neutral world, the drift $\mu(t, s)$ is a function of $\sigma(t, u)$ where $t \le u \le s$.

When Δ is infinitesimal, the discrete version of the model in the *real* world is:

$$\boldsymbol{E}_t\left(\frac{\log f_\Delta(\boldsymbol{t}+\Delta, \boldsymbol{T})}{\Delta} - \frac{\log f_\Delta(\boldsymbol{t}, \boldsymbol{T})}{\Delta}\right) = \mu^*(\boldsymbol{t}, \boldsymbol{T})\Delta$$

$$\mathbf{Var}_t\left(\frac{\log f_\Delta(\boldsymbol{t}+\Delta, \boldsymbol{T})}{\Delta} - \frac{\log f_\Delta(\boldsymbol{t}, \boldsymbol{T})}{\Delta}\right) = \sigma(\boldsymbol{t}, \boldsymbol{T})^2\Delta$$

When Δ is infinitesimal, the discrete version of the model in a *risk-neutral* world is:

$$\widetilde{\boldsymbol{E}_t}\left(\frac{\log f_\Delta(\boldsymbol{t}+\Delta, \boldsymbol{T})}{\Delta} - \frac{\log f_\Delta(\boldsymbol{t}, \boldsymbol{T})}{\Delta}\right) = \mu(\boldsymbol{t}, \boldsymbol{T})\Delta$$

$$\widetilde{\mathbf{Var}_t}\left(\frac{\log f_\Delta(\boldsymbol{t}+\Delta, \boldsymbol{T})}{\Delta} - \frac{\log f_\Delta(\boldsymbol{t}, \boldsymbol{T})}{\Delta}\right) = \sigma(\boldsymbol{t}, \boldsymbol{T})^2\Delta$$

This book does not venture further into the theory; it focuses on a risk-neutral world.

In a risk-neutral world the elements of the binomial trees are as follows under the constraint of $Q(t, S_t) = 1/2$:

$$\frac{\log f_{\Delta}(t+\Delta, T; s_{t+\Delta}) - \log f_{\Delta}(t, T; s_t)}{\Delta} = \mu(t, T, s_t)\Delta - \sigma(t, T, s_t)\sqrt{\Delta} \quad \text{in the up state} \tag{7.18}$$

$$\frac{\log f_{\Delta}(t+\Delta, T; s_{t+\Delta}) - \log f_{\Delta}(t, T; s_t)}{\Delta} = \mu(t, T, s_t)\Delta + \sigma(t, T, s_t)\sqrt{\Delta} \quad \text{in the down state} \tag{7.19}$$

In a multiplication format in a risk-neutral world:

$$f_{\Delta}(t+\Delta, T; s_{t+\Delta}) = \begin{cases} f_{\Delta}(t, T; s_t)e^{[\mu(t,T,s_t)\Delta]\Delta} \times e^{[-\sigma(t,T,s_t)\sqrt{\Delta}]\Delta} & \textit{if } s_{t+\Delta} = s_t u \\ f_{\Delta}(t, T; s_t)e^{[\mu(t,T,s_t)\Delta]\Delta} \times e^{[+\sigma(t,T,s_t)\sqrt{\Delta}]\Delta} & \textit{if } s_{t+\Delta} = s_t d \end{cases} \tag{7.20}$$

Note that in a risk-neutral world, the information of $\mu(t, T)$ is contained in the information of $\sigma(t, s)$, where $t \le s \le T$. The proof is beyond the level of this book. Here it is only necessary to model the process of $\sigma(t, T)$.

How can the volatilities of the forward rates $\sigma(t, T)$ be identified? Some approaches use historical volatility as an estimate of future volatility. Others (for example Black, Derman, and Toy) use the implied volatility from interest rate options. Jarrow (2020) briefly proposes both the historical volatility approach and an implicit volatility approach.

In the historical approach, Jarrow describes two models of volatility: a deterministic volatility model and a nearly proportional volatility function model. The implicit volatility approach uses market prices of interest rate options to derive the volatility structure. Next, we explore the historical approach using the nearly proportional volatility function presented in Jarrow (2020):

$$\sigma(t, T; s_t) = \eta(T - t)\min(\log f(t, T, s_t), M) \tag{7.21}$$

where $\eta() > 0$ is a deterministic function estimated from statistical analysis of historical data and M is a positive large constant. Equation 7.21 implies the larger the forward rate, the larger the volatility. M puts an upper bound on the forward rates.

The following equations of the binomial tree $f(t, T)$ are given with $\sigma(t, T; s_t)$ as inputs. Readers may go to Jarrow's book for the derivations.

Before providing the formulas, define:

$$\mathrm{Cos}\,h(x) = 1/2(e^{x} + e^{-x}) \tag{7.22}$$

The HJM model shows that in a risk-neutral world, the growth rate, $e^{[\mu(t,T,s_t)\Delta]\Delta} = \frac{\mathrm{Cos}\,h([\Sigma_{i=t+\Delta}^{T}\sigma(t,i;s_t)\sqrt{\Delta}]\,\Delta)}{\mathrm{Cos}\,h([\Sigma_{i=t+\Delta}^{T-\Delta}\sigma(t,i;s_t)\sqrt{\Delta}]\,\Delta)}$, is a function of $\sigma(t, s; s_t)$, $t \le s \le T$. Inserting this equation into Equation 7.20 obtains Equation 7.23.

$$f_\Delta(t+\Delta, T; s_{t+\Delta}) = \begin{cases} f_\Delta(t, T; s_t)\dfrac{\operatorname{Cos} h\left(\left[\sum_{i=t+\Delta}^{T}\sigma(t,i;s_t)\sqrt{\Delta}\right]\Delta\right)}{\operatorname{Cos} h\left(\left[\sum_{i=t+\Delta}^{T-\Delta}\sigma(t,i;s_t)\sqrt{\Delta}\right]\Delta\right)} e^{[-\sigma(t,T,s_t)\sqrt{\Delta}]\Delta} & \text{if } s_{t+\Delta} = s_t u \\ f_\Delta(t, T; s_t)\dfrac{\operatorname{Cos} h\left(\left[\sum_{i=t+\Delta}^{T}\sigma(t,i;s_t)\sqrt{\Delta}\right]\Delta\right)}{\operatorname{Cos} h\left(\left[\sum_{i=t+\Delta}^{T-\Delta}\sigma(t,i;s_t)\sqrt{\Delta}\right]\Delta\right)} e^{[+\sigma(t,T,s_t)\sqrt{\Delta}]\Delta} & \text{if } s_{t+\Delta} = s_t d \end{cases} \tag{7.23}$$

The risk-neutral probability is $Q = 1/2$. The next two sections provide concrete examples to generate the forward-rate trees $f(t,3)$, $f(t,2)$, $f(t,1)$, and $f(0,0)$. In particular, Section 7.6.1 generates the initial node of the tree. Section 7.6.2 generates the $\sigma(t, s; s_t)$ tree and the forward-rate trees $f(t, s)$ together.

It is worthy of mention that major interest rate models such as CIR, Vasicek, Ho-Lee, and Hull-White choose to model short-rate processes directly, whereas HJM models the instantaneous forward-rate process, which can be used to derive the short rate's dynamics. In fact, Ho-Lee and Hull-White are special cases of the HJM model with $\sigma(t, T) = \eta$ and $\sigma(t, T) = \eta e^{-\lambda(T-t)}$, respectively. CIR and Vasicek are the first generation of models that are not arbitrage-free. Unlike Hull-White and Ho-Lee, HJM can be generalized to being a multifactor model and it allows free choice of the volatility term structure. However, in the Hull-White, Ho-Lee, Vasicek, and CIR models, the discounted characteristic functions can be derived analytically. So, it offers semi-analytic methods for pricing bonds and derivatives. Another popular type of models, LMM models, utilize discrete forward LIBOR rates that are familiar to traders. There is not space here to discuss all the models. The HJM model serves as a representative model.

7.6.1 Stripping the Initial Forward Rate from Coupon Bonds

Chapter 2 discussed the stripping of the term structure of interest rates from the prices of coupon bonds. The methods of estimating initial forward rates from coupon bond prices are similar. The goal is to strip the coupon bonds to estimate the initial forward rates $f(0, s)$ for use in the HJM model where $0 \le s < T$. The initial forward rates can be easily derived from the initial bond price $P(0, s)$, where $0 \le s \le T$. The process begins by stripping the zero-coupon bond prices, $P(0, s)$, where $0 \le s \le T$, from the coupon bond prices. Equation 7.24 is a straightforward optimization approach for that task. Note that Equation 7.24 is shown with annual coupons, but with minor changes it can be generalized to semiannual coupons:

Choose $P(0, s)$ for $1 \le s \le \max\{T_i: i = 1,\dots,n\}$.

$$\text{Minimize } \sum_{i=1}^{n} \left(\boldsymbol{CB}(\mathbf{0},\ T_i) - C_i \sum_{s=1}^{T_i} \boldsymbol{P}(\mathbf{0},\ s) - \boldsymbol{F} \times \boldsymbol{P}(\mathbf{0},\ T_i)\right)^2 \tag{7.24}$$

In this formula, there are n coupon bonds as inputs, the number of variables is $\max\{T_i: i = 1,\dots,n\}$ since coupons are paid annually. Note that n may be strictly less than max{T_i: $i = 1,\ldots,n$}, the highest maturity of all bonds, (T_i). For example, assume there are three coupon bonds: a one-year coupon bond, a three-year coupon bond, and a five-year coupon bond. The algorithm searches for five values: $P(0,1)$, …, $P(0,5)$. In this case, there is no arbitrage-free strategy to replicate $P(0,1)$ …, $P(0,5)$ using the three coupon bonds. However, if the number of bonds, n, equals the highest maturity of $P(0,t)$, then the system of n equations can be uniquely solved which is equivalent to the bootstrap method in Chapter 2. In that case, the zero coupon bonds can be replicated by the coupon bonds.

The following example illustrates the idea.

Example 7.10: Given the following table of coupon bonds, find $P(0,1)$, $P(0,2)$, $P(0,3)$, $P(0,4)$, and $P(0,5)$ as well as $f(0,0)$, $f(0,1)$, $f(0,2)$, $f(0,3)$, and $f(0,4)$.

Year	Coupon *Ci*	CB(0, *i*)	*F*
1	2.25	101.238	100
2	2.75	103.448	100
3	2	102.941	100
4	2.5	105.853	100
5	1.75	103.64	100

In this case, assuming annual coupons (or easily adjusting the example to assume semiannual coupons) the problem is:

Choose $P(0, s)$ for $1 \le s \le 5$ to

$$\text{Minimize} \sum_{i=1}^{5} (\mathbf{CB}(\mathbf{0}, i) - C_i \sum_{s=1}^{i} P(\mathbf{0}, s) - FP(\mathbf{0}, i))^2.$$

The optimization can be done in an EXCEL spreadsheet using the Solver function (in Step 4):

Spreadsheet 7.16

	A	B	C	D	E	F	G	H
Row 1	Year	Coupon *Ci*	CB(0, *i*)	*F*	*P*(0, *i*)	Partial sum	Mse	*f*(0, *i* – 1)
Row 2					1	0		
Row 3	1	2.25	101.238	100	0.9901022	0.99010222	2.31646E – 09	1.01000
Row 4	2	2.75	103.448	100	0.9802936	1.97039586	2.1898E – 09	1.01001
Row 5	3	2	102.941	100	0.9705898	2.94098568	2.25259E – 09	1.01000
Row 6	4	2.5	105.853	100	0.9609804	3.90196605	2.30207E – 09	1.01000
Row 7	5	1.75	103.64	100	0.9514645	4.85343053	2.27895E – 09	1.01000
Row 8							1.13399E – 08	

Note that the spreadsheet starts with row 1 and column A. The row numbers and the column headers A through H are added for purposes of illustration. Note that columns A–D contain the bond data.

Step 1: Seed the $P(0, i)$ cell column with the top cell = 1 and with 0.9 in E3:E7. Do not copy formulas into column E; only values (which the optimizer will change) individually in each cell. Note that the values do not have to be equal or be 0.9. The values shown above reflect the final solution in column E rather than the input required in this step.

Step 2: Create a partial sum column in column F by putting F2 = 0, F3 = F2 + E3, and copying F3 to the rest of the column (to F7).

Step 3: Create the mean squared error in column G. Set G3 = (C3 – B3 × F3 – D3 × E3)^2 and then copy the rest of the column (except G8). Then set G8 = SUM(G3:G7).

Step 4: Go to EXCEL solver (in the data icon). Set the "target" as \$G\$8, and the cells to be changed as cells \$E\$3:\$E\$7. Then choose Min and then Solve. Column E in the above spreadsheet image is updated with the solution.

Step 5: Finally, since $P(0, t)$ has been calculated, calculate the forward rates $f(0,0), f(0,1), \ldots f(0,4)$ by setting H3 = E2/E3 and copying H3 to cells H7.

Alternatively, R can be used to solve the optimization:

R code 7.1 Strip initial forward rates from coupon bonds

```
#Initialize parameters zb = (P(0,1),…,P(0,5))
zb = c(0.9,0.9,0.9,0.9,.9)
#cb are coupon bonds
#c are coupons
# F face value
cb = c(101.238,103.448,102.941,105.853,103.64)
c = c(2.25,2.75,2,2.5,1.75)
F = c(100,100,100,100,100)
fr = function(zb){(cb[1]-(c[1] + F[1])*zb[1])^2 + (cb
[2]-c[2]*zb[1]-(c[2] + F[2])*zb[2])^2 + (cb[3]-c[3]*zb
[1]-c[3]*zb[2]-(c[3] + F[3])*zb[3])^2 + (cb[4]-c[4]*zb
[1]-c[4]*zb[2]-c[4]*zb[3]-(c[4]  +  F[4])*zb[4])^2  +
+(cb[5]-c[5]*zb[1]-c[5]*zb[2]-c[5]*zb[3]-c[5]*zb
[4]-(c[5] + F[5])*zb[5])^2}
result <- optim(zb,fr)
zb = c(1,result$par)
```

```
f = rep(0,5)
f[1] = zb[1]/zb[2]
f[2] = zb[2]/zb[3]
f[3] = zb[3]/zb[4]
f[4] = zb[4]/zb[5]
f[5] = zb[5]/zb[6]
zb
f
[1] 1.0000000 0.9900872 0.9802761 0.9705829 0.9609755
0.9514659
> f
[1] 1.010012 1.010008 1.009987 1.009998 1.009995
```

Note that using R is quicker, whereas EXCEL is more conducive to learning the intuition. The results match to the fourth digits after the decimals. Having generated the initial nodes of the forward trees, the next section generates the whole tree of forward rates.

Note that forward-rate agreements (FRAs for short) and swaps can also be used to identify zero coupon bond prices. Details can be found in Oosterlee and Grzelak's book, mentioned in the Chapter 3 extensions.

7.6.2 Building the Forward Rate and Volatility Trees

This section assumes $\Delta = 1$. The previous section found the initial node value, $f(0, s)$, where $0 \leq s < T$. The goal of this section is to build $T - 1$ forward trees $f(t,1), \dots f(t, T - 1)$ given $f(0, s)$, $\sigma(t, s)$. Note that all of the trees are not independent. The $f(0,0)$ tree contains only one element. The $f(t,1)$ tree has two columns with $t = 0$ and $t = 1$. The $f(t, T - 1)$ tree has T columns with $t = 0$ to $t = T - 1$.

Begin by inserting $\Delta = 1$ into Equation 7.23:

$$f(t+1, T; s_{t+1}) = \begin{cases} f(t, T; s_t)\dfrac{\operatorname{Cos} h([\Sigma_{i=t+1}^{T}\sigma(t, i; s_t)])}{\operatorname{Cos} h([\Sigma_{i=t+1}^{T-1}\sigma(t, i; s_t)])}e^{-\sigma(t,T,s_t)} & \text{if } s_{t+1} = s_t u \\ f(t, T; s_t)\dfrac{\operatorname{Cos} h([\Sigma_{i=t+1}^{T}\sigma(t, i; s_t)])}{\operatorname{Cos} h([\Sigma_{i=t+1}^{T-1}\sigma(t, i; s_t)])}e^{+\sigma(t,T,s_t)} & \text{if } s_{t+1} = s_t d \end{cases} \tag{7.25}$$

The example continues with $T = 4$.

Start with the $f(0,0)$ tree and end with the $f(t,3)$ tree.

- For the $f(0,0)$ tree, nothing is needed.
- For the $f(t,1)$ tree, to obtain $f(1,1, u)$ and $f(1,1, d)$:

Insert $t = 0$, $T = 1$:

$$f(1, 1, u) = f(0, 1) \times \frac{\text{Cos}\, h(\sigma(0, 1))}{\text{Cos}\, h(0)} \times e^{-\sigma(0,1)}$$

$$f(1, 1, d) = f(0, 1) \times \frac{\text{Cos}\, h(\sigma(0, 1))}{\text{Cos}\, h(0)} \times e^{\sigma(0,1)}$$

Input $f(0,1)$ and $\sigma(0, 1)$ to obtain the tree $f(t,1)$.

- For the $f(t,2)$ tree, to obtain $f(1,2, u)$ and $f(1,2, d)$:
  Insert $t = 0$, $T = 2$ into the following two states:

$$f(1, 2, u) = f(0, 2) \times \frac{\text{Cos}\, h(\sigma(0, 1) + \sigma(0, 2))}{\text{Cos}\, h(\sigma(0, 1))} \times e^{-\sigma(0,2)}$$

$$f(1, 2, d) = f(0, 2) \times \frac{\text{Cosh}(\sigma(0, 1) + \sigma(0, 2))}{\text{Cosh}(\sigma(0, 1))} \times e^{\sigma(0,2)}$$

Insert $t = 1$, $T = 2$, into the following four states:

$$f(2, 2, uu) = f(1, 2, u) \times \frac{\text{Cos}\, h(\sigma(1, 2, u))}{\text{Cos}\, h(0)} \times e^{-\sigma(1,2,u)}$$

$$f(2, 2, ud) = f(1, 2, u) \times \frac{\text{Cos}\, h(\sigma(1, 2, u))}{\text{Cos}\, h(0)} \times e^{\sigma(1,2,u)}$$

$$f(2, 2, du) = f(1, 2, d) \times \frac{\text{Cos}\, h(\sigma(1, 2, d))}{\text{Cos}\, h(0)} \times e^{-\sigma(1,2,d)}$$

$$f(2, 2, dd) = f(1, 2, d) \times \frac{\text{Cos}\, h(\sigma(1, 2, d))}{\text{Cos}\, h(0)} \times e^{-\sigma(1,2,d)}$$

Note that $f(0,2)$ and two σ-trees need to be input for the $f(t,2)$ tree:

Tree 1:
$\sigma(0,1)$

Tree 2:

$\sigma(0,2)\sigma(1,2,u)$
$\sigma(1,2,d)$

- To obtain the f(t,3) tree:

Insert $t = 0$, $T = 3$, into two states:

$$f(1,3,u) = f(0,3) \times \frac{\operatorname{Cos} h(\sigma(0,1) + \sigma(0,2) + \sigma(0,3))}{\operatorname{Cos} h(\sigma(0,1) + \sigma(0,2))} \times e^{-\sigma(0,3)}$$

$$f(1,3,d) = f(0,3) \times \frac{\operatorname{Cos} h(\sigma(0,1) + \sigma(0,2) + \sigma(0,3))}{\operatorname{Cos} h(\sigma(0,1) + \sigma(0,2))} \times e^{\sigma(0,3)}$$

Insert $t = 1$, $T = 3$, into four states:

$$f(2,3,uu) = f(1,3,u) \times \frac{\operatorname{Cos} h(\sigma(1,2,u) + \sigma(1,3,u))}{\operatorname{Cos} h(\sigma(1,2,u))} \times e^{-\sigma(1,3,u)}$$

$$f(2,3,ud) = f(1,3,u) \times \frac{\operatorname{Cos} h(\sigma(1,2,u) + \sigma(1,3,u))}{\operatorname{Cos} h(\sigma(1,2,u))} \times e^{\sigma(1,3,u)}$$

$$f(2,3,du) = f(1,3,d) \times \frac{\operatorname{Cos} h(\sigma(1,2,d) + \sigma(1,3,d))}{\operatorname{Cos} h(\sigma(1,2,d))} \times e^{-\sigma(1,3,d)}$$

$$f(2,3,dd) = f(1,3,d) \times \frac{\operatorname{Cos} h(\sigma(1,2,d) + \sigma(1,3,d))}{\operatorname{Cos} h(\sigma(1,2,d))} \times e^{\sigma(1,3,d)}$$

Insert $t = 2$, $T = 3$, into eight states:

$$f(3,3,uuu) = f(2,3,uu) \times \frac{\operatorname{Cos} h(\sigma(2,3,uu))}{\operatorname{Cos} h(0)} \times e^{-\sigma(2,3,uu)}$$

$$f(3,3,uud) = f(2,3,uu) \times \frac{\operatorname{Cos} h(\sigma(2,3,uu))}{\operatorname{Cos} h(0)} \times e^{\sigma(2,3,uu)}$$

$$f(3,3,udu) = f(2,3,ud) \times \frac{\operatorname{Cos}h(\sigma(2,3,ud))}{\operatorname{Cos}h(0)} \times e^{-\sigma(2,3,ud)}$$

$$f(3,3,udd) = f(2,3,ud) \times \frac{\operatorname{Cos}h(\sigma(2,3,ud))}{\operatorname{Cos}h(0)} \times e^{\sigma(2,3,ud)}$$

$$f(3,3,duu) = f(2,3,du) \times \frac{\operatorname{Cos}h(\sigma(2,3,du))}{\operatorname{Cos}h(0)} \times e^{-\sigma(2,3,du)}$$

$$f(3,3,dud) = f(2,3,du) \times \frac{\operatorname{Cos}h(\sigma(2,3,du))}{\operatorname{Cos}h(0)} \times e^{\sigma(2,3,du)}$$

$$f(3,3,ddu) = f(2,3,dd) \times \frac{\operatorname{Cos}h(\sigma(2,3,dd))}{\operatorname{Cos}h(0)} \times e^{-\sigma(2,3,dd)}$$

$$f(3,3,ddd) = f(2,3,dd) \times \frac{\operatorname{Cos}h(\sigma(2,3,dd))}{\operatorname{Cos}h(0)} \times e^{\sigma(2,3,dd)}$$

Note that $f(0,3)$ and three σ trees are needed as inputs to the $f(t,3)$ tree:

Tree 1:
$\sigma(0,1)$

Tree 2:

$\sigma(0,2)$	$\sigma(1,2,u)$	
	$\sigma(1,2,d)$	

Tree 3:

$\sigma(0,3)$	$\sigma(1,3,u)$	$\sigma(2,3,uu)$
		$\sigma(2,3,ud)$
	$\sigma(1,3,d)$	$\sigma(2,3,du)$
		$\sigma(2,3,dd)$

It is clear that the $f(t,s)$ trees are dependent. Note that the $f(t,3)$ tree depends on all σ trees.

Next, build the σ trees:

$$\sigma(t,T;s_t) = \eta(T-t)\min(\log f(t,T,s_t), M)$$

For the four-period tree, there is no need for a big M.

Note that the $f(t, s)$ tree is needed for $\sigma(t, s; s_t)$, where $0 \leq s < T$.

$f(t,1)$ tree:

$$\sigma(0, 1) = \eta(1) \times \log f(0, 1)$$

$f(t,2)$ tree:

$$\sigma(0, 2) = \eta(2) \times \log f(0, 2)$$

$$\sigma(1, 2, u) = \eta(1) \times \log f(1, 2, u)$$

$$\sigma(1, 2, d) = \eta(1) \times \log f(1, 2, d)$$

$f(0,3)$ tree:

$$\sigma(0, 3) = \eta(3) \times \log f(0, 3)$$

$$\sigma(1, 3, u) = \eta(2) \times \log f(1, 3, u)$$

$$\sigma(1, 3, d) = \eta(2) \times \log f(1, 3, d)$$

$$\sigma(2, 3, uu) = \eta(1) \times \log f(2, 3, uu)$$

$$\sigma(2, 3, ud) = \eta(1) \times \log f(2, 3, ud)$$

$$\sigma(2, 3, du) = \eta(1) \times \log f(2, 3, du)$$

$$\sigma(2, 3, dd) = \eta(1) \times \log f(2, 3, dd)$$

The inputs to the $\sigma(t, 1)$ tree are $\eta(1)$ and $f(0,1)$; the inputs to the $\sigma(t, 2)$ tree are $\eta(1)$, $\eta(2)$, $f(0,2)$, and $f(0,1)$; and, finally, the inputs to the $\sigma(t, 3)$ tree are $\eta(1)$, $\eta(2)$, $\eta(3)$, $f(0,3)$, $f(0,2)$, and $f(0,1)$. There are a seven total direct inputs to the σ trees and the forward-rate trees, $f(t, s)$: $\eta(1)$, $\eta(2)$, $\eta(3)$, $f(0,3)$, $f(0,2)$, $f(0,1)$, and $f(0,0)$. For the one-factor HJM model, there are inputs of the $Q(t)$ risk-neutral probability tree that have $1 + 2 + 4 = 7$ elements. Thus there are 14 parameters inputs to the one-factor HJM model when $T = 4$.

The next section provides a concrete example.

7.6.3 Building a Forward Tree and Valuing Interest Rate Derivatives

Using the formulas of forward rates and volatility trees from the previous section, this section completes the framework for valuing derivatives in Example 7.11.

Example 7.11: Assume that $f(0,3) = f(0,2) = f(0,1) = f(0,0) = 1.01$ as calibrated in the H column of the spreadsheet in Section 7.6.1. Following Jarrow's book, assume that the historical data implies $\eta(1) = 0.11765$, $\eta(2) = 0.08825$, and $\eta(3) = 0.06865$.

a. Generate four trees: $f(0,0)$, $f(t,1)$, $f(t,2)$, and $f(t,3)$.
b. Generate the $r(3)$ tree and the $Q(3)$ tree so that any derivative can be valued.

Solution of a. Create an EXCEL spreadsheet in the following format:

The first block is input of $\eta(1)$, $\eta(2)$, $\eta(3)$.

Lay out three σ trees horizontally. Next, layout the forward tree horizontally. Begin by building three parent nodes from the forward trees to the σ trees: $f(0,1), f(0,2), f(0,3) \rightarrow \sigma(0,1), \sigma(0,2), \sigma(0,3)$.

Then build the first generation of the three forward trees: $f(1,1,u)$, $f(1,1,d)$, $f(1,2,u)$, $f(1,2,d)$, $f(1,3,u)$, $f(1,3,d)$. Then build the corresponding first generation of each of the two σ trees: $f(1,2,u), f(1,2,d) \rightarrow \sigma(1,2,u), \sigma(1,2,d)$, $f(1,3,u), f(1,3,d) \rightarrow \sigma(1,3,u), \sigma(1,3,d)$.

Next, build the second generation of the two forward trees: $f(2,3,uu)$, $f(2,3,ud)$, $f(2,3,du)$, $f(2,3,dd)$, $f(2,2,uu)$, $f(2,2,ud)$, $f(2,2,du)$, $f(2,2,dd)$ and build the corresponding second generation of one σ tree: $f(2,3,uu)$, $f(2,3,ud)$, $f(2,3,du)$, $f(2,3,dd) \rightarrow \sigma(2,3,uu)$, $\sigma(2,3,ud)$, $\sigma(2,3,du)$, $\sigma(2,3,dd)$.

Finally, build the third generation of one forward tree: $f(2,2,uuu)$, $f(2,2,uud)$, $f(2,2,udu)$, $f(2,2,udd)$, $f(2,2,duu)$, $f(2,2,dud)$, $f(2,2,ddu)$, $f(2,2,ddd)$.

Here are the results with the convention that cells with (c) have formulas inserted rather than copied from other cells.

Spreadsheet 7.17

$\eta(1)$		$\eta(2)$			$\eta(3)$			Flat forward rate
0.11765(c)		0.08825(c)			0.06865(c)			1.01(c)
$\sigma(0,1)$		$\sigma(0,2)$	$\sigma(1,2,u)$		$\sigma(0,3)$	$\sigma(1,3,u)$	$\sigma(2,3,uu)$	
							$\sigma(2,3,ud)$	
			$\sigma(1,2,d)$			$\sigma(1,3,d)$	$\sigma(2,3,du)$	
							$\sigma(2,3,dd)$	
0.0011707(c)		0.0008781	0.0010675(c)		0.0006831	0.0008180	0.0009944(c)	
							0.0011869	
			0.0012741			0.0009385	0.0011410	
							0.0013618	
$f(0,1)$	$f(1,1,u)$	$f(0,2)$	$f(1,2,u)$	$f(2,2,uu)$	$f(0,3)$	$f(1,3,u)$	$f(2,3,uu)$	$f(3,3,uuu)$
								$f(3,3,uud)$
				$f(2,2,ud)$			$f(2,3,ud)$	$f(3,3,udu)$
								$f(3,3,udd)$
	$f(1,1,d)$		$f(1,2,d)$	$f(2,2,du)$		$f(1,3,d)$	$f(2,3,du)$	$f(3,3,duu)$
								$f(3,3,dud)$
				$f(2,2,dd)$			$f(2,3,dd)$	$f(3,3,ddu)$
								$f(3,3,ddd)$
1.01	1.0088190(c)	1.01	1.0091149(c)	1.0080388	1.01	1.0093120(c)	1.0084879	1.0074861
								1.0094918
				1.0101933(c)			1.0101391	1.0089416
								1.0113394
	1.0111837(c)		1.0108887(c)	1.0096023		1.0106918(c)	1.0097453	1.0085945
								1.0108987
				1.0121784			1.0116425	1.0102667
								1.0130220

Note that $r(0) = f(0,0)$, $r(1) = f(1,1)$, $r(2) = f(2,2)$, and $r(3) = f(3,3)$. Merge the columns of the forward trees, create the short-rate trees $r(t)$, and set $Q(t) = 1/2$. The resulting framework is ready to price almost any derivative. It can also generate $P(t,4)$ as was used in all previous sections.

Chapter Summary

This chapter detailed the HJM model and demonstrated its usefulness in pricing interest rate derivatives. The HJM model is an arbitrage-free model of fixed-income valuation derived in a risk-neutral framework that models the instantaneous forward rate, $f(t, T_1) = \lim_{T_2 \to T_1} f(t, T_1, T_2)$, as a d-dimensional Weiner process. This chapter focused on the single-factor HJM model to demonstrate applications of the model to fixed-income derivative pricing.

Similar to Jarrow (2020), this chapter illustrates the one-factor HJM model in a binomial tree setting. To avoid abstract theory, this chapter began with generalizing the risk-neutral probability in earlier chapters:

$$Q = \frac{e^{r\Delta t} - d}{u - d}$$

Note that an underlying asset needs to be specified to identify the up factor u and down factor d. A short rate $r(t)$ is also needed to column by column as discount values in earlier chapters. Here, the underlying asset tree is the bond price evolvement tree $P(t, T)$. So, it becomes intuitively clear that two trees (with only node values but not including probability measures), $P(t, T)$ tree and $r(t)$, are necessary without relying on a proof from the martingale property. Additionally, through examples, it is shown that it is equivalent to use a zero coupon bond $P(0, T)$ and money market fund $m(t)$ to hedge any derivative. An attractive feature of the HJM model is that, unlike other seminal interest rate models, the prices of derivatives and the specification of the hedge are developed together. Furthermore, the hedging approach here is similar to earlier approaches in that a bond is used instead of a stock as the underlying asset and a money market fund is used instead of a constant risk-free rate to hedge a call option.

In particular, after understanding that two trees with node values only are necessary for derivative pricing and hedging, this chapter adopts Jarrow's (2020) examples to price and hedge the options on bonds, cap, floor, swap, forward, and futures on bonds. During the process, this chapter clearly illustrates why some derivatives do not require the HJM model for valuation. This chapter also clarifies why some hedges need to focus only on derivative values at any tree node, while others need to focus on derivative values and derivative cash flows at any tree node. For improved accessibility, this chapter uses concrete, hands-on EXCEL examples rather than a theoretical approach, using martingale properties and just numbers in a binomial tree setting as in Jarrow (2020). Although this chapter does not prove theory, it offers intuition accessible to undergraduates about the otherwise-abstract HJM models. The end of the chapter briefly discusses the estimation of parameters, including estimation of volatilities for the forward rate's evolvement so it completes the demonstration of the model.

Extensions and Further Reading

This chapter covers the HJM one-factor model. The HJM model is a term structure model that allows several factors to be used to describe the process of the instantaneous forward rate. Monte Carlo simulation is especially useful in real-world applications because it can be used in applications where trees grow exponentially.

From a theoretical perspective, the following book is useful for exploring the details of the HJM model:

> *Modeling Fixed Income Securities and Interest Rate Options* by Robert Jarrow, Third edition, CRC Press.

This book covers derivatives with underlying assets including equities, equity indexes, interest rates, and bonds. It does not cover underlying assets such as foreign exchange rates. However, the models on derivatives with underlying asset foreign exchange rates are presented in the Black-Scholes and interest rate models frameworks. Readers can explore the details in *Mathematical Modeling and Computation in Finance: With Exercises and Python and MATLAB Computer Codes* by Cornelis W. Oosterlee and Lech A. Grzelak.

As previously noted, not all models can be fully covered in this book. However, this book provides a wide spectrum of valuable models. It is important to become familiar with a variety of models. Do not "fall in love" with your favorite model. Fundamental thinking and the implementation of relevant skills are essential – and they are best acquired through a broad knowledge of diverse models and methods.

End-of-Chapter Problems (Fundamental Problems)

1. Assume that the time period is $\Delta = 1$ and all interest rates are deterministic in each period. Using five years as the time period, what is the minimal set of information needed to obtain all information about zero coupon bond prices, $P(t, s)$, where $0 \le t < s \le 5$; forward rates, $f(t, s)$ where $0 \le t \le s \le 4$; short rates, $r(t)$ where $0 \le t \le 4$; and money market funds, $m(t)$ where $1 \le t \le 5$. Show all work as in Example 7.1.
2. Assume that four-year zero coupon bond tree, $P(t,4)$, is given and the short-rate tree, $r(t)$, is given as in Example 7.3.
   a. Build a synthetic two-year bond tree $P(t,2)$ using the four-year zero coupon bond and money market fund.

b. Find the number of units in a four-year zero coupon bond, $P(0,4)$, and the number of units in a money market fund, $m(t)$, needed at each node to replicate this synthetic bond $P(0,2)$.

3. Assume that the four-year zero coupon bond tree is given as $P(t,4)$ and the short-rate tree, $r(t)$, is given as in Example 7.3. In Example 7.3, $P(0,3)$ is shorted by going long a synthetic (replicating) $P(0,3)$ portfolio using the four-year zero coupon bond, $P(0,4)$, and money market fund $m(t)$. Find the hedge ratio at time $t = 0$. Later, the duration hedge is used to hedge $P(0,3)$ by $P(0,4)$. The hedge ratio is different at time $t = 0$. Show the duration hedge will have a risk (up state and down state are different) at $t = 1$. Assume the evolving of the market follows the HJM framework and that the forward curve is flat at 1.01. Therefore, there is no difference whether duration or modified duration is used to form the hedge.

4. Given a $P(t,4)$ tree and the $r(t)$ tree over a four-year period as in Example 7.3:
   a. Find the value of a two-year European put option on a three-year zero coupon bond with a strike of $K = 0.991$.
   b. What is the put's replicating strategy?

5. Given a $P(t,4)$ tree and the $r(t)$ tree over a four-year period as in Example 7.3. Assume that if the option is exercised at time t, the option holder receives the next coupon at time t + 1, not at time t.
   a. Find the value of a two-year American call option on a three-year 1% coupon bond paid every year with a strike of $K = 0.99$.
   b. What is the call's replicating strategy using the underlying coupon bond?

6. Given a $P(t,4)$ tree and the $r(t)$ tree as before, consider a three-year floor at strike, $K = 1.02$:
   a. Determine the price of the floor.
   b. Use the bond tree $P(0,4)$ to form a hedge.

7. Given: $P(0,3) = 0.9705901$, $P(0,2) = 0.9802960$, and $P(0,1) = 0.9900990$. Note that these are consistent with the $P(t,4)$ tree and the $r(t)$ tree given in Example 7.3. Also, the $P(t,4)$ tree and the $r(t)$ tree are given as before.
   a. Find the price of a four-year swap that pays annually and has $L = 1$.
   b. Replicate the swap using the zero coupon bond in $P(0,4)$.

8. Finish the details in Example 7.8.

9. Given: $P(0,3) = 0.9705901$, $P(0,2) = 0.9802960$, and $P(0,1) = 0.9900990$. Note that these are consistent with the $P(t,4)$ tree and the $r(t)$ tree given in Example 7.3. Also the $P(t,4)$ tree and the $r(t)$ tree are given as before.
   a. Find the forward price, $F(0,3,4)$.
   b. Hedge the forward using $P(0,4)$.

10. Finish the details in Example 7.9.
11. Given $P(0,3) = 0.9705901$, $P(0,2) = 0.9802960$, and $P(0,1) = 0.9900990$. Note that these are consistent with the $P(t,4)$ tree and the $r(t)$ tree given in Example 7.3. Also, the $P(t,4)$ tree and the $r(t)$ tree are given as before.
   a. Find the future price, $\mathcal{F}(0,3,4)$.
   b. Hedge the futures using $P(0,4)$.

12. Assume that $f(0,2) = f(0,1) = f(0,0) = 1.02$. Follow the approach of Jarrow (2020) and assume that the historical data implies $\eta(1) = 0.11$, $\eta(2) = 0.09$.
   a. Generate three trees: $f(0,0)$, $f(t,1)$, $f(t,2)$.
   b. Generate the $r(2)$ tree and the $Q(2)$ tree so that any derivative can be valued.

Note

1 This chapter generally uses the same number of time steps (4) as Jarrow (2020) and the same selections of maturity times (e.g., 1, 2, 3 or 4) for most of the financial contracts in the examples due to the difficulty of depicting large non-recombining trees in a book format and therefore the limited number of time steps (4). This chapter also uses Jarrow's parameter values that he based on historical data, since estimation of those values is not a goal of this brief analysis. Although parameters and maturity choices here follow Jarrow (2020), this chapter departs substantially from Jarrow by using EXCEL spreadsheets, as well as other perspectives.

Reference

Robert, A. Jarrow. (2020). *Modeling fixed income securities and Interest rate options (Third edition)*. CRC Press.

8

Credit Risk and Credit Derivatives

This chapter first examines various default probabilities and how to extract default probabilities from market prices, and then it introduces single-name credit derivatives, in particular, credit default swaps (CDSs). Valuation of credit derivatives is discussed both from the analytical perspective and from the simulation perspective. This chapter also discusses multiple-name credit derivatives, primarily collateralized debt obligations (CDOs). First, CDOs are discussed with the assumption of uncorrelated bond defaults. Then, the effects of correlation and the simulation of the prices of CDOs with positively correlated bond defaults are considered. Finally, CDO squareds are briefly discussed. The chapter concludes with a discussion of credit derivatives and the 2007–2008 global financial crisis.

8.1 Default Probabilities and Extract Default Probabilities from Market

Default risk is the chance that companies or individuals will fail to make the required payments on their debt obligations. ***Credit risk*** is often interpreted as being synonymous with default risk, but can also be used to include broader issues such as the potential economic effects of changes in credit spreads or credit ratings. A ***credit rating*** is an evaluation of the credit risk of a prospective debtor, which is an implicit forecast of the likelihood of the debtor defaulting. Credit ratings are important metrics within the financial industry. Rating agencies analyze the credit risk of most publicly traded credit-based securities. Rating agencies are for-profit firms that in recent decades have been primarily paid by firms to rate their new securities rather than being paid by investors subscribing to their publications (as was common decades ago). In the United States, the ***SEC (Securities and Exchange Commission)*** recognizes several ***NRSROs (Nationally Recognized Statistical Rating Organizations)*** that may provide information that may be relied upon for regulatory purposes. The credit rating industry has consolidated into three major rating agencies: Moody's, S&P, and Fitch. This book uses the notation of Moody's:

Aaa: highest rating with almost no chance of default.
Aa1–Baa3: investment grade bonds.

Ba1–Caa3: high yield or junk bonds.
C / (sometimes referred to as D): lowest rating; used for bonds in default.

Note that Aa indicates greater safety than A and similarly Baa better safety than Ba. Most rating agencies provide further detail. For example, the highest Moody's rating within the Baa category is Baa1 and the lowest rating within the Baa category is Baa3.

Modeling the probability of default is the starting point for credit rating and credit risk modeling. The default probability is one of the most important factors in credit derivative pricing. This section defines various default probabilities and discusses their relations. If one knows the value of one type of default probability, it can be used to derive the rest.

8.1.1 Types of Default Probabilities

This section introduces five default probabilities from a mathematical perspective. It is assumed that default can only occur once. Let t be a positive real number which represents time t. The five key probabilities related to default are:

1. ***Cumulative default probability P(t)***:

$$P(t) = P(\textbf{default before } t) \tag{8.1}$$

$P(t)$ is the chance that default occurs before time t. $P(5)$, for example, would be the probability that the corporation defaults before year 5. $P(t)$ is a cumulative probability function.

2. ***Cumulative survival probability*** $Q(t)$:

$$Q(t) = P(\textbf{survival until } t) \tag{8.2}$$

$Q(t)$ is the chance that the entity survives through time t. Note that survival is the opposite of default. Therefore,

$$P(t) = 1 - Q(t) \tag{8.3}$$

3. ***Conditional default probability***:

$$P_{t+\Delta t\,|\,t} = P(\textbf{default between}(t,\, t+\Delta t)\,|\,\textbf{no default before } t)$$

$$= \frac{P(\textbf{default happens } (t,\, t+\Delta t))}{P(\textbf{no default happens before } t)}$$

$$= \frac{P(t + \Delta t) - P(t)}{1 - P(t)}$$

$$P_{t+\Delta t \mid t} = \frac{Q(t) - Q(t + \Delta t)}{Q(t)} \tag{8.4}$$

4. ***Default intensity*** *λ(t)*:
   Default intensity, also called the ***hazard rate***, is denoted by $\lambda(t)$:

$$\lambda(t) = \lim_{\Delta t \to 0} \frac{P_{t+\Delta t \mid t}}{\Delta t} \tag{8.5}$$

By simple calculus, the relation between $\lambda(t)$ and $Q(t)$ is:

$$Q(T) = Q(t)e^{-\int_t^T \lambda(t)dt} \tag{8.6}$$

In particular, if $\lambda(t)$ is constant for $(t, t + \Delta t)$, then

$$\lambda = \frac{P_{t+\Delta t \mid t}}{\Delta t}$$

Proof
By definition,

$$\lambda(t) = \lim_{\Delta t \to 0} \frac{P_t + \Delta t \mid t}{\Delta t}$$

Insert Equation 8.4 into the above equation,

$$\lambda(t) = \lim_{\Delta t \to 0} \frac{\frac{Q(t) - Q(t + \Delta t)}{Q(t)}}{\Delta t}$$

Switching $Q(t)$ and Δt,

$$\lambda(t) = \lim_{\Delta t \to 0} \frac{\frac{Q(t) - Q(t + \Delta t)}{\Delta t}}{Q(t)}$$

Using the definition of the derivative,

$$\lambda(t) = \frac{-\frac{dQ}{dt}}{Q(t)}$$

Multiply each side by $-dt$:

$$\frac{dQ}{Q} = -\lambda(t)dt$$

Equivalently,

$$dLnQ = -\lambda(t)dt$$

Integrate both sides from t to T,

$$LnQ(T) - LnQ(t) = -\int_t^T \lambda(t)dt$$

Solve for $Q(T)$,

$$Q(T) = Q(t)e^{-\int_t^T \lambda(t)dt}.$$

In particular, if $t = 0$ and $\lambda(t)$ is the constant λ, then

$$\boldsymbol{Q(T) = Q(0)e^{-\lambda T} = e^{-\lambda T}} \tag{8.7}$$

Now, letting t be a positive integer that represents the time gives the following probability related to default:

5. ***Non-cumulative default probability*** P_t:

$$\boldsymbol{P_t = P(\text{default probability during } t\text{th year})} \tag{8.8}$$

The non-cumulative default probability, P_t, is the probability that default will occur in year t. So, the cumulative default probability is the sum of the non-cumulative default probabilities for each year less than t:

$$\boldsymbol{P(t) = P_1 + P_2 + \dots + P_t} \tag{8.9}$$

Equivalently:

$$\boldsymbol{P_t = P(t) - P(t-1) = Q(t-1) - Q(t)} \tag{8.10}$$

For a low credit-risk investment, P_t tends to be an increasing function of t since a firm currently in good financial health will typically take several years to decline in health to the point of default. For a poor grade investment, P_t is likely to be a decreasing function of t since most distressed firms that have survived the next few years are likely to have recovered financial strength.

Consider two Examples: 8.1 and 8.2.

Example 8.1: Given $P_3 = 10\%$ and $P(2) = 30\%$, find $P_{3|2}$ and λ between years 2 and 3 assuming $P_{3|2}$ and λ are constants between year 2 and year 3.

$$P_{3|2} = P(\text{default between year 2 and year 3} \,|\, \text{no default before year 2})$$
$$= \frac{P_3}{Q(2)} = \frac{P_3}{1 - P(2)} = \frac{10\%}{1 - 30\%} = 14.29\%$$

If λ is constant between year 2 and year 3,

$$\lambda = \frac{P_{3|2}}{3 - 2} = 14.29\%.$$

Example 8.2: Assuming that hazard rate $\lambda = 1\%$, find $Q(3)$, $P(3)$, P_3, and $P_{3|2}$.

Given $Q(3) = e^{-0.01\times 3} = 0.9704$, then

$$P(3) = 1 - Q(3) = 0.02955,$$

$$P_3 = Q(2) - Q(3) = e^{-0.01\times 2} - e^{-0.01\times 3} = 0.009753,$$

$$P_{3|2} = \frac{P_3}{Q(2)} = 0.0099501 \,(\text{or approximately } \lambda\Delta t = \lambda = 1\%).$$

In the following sections, the above five default probabilities are calculated based on three types of security prices: bond prices, asset swap prices, and credit default swap prices. Note that these default probabilities are derived from three types of security prices, and are risk-neutral estimates. In order to do so, the expected cash flows are found, assuming a risk-neutral world. Recall that in Chapter 1's Mathematics Review, the expected value, $E(X) = \mu_X = \sum_x xp(x)$, is the *probability weight* × *value*. In this chapter, cash flows occur at different times; therefore, an expected cash flow is calculated using the following formula to adjust for the time value of money:

$$\textbf{Expected cash flow} = \textstyle\sum_t \textbf{probability weight} \times \textbf{cash flow} \times \textbf{time discount factor} \quad (8.11)$$

The time discount factor is less than one and adjusts cash flows for the time value of money.

In finance, an expected cash flow (for example an expected loss) and an ***expected cash flow expressed as a present value*** are different. The first one does not account for time value of money; the second one does. In the rest of our chapter, we simplify the notation by referring to expected cash flows/loss discounted for the time value of money as simply ***expected cash flows/loss***. So, the above formula (containing a time discount factor) is different from the formula typically used in mathematics because it includes a time discount factor.

In the following three sections, the default probabilities are extracted from a zero coupon bond, a coupon bond, and an asset swap, respectively, assuming a risk-neutral world. Unlike Chapter 2 and Chapter 7, where the bonds are mostly default-risk-free securities such as Treasury bills, notes, and bonds, the bonds in this chapter are mostly defaultable (i.e., credit risky) bonds.

8.1.2 Deriving Default Probabilities from *N*-year Risk-Free Zero Coupon Bonds and Defaultable Zero Coupon Bonds

Recall from Chapter 2 that the T-year risk-free bond price at time t, $B(t,T)$, which may be a zero-coupon bond or non-zero-coupon bond, is a function of the risk-free rate, r. $B(t,T)$ is calculated by discounting the bond's face value (and coupons if any) by the risk-free rate. The T-year *risky bond* (often called a ***defaultable bond*** in this chapter to emphasize the default probability as the source of the risk rather than only interest rate uncertainty) price at time $t, \hat{B}(t, T)$, is a function of the bond yield y. $\hat{B}(t, T)$ is calculated by discounting the face value (and coupons if any) by the bond yield y. For a given face value and coupon rate, the risk-free rate, r, assuming constant, and the risk-free bond price, $B(t,T)$, can be viewed as a pair. If either the price or the risk-free rate is known, then the other can be calculated. Similarly, if the risky bond price $\hat{B}(t, T)$ is known, then y can be calculated and vice versa.

In order to derive a default probability from a defaultable bond price, one must know the bond's recovery rate, R. The ***recovery rate*** is the amount of money the bondholder will recover in the event of default, expressed as a rate. The recovery rate may be viewed as the value of the bond immediately after the issuer defaults expressed as percent of its face value. A typical recovery rate for a corporate bond is approximately 40%. The higher the anticipated recovery rate, the higher the current bond price and the lower the yield.

The objective is to find one of the five types of default probabilities using the available defaultable bond information. That default probability can then be converted into any of the other default probabilities (Figure 8.1).

This section begins with the calculations for defaultable zero coupon bonds and then performs the calculations for defaultable non-zero-coupon bonds in the next section.

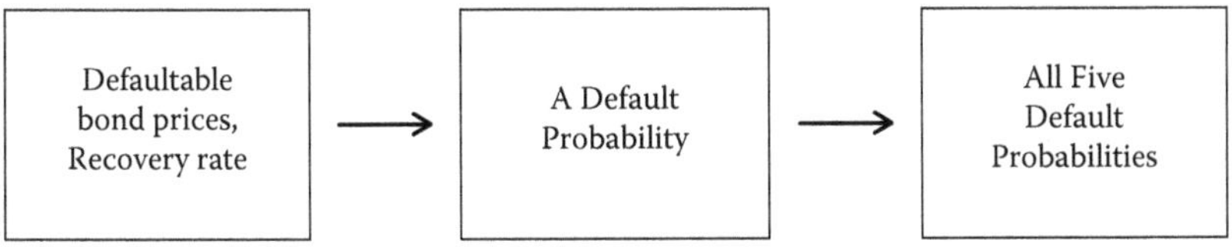

FIGURE 8.1
Extract Default Probabilities from Defaultable Bond Prices.

Given that a T-year risk-free zero coupon bond has an interest rate, r, the T-year defaultable zero coupon bond has a yield $y = r + s$, where s is the ***yield spread***. Given the defaultable bond's recovery rate, R, and its yield spread s above the risk-free rate, it can be shown that the default intensity can be approximated by:

$$\lambda = \frac{s}{1 - R} \tag{8.12}$$

In this equation, it is assumed that λ is constant and, for simplicity, that the recovery cash flow occurs at maturity. In practice, bond defaults are possible throughout a bond's life and recoveries usually take several years from the point of default. The following fundamental equation is needed to derive the above equation.

8.1 Fundamental Equation for a Defaultable Bond

$$\sum \textbf{probability weight} \times \textbf{loss} \times \textbf{time discount factor}$$
$$= \boldsymbol{B(0, T) - \hat{B}(0, T)} \tag{8.13}$$

where $B(0, T)$ is the T-year risk-free bond price today and $\hat{B}(0, T)$ is a T-year defaultable bond price today. It is assumed that all calculations are in a risk-neutral world.

The left side of Equation 8.13 is the expected loss expressed in a risk-neutral world while the right side is the bond price difference, which is a reflection of the loss in the market.

In this case, only zero coupon bonds are considered, so $B(0, T)$ is the T-year risk-free zero coupon bond price today and $\hat{B}(0, T)$ is a T-year defaultable zero coupon bond price today.

Assume one year as the unit of time. The *probability weight* is the non-cumulative default probability in each year, the loss in the event of default is $L(1-R)$ where L is the bond's face value (principal), R is the bond's recovery rate, and the time discount factor is e^{-ri} for year i.

Let P_i be the non-cumulative default probability in year i. From Equations 8.7 and 8.10, $P_i = Q(i-1) - Q(i) = e^{-\lambda(i-1)} - e^{-\lambda i}$, given default intensity λ as a constant, the loss given default is $L(1-R)$, and e^{-rT} is the time discount factor since the recovery value is assumed to be paid at the bond's scheduled maturity date (T).

$$\text{Expected loss} = (e^{-\lambda 0} - e^{-\lambda 1}) \times L(1-R) \times e^{-rT} + (e^{-\lambda 1} - e^{-\lambda 2}) \times L(1-R) \times e^{-rT} + \ldots + (e^{-\lambda(T-1)} - e^{-\lambda T}) \times L(1-R) \times e^{-rT}$$

Through simplification,

$$\text{Expected loss} = (e^{-\lambda 0} - e^{-\lambda T})L(1-R) \times e^{-rT}$$

$$\text{Bond price difference} = B(0,T) - \hat{B}(0,T) = Le^{-Tr} - Le^{-yT} = Le^{-Tr} - Le^{-(r+s)T}.$$

Thus,

$$(e^{-\lambda 0} - e^{-\lambda T})L(1-R) \times e^{-rT} = Le^{-rT} - Le^{-(r+s)T}.$$

Which is equivalent to

$$(1 - e^{-\lambda T})(1-R) = 1 - e^{-sT}.$$

If λT and sT are small, then using a Taylor expansion, $e^x \approx 1 + x$, to simplify:

$$(1 - 1 + \lambda T)(1-R) \approx 1 - 1 + sT$$

$$\lambda = \frac{s}{1-R} = \frac{y-r}{1-R}$$

and, $s = \lambda(1-R)$.

Note that the default intensity is a function of bond yield spread s and recovery rate R. The higher the bond yield spread, the higher the probability of default; the lower the bond recovery rate, the higher the bond yield spread. This formula is a rough approximation; however, it is used in the financial industry.

8.1.3 Deriving Default Probabilities from Coupon Bond Prices

The fundamental equation from the previous section can be applied for the case of coupon bonds; namely, that the expected loss is equal to the difference in prices between the T-year defaultable and otherwise equivalent risk-free bonds. The default probability can be derived from the coupon bond price and its recovery rate. An example similar to Hull's is used to illustrate the idea.

Example 8.3: Assume the risk-free rate is $r = 2.5\%$, a three-year defaultable coupon bond with a coupon of 6% paid annually (at the end of each year) has yield $y = 4.5\%$, and the recovery rate of the risky bond $R = 40\%$. Assume that default can only occur in the middle of a year and the recovery amount is paid immediately at default. Find the default probability P for each year assuming that P is constant ($P_1 = P_2 = P_3 = P$).

For convenience, assume that each bond's face value (principal) is 100, although it could be any face value, L. Equation 8.13 is used here similar to its use in Section 8.1.2.

$$\sum \text{probability weight} \times \text{loss} \times \text{time discount factor} = B(0, T) - \hat{B}(0, T)$$

Assuming that each probability can only occur in the middle of each year, there are three terms, each with a probability weight of P. Now consider the loss if default occurs in the middle of the first year. Without default, the bond buyer is going to receive \$6 (coupon rate×face value) at the end of the first, second, and third years in addition to the face value at the end of the third year. At time $t = 0.5$, the present value of the bond's future cash flows in a risk-neutral world is $B(0.5,3)$. In the case of default, the bondholder is only going to receive $100 \times R$. The difference is the loss: $B(0.5,3)-100R$, if default occurs at $t = 0.5$. Similarly, if the default occurs at $t = 1.5$, the loss is $B(1.5,3)-100R$, and if the default occurs at $t = 2.5$, the loss is $B(2.5,3)-100R$. The expected loss (the left side of the previous equation) is the probability-weighted sum of the loss, each multiplied by its time discount factor:

$$P \times (B(0.5, 3) - 100R) \times e^{-r\times 0.5} + P \times (B(1.5, 3) - 100R) \times e^{-r\times 1.5} + P \times (B(2.5, 3) - 100R) \times e^{-r\times 2.5}$$

where r is risk-free interest rate. The right side in the following equation is the difference of the prices of the defaultable and risk-free bonds.

$$P \times (B(0.5, 3) - 100R) \times e^{-r\times 0.5} + P \times (B(1.5, 3) - 100R) \times e^{-r\times 1.5} + P \times (B(2.5, 3) - 100R) \times e^{-r\times 2.5} = B(0, 3) - \hat{B}(0, 3)$$

Note that:

$$B(0.5, 3) = 6 \times e^{-r\times 0.5} + 6 \times e^{-r\times 1.5} + 106 \times e^{-r\times 2.5}$$

$$B(1.5, 3) = 6 \times e^{-r\times 0.5} + 106 \times e^{-r\times 1.5}$$

$$B(2.5, 3) = 106 \times e^{-r\times 0.5}$$

$$B(0, 3) = 6 \times e^{-r\times 1} + 6 \times e^{-r\times 2} + 106 \times e^{-r\times 3}$$

$$\hat{B}(0, 3) = 6 \times e^{-y\times 1} + 6 \times e^{-y\times 2} + 106 \times e^{-y\times 3}$$

where y is the yield of the defaultable bond.

Finally, we can solve for P because in the equation P is the only unknown. It is very easy to do the above process in a spreadsheet.

Spreadsheet 8.1

R1	Table 1											
R2	Col.B	Col.C	Col.D	Col.E	Col.F	Col.G	Col.H	Col.I	Col.J	Col.K	Col.L	Col.M
R3	T	*D* factor	Cash	100×*R*	*B*(*t*,*T*)	Loss	Expected loss	P		*r*	*y*	*R*
R4	0.5	0.98758	6	40	111.2824	71.28242	2.17137267	0.03084		0.025	0.045	0.4
R5	1.5	0.96319	6	40	108.0241	68.02408	2.02095772	0.03084				
R6	2.5	0.93941	106	40	104.6832	64.68325	1.87425653	0.03084				
R7							6.06658693					

Above cell is Expected Loss=Sum of column G

R10	Table 2			
R11	T	Cash	*B*(0,3)	$\tilde{B}$(0,3)
R12	1	6	5.85186	5.735985
R13	2	6	5.70738	5.483587
R14	3	106	98.3408	92.61389
R15		Sum	109.9	103.8335
R16				

R17 Difference of $B(0,3)-\tilde{B}(0,3)$

R18 6.066587

R20 Target Cell=Difference−Expected Loss

R21 2.31E−14

The solution can be found using EXCEL's Solver or a similar function in other spreadsheet software:

Step 1. Build up Table 1

First, enter values in columns A through G: column B values are times 0.5, 1.5, 2.5; column C is $e^{-r\times 0.5}$, $e^{-r\times 1.5}$, $e^{-r\times 2.5}$; column D contains cash flows; column E is the recovery amount $40; column F is $B(0.5,3)$, $B(1.5,3)$, and $B(2.5,3)$; column G is the loss $B(t,3)$–40. Columns K through M have the inputted risk-free rate, yield, and recovery rate. In fact, they are not columns, just three numbers. Then create column H as: column I times column G times column C. Leave the first cell (I4) of column I empty and enter formulas to make I5 equal to I4, and I6 equal to I4. Then form H7 as the sum of column H.

Step 2. Build up Table 2

Column B contains years 1, 2, 3; column C contains cash flows; column D contains the present value of $6 in year 1, present value of $6 in year 2, and present value of $106 in year 3, all discounted by the risk-free rate; column E contains the present values of $6 in year 1, present value of $6 in year 2, and present value of $106 in year 3 discounted by the bond yield. Sum columns D and E in Table 2 to get cell D15 as $B(0,3)$ and cell E15 as $\hat{B}(0, 3)$.

Step 3. Calculate several important cells

Enter the formula in B18 as the difference of $B(0, 3) - \tilde{B}(0, 3)$, then make B21 as the difference of H7 and B18 which is the difference of Expected Loss (left-hand side of the Equation 8.13) and the bond price difference (right-hand side of Equation 8.13).

Step 4. Open Solver-type function

Set the target cell B21=0 (Left side=Right side) by adjusting cell I4, the probability.

The Solver finds the solution is P=0.03084, which is shown in I4 (and the cells below I4).

Example 8.4 examines a longer-term defaultable bond with similar default risk to the default risk in the previous bond example (Example 8.3) over its first three years, namely, assume $P_1 = P_2 = P_3 = 0.03084$ as before.

> **Example 8.4:** Assume the risk-free rate is $r = 2.5\%$, that a five-year risky coupon bond with a coupon of 5% paid annually has yield: $y = 4.75\%$, and the recovery rate of the risky bond $R = 40\%$. Assume that default can only occur in the middle of each year and the recovery amount is paid immediately at default. Find the default probability $\hat{P}$ for years 4 and 5 using the P calculated in the previous problem for the first three years. Assume $\hat{P}$ is constant, $P_4 = P_5 = \hat{P}$, and that $P_1 = P_2 = P_3 = P$, which is constant and is equal to the value calculated in the previous example.

Inserting into the bond pricing equation used in the previous example,

$$\underline{P} \times [\underline{B(0.5, 5) - 100R}] \times \underline{\quad e^{-r\times 0.5}} + \underline{P} \times [\underline{B(1.5, 5) - 100R}] \times \underline{\quad e^{-r\times 1.5}}$$
$$+ \underline{P} \times [\underline{B(2.5, 5) - 100R}] \times \underline{\quad e^{-r\times 2.5}} +$$
$$\underline{\hat{P}} \times [\underline{B(3.5, 5) - 100R}] \times \underline{\quad e^{-r\times 3.5}} + \underline{\hat{P}} \times [\underline{B(4.5, 5) - 100R}] \times \underline{\quad e^{-r\times 4.5}}$$
$$= B(0, 5) - \hat{B}(0, 5)$$

Note that in this equation, the coupon is $5. It does not need to be same as the coupon rate in Example 8.3.

There is now one unknown $\hat{P}$, which can be solved given the default probability from the first three years found using the price of the three-year bond. So, using two bonds of different maturities, we can find the two default probabilities relating to different time periods. Extending this example, if there are N bonds with maturities ranging from 1 to N the default probabilities can be calculated for each year.

If it is assumed instead that the bond default intensity, λ, is constant, then one bond price can be used to solve for one default intensity, λ. Using multiple bond prices, piecewise constant default intensities can be found.

8.1.4 Deriving Default Probabilities from Asset Swaps

This section discusses another product related to interest rates: asset swaps. Asset swaps have a specified rate, s, which is negotiated when the contract is initialized. The ***asset swap rate***, s, is also called the ***price of the asset swap***. The asset swap price, s, can be used to derive all of the five default probabilities. The price of an asset swap can be derived from bond prices (Figure 8.2).

Asset swaps are financial derivative contracts between two parties. In its simplest form, the holder of an asset delivers cash flows from that asset to a counterparty who, in return, delivers a floating interest rate. The two parties agree to exchange these cash flows over the life of the swap, which often matches the maturity of the asset.

The following diagram illustrates the cash flow of an asset swap between two parties A and B where the asset is a coupon bond. Each year contains n periods (Figure 8.3).

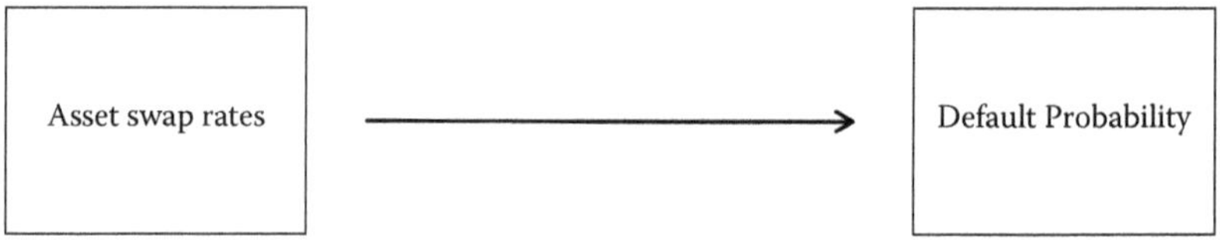

FIGURE 8.2
Extract Default Probability from Asset Swap.

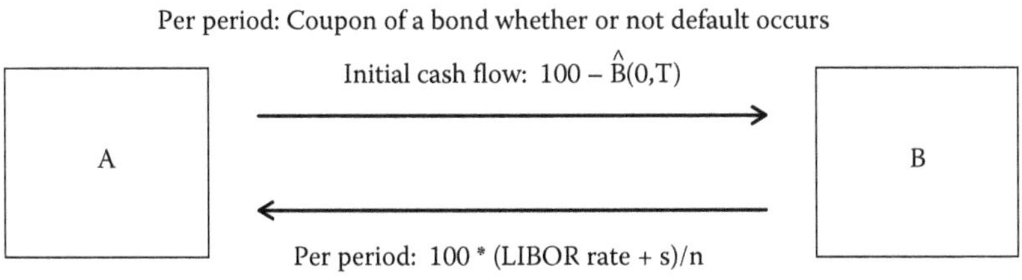

FIGURE 8.3
Asset Swap Cash Flows.

The LIBOR rate in the above diagram is a floating interest rate, discussed in Chapter 2, that fluctuates based on supply and demand for short-term, very low risk capital.

For a particular defaultable bond, today's price, $\hat{B}(0, T)$, is known and the coupon rate is specified. Therefore, the cash flows that A promises to pay to B are known today. Note that A pays B the coupon regardless of whether the bond defaults. The cash flows from B to A are based on the sum of a variable rate (LIBOR) and the asset swap rate s. The asset swap rate, also called the price of an asset swap, is calculated and specified in the swap contract at initialization. The only uncertainty in the cash flows between A and B comes from fluctuations in the LIBOR rates. Investors often use asset swaps to exchange a fixed rate of interest with a floating interest rate. Note that a total return swap is used to hedge credit risk while an asset swap is used to hedge risk free interest rate risks. Asset swaps are similar to interest rate swaps where investors swap a fixed rate with a floating rate. The primary difference is that an asset swap is based on a defaultable bond while an interest rate swap is based on riskless fixed interest rates. Interest rate swaps focus on transferring interest rate risk. Asset swaps use the coupons of the defaultable bond to transfer the fixed interest rate to floating rate. Note that unlike in the case of a credit default swap, which is discussed later, there is no default risk transfer between the two parties of the asset swap. Unlike the case of total return swap in Chapter 5, there is no credit risk transferred between the two parties of the asset swap. Though the asset swap rate can be used to derive the default probability, the asset swap itself cannot be used to hedge the default risk or credit risk.

Next, we explore how to find an asset swap rate s. In a swap contract, the key equation is:

8.2 Fundamental Equation for Swap

Expected payoff of A to B = Expected payoff of B to A (8.14)

(with expected payoffs expressed as present values in a risk-neutral world)

First, defining s as the annual rate for the swap contract and assuming that the notional amount of the swap is \$100 and the period of the swap is in half-years (since most U.S. coupon bonds pay semiannual coupons), define V as the present value of cash flows directly related to s:

$$V = s/2 \times 100 \times e^{-r\times 0.5} + s/2 \times 100 \times e^{-r\times 1} + ... + s/2 \times 100 \times e^{-r\times T}$$

The annual swap rate is divided by 2 because usually asset swaps are quoted as an annual rate that is paid semiannually.

In order to derive the asset swap's price s from the corresponding bond prices, assume that A pays \$100 to B at the swap's end and B pays \$100 to A at the swap's end. The cash flows of the swap differ between a fixed coupon and a floating-rate coupon in the following discussion.

Note that the present value of the promised payments of A to B is the sum of the value of a riskless bond that pays the same coupon as the defaultable bond (if \$100 is included as assumed), which is denoted by $B(0,T)$, plus initial cash flows, which are the excess of the principal value of a defaultable bond over its price: $100 - \hat{B}(0, T)$. Together, they form the sum: $B(0, T) + 100 - \hat{B}(0, T)$.

Before discussing the present value of the expected payoff of B to A, note the following:

Fact: The price of a floating-rate bond equals its face value when the coupon is reset.

Explanation: If a bond's coupon rate is continuously set to the LIBOR rate and if the bond's face value is \$100, then the market price of the bond will be steady at \$100. This is because owning the bond offers the same interest income as having the bond's face value inside a bank account that offers LIBOR. If the coupon rate is reset non-continuously, then the bond's value converges to its face value at each coupon reset date and offers the same value and risk as having a bond with a time-to-maturity equal to the time-to-reset.

The present value of the expected payoff of B to A has two parts: the present value of a floating rate bond of \$100 as assumed, which is PV(A floating bond), plus the present value of the asset swap rate, V. Note that PV (A floating bond) = 100. So, the present value of the expected payoff of B to A is $100 + V$. Setting the two payoffs equal in value:

$$B(0, T) + 100 - \hat{B}(0, T) = 100 + V.$$

As discussed above, the reason that a floating-rate bond has a price equal to its principal amount, say $100, can also be seen by viewing the bond as being equivalent to having $100 in a bank account that regularly adjusts to paying a competitive interest rate (equal to the floating rate). When the floating rate is a competitive market rate for money, the value of the floating rate bond will equal its principal amount.

Note that the principal or notional amount of the swap's underlying bonds due at the maturity of the swap (e.g., $100) cancel out. Party A gives B $100 at the end, and B gives A $100. Cancelling the $100 in above equation leaves the relation between the asset swap and its corresponding defaultable bond:

$$B(0, T) - \hat{B}(0, T) = V. \tag{8.15}$$

8.3 Fundamental Equation for an Asset Swap

Let s be the asset swap rate, $B(0,T)$ the risk-free bond price (which has the same semiannual coupon as the defaultable bond), and $\hat{B}(0, T)$ the defaultable bond price.

$$B(0, T) - \hat{B}(0, T) = s/2 \times 100 \times e^{-r\times 0.5} + s/2 \times 100 \times e^{-r\times 1} + \ldots + s/2 \times 100 \times e^{-r\times T} \tag{8.16}$$

$B(0,T)$ is a function of the risk-free rate and $\hat{B}(0, T)$ is a function of the bond yield y. Given the riskless and risky bond prices, the asset swap's price can be found (i.e., s can be found). Conversely, if the asset swap rate, s, is known, the bond yield can be calculated. The calculations can also be approximated using $s \approx y - r$. While the asset swap rate, s, is approximately equal to the bond yield spread, y–r, the accurate form of the relation is Equation 8.16.

Now consider how to find the default probability given s. By Equation 8.16, given s, then $B(0, T) - \hat{B}(0, T)$ is known. From Examples 8.3 and 8.4, the default probability can be solved using $B(0, T) - \hat{B}(0, T)$. Example 8.5 demonstrates calculating the default probability.

Example 8.5: Consider an asset swap based on a five-year defaultable coupon bond with a 6% coupon rate paid semiannually and recovery rate $R = 40\%$. Assume the five-year risk-free interest rate is constant at $r = 5\%$ and that the five-year asset swap rate $s = 150$ basis points, paid semiannually. Assuming that any default occurs in the middle of a year and the recovery amount is paid immediately at default, find the default probability P for each year (assume P is constant, $P_1 = P_2 = P_3 = P_4 = P_5 = P$).

By combining Equation 8.13 and Equation 8.16,

$$\underline{P}\,[\underline{B(0.5, 5) - 100R}]\,\underline{e^{-r\times 0.5}} + \underline{P}\,[\underline{B(1.5, 5) - 100R}]\,\underline{e^{-r\times 1.5}}$$
$$+ \underline{P}\,[\underline{B(2.5, 5) - 100R}]\,\underline{e^{-r\times 2.5}} +$$
$$\underline{P}\,[\underline{B(3.5, 5) - 100R}]\,\underline{e^{-r\times 3.5}} + \underline{P}\,[\underline{B(4.5, 5) - 100R}]\,\underline{e^{-r\times 4.5}}$$
$$= B(0, 5) - \hat{B}(0, 5) = s/2 \times 100 \times e^{-r\times 0.5} + s/2 \times 100 \times e^{-r\times 1} + \ldots + s/2$$
$$\times 100 \times e^{-r\times 5}$$

Inserting for r and s to find P:

$$\frac{s}{2} \times 100 \times e^{-r\times 0.5} + \frac{s}{2} \times 100 \times e^{-r\times 1} + \ldots + \frac{s}{2} \times 100 \times e^{-r\times 5} = 6.553372$$

$$\underline{P} \times [\underline{B(0.5, 5) - 100R}] \times \underline{e^{-r\times 0.5}} + \underline{P} \times [\underline{B(1.5, 5) - 100R}] \times \underline{e^{-r\times 1.5}}$$
$$+ \underline{P} \times [\underline{B(2.5, 5) - 100R}] \times \underline{e^{-r\times 2.5}} +$$
$$\underline{P} \times [\underline{B(3.5, 5) - 100R}] \times \underline{e^{-r\times 3.5}} + \underline{P} \times [\underline{B(4.5, 5) - 100R}] \times \underline{e^{-r\times 4.5}}$$
$$= 288.48P$$

Solve $P = 2.27\%$.

So far, default probabilities have been extracted from market values assuming a risk-neutral world. An important question is whether or not the extracted default probability must be the same as the real-world default probability? The answer is "no." The next section examines this issue.

8.1.5 Statistical, Historical, and Risk-Neutral Default Probabilities

A ***statistical default probability*** is an unbiased indication of the likelihood of default. In practice, it is generally unobservable and is often referred to as a real-world default probability or P-measure. Practitioners often estimate these unknown probabilities by observing past or ***historical default probabilities***. For example, long-term observation of the corporate bond market may indicate that an average of 0.5% of the bonds of a particular credit rating defaulted each year.

Practitioners often use these historic observations of default tendencies as the basis for predicting the likelihood of future defaults for bonds within that rating class. However, it is reasonable to believe that during harsh economic times the actual (i.e., statistical) probability of default is higher than the historical experience, with the opposite holding during periods of sustained economic improvement.

Since default probability is a driver of bond prices and bond yields, investors form expectations of default probabilities and use those expectations to select bonds. It may appear to be a simple procedure to infer bond prices from estimated default probabilities, or conversely, to infer default probabilities from observed bond prices. But the analysis is made more complex by the need to adjust values for risk and investor aversion to bearing risk.

Consider the following example that is made much simpler but unrealistic by assuming that the riskless interest rates for all maturities are zero. Suppose that an investor observes a zero coupon one-year bond with a principal or face value amount of $100, and believes that there is an x% likelihood that the bond will be paid in full ($100) and a 100 – x% likelihood of a default with no recovery ($0). To the extent that investors are risk averse and that the bond contains systematic risk, the bond should have a market price of *less than* x% × $100 in order to compensate investors for bearing the risk of default (i.e., to provide an expected return premium for bearing risk). Now consider a concrete example: suppose that an analyst observes another one-year defaultable bond that offers no recovery in the event of default. If the bond has a price of $49 the analyst can conclude either that the bond has a default probability of more than 50% or that the market is demanding a risk premium (or a combination of both). For example, the $49 price can be attributed entirely to a default probability of 51% or perhaps to a default probability of 50% with a risk premium of (roughly) 2% for bearing the default risk of the bond (i.e., in the second case the bond's expected payoff is $50 and is 2.04% higher than the $49 price indicating the exact risk premium).

Here is the point: risky bond prices are driven by two unobservable values: true default probabilities and risk premiums. It is generally impossible to separate the two effects. As in the above example, the analyst does not know whether the $49 bond price is due to the default probability being over 50% or at least in part due to a risk premium being demanded by risk-averse investors.

Fortunately, financial economists developed a solution. The breakthrough in finance is this process: (1) observe bond prices (and bond yields), (2) apply the risk-neutral pricing approach by assuming that all risk premiums are zero, and (3) use the bond data to estimate *risk-neutral* probabilities of default. A ***risk-neutral default probability*** (i.e., a Q-measure) is the default probability that explains defaultable bond prices relative to riskless bond prices under the assumption that all risk premiums are zero (i.e., all investors are risk neutral).

Let's return to the previous simplified example. An analyst can use the $49 bond price to infer that the *risk-neutral* probability of default is 51% (still assuming zero recovery and zero riskless interest rates):

$$\$49 = (P \times \$0) + (1-P) \times \$100$$

$$P = \$51/\$100 = 51\%$$

where P is the risk-neutral probability of default.

The risk-neutral default probability is an *upward-biased estimate* of the statistical default probability with the extent of the bias driven by: (1) the (unobservable) degree of investor risk aversion and (2) an assumption that default is more likely to occur when the overall performance of the economy and market portfolio is poor. The power of the risk-neutral approach is that the risk-neutral default probabilities on bonds of a particular rating class inferred from one market can be used to value similarly rated bonds in other markets or to value otherwise similar bonds with other recovery rates and coupons. Risk-neutral pricing provides tremendous abilities to establish *relative* pricing relations that must hold to prevent arbitrage.

Actual/historical or statistical default probabilities are often termed ***real-world default probabilities*** in finance to distinguish them clearly from risk-neutral default probabilities. Financial derivative valuation can be viewed as dealing with two parallel worlds: a risk-neutral world and a real world. The risk-neutral world is mathematically tractable but uses biased probabilities by assuming all risk premiums are zero and allowing the effects of risk premiums to be captured by these pseudo probabilities.

> **Example 8.6:** Assume that A-rated bonds have a 10-year historical *cumulative* default probability of 0.5% and an estimated recovery rate (R) of 40.0%, A-rated ten-year zero coupon bonds have a yield of 6%, and the risk-free rate 5.4%. Find the historical default density and the risk-neutral default intensity.

Using the historical cumulative default probability:

$$P(10) = 1 - e^{-10\lambda}$$

$$0.005 = 1 - e^{-10\lambda}.$$

Therefore, the historical default intensity is:

$$\lambda = \frac{Ln(1 - 0.005)}{-10} = 0.0005013.$$

Now, find the risk-neutral default intensity, λ:

$$s = y - r = 0.060 - 0.054 = 0.006$$

$$\lambda = \frac{s}{1-R} = \frac{0.006}{0.60} = 0.010.$$

The risk-neutral default probability, 0.010, is much larger than the historical default probability, 0.0005013. In a risk-neutral world, a bond with a recovery rate of 40% and offering 0.6% higher annual yield than a riskless bond must have a 1% annual probability of defaulting on 60% of its value to be equally attractive to the riskless bond (as viewed by risk-neutral investors). It is not possible to know if the market perceives the actual statistical probability of default to be closer to the historical default rate or the risk-neutral rate because it is not known how risk-averse investors are in the real world or what the size is of the risk premium they demand. Market yields on risky bonds must exceed riskless yields in order to induce risk-averse investors to bear the default risk. However, it is not possible (without further information) to decompose the yield spreads into the portion attributable to risk aversion and the portion attributable to expectations of future default rates (or recovery rates) and how they differ from observed historical rates.

This completes the section on default probabilities and extracting risk-neutral default probabilities from the market. The next section examines credit derivatives and the use of credit derivatives to hedge default risk.

Section Summary

1. Five different default probabilities were discussed: the cumulative default probability $P(t)$, cumulative survival probability $Q(t)$, non-cumulative default probability P_t, conditional probability $P_{t+\Delta t\,|\,t}$, and default intensity $\lambda(t)$. Knowledge of one enables computation of all.

$$P(t) = 1 - Q(t)$$

$$P_{t+\Delta t\,|\,t} = \frac{Q(t) - Q(t+\Delta t)}{Q(t)}$$

$$Q(T) = Q(t)e^{-\int_t^T \lambda(t)dt}$$

$$Q(T) = e^{-\lambda T}$$

If t is an integer, we have

$$P(t) = P_1 + P_2 + \dots + P_t$$

$$P_t = P(t) - P(t-1) = Q(t-1) - Q(t).$$

2. Risk-neutral default probabilities can be calculated from prices of various financial products. Statistical default probabilities (P-measures) are often approximated by historical default probabilities of corporations with similar bond ratings. Risk-neutral default probabilities (Q-measures) are higher than statistical default probabilities in a risk-averse world (and typically higher than historical probabilities) because risk-neutral pricing models adjust for aversion to risk by assigning higher probabilities to unattractive outcomes. This assumes that unattractive outcomes for the bond are correlated with unattractive outcomes for the total market.
3. As an approximation, the following relation can be used to estimate a bond's risk-neutral default intensity from a zero coupon bond yield, y, or the bond yield spread, s:

$$\lambda = \frac{s}{1-R} = \frac{y-r}{1-R}$$

where r is the risk-free rate and R is the recovery rate.
4. The following equation can be used to estimate a default probability from a risky bond price and a riskless bond price:

$$\sum \text{probability weight} \times \text{loss} \times \text{time discount factor} = B(0,T) - \hat{B}(0,T)$$

where $B(0, T)$ is the T-year risk-free bond price today and $\hat{B}(0, T)$ is a T-year risky bond price today.
5. The relation between an asset swap rate and relative bond prices is:

$$B(0,T) - \hat{B}(0,T) = s/2 \times 100 \times e^{-r \times 0.5} + s/2 \times 100 \times e^{-r \times 1} + \dots + s/2 \times 100 \times e^{-r \times T}$$

where s is the asset swap rate and the payment is semiannual.
6. The following equation can be used to estimate the default probability from an asset swap rate s:

$$\sum \text{probability weight} \times \text{loss} \times \text{time discount factor}$$
$$= s/2 \times 100 \times e^{-r\times 0.5} + s/2 \times 100 \times e^{-r\times 1} + \ldots + s/2 \times 100 \times e^{-r\times T}.$$

Demonstration Exercises

Assume that risk-free interest rates are flat at $r = 2.5\%$, a three-year risky coupon bond with a coupon of 4% paid annually has a yield of 4.5% and a recovery rate $R = 40\%$, that any defaults occur in the middle of a year and that the recovery amount is paid immediately at default. Find the default intensity, λ, by assuming that λ is constant.

By the relation above:

$$P_t = Q(t-1) - Q(t) = e^{-\lambda(t-1)} - e^{-\lambda t}.$$

By the following fundamental equation:

$$\sum \text{probability weight} \times \text{loss} \times \text{time discount factor} = B(0, T) - \hat{B}(0, T)$$

$$\underline{(e^{-\lambda 0} - e^{-\lambda 1})} \times \underline{[B(0.5, 3) - 100R]} \times \underline{e^{-r\times 0.5}} + \underline{(e^{-\lambda 1} - e^{-\lambda 2})} \times \underline{[B(1.5, 3) - 100R]} \times \underline{e^{-r\times 1.5}}$$
$$+ \underline{(e^{-\lambda 2} - e^{-\lambda 3})} \times \underline{[B(2.5, 3) - 100R]} \times \underline{e^{-r\times 2.5}} = B(0, 3) - \hat{B}(0, 3)$$

where

$$B(0.5, 3) = 4 \times e^{-r\times 0.5} + 4 \times e^{-r\times 1.5} + 104 \times e^{-r\times 2.5}$$

$$B(1.5, 3) = 4 \times e^{-r\times 0.5} + 104 \times e^{-r\times 1.5}$$

$$B(2.5, 3) = 104 \times e^{-r\times 0.5}$$

$$B(0, 3) = 4 \times e^{-r\times 1} + 4 \times e^{-r\times 2} + 104 \times e^{-r\times 3}$$

$$\hat{B}(0, 3) = 4e^{-y1} + 4e^{-y2} + 104e^{-y3}.$$

Since λ is the only unknown, the solution can be found via trial and error.

8.2 Single-Name Credit Derivatives

A ***credit derivative*** is a financial contract between two parties where the cash flows depend on whether or not a credit event such as default occurs. There are several types of credit derivatives. This chapter discusses three such products: credit default swaps (CDSs for short), collateralized debt obligations (CDOs for short), and CDO squared. The reader can find additional credit derivatives in comprehensive books on credit derivative products including Dominic O'Kane's book *Modelling Single-name and Multi-name Credit Derivative*. This section begins with single-name credit default swaps.

8.2.1 Credit Default Swaps (CDSs)

A ***credit default swap (CDS)*** is a single-named credit derivative with outcomes that depend on whether a default event occurs for a single corporation (i.e., name). CDSs provide cost-effective management of exposure to default risk. CDS valuation reflects the fundamental concepts of credit derivative valuation.

The starting point to analyze or value any credit derivative is to understand its cash flows. The cash flows of a CDS between two parties A and B are diagramed below, assuming that each year contains four payment periods; that the ***CDS rate (CDS spread or CDS price)*** is s, quoted annually; and that L is the notional principal (Figure 8.4).

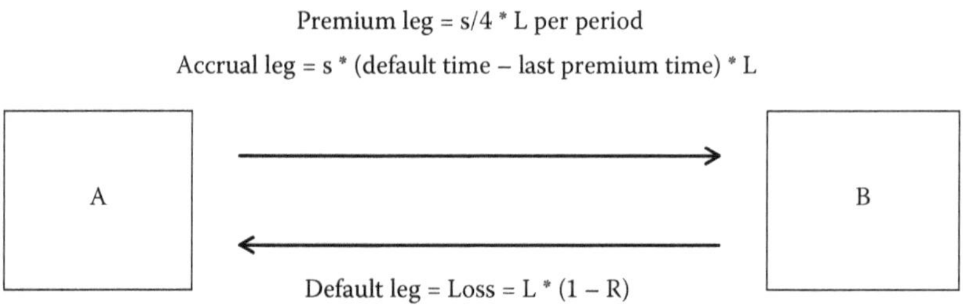

FIGURE 8.4
Credit Default Swap Cash Flows Between Parties A and B.

Party A is the ***protection buyer*** and B is the ***protection seller***. The ***premium leg*** is a periodic payment made by A to B each period before a default occurs. A does not pay the premium leg when or after a default occurs. The accrual leg and ***default leg*** are one-time payments that are made only if a default occurs.

A CDS is very much like a car insurance contract. Party A, the protection buyer, is also referred to as the ***CDS buyer***, just like a car driver seeking accident protection is a car insurance buyer. If a default occurs (or a car accident occurs in the insurance analogy), the insurance buyer, Party A, will receive compensation (a cash flow) from the protection seller (Party B or an insurance company) for the loss. In the case of a CDS that cash flow received in the event of a default is $L \times (1 - R)$. Party A pays Party B a premium leg in order to receive this protection against credit losses much like a consumer pays insurance premiums to receive protection against losses from accidents.

One difference between a car insurance contract and a CDS is that a CDS provides protection against a single default (and then ceases to exist) while car insurance companies often continue coverage after an accident. In a CDS, if the default occurs in the middle of a period, the CDS seller does not receive the full premium leg at the end of the period even though the seller protected the buyer for a half period. However, there is an ***accrual leg*** payment that compensates the seller from the time of the last premium paid to the default time, which is termed the ***accrual time***.

Having cash (i.e., a very short-term riskless bond) and entering a CDS on ABC Corporation as a protection seller is equivalent to owning a risky bond issued by ABC Corporation. Owning a risky bond and entering a CDS as a protection buyer is like owning cash. So, why do CDSs exist if they provide risk exposures similar to cash positions and risky bonds positions? First, CDS contracts are cost-effective securities (in terms of transactions costs) with which to manage credit or default risk. On the other hand, using the cash or spot bond markets to trade corporate bonds tends to be less cost-effective and typically has less liquidity. Another disadvantage of using cash positions (i.e., long or short positions in cash bonds rather than CDSs) is that short positions in corporate bonds may be difficult, cost ineffective, or even impossible to establish.

The next section calculates the price (spread) of a CDS analytically by making some assumptions such as default intensity being a constant λ and default only occurring in the middle of the period (which of course is unrealistic in real life but easier to model).

8.2.2 CDS Valuation Using the Analytical Method

Equation 8.14 in Section 8.1 depicts that in a risk-neutral world the expected payoff of A to B will equal the expected payoff of B to A since financial derivatives are zero-sum contracts. Equation 8.17 applies this relation for the case of a CDS. As before, all expected values are expressed as present values in a risk-neutral world.

8.4 Fundamental Equation for CDS Pricing

$$\mathbf{E(Premiumleg) + E(Accrualleg) = E(Defaultleg)} \quad (8.17)$$

where E(leg) is the expected value of the cash flows of a leg under risk-neutral measures and is equal to the sum of the probability-weighted cash flows multiplied by time discount factors.

Assuming quarterly payments paid at the end of each quarter, that default occurs in the middle of a quarter, that default payments and accrual payments are also paid in the middle of a quarter, and that the interest rate, r, is constant, then:

$$\text{E(Premium leg)} = \sum_{i=1}^{n} Q(t_i) \times \frac{Ls}{4} \times D(0, t_i)$$

$$\text{E(Accrual interest leg)} = \sum_{i=1}^{n} (Q(t_{i-1}) - Q(t_i)) \times \frac{Ls}{8} \times D(0, (t_i + t_{i-1})/2)$$

$$\text{E(Default leg)} = \sum_{i=1}^{n} (Q(t_{i-1}) - Q(t_i)) \times L(1 - R) \times D(0, (t_i + t_{i-1})/2)$$

where $n = 4T$ is the number of periods, t_i is the time when a premium is due, $Q(t)$ denotes the survival probability at t, R is the recovery rate, L is the notional principal, s is the CDS spread, and $D(0,t)$ is the time discount factor where $D(0, t_i) = e^{-rt_i}$.

Note that the ith premium is paid only if the corporation survives until t_i, so the probability weight is the survival probability $Q(t_i)$. On the other hand, accrual and default legs are paid only if default occurs in the period $[t_{i-1}, t_i]$, so the probability weight is the ith period default probability: $Q(t_{i-1}) - Q(t_i)$.

The contract matures (expires or settles) in T years, there are $4T$ periods and the summation has $4T$ terms. Each term has three factors: the probability weight, cash flow, and time discount factor.

$$\text{E(Premium leg)} = \sum_{i=1}^{n} Q(t_i) \times \frac{Ls}{4} \times D(0, t_i) =$$

____×_______×________+_______×_______×________+

...+_______×___________×___________

(There are $n = 4T$ periods and T is the time to maturity in years.)

$$\text{E(Accrual interest leg)} = \sum_{i=1}^{n} (Q(t_{i-1}) - Q(t_i)) \times \frac{Ls}{8} \times D(0, (t_i + t_{i-1})/2) =$$

____×______×_______+______×______×_______+
...+_______×_________×__________

(There are $n = 4T$ periods and T is the time to maturity in years.)

$$\text{E(Default leg)} = \sum_{i=1}^{n} (Q(t_{i-1}) - Q(t_i)) \times L(1 - R) \times D(0, (t_i + t_{i-1})/2) =$$

____×______×_______+______×______×_______+
...+_______×_________×__________

(There are $n = 4T$ periods and T is the time to maturity in years.)

It is easy to see that s can be factored from the premium leg and the accrual leg, and that L can be canceled from both sides, so we can derive s as a quotient with the default leg's n terms in the numerator and with the premium leg's n terms and accrual interest's n terms in the denominator.

8.5 CDS Spread Formula

Assume that CDS premiums are paid quarterly at the end of each quarter, default can only occur in the middle of a premium payment period, and that default and accrual legs are paid in the middle of the period. The CDS spread s can be calculated as follows:

$$\mathbf{S} = \frac{\sum_{i=1}^{n} (Q(t_{i-1}) - Q(t_i)) \times (1 - R) \times D(0, (t_i + t_{i-1})/2)}{\sum_{i=1}^{n} Q(t_i) \times \frac{1}{4} \times D(0, t_i) + \sum_{i=1}^{n} (Q(t_{i-1}) - Q(t_i)) \times \frac{1}{8} \times D(0, (t_i + t_{i-1})/2)} \quad (8.18)$$

where $n = 4T$ is the number of periods, t_i is the premium payment time, $Q(t)$ denotes the survival probability at t, R is the recovery rate, and $D(0,t)$ is the time discount factor.

Consider the following example.

Example 8.7: Suppose that the risk-free zero coupon curve is constant at 5% per annum with continuous compounding. Consider a five-year CDS. Assume that premium payments are made annually at the end of each year, default can only occur in the middle of a year, and the default leg and accrual leg are paid immediately after default. Given the default intensity $\lambda = 1\%$ and that the recovery rate is 40%, find the CDS spread s.

Note that the quarterly premium payment factors 1/4 and 1/8 in Equation 8.18 are replaced by 1 and ½ for annual payments, respectively.

$$s = \frac{(1 - e^{-\lambda 1}) \times (1 - 0.4) \times e^{-0.5\times r} + (e^{-\lambda 1} - e^{-\lambda 2}) \times (1 - 0.4) \times e^{-1.5\times r} + \ldots + (e^{-\lambda 4} - e^{-\lambda 5}) \times (1 - 0.4) \times e^{-4.5\times r}}{(e^{-\lambda 1} \times 1 \times e^{-1\times r} + \ldots + e^{-\lambda 5} \times 1 \times e^{-5\times r}) + (e^{-\lambda 0} - e^{-\lambda 1}) \times 1/2 \times e^{-0.5\times r} + \ldots + (e^{-\lambda 4} - e^{-\lambda 5}) \times 1/2 \times e^{-4.5\times r})}$$

The CDS spread, s, is found by inserting the known values and calculating. In the next example, a spreadsheet is used to calculate s.

Example 8.8: Using a spreadsheet to find the CDS spread s in Example 8.7.

Spreadsheet 8.2

r	0.05
T	5
Default occurs halfway each year	
Lambda	0.01
R	0.4
Payment made annually	

				probability weighted sum
Premium leg				
Year	**Survival probability**	**Premium**	**Discount rate**	
1	0.990049834	s	0.95122942	0.94176453
2	0.980198673	s	0.90483741	0.88692043
3	0.970445534	s	0.86070797	0.83527021
4	0.960789439	s	0.81873075	0.78662786
5	0.951229425	s	0.77880078	0.74081822
Sum				4.19140126
Accrual leg				
Year	**Default probability**	**Accrual**	**Discount rate**	
0.5	0.009950166	s/2	0.97530991	0.00485224
1.5	0.00985116	s/2	0.92774348	0.00456967
2.5	0.00975314	s/2	0.88249690	0.00430355
3.5	0.009656094	s/2	0.83945702	0.00405293
4.5	0.009560015	s/2	0.79851621	0.00381691
Sum				0.02159533
Default leg				
Year	**Default probability**	**Discount rate**		
0.5	0.009950166	0.97530991	0.00582269	
1.5	0.00985116	0.92774348	0.00548361	
2.5	0.00975314	0.88249690	0.00516426	
3.5	0.009656094	0.83945702	0.00486352	
4.5	0.009560015	0.79851621	0.00458029	
Sum			0.02591439	

CDS spread=default sum/(premium sum+accrual sum)
=0.02591439/(4.19140126+0.02159533)
=0.006151061

The CDS spread, s, is 0.00615.

The next section uses simulation to price a CDS. Note that with simulation it is unnecessary to assume that default occurs in the middle of the period.

8.2.3 CDS Valuation Using Simulation

CDS pricing using simulation requires a specification of random time when default occurs. This section begins by reviewing some probability distributions which can be used to model default time.

Mathematics Review 8.1 ***Poisson Distribution***: A discrete random variable X follows a Poisson distribution if

- $X = k$ then there are k events occurring in a fixed interval of time or space.
- Events occur with a known average rate λ per unit of time or space.
- In the small time interval Δt, $P(X = 1 \text{ in time } \Delta t) = \lambda\Delta t$ and $P(X = 0 \text{ in time } \Delta t) = 1 - \lambda\Delta t$.
- These events occur independently.

The probability function of X is:

$$P(X = k) = \frac{e^{-\lambda t}(\lambda t)^k}{k!}.$$

Exponential Distribution: A continuous random variable Y follows an exponential distribution if

- Y models the time between events in a Poisson process.
- It is a process in which events occur continuously and independently at a constant average rate λ.

The cumulative distribution function of Y is:

$$P(Y \le t) = 1 - e^{-\lambda t}$$

where λ is the average rate of occurrence.

In modeling default events, it is often assumed for simplicity that the default intensity is a constant λ. Recall in the Equation 8.7, the cumulative survival probability before t is: $Q(t) = e^{-\lambda t}$ and in Equation 8.3, the cumulative default probability before t is: $P(t) = 1 - Q(t) = 1 - e^{-\lambda t}$, where λ is the default intensity. So, the default time can be modeled as a continuous random variable τ with an exponential distribution.

The cumulative distribution function is $F(t) = P(\tau \le t) = 1 - e^{-\lambda t}$, where λ is the default intensity. Given the independence assumption in the exponential distribution, it is easy to see that the default intensity, λ, is the average rate of occurrence parameter λ in a Poisson distribution. In addition, it is easy to simulate an exponential distribution random variable and so we use the simulated exponential random variable as the default time. For each default time τ that is generated, the default leg, premium leg, and accrual leg for the corresponding path can be found.

Default leg = $L \times (1 - R) \times$ time discount factor:

$$\text{Premium leg} = \sum_{i=1}^{\text{number of quarters before default}} L \times \frac{s}{4} \times \text{time discount factor},$$

$$\text{Accrual leg} = L \times s \times (\text{default time } \tau - \text{last premium payment time}) \times \text{time discount factor}.$$

Note that the default leg and accrual leg are paid only once in each path and are paid at the point where the default occurs. Premium payments are paid quarterly until default occurs. The simulation is repeated to a total of m paths. The expected default leg, expected premium leg, and expected accrual leg can be calculated by taking arithmetic averages of each leg across the simulated paths. Note that even though s is an unknown in the accrual leg and premium leg, it can be factored and therefore we can solve for s. Equation 8.18 can be modified to produce a simulation version of the fundamental equation for a CDS price.

8.6 Fundamental Equation for CDS Pricing (Simulation Version)

We assume that the premium is paid quarterly. The accrual and default legs are paid immediately after a default, and the default intensity λ is assumed constant. We simulate default time τ_j where $1 \le j \le m$ for m paths, each with an independent exponential distribution and with default intensity λ.

We can use the following equation to solve for s. Assuming m paths are generated:

$$\frac{\sum_{j=1}^{m} \textbf{Premium leg}_j}{m} + \frac{\sum_{j=1}^{m} \textbf{Accrual leg}_j}{m} = \frac{\sum_{j=1}^{m} \textbf{Defualt leg}_j}{m} \tag{8.19}$$

where index j is the value corresponding to the jth path. In each path, we have

$$\text{Premium leg}_j = \sum_{i=1}^{\text{number of quraters before default}} L \times \frac{s}{4} \times D(0, t_i)$$

where t_i is the end of ith quarter.

$$\text{Accrual leg}_j = L \times s \times (\tau_j - \text{last quarter before default}) \times D(0, \tau_j)$$

$$\text{Default Leg}_j = L \times (1 - R) \times D(0, \tau_j).$$

Note that the difference between Equations 8.17 and 8.19 is that Equation 8.19 (i.e., using simulation) does not involve probability weights.

The following example demonstrates coding with R since R can be an efficient approach with which to perform simulations.

Example 8.9: Consider a five-year CDS. Assume that the risk-free rate is 5%, the recovery rate is 40%, and the default intensity is 1%. Assume the premium is paid at the end of each year (it is very easy to adjust to quarterly). Assuming that the default leg and accrual leg are paid immediately after default, find the CDS spread s.

R-Code 8.1 Find a CDS Spread Using Simulation

```
#Simulate 100,000 paths and save the default leg, accrual leg and premium leg in
#vectors D,A,P
r=0.05
R=0.4
D=rep(0,100000)
A=rep(0,100000)
P=rep(0,100000)
for(i in 1:100000)
{
t=rexp(1,0.01)
if (t>5)
{
D[i]=0
A[i]=0
P[i]=0
for(j in 1:5)
{
P[i]=P[i]+exp(-r*j)
}
}
if (t<5)
{
m=floor(t)
D[i]=(1-R)*exp(-r*t)
A[i]=(t-m)*exp(-r*t)
P[i]=0
for(j in 1:m)
{
P[i]=P[i]+exp(-r*j)
}
}
}
S=mean(D)/(mean(A)+mean(P))
S
```

The estimate of s obtained by the authors in executing a single simulation was 0.006131655, which is close to the analytical solution found for s in the previous section.

In further analysis of the spread, s, it seems that

$$s \approx \lambda(1 - R).$$

In our case, $s \approx 0.01 \times (1 - 0.4) = 0.006$. This makes sense since the default is a Poisson arrival with a constant frequency of λ, so the maturity of the contract should not matter. Of course, the risk-free rate causes the estimate to deviate slightly from 0.006.

8.2.4 Calibrating Piecewise Constant Default Probabilities from Multiple CDS Spreads

In Section 8.1, bond prices and asset swap rates were used to find default probabilities. Recall that in Sections 8.2.2 and 8.2.3 default intensity, λ, was used to solve for the CDS spread via both the analytical method and the simulation method. This section examines the opposite task: given a CDS spread, s, solve for the default probability analytically.

Inserting $Q(t) = e^{-\lambda t}$ into Equation 8.18 generates a simplified relation between the default intensity λ and the CDS spread s. If λ is known, the CDS spread s can be found. Conversely, given a CDS spread, s, the default intensity λ can be found. EXCEL Solver or a similar spreadsheet function can be used to find λ under the analytic approach. In the simulation approach, the trial-and-error search method is used. Specifically, values of λ are selected until the simulated CDS spread matches the market quote. Furthermore, given several CDS spreads, $Q(t)$ can be found while λ is piecewise constant. This is demonstrated in detail in the end-of-chapter project.

Note that it is possible to estimate a default probability using either a CDS spread, bond prices, or an asset swap rate. Each input (i.e., each market) may lead to a different estimate of the default probability. These differences can be used to signal arbitrage opportunities that take opposite exposures to the implied risk-neutral default probabilities in the CDS, asset swap and bond markets. For example, if the CDS market implies a much higher default probability than the bond market or the asset swap market, then an arbitrager could establish a short position in a CDS (selling the credit protection which is a synthetic equivalent to a long position in the defaultable bond) with the higher implied default probability and a short position in bonds within those markets with lower implied default probabilities. Properly hedged for risk, the arbitrager will generate a profit if the implied default probabilities between the two markets converge. Note that the two positions are hedged. If the default risk increases (decreases), although a short position in as a CDS protection seller loses (gains) money, a short position in bond market gains (losses) money.

The more accurate the analysis by arbitragers and the lower their transactions costs, the more that the actions of arbitragers drive the markets prices toward values that are in equilibrium with each other (i.e., the implied default probabilities for each credit name are equated across markets). That competition facilitates risk management by other market participants such as financial institutions who can be confident that they are obtaining competitive prices while using whichever market is more convenient or has lower transactions costs.

This completes the analysis of single-name credit derivatives. The next section analyzes multi-name credit derivatives.

It may seem that the CDO material is outdated. Why do we cover the next two sections? Some of the financial instruments (especially CDOs) were much less commonly used in the immediate aftermath of the global financial crisis. But that has changed. CDOs are making a comeback now as they did in the 1990s after that structured product crisis. Uzialko[1] notes:

"Now, CDOs are making a comeback. While the market is still a fraction of what it once was – today it stands at roughly $70 billion compared to more than $200 billion pre-crisis – major institutions like Citigroup and Deutsche Bank have skin in the CDO game once again."

Uzialko's number is somewhat outdated. More recently, the *Financial Times* claimed that investors are "flocking back to a complex debt derivative product blamed for amplifying losses in the financial crisis" and that "The volumes have grown nicely."[2]

The claim that the CDO material (including the copula method) is outdated may actually be an outdated claim. To borrow a catchphrase from the *Poltergeist*: "They're baaack!"

Another reason is that CDO models provide good practice of simulation with correlation. In addition, without understanding CDOs, we can't understand the 2007–2008 financial crash.

Summary of Single-Name Credit Derivatives

This section introduced important concepts about credit derivatives pricing. The fundamental equation balances the present values of the cash flows on each side of a derivative to be equal. In the case of CDSs, this means: E(Premium leg)+E(Accrual leg)=E(Default leg). The equation simply requires that the discounted expected value of cash flows from A to B be equal to the discounted cash flows from B to A. By assuming a risk-neutral world (i.e., a risk-neutral framework), the values can be calculated by: probability weighted × cash flow × time discount factor, where the probabilities are risk neutral rather than statistical. Simulation can also be used to value the credit derivatives. The difference is that in the simulation approach the probability weights are not explicit in the computation of the final value: the value is an arithmetic average of the values (cash flow × time discount factor) from each path. The probability distribution governs the simulation paths. Therefore, the probabilities of various outcomes are implied by the numbers of simulated paths.

Demonstration Exercises

Suppose that the risk free zero-coupon curve is flat at 5% per annum with continuous compounding. Consider a five-year CDS. Assuming that default can only occur at the midpoint of each year, that the CDS spread is 150 basis points, that the recovery rate is 40%, and that payment is made annually at the end of a year, find the default intensity λ.

$$s = \frac{\begin{array}{c}(1 - e^{-\lambda 1}) \times (1 - 0.4) \times e^{-0.5 \times r} + (e^{-\lambda 1} - e^{-\lambda 2}) \times (1 - 0.4) \\ \times\, e^{-1.5 \times r} + \ldots + (e^{-\lambda 4} - e^{-\lambda 5}) \times (1 - 0.4) \times e^{-4.5 \times r}\end{array}}{\begin{array}{c}(e^{-\lambda 1} \times e^{-1 \times r} + \ldots + e^{-\lambda 5} \times e^{-5 \times r}) + (e^{-\lambda 0} - e^{-\lambda 1}) \times 0.5 \\ \times\, e^{-0.5 \times r} + \ldots + (e^{-\lambda 4} - e^{-\lambda 5}) \times 0.5 \times e^{-4.5 \times r})\end{array}}$$

Given s, EXCEL Solver can be used to find λ.

8.3 Collateralized Debt Obligations (CDOs) and a CDO Model with Independent Bond Defaults

The previous section discusses the single-name credit derivative CDS. At the end of the previous section, a rationale was offered indicating why this book includes the next two sections. This section discusses multi-name credit derivatives, namely collateralized debt obligations. Generally speaking, a ***collateralized debt obligation*** (***CDO*** for short) is a type of structured product that divides the cash flows from a portfolio of fixed-income assets such as bond portfolios into ***tranches*** with different characteristics, typically different risk exposures. In other words, a CDO *structures* the risk of a fixed-income portfolio into various tranches much like the capital structure of a major corporation structures the risk of its assets into various security classes with different risk exposures such as senior debt, junior debt, and equity. Note that tranche investors receive a sequence of coupon payments and bear the losses due to defaults. The different tranche investors have different risk exposures related to defaults and receive different levels of coupon payments. Therefore, ***tranche investors*** offer default protection and are therefore "protection sellers". Sometimes the term "tranche investor" is shortened to "tranche" such that a tranche would be said to be selling default protection.

8.3.1 Overview of CDO Tranches and Structuring of Cash Flows

The concept of a CDO has been applied to virtually every type of fixed income portfolio resulting in collateralized loan obligations (that structure bank loan portfolios), collateralized mortgage obligations (that structure mortgage portfolios), and so forth. This section focuses on CDOs that structure a portfolio of corporate bonds into a number of tranches ranging from the most ***senior tranche*** that usually has the lowest risk and highest available credit rating (AAA) to the tranche with the highest risk and lowest seniority (the ***equity tranche***). A ***mezzanine tranche*** has risk and reward in the middle. Here the risk we refer to the risk of longing a tranche since they are protection sellers. A CDO of corporate bonds is often referred to as a *multi-name instrument* because the CDO's asset portfolio contains credit risk exposures (i.e., default risk exposures) to many corporations. The diversification offered by the bond portfolio attracts investors who wish to invest in the CDO's tranches and who select a tranche that best matches the investor's desired risk exposure. In cash CDOs, the proceeds from selling the tranches to investors are used to finance the acquisition of the CDO's collateral portfolio.

Consider the following highly simplified CDO structure that will serve as the foundation for a model. A sponsor establishes the CDO structure by assembling positions in the bonds of 100 corporations and pools them together into a collateral pool or portfolio. To model the structure with simplicity, assume that each of the 100 corporate bonds in the collateral pool has a principal value of \$1 for a total principal amount of \$100. Further assume that there are only three investor tranches issued against the collateral bond portfolio: a \$75 senior tranche, a \$20 mezzanine tranche, and a \$5 junior/equity tranche.

Figure 8.5 depicts a balance sheet view of a CDO with assets being financed by tranches of securities. The junior or equity tranche is placed to the center, much like in a corporate balance sheet, to indicate that it has the lowest priority to value created by the assets. Cash generated by the collateral portfolio (after managerial expenses) is distributed to the tranches with the highest priority being to meet the scheduled payments to the senior tranche. Next in line for cash is the mezzanine tranche, with remaining cash paid to the holders of the junior or equity tranche.

The structure of tranches prioritizes the crediting of coupons and principal payments from the collateral pool to the tranche investors. Coupons are credited first to the senior-most tranche until the tranche's coupon rate has been reached and then cascade to the less senior tranche's in a process

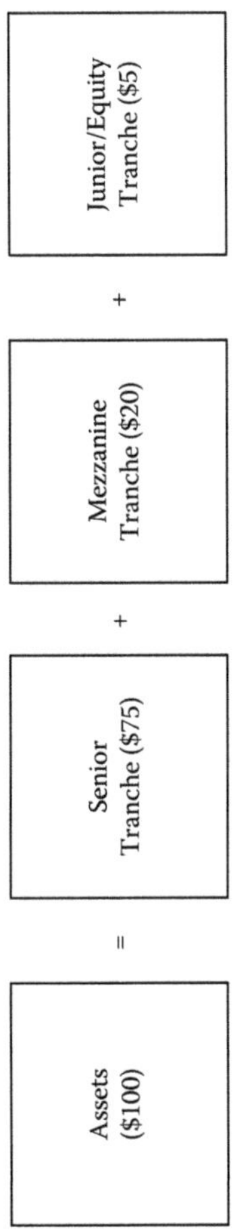

FIGURE 8.5
A Balance Sheet View of a Simplified CDO.

known as a cash waterfall. Similarly, principal payments are credited first to the most senior tranches.

The tranches also structure the prioritization of bearing losses from defaults. Briefly, default losses are assessed against junior-most tranches first, with more senior tranches protected from default losses until all tranches with lower seniority have been wiped out. Figure 8.6 indicates the exposures to default losses using the concept of attachment points and detachment points.

Tranche Name	Size	Attachment Point	Detachment Point
Senior tranche	75% or $75	25%	100%
Mezzanine tranche	20% or $20	5%	25%
Junior/Equity tranche	5% or $5	0%	5%

FIGURE 8.6
Prioritization of Protection from Default Losses.

The attachment point is the loss level that must occur to the portfolio in order for the corresponding tranche investors to begin to bear losses. For example, the least-senior tranche (the junior or equity tranche) begins to suffer losses from the very first defaults in the collateral portfolio so its attachment point is 0%. The detachment point is the loss level that, if reached, completely wipes out a tranche. For example, the junior tranche becomes worthless in this highly simplified model if the portfolio loses 5% of its value due to defaults since the initial value of the junior tranche only represents 5% of the value of the total structure and because the junior tranche is the first tranche to bear losses. Note that the senior tranche does not bear losses until 25% of the portfolio value has been lost due to defaults, and is not completely wiped out until the portfolio is completely wiped out (i.e., a detachment point of 100%). The various CDO tranches are usually rated with the most senior tranches receiving the coveted rating of AAA (or Aaa).

To illustrate, the collateral pool of corporate bonds might have an average fixed coupon rate of, say, 6%. The senior tranches, with their superior priority to cash flows, might be large and AAA-rated while offering investors only a 3% coupon rate due to their role as 'protection sellers' (i.e., the senior tranche sacrifices coupon in order to have more junior tranches bear the first losses). The more junior tranches, as protection sellers, might be small, unrated, and high coupon (e.g., 20%). If the bonds held as collateral are floating rate, the coupons of the CDO tranches will tend to be floating rate and are quoted in terms of a spread to LIBOR. Regardless, interest rate swaps can be used to convert fixed-rate products to floating-rate products and vice versa. For modeling purposes, fixed-rate products are assumed in the rest of the book.

8.3.2 Economic Roles and Risks of CDOs

The primary economic role of CDOs of corporate bonds is offering investors diversified exposure to a portfolio of credit risks through tranches that offer differing levels of default risk and credit ratings. The variety of risk exposures is formed by structuring numerous tranches of securities with varying degree of credit protection. Investment-grade CDO tranches often allow insurance companies to skirt regulations by diversifying into non-investment grade debt using credit-protected tranches on well-diversified bond portfolios.

In theory, CDOs cannot increase the aggregate level of risk in an economy because CDOs merely serve as conduits that flow the existing risks from a bond portfolio to investors through various tranches. In fact, since CDOs facilitate diversification, there is a valid argument that CDOs lower the aggregate level of diversifiable risk borne by investors.

CDOs allow investors to diversify broadly across credit risks while being able to select their desired level of exposure to the risk of high default rates. The foundation to understanding the risks of CDOs is to carefully study the effects of various levels of default losses from the CDO bond portfolio on the values of the CDO's various tranches. That foundation begins with an accounting view of the allocation of cash and defaults to the tranches. An analyst uses a computer model to mechanically apply the CDO structure's cash-flow priorities to credit tranches with cash flow received from the portfolio and debit tranches for default losses experienced in the collateral portfolio.

Obviously, high default rates hurt all tranches, especially junior tranches that bear the first losses. It is less obvious how *correlations* among defaults affect different tranches. As is discussed in detail in a later section, the structuring of CDO tranches exposes senior tranches to the risk that correlations between the defaults of different corporations will be high, while junior tranches are more exposed to the risk that correlations between the defaults of different corporations will be low. In addition, the *recovery rate* also matters. In a low-recovery environment, senior tranches are more vulnerable to losses. How do rating agencies and prospective investors of tranches assess the risks and rewards of the various tranches given so much uncertainty regarding the number and timing of actual defaults in the CDOs bond portfolio? Analysis of these surprisingly complex relationships requires sophisticated financial models primarily based on simulation, as discussed in the next several sections.

There are two major kinds of CDOs: Cash CDOs and synthetic unfunded CDOs. A ***cash CDO*** is a CDO that holds debts securities in the collateral portfolio with payoffs, either coupons or principal payments, to the investors, which come from the actual cash flows of the assets in the pool. A ***synthetic unfunded CDO*** is a CDO that does not actually hold the referenced debt securities but rather obtains exposure to a credit risks through credit derivatives. Synthetic unfunded CDO's payoffs are driven by credit events and financial derivatives on the referenced entities. Cash CDOs are discussed in Section 8.3.3 and synthetic unfunded CDOs are discussed in Section 8.3.4. Cash CDOs are popularly used in practice. Modeling is often based on synthetic unfunded CDOs for simplicity.

8.3.3 An Example of a Cash CDO Structure

This section examines an example of a cash CDO.

Assume that a CDO is formed with a five-year time horizon and 100 equally sized positions ($1 of principal) in corporate bonds. Assume that the CDO has three tranches: a junior tranche with a principal amount of $5, a mezzanine tranche with a principal amount of $20, and a senior tranche with a principal amount of $75. Assume that each corporate bond has the same default probability each year (5%) and the same anticipated recovery rate of 40% in the event of default. Further assume that any defaults occur in the middle of a period. This assumption is relaxed in later sections.

The cash flows of all tranches are as follows: The junior tranche investors pay $5 to purchase the CDO's junior tranche. Mezzanine tranche investors pay $20 to purchase the CDO's mezzanine tranche with senior tranche investors paying $75 to purchase all of the CDO's senior tranche. The investors in each tranche receive a fixed coupon (or a floating rate plus a spread) based on the remaining principal for that tranche. The principal amounts of some or all of the tranches decline as the collateral bond portfolio experiences losses from defaults. The following scenario illustrates the mechanics of reducing the principal amounts for default losses.

Default scenario analysis:

Suppose that in the first year, 7 of the 100 corporate bonds default. Using the $1 per bond initial principal and the assumed recovery rate of 40%, the total first-year loss is $7 × 0.60 = $4.20. The remaining principal for the junior tranche falls to $5.00 – $4.20 = $0.80. Interestingly, for this cash CDO, the $2.80 recovered principal on the seven defaulted bonds serves as a principal repayment to the most-senior tranche. Therefore, the principal amount of the most-senior tranche is reduced by $2.80. Both the senior and junior tranches' future coupons are reduced as the principal amounts of the tranches are reduced. From a balance sheet view of the CDO, the assets fall by $7.00 due to the seven bonds ceasing to exist. The senior tranche drops $2.80 because the 40% that was recovered was distributed to the investors in that tranche, and the junior tranche drops by $4.20 to reflect the responsibility of the junior-most tranche to bear the initial default losses.

As bonds in the collateral portfolio continue to default, the most-junior tranche is marked down for the unrecovered principal amounts while the senior-most tranches continue to be paid down using the recovered portion of the principal amounts from defaulted bonds. When the expected default rates are due to high probabilities of very high default rates and the recovery rates are potentially very low, the senior tranche has a relatively high-risk exposure. Interestingly, the most-junior tranches suffer most when moderate default rates are highly likely (and low recovery rates). Due to their call-option-like exposure, junior tranche investors prefer a highly uncertain future (with substantial chances of low default rates) to a certain path of moderate defaults (and low recoveries) because, as bearers of first losses, moderately poor asset performance will lead to total losses among junior tranches. This issue is detailed further in subsequent analyses.

There are three key parameters that drive the losses to the various tranches. First, the ***expected default rate*** of the collateral bonds is a crucial determinant of the average losses that are expected each year from defaults in the bond portfolio. Second, the ***recovery rate*** determines the losses from each default – the lower the recovery rate, the larger the losses and the faster the junior tranche is wiped out. Third, the ***correlation of defaults*** drives the relative risks of the various tranches. It is not easy to understand why default correlations matter. However, the effects of default correlations will be carefully explained in later sections. The health of an economy tends to experience cycles. There are recessions during which credit default rates can be very high and periods of prosperity during which corporate bond defaults are relatively rare. CDO simulation models often include a default correlation parameter that can be used to model the extent to which defaults tend to occur in clusters or independently. Therefore, the assumed default correlation parameter effects the relative risks and values of the various tranches of each CDO.

The next section examines a synthetic unfunded CDO.

8.3.4 Synthetic Unfunded CDO Structures

The above discussion of CDOs only touches on the complexities and varieties of CDO structures used in practice. Up to this point the focus has been on *cash* CDOs, where investors purchase CDO tranches with cash that is used to purchase and hold a portfolio of corporate bonds in the collateral portfolio.

CDO modeling often focuses on *synthetic unfunded* CDO structures. A CDO is synthetic when the CDO's portfolio consists of short positions in credit default swaps (i.e., serves as credit risk protection sellers) to obtain default risk exposure rather than buying risky bonds with cash. Thus, a hypothetical synthetic CDO structure might sell credit protection on 100 names (i.e., corporate bonds) using CDSs and invests cash (if any) in low-risk interest-bearing securities. A CDO structure is *unfunded* when the risk exposures to the tranches are obtained through CDSs with the tranche investors and without cash collateral. Thus, in an unfunded synthetic CDO there are no initial cash flows – the structure is simply a set of CDSs.

For the purposes of modeling CDOs, it is useful to create highly abstracted scenarios that avoid many complexities. So, although cash CDOs are common, CDO modeling is often performed using highly simplified synthetic unfunded CDO structures. Synthetic unfunded CDO structures allocate default losses from the CDO bond portfolio to the CDO's various tranches based on their seniority and their notional principal values. To avoid confusion, the two parties of each tranche are referred to as the ***tranche protection seller*** (the investor in the tranche) and the ***tranche protection buyer***. The junior tranches serve as protection sellers and therefore receive relatively large spreads from the protection buyers who pay the spreads for credit protection. The tranche protection buyers' payments are typically based on floating interest rates (LIBOR-based) rather than fixed rates and corresponding tranche spreads. But for illustration purposes, we use fixed rates.

The obligation of the junior tranche protection seller is to pay the unrecoverable loss to the junior tranche protection buyer when defaults occur. After this insurance-like payment, the junior tranche notional principal decreases by the loss amount. This process continues potentially until the junior tranche is wiped out. Then, the mezzanine tranche protection seller begins to pay mezzanine protection buyers the losses due to further defaults. At that time, the notional principal of the mezzanine tranche decreases by the loss amount. If the mezzanine tranche is wiped out, the senior tranche protection seller begins to pay the senior protection buyers. The protection buyers of all tranches (i.e., the counterparties to the CDSs) pay premiums to the tranche investors based on any remaining notional principal before they are wiped out. In practice, similar to the case of CDSs, there are accrual interest payments in addition to default payments and premium payments. Next, a simplified CDO model is used to illustrate the concepts.

8.3.5 A Simplified Simulation based CDO Model with Independent Bond Defaults

This section investigates a simulation-based CDO model for a synthetic unfunded CDO with fixed coupon rates (which are called ***CDO spreads or CDO prices***). Defaults are assumed to be independent in this section for simplification, although they are not independent in reality. The cash flows of various tranches are analyzed using the simulated defaults.

Example 8.10: Consider a five-year, synthetic unfunded CDO structure that sells protection on 100 names (corporate CDSs) each with a \$1 notional principal amount. Assume that each corporation (or name or credit) has a default probability of 5% in each year and a fixed recovery rate of 40%. Initially, assume that all defaults occur in the middle of the period (year). Suppose that the CDO prices (spreads) of the tranches are as follows: the senior tranche (representing 75% of the structure) has a spread = 1%, the mezzanine tranche (representing 20% of the structure) has a spread = 3%, and the junior tranche (representing 5% of the structure) has a spread = 30%. In practice, these spreads (CDO prices) are determined by supply and demand in the marketplace and typically by analysts using mathematical models such as a copula model. The spreads are the potential compensation in the coupon rates received by each tranche for bearing default risk (i.e., offering credit protection to the counterparties). Assume that all defaults occur independently.

In order to model the CDO's cash flows, the potential defaults need to be modeled through time. A simulation of defaults is used to model the various outcomes of defaults. In this example, all defaults are assumed to be statistically independent. This initial assumption is unrealistic and is relaxed in the next section. The reason it is assumed here is to provide a simplified illustration of the cash flows. Before exploring the simulation, two probability distributions are reviewed in Mathematics Review 8.2.

Mathematics Review 8.2 Bernoulli Distribution: A discrete random variable X follows a ***Bernoulli distribution*** if

- $P(X=1) = p$ and $P(X=0) = 1 - p$

Binomial Distribution: A discrete random variable X follows a ***binomial distribution*** if

- There are n trials and each trial has a probability of p to be 1 (success) and has a probability of $1 - p$ to be 0 (i.e., each trial is a Bernoulli distribution).
- X is the number of successes during the n trials.
- All trials are independent.

The probability distribution of X is:

$$P(X = k) = \binom{n}{k} p^k (1 - p)^{n-k}$$

where n and p are parameters for the binomial distribution.

1. ***Simulation of independent defaults***:
   Since defaults are initially assumed to occur in the middle of a year, the possible default times are fixed, namely the middle of each year. It

is necessary to simulate the number of defaults that occur in each path for all 100 corporations. Only one way of performing a simulation is illustrated, although there are several ways to perform simulations. As the expected default probability is 5% in each year and the CDO is for five years, the cumulative default probability for each corporate bond is 25%. Each corporation's default can be modeled as a Bernoulli distribution with 1 for default and 0 for no default. Assuming the defaults are independent (very unlikely in real life), the 100 potential corporate defaults are modeled as a binomial distribution.
First, simulate 100 independent Bernoulli distributions with $p = 0.25$, with $x = 1$ for default and $x = 0$ for no default. Each x specifies if a corporation defaults or not. The result is a 100-dimension vector with values 1 representing default and 0 representing no default. However, cash flows are related to the time period in which defaults occur, so further modeling is needed regarding the default year. Defaults are equally likely to occur in each of the five years, so the program simulates 100 rolls of a five-faced dice. The dice has five equally likely faces of 1, 2, 3, 4 or 5, each representing the default year if default occurs, creating the random variable y. Multiplying x and y creates another vector with a dimension of 100 representing the 100 possible values, with 1–5 representing defaults in the five possible years and 0 representing no default. Here is the R-code and the resulting vector from a sample trial.

R-Code 8.2 Generate One Path of Default Scenarios for 100 Corporations over Five Years

```
#R code simulating default year 1, 2, 3, 4, 5 and 0 for no default for 100 bonds:
x= rbinom(100,1,0.25)
y=sample(c(1,2,3,4,5),100,replace=TRUE)
z=x*y
z
R-Result
[1] 5 0 1 0 0 5 1 0 1 0 4 0 0 0 0 0 0 0 0 0 0 0 0 0 4 0 0 2 0 0 0 0 0 0 4
0 0
[38] 0 5 0 0 0 0 3 2 0 0 3 3 4 0 0 0 2 0 0 0 5 0 0 0 0 0 0 0 0 0 0 4 0 5
0 0 0
[75] 0 2 0 0 5 2 0 0 0 0 0 0 0 0 0 0 0 0 0 0 0 5 0 1 4 0
```

The next step is to tabulate the number of defaults in each year for all 100 corporations combined:

Year	Defaults
1	4
2	5
3	3
4	6
5	7
Total	25

2. ***Analysis of total cash flows for tranches using above simulated independent defaults***:
Recall the fundamental equation for a CDS price (Equation 8.17):

$$E(\text{Premium leg}) + E(\text{Accrual leg}) = E(\text{Default leg}).$$

In the case of an unfunded synthetic CDO, for each year the left side of the above equation represents the cash flows (premiums) *received* by all of the tranches investors from the CDO counterparties. Note that the premium leg captures the full-year payments and the accrual leg captures the half-year payments (all defaults are assumed to occur at mid-year in this example). The right side of the equation captures all the payments *made* from the tranche investors to the counterparties of the CDO (as a result of defaults). As a risk-neutral model, Equation 8.17 sets the expected cash flows between the two parties (the tranche investors and the external counterparties) equal to each other.

This section details the cash flows for each tranche using the three above legs. The simulated number of defaults in each of the five years of the CDO's life determines each tranche's cash flows. In this section, all tranches (i.e., the entire CDO) are briefly analyzed together. Then, each specific tranche is analyzed one at a time afterward.

The total number of defaults in each year and the recovery rate are used to calculate the payouts from the tranche investors. The recovery rate, R, is assumed in this example to be 0.40, so the loss given default, $1 - R$, is 0.60. The loss to (i.e., payout from) the total structure in the above simulation path through all five years is \$15 (found as 25 defaults × \$1 principal × 0.60). The first \$5 of the \$15 default losses fall on the most-junior tranche (that serves as a protection seller) and wipes out that tranche. The next \$10 of default losses fall on the mezzanine tranche.

The \$15 of total default losses is viewed as forming the default legs (the center side of Equation 8.17) paid by the junior tranche and mezzanine tranche as protection sellers and is discussed later. Note that each tranche receives compensation for bearing default risk, which is found as the product of its spread and its remaining size, and is called the premium leg (and in the case of partial years, the accrual leg).

Before defaults begin to affect the senior and mezzanine tranche, the senior tranche receives \$0.75 (\$75 principal × 1% spread) annually, and the mezzanine tranche receives \$0.60 (\$20 × 3%) annually. Unlike the case of a cash CDO, in a synthetic unfunded CDO, it is not necessary to reduce the senior tranche's principal due to the recoveries obtained in the defaults borne by the other tranches because there are no bonds actually being held and therefore there are no actual cash recoveries from defaults. All cash flows (i.e., premiums and defaults) are driven by the CDSs.

To review the net cash flows of the structure, start with the junior tranche. The junior tranche investors pay nothing to establish their position at the beginning of the contract but the junior tranche receives interest based on the tranche's \$5 notional principal and 30% spread. Once defaults occur, the junior tranche investors must pay the defaulted amount to the protection buyers. The tranche's notional principal begins to decline with each default and it begins receiving interest on only the remaining notional principal. Once the junior tranche is wiped out (or at the end of contract), the junior tranche's payments for defaults and receipts of interest are ended.

Next, analyze the cash flows year by year for all three tranches.

3. ***The junior tranche cash flow year by year using above simulated independent defaults***:

Year 1: The cash flow for the junior tranche in the first year has three components or legs: the default leg (i.e., loss), the premium leg, and the accrual leg. The three legs to the junior tranche are illustrated below (Figure 8.7).

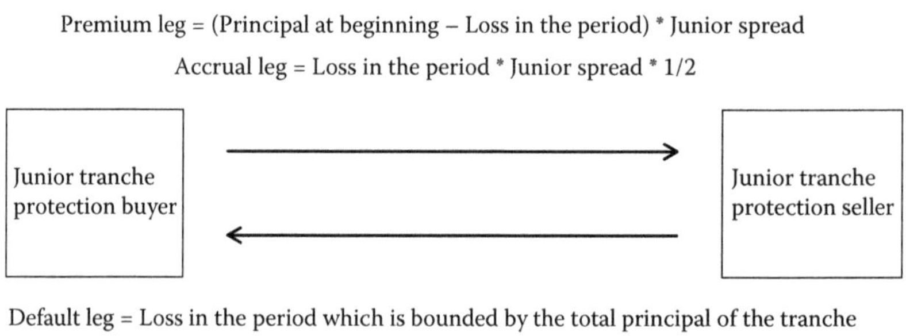

FIGURE 8.7
Junior Tranche Cash Flows.

At the end of the first year, the *default leg* is \$2.40 (four defaults × \$0.60 losses per default). If there had been no default losses, the premium leg would simply be the notional principal (\$5.00) times the spread (0.30). However, when there are losses, the premium leg is based on the tranche values less losses from defaults. Since this example assumes that default occurs in the middle of the period, there is an accrual leg payment for half of the period that approximates the spread earnings during the year on those defaults that occurred later in the year. The spread earnings to the junior tranche from selling credit protection based on end-of-period values form the *premium leg* and would be 30% of the tranche's size at the end-of-the-period.

In summary, the amount of principal left in the junior tranche at the end of one year is (5.00 – 2.40) = 2.60, which is then multiplied by the junior tranche spread (0.30) to form the premium leg = 0.78. The *accrual leg* adds to the junior tranche the quantity $2.4 \times 0.30 \times 1/2 = 0.36$, which is the premium on defaulted names accrued prior to their default. The premium leg and accrual leg together can be viewed as adjusting the total premium payments to reflect an average of the beginning and end-of-period values. In fact, using the start-of-period tranche values produces premium payments of $5.00 \times 0.30 = 1.50$, using the end-of-period tranche values produces premium payments of $2.60 \times 0.30 = 0.78$, and using the midpoint $(5.00 + 2.60)/2 = 3.80$ produces a premium of $3.80 \times 0.30 = 1.14$ which is the average of 0.78 and 1.50. So, the model can be viewed as assuming that the spread is paid based on the average of the year's beginning and ending notional principals.

In the second period (year 2) of the simulation, there were five defaults, each with a principal of $1 for a total loss after the 40% recovery of $3.00. Note that this loss wipes out the junior tranche since it had only $2.60 left in principal after the previous year's losses. So the junior tranche protection seller pays $2.60 to its protection buyer and has no further role in this path. There is an accrual leg payment of $\$2.6 \times 1/2 \times 0.3 = 0.39$ from the junior tranche protection buyers to junior tranche protection sellers as a result of the spread being paid through mid-year (i.e., the accrual leg).

4. ***The mezzanine tranche cash flow year by year using the above simulated independent defaults***:

In year 1, the mezzanine tranche protection buyers/sellers do not gain or lose from the defaults (because all of the default losses were absorbed by the junior tranche) and so the mezzanine tranches's premium legs are formed by the product of their principal amounts and their premium rates ($20 \times 0.03 = 0.60$). While the four defaults caused a reduction in the principal of the junior tranche from $5.00 to $2.60, they left the mezzanine and senior tranches at their original principal values.

In year 2, with the junior tranche going away, the mezzanine tranche as protection seller to a protection buyer begins to bear default losses. The $3.00 second-year loss leaves $0.40 to be borne by the mezzanine tranche after the junior tranche bore $2.60. This $0.40 loss serves as the default leg for the mezzanine tranche paid to the protection buyer in year 2 (Figure 8.8).

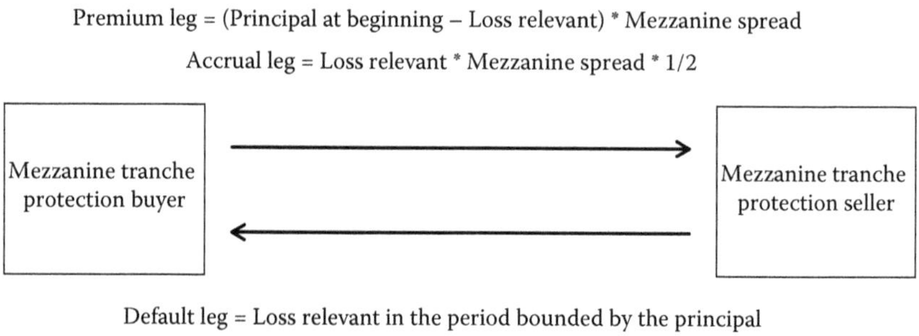

FIGURE 8.8
Mezzanine Tranche Cash Flows.

The mezzanine tranche protection buyer pays a premium leg based on the principal left which is equal to: principal at beginning – relevant loss = 20.00 – 0.40 = 19.60. The premium leg is therefore: principal left × spread = (20.00 – 0.40) × 0.03 at the end of second year. The accrual leg (again, reflecting spreads through mid-year on defaults) is 0.40 × 0.03 × 1/2.

In years 3–5, the mezzanine continues to decline in principal value, with a declining premium leg and a non-zero accrual leg. For example, from the simulation there are three defaults in the third year for a loss of $1.80, so the mezzanine tranche as protection seller pays $1.80 for the default leg.

5. ***The senior tranche cash flow year by year using the above simulated independent defaults***:

The senior tranche does not have any default legs since the mezzanine tranche survives the five-year simulated path. The senior tranche receives a premium (principal × spread = $75.00 × 0.01 = 0.75) every period and has no accrual leg or default leg until and unless 42 corporations have defaulted. The senior tranche premium leg, accrual leg, and default leg are illustrated in the following diagram, even though in this simulation, there is no loss to the senior tranche since the total loss to the entire structure is only $15 (Figure 8.9).

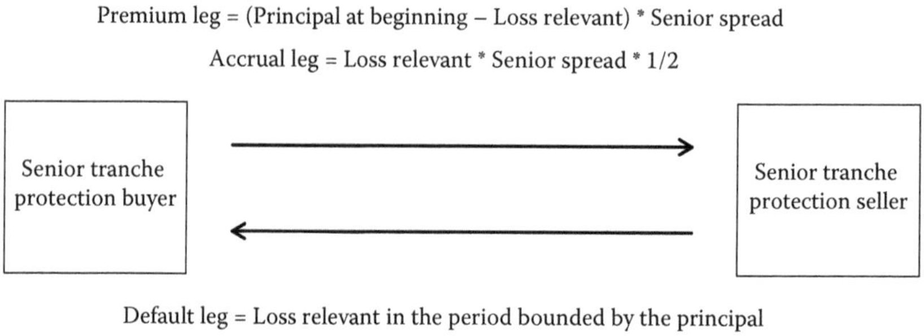

FIGURE 8.9
Senior Tranche Cash Flows.

Having investigated the cash flows in this simulation, consider the following question:

6. ***How safe is the senior tranche?***

Now consider how likely it is that the senior tranche will be affected by paths with numerous defaults. If 42 defaults occurred, there would be a loss of \$0.60 × 42 = \$25.20 and the senior tranche would begin to lose principal. Since the defaults of each name – in this example – are assumed to be independent of each other, the number of defaults in five years follows the binomial distribution with probability 0.25. Here is the calculation of the probability that the senior tranche will *not* experience default losses:

$$P(41 \text{ corporations or less}) = \sum_{i=0}^{41} \binom{100}{i} 0.25^i 0.75^{100-i} = 0.999852926.$$

R-code:

```
a=pbinom(41,100,0.25)
```

The probability that the senior tranche will *not* experience default losses is 0.9998529, meaning that there is a probability of about 0.01% that the senior tranche will be at least slightly impaired.

It appears that the senior tranche holders are quite safe as they collect premiums with little chance of bearing defaults. Many investors and rating agencies thought just that prior to the global financial crisis of 2007–2008. However, note that *the above simulation assumed that all of the corporate bonds experienced defaults independently of each other*. In practice, default levels follow a cycle and tend to exhibit positive correlation. The next section explores CDO default correlation in detail and its role in the financial crisis.

Before exploring the modeling of default correlations, examine an extreme case where the default correlation between corporate bonds is 1. If the default correlation is 1, the situation is as if there is only one corporation. In other words, if 1 corporation defaults, then all 100 corporations default. Even though the default probability is only 5%, that implies that within five years, there is a 25% chance that all bonds will default.

What occurs if all corporations default in the second year with a principal loss equal to 0.6 × 100 = \$60? The senior tranche protection seller has to pay \$35 of the \$60 loss as a default leg to the senior tranche protection buyer because the other \$25 of defaults have wiped out the junior and mezzanine tranches. In this case, the 75% probability of earning a 1% spread (\$0.75 per year) for five years on the senior tranche is insufficient to compensate the senior tranche protection seller's risk of paying \$25 with a 25% probability. This example shows us how important the correlation is and motivates the next section.

The most likely correlation is between 0 and 1. The next section uses a powerful model, the copula method, to model the correlation. The model shows that an appropriate price (spread) for each tranche is a function of the correlation coefficient and a function of the recovery rate in addition to a function of the expected default rate. The sensitivity is especially large for the senior tranche. If default correlation is ignored (i.e., assumed to be zero), the risk and price of the senior tranche generated by the model will be wrong. Most investors prior to the financial crisis thought that senior CDO tranches were safe and the rating agencies rated many senior tranches as AAA. But high-default correlations compounded by a low recovery rate were a major reason for the CDO losses of even the seniormost tranches during the 2007–2008 financial crisis. The crisis teaches a very important lesson: model assumptions in mathematical finance are very important and need to be understood by all parties involved in decision making.

8.4 Collateralized Debt Obligations (CDOs) Model with Positively Correlated Bond Defaults

As discussed in the previous section, correlation among bond defaults plays an important role in the relative pricing of a CDO's tranches. The binomial distribution cannot be used to model correlated defaults. In mathematical models, the *copula* method is used to model correlated distributions.

8.4.1 The Copula Method and the Mathematical Background for CDO Pricing

In order to value CDO tranches (i.e., estimate spreads for different tranches), it is necessary to model multiple defaults for different corporations taking into account correlation between the defaults. As in the previous CDS simulation, each individual corporation's default time can be modeled as an exponentially distributed random variable τ_i. This section forms a joint distribution of the 100 corporate default times with a specified correlation. Throughout the analyses that follow, bond defaults are assumed to be positively correlated, which is very consistent with historical evidence.

Mathematics Review 8.3 includes how to model correlated normal distributions. It begins by reviewing how to simulate n standard normal distributions with a pairwise correlation ρ.

> **Mathematics Review 8.3** Let $M, Z_1, \dots Z_n$ be $n + 1$ independent standardized normal random variables and
>
> $$X_i = \sqrt{\rho}M + \sqrt{1-\rho}Z_i.$$
>
> Then, $X_1, \dots X_n$ are n standardized normal distributions with pairwise correlation ρ where
> $0 \le \rho \le 1.$

We begin with a proof of the results in Mathematics Review 8.3.

First, note that $X_i \sim N(0, 1)$. X_i follows a normal distribution since it is a linear combination of independent normal distributions.

$$E(X_i) = \sqrt{\rho}E(M) + \sqrt{1-\rho}E(Z_i) = 0$$

$$\mathrm{Var}(X_i) = (\sqrt{\rho})^2\mathrm{Var}(M) + (\sqrt{1-\rho})^2\mathrm{Var}(Z_i) = 1$$

Next, note that the correlation between X_i and X_j is ρ, assuming that $i \ne j$:

$$\mathrm{Cov}(X_i, X_j) = \mathrm{Cov}(\sqrt{\rho}M + \sqrt{1-\rho}Z_i, \sqrt{\rho}M + \sqrt{1-\rho}Z_j).$$

By expanding:

$$= \rho\mathrm{Cov}(M, M) + \sqrt{\rho}\sqrt{1-\rho}\mathrm{Cov}(M, Z_j) + \sqrt{1-\rho}\sqrt{\rho}\mathrm{Cov}(Z_i, M) + (1-\rho)\mathrm{Cov}(Z_i, Z_j)$$

$$= \rho\mathrm{Cov}(M, M) + 0 + 0 + 0$$

$$= \rho\mathrm{Var}M$$

$$= \rho$$

$$\mathrm{Correlation}\ (X_i, X_j) = \frac{\mathrm{Cov}(X_i, X_j)}{\mathrm{std}(X_i)\mathrm{std}(X_j)} = \mathrm{Cov}(X_i, X_j) = \rho.$$

The variables $X_1, X_2, \dots, X_n$ are n standardized normal random variables with a pairwise correlation of ρ (i.e., there are n correlated normal distributions).

Correlated exponential distributions are obtained by transforming normal distributions to exponential distributions. By ***Sklar's theorem***: *Any multivariate joint distribution can be written in terms of univariate marginal distribution* *functions and a copula, which describes the dependence structure between the variables*. The marginal distribution in this case is the exponential distribution and the copula generated is called a Gaussian copula or a normal distribution copula. A ***copula*** is a multivariate cumulative distribution function for which the marginal probability distribution of each variable is uniform. A ***Gaussian copula*** is a multivariate cumulative distribution over the unit cube, which is constructed from a multivariate normal distribution by using the ***Probability Integral Transformation***. The correlation, ρ, in the normal distributions is called the ***copula correlation*** in a copula model. The correlation in the copula model emanates from the correlation at the normal distribution level, it is not correlation from the exponential distribution. Therefore, ρ is described as a copula correlation, not a default-time correlation.

In constructing correlated exponential distributions, or multivariate joint distribution functions with a marginal exponential distribution, the following transformation between different distributions is important.

Start with two Probability Integral Transformation theorems: one transforms an arbitrary continuous distribution to a uniform distribution and the other transforms the uniform distribution to any given continuous distribution with c.d.f. $F(x)$:

Mathematics Review 8.4 Probability Integral Transformation Theorem to the Uniform Distribution

Let X be a random variable that has a continuous cumulative distribution function $F(x)$ and let $Y = F(X)$. The distribution of Y is the uniform distribution on the interval [0,1].

Mathematics Review 8.5 Probability Integral Transformation Theorem from the Uniform Distribution

Let X have the uniform distribution on the interval [0,1], and let $F(x)$ be a continuous cumulative distribution function with quantile function $F^{-1}(x)$. Then, $Y=F^{-1}(X)$ has a cumulative distribution function of F.

So, for any random variable X, if its own c.d.f., F, is applied to the random variable X, then the new random variable, $F(X)$, is uniformly distributed. If X is a uniformly distributed random variable, then if a c.d.f. inverse exists, F^{-1}, and is applied to X, the new random variable $F^{-1}(X)$ has a c.d.f. of F.

So, consider a random variable X with a c.d.f. $F(x)$. We can generate $Y = F(X)$ as a uniform distribution, and then apply another c.d.f. inverse G^{-1} to Y to obtain a random variable $Z = G^{-1}(F(X))$, which has a distribution of $G(x)$. So the random variable X with c.d.f. $F(x)$ is transformed to a random variable $Z = G^{-1}(F(X))$ with c.d.f. $G(x)$. Note that we follow the common convention where capital letters represent random variables and lowercase letters represent the independent variables in cumulative distribution functions.

Consider X_i, which in Mathematics Review 8.3 was modeled as: $\sqrt{\rho}M + \sqrt{1-\rho}Z_i$. Note that the first random factor, M, introduces a random effect that is common to each X_i, while the second random effect, Z_i is unique to X_i. In the copula model, M represents the shared tendencies of default to occur across all corporations in the market while X_i represents risks of default unique to a particular corporation, i. Thus, the copula method allows for correlation among defaults to be driven by their common correlation that enters the model through the market factor M, and is scaled by the parameter ρ, as well as idiosyncratic defaults through the Z_i's.

We can model two positively correlated normal distributions (in this case, using Z_i and M) as in Review 8.3, use the Probability Integral Transformation theorem in Review 8.4 to obtain correlated uniform distributions, and then use Review 8.5 to obtain correlated exponential distributions. This method is called the ***Gaussian copula method*** and was pioneered by David Li.[3] Given marginal distributions and given a copula function, an analyst can build up a joint distribution with a correlation ρ (the copula correlation from the normal distribution). The copula correlation is not the default correlation in the exponential distribution. The formula for time to default is $\tau_i = G^{-1}(F(X_i))$, where τ_i is the ith corporation's default time, G is the exponential distribution c.d.f. function, and F is the normal distribution c.d.f. function.

The next section demonstrates computation of CDO prices based on the copula model.

8.4.2 CDO Pricing Based on Simulation and the Copula Model

This section provides an algorithm for CDO pricing.

The ideas from the previous section can be used to generate n correlated random variables as part of a simulation to model the times to default, if any, for n corporations.

Here are the steps to generate correlated default times for n corporations.

Step 0: Based on Mathematics Review 8.3, generate $X_1, \ldots, X_n$, n standardized normal random variables with pairwise correlation ρ:

$$\boldsymbol{X_i = \sqrt{\rho}M + \sqrt{1-\rho}Z_i} \tag{8.20}$$

where $M, Z_1, \ldots Z_n$ are $n+1$ independent standardized normal random variables and ρ is the copula correlation.

Step 1: Define n uniform random variables:

$$Y_i = N(X_i) \tag{8.21}$$

where N is the cumulative distribution function of $N(0,1)$: $N(x) = \int_{-\infty}^{x} \frac{1}{\sqrt{2\pi}} e^{-(1/2)\times x^2} dx$.

By Review 8.4, it is not hard to prove that Y_i follows uniform (0,1).

Step 2: Define n correlated default times:

$$\tau_i = F^{-1}(Y_i) \tag{8.22}$$

where $1 - e^{-\lambda t} = F(t)$.

By Review 8.5, it is not hard to prove that τ_i follows an exponential distribution. Note that the τ_i are correlated exponential distributions, which model defaults well.

The above steps model correlated defaults. Next, the simulation involves projecting cash flows. If it is assumed that premiums are paid at the end of a quarter, that defaults and accruals are paid as they occur, and that default intensity, λ, is a constant, simulation can be used to find the CDO spreads for each tranche.

Algorithm

1. First simulate m paths with each path containing the default time, if any, for $n = 100$ corporations, which is represented by the row vector $(\tau_1, \ldots, \tau_{100})$, where τ_i is the default time for the ith corporation's debt.
2. Estimate each tranche's cash flows separately. The cash flows are found for a particular simulation path given the default vector, the recovery rate, the tranche's spread (premium), and the initial premium amount. The cash flows depend on the defaults relevant to the tranche being analyzed. For each path, j, calculate the default leg, premium leg, and accrual leg as in the previous sections. Here are the formulas for each leg, where τ_i is the ith corporation's default time and t_i is the end of the ith quarter.

$$\textbf{Default leg}_j = \sum\nolimits_{\textbf{all relevant defaults}} (\mathbf{1} - \boldsymbol{R}) \times \boldsymbol{D}(0,\ \tau_i) \tag{8.23}$$

$$\textit{\textbf{Premium leg}}_j = \sum\nolimits_{i=1}^{\textbf{number of quarters before trancheis wiped out}} \textit{\textbf{Principal left in ith quarter}} \times \frac{s}{4} \times \boldsymbol{D}(0,\ \boldsymbol{t}_i) \tag{8.24}$$

$$\textit{Accrual leg}_j = \sum\nolimits_{\text{all relevant defaults}} (1 - R) \times s \times (\tau_i - \textit{the quarter before this default}) \times D(0, \tau_i) \quad (8.25)$$

3. Note that the above formulas share similarity with those in Equation 8.19 for pricing CDSs using simulation. Next, take the arithmetic average of the cash flows for the three legs and the m paths.

$$\frac{\sum_{j=1}^{m} \textit{Premium leg}_j}{m} + \frac{\sum_{j=1}^{m} \textit{Accrual leg}_j}{m} = \frac{\sum_{j=1}^{m} \textit{default leg}_j}{m} \quad (8.26)$$

Note that the premium and accrual legs are functions of s, the spread. The spread can be factored out. Solve for s for the tranche. Similarly, s can be solved for the other tranches. Note that the correlation ρ is a parameter that is used to simulate the default time vectors, so the spread is a function of ρ.

8.4.3 CDO Spreads as a Function of the Correlation Coefficient

A graph of the CDO spreads as a function of the copula correlation, ρ, can be formed by solving for the spreads over different values of ρ. The following example is from the honors thesis of Lafayette College undergraduate student Jinjin Qian '08. Her example assumes a risk-free interest rate of 0% and a default intensity $\lambda = 1\%$. She modeled a five-year CDO contract with 125 corporate bonds, five tranches, and a recovery rate of 40%. The five tranche widths most junior to most senior are:

0% – 3% (Most Junior), 3% – 6%, 6% – 12%, 12% – 22%, and 22% – 100% (Most Senior).

The leftmost tranche above is the first to bear default losses. The right most tranche, the most-senior tranche, begins suffering default losses when 22% of the initial value of the assets has been wiped out. The senior tranche does not become worthless unless 100% of the assets are wiped out (which could only happen if the recovery rate were 0% and all bonds defaulted).

Figure 8.10 depicts the spreads that markets should require for each of these tranches given the parameters discussed above and the copula correlation given on the horizontal axis.

Note that when ρ is small, the junior tranche is expensive (i.e., it has a large spread) and the senior tranche requires only a very small spread. This is because when defaults have little or no correlation, annual realized default rates tend to be near to the default intensity ($\lambda = 1\%$). The low but

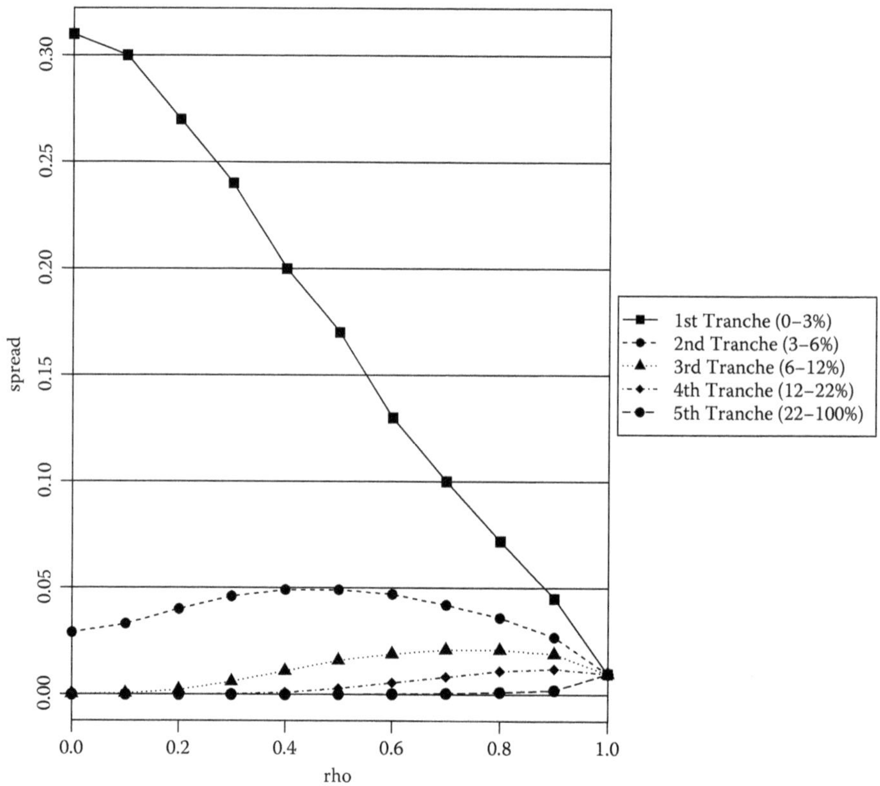

FIGURE 8.10
Relation Between CDO Tranche Spreads and Copula Correlation ρ.

relatively constant default rates are very likely to completely wear away the junior-most tranche but leave the mezzanine and senior tranches unscathed.

As ρ becomes larger, defaults are more likely to cluster. Clustering helps the junior-most tranches because they can perform well in the scenarios where there are few defaults. On the other hand, the mezzanine and senior tranches are vulnerable to very high default rates caused by correlations (because with high correlation, the defaults will cluster to the point that the losses invade their principals). If $\rho = 1$, the spreads converge because each path will either have 125 defaults with maximum losses (wiping out all) or no defaults and no losses.

Because ρ is so important, it is vital that analyses of defaults include tests over a wide spectrum of potential values for ρ, not just tests using an *expected* value for ρ. The same applies to the recovery rate. In the above analysis, we assume that the recovery rate is constant at 40%. In fact, when the economy enters a recession, the default rate rises, the correlation of defaults is high, and the recovery rate declines (as the housing market declines). The next section analyzes the effect of the recovery rate.

8.4.4 Recovery Does Matter in CDO Spreads and in Global Financial Crisis

In all of the above analyses, the recovery is assumed constant at 40%, meaning that if all bonds defaulted the structure would still only lose 60%. If the recovery becomes low, the senior tranche becomes risky. In an environment of a low recovery rate, the senior tranche spread should be much higher than that in an environment of a high recovery rate. Before the global financial crisis, the copula model was applied over a wide range of default outcomes, but was applied only over a narrow range of correlations and a narrow range of recovery rates (or with recovery rates that were not linked to realized default rates).

Both correlations and recovery rates are important to the estimation of risk and prices. One problem in the global financial crisis was that the market did not understand the implicit correlation and recovery assumptions used in valuing credit derivatives under the copula model. Most senior tranches were rated as AAA. However, AAA ratings were only reasonable if default correlation is sure to be no more than moderate and recovery rates sure to be substantial. When a CDO contains subprime mortgages, the story is especially different. Pools with subprime mortgages should have been anticipated to have potentially high default rates, high default correlations, and low recovery rates *all at the same time*. It is likely that many analysts did not do sensitivity tests for highly correlated defaults and very low recovery rates even though common sense dictates that correlated defaults and low recovery rates should be expected in a major real estate crisis.

The credit risk of CDO tranches was underestimated, which provided inexpensive financing for risky mortgages such as subprime mortgages. Mortgage lenders issued subprime mortgages freely because the mortgages could be sold into CDO pools – relaxing natural market controls on how many subprime mortgages were originated. Financial institutions were eager to collect higher yields from being protection sellers because they trusted the AAA ratings that were assigned to senior tranches. Borrowers, able to leverage highly using subprime mortgages, rushed to acquire real estate at higher and higher prices in "hot" markets. When real estate prices declined and an economic slowdown occurred, many subprime borrowers could not afford the payment and began to default, causing real estate prices to spiral downward. Thus, high default rates and low recovery rates hit the mortgage market simultaneously.

The bottom line is that the risks of CDO tranches with underlying subprime mortgage exposure were severely underestimated. The complexity of the structures and the analytical models obfuscated the limitations of the models as applied. In addition to the above problems, a product known as CDO squareds has been identified as contributing to the financial crisis. CDO squareds are discussed in the next section.

8.5 CDO Squareds

A ***CDO squared*** is a financial derivative (i.e., a CDO) with underlying assets containing tranches of other CDOs. Although CDO squareds are no longer popular or even barely exist, they were quite popular before the 2007–2008 global financial crisis. In fact, CDO squareds on subprime mortgages were one of the most problematic types of securities during the depth of the global financial crisis (due not only to high losses but also due to their complexity, which caused many investors to be unclear as to their value during the crisis). Models of CDO squareds can be used to analyze leveraged positions in CDO tranches – a practice that is common. Therefore the underlying mathematics remains relevant. Consider the following example of a CDO squared.

Assume that there are 40 mortgage *pools* inside 40 CDOs, each containing 1,000 mortgages. Thus, each CDO has one of the mortgage pools. Each of the CDOs is structured with three tranches: a junior tranche, a mezzanine tranche, and a senior tranche with tranche widths of 5%, 20%, and 75%, respectively. The pools and CDOs are illustrated in Figure 8.11.

Next, consider the case in which all 40 mezzanine tranches from the CDOs are used as the collateral assets for one or more CDO squareds. The use of all 40 of the mezzanine tranches as collateral for a CDO squared is illustrated in Figure 8.12.

The CDO squared has a total principal amount of $200 × 40 = $8,000. The CDO squared's senior tranche represents $8,000 × 0.75 = $6,000 of the structure while the mezzanine tranche is worth $1,600 and the junior tranche is worth $400. The diagram illustrates that the CDO squared is structured just like a regular CDO. The difference is that the assets of the CDO squared are CDO tranches rather than mortgages or corporate bonds.

Now consider what happens when the mortgages underlying the 40 CDOs begin to default. Assume that there are three types of outcomes to the 40 CDOs: 500 mortgages default inside 20 of the CDOs, 250 mortgages default inside 10 of the CDOs, and 100 mortgages default inside the remaining 10 CDOs.

Assuming a mortgage recovery rate $R = 0.6$, what will happen to the mezzanine tranches of the 40 CDOs (differentiated by loss outcomes), and what will that do to the tranches of the CDO squared? Here are the losses to the mezzanine tranches of the 40 CDOs based on the three types of default outcomes:

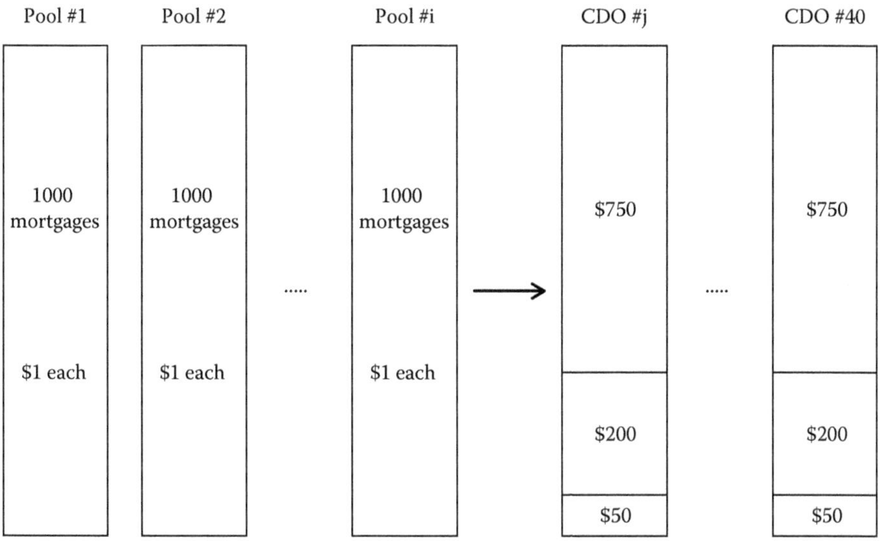

FIGURE 8.11
Subprime Mortgages Into CDOs.

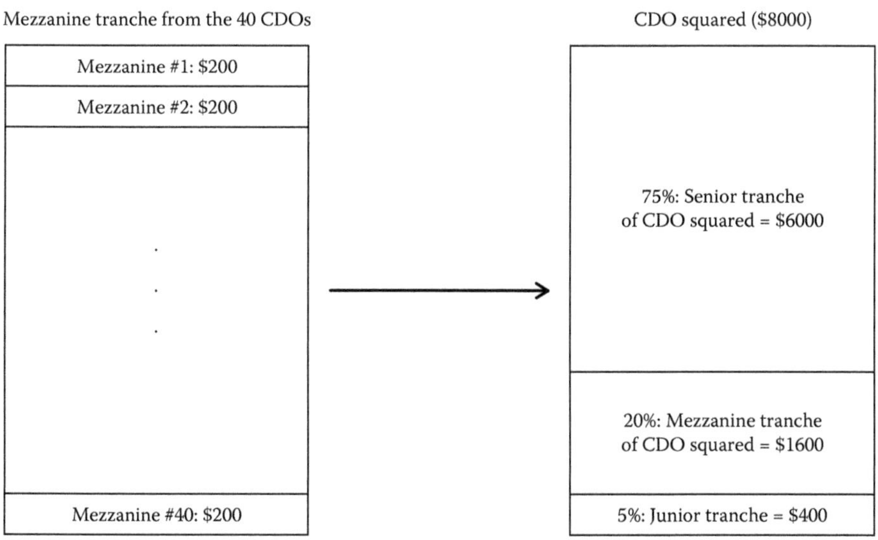

FIGURE 8.12
Subprime Mortgages CDOs into CDO Squared.

First type of CDO loses 0.4 × 500 = 200, and so its mezzanine tranche loses \$150.

Second type of CDO loses 0.4 × 250 = 100, and so its mezzanine tranche loses \$50.

Third type of CDO loses 0.4 × 100 = 40, and so its mezzanine tranche loses 0.

Therefore, the collateral pool of the CDO squared loses (\$150 × 20) + (\$50 × 10) = \$3,500. The consequence of this loss is that the CDO squared loses all of its junior tranche (\$400), loses all of its mezzanine tranche (\$1,600), and loses 3,500 − 1,600 − 400 = \$1,500 of its senior tranche.

So, even though the senior tranches of each regular CDO were safe (and the mezzanine tranches of the 40 CDOs had only partial losses or no losses at all), the senior tranche of the CDO squared was not safe. The senior most CDO squared tranche loses 1,500/6,000 = 25% of its principal.

The detailed algorithm for pricing CDO squareds is beyond the scope of this chapter. In order to price a CDO squared, as noted in Hull's book, two correlations must be modeled: ***within-pool correlation*** (which is the default copula correlation between the individual corporations/mortgages within a pool), and ***between-pool correlation*** (which is the default copula correlation between the pools).

The formula to generate the correlated normal distribution expands from one M-type source of randomness (M) for regular CDOs to two for CDO squareds (M_{bp} and M_{wp}):

$$\sqrt{\alpha\rho}\,M_{\text{bp}} + \sqrt{(1-\alpha)\rho}\,M_{\text{wp},j} + \sqrt{1-\rho}\,Z_{ij}$$

where ρ is the within pool correlation and α is the between pool correlation.

Why were CDO squareds so popular? The popularity of *regular* CDOs is because a pool of risky assets (such as subprime mortgages) can be financed in large part by issuing senior tranches that are AAA-rated. In the years leading to the financial crisis, lots of high-yielding subprime mortgages were issued using the money obtained from issuing the AAA-rated tranches of regular CDO structures. The senior tranches could be easily sold at very low yields to institutions eager for highly rated investments that yielded even just a little more than the riskless rate. But rating agencies tended to limit the size of the AAA-rated senior tranches to about 75% of the financing. The CDO *squared* concept allowed financial engineers to take many of the mezzanine and junior tranches (that were not AAA-rated and therefore could not be financed directly as low-cost AAA-rated tranches) and collect them into a CDO squared which could then receive AAA-ratings on its senior tranche. Thus, CDO squareds indirectly allowed a higher percentage of the principal amounts of subprime mortgages to be financed with AAA-rated funding. In fact, the process could continue to multiple layers of CDOs – each of which increased the amount of AAA-rated financing sources that could be used to fund the issuance of new subprime mortgages.

Practitioners who lacked a deep understanding of the mathematics of the copula model could not see the deficiencies in the way the model was being applied.

This section completes the chapter's coverage of three major types of credit derivatives. There are other credit derivatives such as forward-starting CDSs, options on CDSs, first to default swaps, etc. which are beyond the scope of this book. Readers may wish to learn about these products in Dominic O'Kane's book *Modeling Single-name and Multi-name Credit Derivatives.*

8.6 The Financial Crisis and Credit Derivatives

Financial derivatives are widely believed to have played an important role in causing or exacerbating the global financial crisis of 2007–2008. This chapter has explored the derivatives at the heart of the crisis: CDOs. The CDO structures that caused the biggest harm contained subprime mortgages in their collateral pools. The ***subprime mortgages*** were residential mortgages (i.e., mortgages on houses) that had relatively high default risk relative to prime mortgages. The subprime mortgages that caused the biggest losses tended to be loans on houses in "hot" real estate markets – those metropolitan areas in which housing prices had soared the most in the years leading up to the crisis. These regions with high real estate price volatility saw the values of the mortgaged properties fall dramatically as the crisis unfolded, which led to expectations of high default rates, high default correlation, and eventually low recovery rates.

As applied, the copula model simulates random dispersion in the number of defaults that occur around the expected default rate. But the model as commonly applied did not allow some of the parameters to vary randomly – especially the recovery rate and the correlation. There is an expression in finance that in times of financial crisis "correlations go to one" – meaning that lots of asset classes that appeared to offer diversification benefits from their lack of return correlation prior to a crisis transform into having highly correlated returns in a crisis as their prices collapse together. A major problem with CDO modeling prior to the crisis was that recovery rates and correlations were not allowed to take on extreme values in the models at the same time that the default rate soared. In other words, the model's randomness allowed actual default rates to differ widely from the expected default rate (which was inserted as a parameter) but it did not allow recovery rates to correlate negatively with those default rates. It is common sense that in a real estate market with soaring rates of defaults (and foreclosures) there will be plummeting recovery rates due to depressed real estate prices, which in turn can repeat the cycle.

Another key issue is that rating agencies and investors tended to believe that the market prices of CDOs would adhere to model prices based on realistic inputs. But, in times of crisis market prices can be driven far below model values due to forced liquidations of portfolios caused by sell orders from distressed investors.

Another factor is that the market prices of CDOs containing subprime mortgages are based on investor expectations, which in times of crisis can be far outside the range of outcomes anticipated during periods of economic health. Thus, CDO prices dropped not just to reflect actual mortgage default rates, but also to reflect the potential for higher future default rates and lower future recovery rates.

This chapter discussed the vulnerability of the senior-most tranches to extraordinarily high default rates, correlations and extraordinarily low recovery rates. As the likelihood of experiencing these worst-case scenarios increased, the AAA-rated senior CDO tranches dramatically fell in price. These AAA-rated tranches had been purchased and held by conservative investors with low tolerances for losses and risks. The rush to exit these tranches generated liquidations that accelerated the price declines of these residential mortgage-backed securities. Senior CDO tranches began to be priced at levels indicating expectations of ridiculously high default rates and extremely low recovery rates. The complexity of the products exacerbated the problem as few investors understood the true risks of the collateral assets and the arcane mechanics of the CDO structures (especially CDO squareds) well enough to purchase distressed tranches on speculation and offset the sell-off caused by liquidations by conservative investors.

This chapter has detailed the complexity of assessing the risks of CDOs with collateral pools containing debt securities subject to default risk. In particular, the most senior tranches are vulnerable to: (1) high default rates and high default correlations and (2) low recovery rates due to falling asset prices. The financial crisis revealed the inability of investors – and even rating agencies – to understand the mathematics of the models including the copula method and its assumptions.

Two observations should be noted about the net benefits to financial derivatives even after the fiascos of the recent financial crisis. First, the ability of financial derivatives to devastate an economy demonstrates the power of these products. Financial derivatives are used for important economic activities such as risk management and hedging – they are not just devices for whimsical speculations. If they were purposeless gambling devices their demise would not have so badly devastated the entire economy. Financial derivatives themselves are zero-sum games in the short run; they are merely vehicles for transferring risk. Each loss they convey to one side of the contract tends to be a profit to the other side of the contract. In other words, the CDO losses were primarily driven by declining mortgage values, not by the CDO structures themselves. Second, in the decades leading up to the financial crisis, financial derivatives were important drivers of the economy's success.

In particular they generated tremendous cost savings within the real estate financing industry that was passed on to borrowers in the form of relatively low loan rates. Modern economies benefited greatly from the highly efficient capital markets that evolved in the previous decades as a result of financial engineering innovations such as financial derivatives.

Chapter Summary

This chapter discusses default probabilities and investigates how to derive default probabilities from bond and asset swaps. Then, the chapter overviews the management of credit risk in general and default risk in particular. Credit derivatives are introduced for hedging credit risk and default risk. The potential role of credit derivatives in the global financial crisis is discussed.

Credit derivative contracts have two parties. The two parties swap cash flows based on credit events. Analysts need to clearly understand the cash flows being exchanged between the two parties so that they value the derivatives and model their risks.

CDOs are often modeled with the copula method, which has been used effectively to capture the tendency of actual default rates to differ randomly from expected default rates. The cash flows to CDOs or CDO squareds also depend on default correlation. In fact, CDO prices (spreads) are very sensitive to copula correlations. Recovery rates serve as another important parameter in CDO pricing. So correctly modeling the copula correlation and the recovery rate is vital for valuing CDO tranches. CDO models were not used effectively to capture extremes in these important parameters (i.e., recovery rates and correlations) in the years leading up to the 2007–2008 global financial crisis.

Demonstration Exercises

1. Consider a CDO with 100 bonds (i.e., credit names or corporations) with $1 principal for each bond. Assume the recovery rate is 20%. There is an equity tranche, a mezzanine tranche and a senior tranche within the CDO structure. The attachment for the equity tranche is 0% and the detachment is 5%. The attachment for the mezzanine tranche is 5% and the detachment is 25%. The attachment for the senior tranche is 25% and the detachment is 100%. Assume that

the CDO has five years to termination, that default occurs at the middle of the year, that the default and the accrual legs are paid at the time of default, that the premium is paid at the end of the year, that the equity tranche spread is 30%, that the mezzanine tranche spread is 3%, that the senior tranche spread is 1% and that there are three defaults each year. Calculate the mezzanine tranche premium leg, the default leg, and the accrual leg in the third year.
First-year total loss: $(1 - 0.2) \times 3 = 2.4$
Second-year total loss: $(1 - 0.2) \times 3 = 2.4$
Third-year total loss: $(1 - 0.2) \times 3 = 2.4$
$2.4 + 2.4 + 0.2 = 5$ is the loss to the junior tranche.
Remaining loss charged against the mezzanine tranche is $2.4 - 0.2 = 2.2$
Default leg: $2.20 paid by the protection seller to the protection buyer at 2.5 years.
Premium leg: $(20 - 2.2) \times 0.03 = 0.534$ paid to the protection seller at the end of year 3.
Accrual leg: $2.2 \times 1/2 \times 0.03 = 0.033$ paid to the protection seller at 2.5 years.

2. Pool A contains 1,000 mortgages with $1 principal for each mortgage. It is structured into a CDO with three tranches: 5% junior, 20% mezzanine, and 75% senior. Pool B also contains 1,000 mortgages with $1 for each mortgage. It is also structured into a CDO with 5% junior, 20% mezzanine, and 75% senior tranches. Now the two mezzanine tranches of Pools A and B are put together into an asset pool that is further structured into a CDO squared with 5% junior, 20% mezzanine, and 75% senior tranches. Assume that 20% of the mortgages in both pools default and the recovery value is 0.6.
   a. What is the loss for each pool?
   b. What percent of the junior tranche of each CDO is lost?
   c. What percent of the mezzanine tranche of each CDO is lost?
   d. What percent of the senior tranche of each CDO is lost?
   e. What percent of the junior tranche of the CDO squared is lost?
   f. What percent of the mezzanine tranche of the CDO squared is lost?
   g. What percent of the senior tranche of the CDO squared is lost?

It is clear that Pools A and B are structured into $50 junior, $200 mezzanine, and $750 senior tranches. The CDO squared then pools $200 + $200 of mezzanine tranches from the two pools together as its $400 asset pool. The CDO squared has three tranches: $20 in the junior, $80 in the mezzanine, and $300 in the senior tranche.

a. 20% × $1,000 × (1 − 0.6) = $80
b. $80 CDO, loss > junior principal = $50, so 100% of each junior tranche is lost.
c. $80 − $50 = $30 to be absorbed by each mezzanine tranche; ($30/$200) = 15% lost.
d. The senior tranche is untouched: 0% loss.
e. When the two pools are put together, the total loss is $60 which is greater than the junior tranche principal ($20), so 100% of the junior tranche is lost.
f. Relevant loss to the mezzanine tranche is ($60 − $20)/80 = 50%, so 50% is lost.
g. The senior tranche of the CDO squared is not touched: 0%.

Extensions and Further Reading

CDOs became unpopular after the global financial crisis of 2007–2008. However, structured products in general and tranching of cash flows based on credit risk are still very important. There are many credit derivatives that were not covered in this chapter due to space limitations. There are also many credit derivative pricing models such as Jarrow–Turnbull credit risk models that were not covered. The reasons to study the copula model include that the model is cited as a cause of the 2007–2008 financial crisis and offers important lessons from that history. Another reason is that the copula model is an example of an interesting programming implementation. This book focuses on teaching skills, not just models.

Interested readers can read the book:

Modeling Single-name and Multi-name Credit Derivative by Dominic O'Kane

End-of-Chapter Project

Calibrating A Piecewise Survival Curve Using CDS Market Quotes

Perfectly efficient financial markets do not have persistent arbitrage opportunities. In practice, reasonably efficient markets may have temporary arbitrage opportunities, but they vanish quickly as market prices tend toward equilibrium values. One source of potential arbitrage occurs when inconsistent information is used in valuing securities in different financial markets. For example, the bond market (i.e., the cash market for bonds), asset swap market, and CDS market may at times trade based on different information, leading to inconsistent security prices. Arbitragers extract information implied from prices within each market and compare the information for consistency based on models. If two or more markets imply inconsistent information on the same underlying assets (e.g., expected default rates of similar bonds) it is an arbitrage opportunity for traders to establish long positions in the relatively undervalued assets, establish short positions in the relatively overvalued assets, and hold the portfolio until the prices converge (i.e., begin trading based on consistent information).

Calibrating data from a particular financial market to derive information that can be used in other similar markets is an important skill based on financial mathematics. Through this project, readers practice the skill of calibration. This project requires readers to understand the mathematical models and to be familiar with solving complicated equations. Readers may use Mathematica, EXCEL, or other software. This project adapts an idea from Chapter 8 in Dominic O'Kane's book on deriving the CDS survival curve, which is equivalent to deriving piecewise risk-neutral default intensities.

From this chapter, readers can use pricing information from CDS markets to ascertain implied information regarding corporate default. In particular, given several CDS spreads, the piecewise risk-neutral default intensity for a particular corporation can be found. The implied information regarding corporate default derived from CDS market should be consistent with the implied information from the bond market and asset swap market. Any inconsistencies may indicate an arbitrage opportunity.

1. *Review of CDS Prices*

   Given survival probability $Q(t)$ (which is the probability that an entity survives until time t), the constant risk-free discount rate, r, and the expected recovery rate, R, find the CDS contract price i.e., spread) s. By Equation 8.18,

$$s = \frac{(1-R)\sum_{i=1}^{n}(Q(t_{i-1})-Q(t_i))D\left(0, \frac{t_{i-1}+t_i}{2}\right)}{\sum_{i=1}^{n}\left[Q(t_i)\times\frac{1}{4}\times D(0, t_i)+(Q(t_{i-1})-Q(t_i))\times\frac{1}{8}\times D\left(0, \frac{t_{i-1}+t_i}{2}\right)\right]}$$

where $D(0,t)$ is a time discount factor, which is e^{-rt} if r is constant. If r is not constant, it can be a deterministic function of time t. More generally, replace ¼ by $\Delta(t_{i-1}, t_i)$:

$$s = \frac{(1-R)\sum_{i=1}^{n}(Q(t_{i=1})-Q(t_i))D\left(0, \frac{t_{i-1}+t_i}{2}\right)}{\sum_{i=1}^{n}\left[Q(t_i)D(0, t_i)+(Q(t_{i-1})-Q(t_i))\frac{1}{2}D\left(0, \frac{t_{i-1}+t_i}{2}\right)\right]\Delta(t_{i-1}, t_i)}.$$

To simplify the formula in the numerator, use $\frac{1}{2}(D(0, t_{i-1})+D(0, t_i))$ to replace $D\left(0, \frac{t_{i-1}+t_i}{2}\right)$. Note that the numerator comes from the default leg. Instead of using quarters as an interval for default, any interval can be used. So the numerator is:

$$\frac{1-R}{2}\sum_{j=1}^{K}(Q(t_{j-1})-Q(t_j))(D(0, t_{j-1})+D(0, t_j)).$$

To simplify the denominator, assume the accrual interest is paid at the end of quarter instead of the middle of a quarter. So, the denominator is:

$$\sum_{i=1}^{n}[Q(t_i)D(0, t_i)+(Q(t_{i-1})-Q(t_i))\frac{1}{2}D(0, t_i)]\Delta(t_{i-1}, t_i)$$

which is equivalent to:

$$\sum_{i=1}^{n}D(0, t_i)(Q(t_{i-1})+Q(t_i))\frac{1}{2}\Delta(t_{i-1}, t_i).$$

Hence,

$$s = \frac{\frac{1-R}{2}\sum_{j=1}^{K}(Q(t_{j-1})-Q(t_j))(D(0, t_{j-1})+D(0, t_j))}{\sum_{i=1}^{n}D(0, t_i)(Q(t_{i-1})+Q(t_i))\frac{1}{2}\Delta(t_{i-1}, t_i)} \quad (1)$$

Equation (1) is easier than Equation 8.18. If $D(0,t)$, s, and R, are known, they can be used to calibrate the CDS Survival curve, $Q(t)$ $(0 \le t \le T)$ (where T is the maturity of the longest contract) by using many CDS contracts.
Note that Equation (1) above is used as the relation between the survival curve and the CDS spread s for this project.

2. ***Review the Interpolation Scheme***:
   Interpolation formula:
   If $Q(t_{n-1})$ and $Q(t_n)$, are known, they can be used to find $Q(t_*)$, where $t_* \in [t_{n-1}, t_n]$.
   Specifically,

$$Q(t_*) = Q(t_{n-1})e^{-(t_*-t_{n-1})\lambda} \tag{2}$$

So λ can be found by equation (2):

$$\lambda = \frac{Ln(Q(t_n)/Q(t_{n-1}))}{-(t_n - t_{n-1})} \tag{3}$$

which can then be inserted into (2)

$$Q(t_*) = Q(t_{n-1}) \times \left(\frac{Q(t_n)}{Q(t_{n-1})}\right)^{\frac{t_*-t_{n-1}}{t_n-t_{n-1}}} \tag{4}$$

Equation (4) gives the survival probability at any time between the interval $[t_{n-1}, t_n]$.

3. ***Calibrating the CDS Curve***:
   In the following project, assume the LIBOR curve is flat at 5%, i.e., $D(0, t) = e^{-0.05t}$, $R = 40\%$, monthly time steps are used for K, and $\Delta(t_{i-1}, t_i) = \text{constant} = 0.25$.
   Given the following CDS curve, find $\{Q(T_0), Q(T_{0.5}), Q(T_1), Q(T_2), Q(T_3), Q(T_4), Q(T_5), Q(T_7), Q(T_{10})\}$:

	Six months	One year	Two years	Three years	Four years	Five years	Seven years	10 years
S	145 bp	145 bp	160 bp	175 bp	190 bp	220 bp	245 bp	270 bp

Note: bp stands for basis points where (100 bp = 1% = 0.01).[4]
Note that the longest maturity is 10 years, so $Q(t)$ cannot be found for $t > 10$. There are eight CDS spreads. So, eight different risk-neutral default intensities can be found corresponding to eight time periods $[T_0, T_{0.5}]$, …, $[T_7, T_{10}]$. This is equivalent to having the eight $Q(t)$ at the ending points of the time intervals. Once these eight $Q(t)$ are found, the survival curve can be found by Equation (4).

Algorithm:

Given $\{T_0, T_{0.5}, \dots, T_{10}\} = \{0, 0.5, 1, 2, 3, 4, 5, 7, 10\}$ and $\{S_{0.5}, \dots, S_{10}\} = \{0.0145, 0.0145, 0.016, 0.0175, 0.019, 0.022, 0.0245, 0.027\}$, note that $\Delta(t_{i-1}, t_i) = \text{constant} = 1/4$ in all calculations.

1. Set $Q(T_0) = 1$.
2. Assume that $Q(T_{0.5}) = x$. Use the spread $S_{0.5} = 0.0145$ given above. Equation (1) contains $Q(T_{1/12}), Q(T_{2/12}), \dots, Q(T_{6/12})$, which need to be expressed as functions of x. By Equation (4), $Q(T_{1/12}) = Q(T_0)\left(\frac{Q(T_{0.5})}{Q(T_0)}\right)^{\frac{\frac{1}{12}-0}{0.5-0}} = x^{1/6}$, similarly, $Q(T_{5/12}) = x^{5/6}$. Now Equation (1) only has one unknown x, which can be easily found.
3. Given $Q(t)$ for $0 \le t \le 0.5$ and, in particular, $Q(T_{0.5})$, assume $Q(T_1) = x$ is the new unknown. Equation (1) contains $Q(T_{1/12})$, $Q(T_{2/12}), \dots, Q(T_{6/12})$ which are known and $Q(T_{7/12}), Q(T_{8/12}), \dots, Q(T_1)$ which are not yet known. Next, express them as a function of x. By Equation (4),
$$Q(T_{7/12}) = Q(T_{0.5})\left(\frac{Q(T_1)}{Q(T_{0.5})}\right)^{\frac{\frac{7}{12}-0.5}{1-0.5}} = Q(T_{0.5})\left(\frac{x}{Q(T_{0.5})}\right)^{1/6},$$
similarly, $Q(T_{11/12}) = Q(T_{0.5})\left(\frac{x}{Q(T_{0.5})}\right)^{5/6}$. Now Equation (1) only has one unknown x. By inserting $S_1 = 0.0145$, solve for x.
4. By continuing this procedure, solve for $Q(t)$ at $t = 2, 3, 4, 5, 6, 7, 10$. This is called calibrating the CDS curve. It is a very popular practice in the financial industry and is another form of bootstrapping in addition to the bootstrapping of the term structure of spot interest rates previously detailed in Chapter 2.

End-of-Chapter Problems (Fundamental Problems)

1. Assuming that the default intensity is constant $\lambda = 1\%$ for five years, find each year's default probabilities P_1, P_2, P_3, P_4, P_5. Are they increasing or decreasing? Intuitively describe your answer.
2. Assuming the default probability in each of the five years is a constant 1% ($P_1 = P_2 = P_3 = P_4 = P_5 = 1\%$) and the default intensity

is constant for each year, find $\lambda_1, \lambda_2, \lambda_3, \lambda_4, \lambda_5$. Are the default intensities increasing or decreasing? Intuitively describe your answer.

3. Suppose that the default intensity is 1% for each year for a four-year period.
   a. Find the cumulative default probability in the first three years.
   b. Find the survival probability until the end of the fourth year.
   c. Find the default probability in the third year.
4. Assume that $Q(0.5) = x$ and that λ is constant during the time [0,0.5]. Find the survival probabilities $Q(1/12)$, $Q(2/12)$, $Q(3/12)$, and $Q(5/12)$ as a function of x.
5. Assume that $Q(0.5) = x$, $Q(1) = y$ and that λ is constant during the time [0.5,1]. Find the survival probabilities $Q(7/12)$, $Q(8/12)$, $Q(9/12)$, and $Q(11/12)$ as a function of x and y.
6. If the risk-free rate is flat at 3%, the three-year zero coupon risky bond has a yield of 9%, and the recovery rate is 40%, find the default intensity.
7. If the risk-free interest rate curve is flat at $r = 4\%$, the three-year risky coupon bond with a coupon of 4% paid annually has a 5% bond yield, the risky bond recovery rate $R = 40\%$, and default occurs in the middle of a year and the recovery amount is paid immediately at default, find the default probability P each year (assume P is constant, $P_1 = P_2 = P_3 = P$).
8. If the risk-free interest rate curve is flat at $r = 4\%$, the three-year risky coupon bond with a coupon of 4% paid annually has a 5% bond yield, the risky bond recovery rate $R = 40\%$, default occurs in the middle of a year and the recovery amount is paid immediately at default, find the default intensity λ (assume λ is constant).
9. If the risk-free interest rate curve is flat at $r = 4\%$, the three-year risky coupon bond with a coupon of 4% paid annually has a 5% bond yield, a five-year risky bond with a coupon of 5% paid annually has a 6% bond yield, both risky bonds have recovery rate $R = 40\%$, default occurs in the middle of a year and the recovery amount is paid immediately at default, find the default intensities λ_1 and λ_2 (assume λ_1, the first three-year default intensity, is constant and λ_2 the last two-year default intensity, is constant).
10. Assume the risk-free interest rate curve is flat at $r = 5\%$. Consider an asset swap that is based on a three-year risky coupon bond with a coupon of 6% paid annually. Assume that the three-year asset swap rate s = 100 basis points, paid semiannually, and that the risky bond recovery rate $R = 40\%$. Assuming that default occurs in

the middle of a year and the recovery amount is paid immediately at default, find the default probability P each year (assume P is constant, $P_1 = P_2 = P_3 = P$).

11. Assume the risk-free interest rate curve is flat at $r = 5\%$. Consider an asset swap that is based on a three-year risky coupon bond with a 6% coupon paid annually. Assume that the default probability is 1% per year and the risky bond recovery rate $R = 40\%$. Find the three-year asset swap rate spread paid semiannually (assume that default occurs in the middle of a year).

12. A corporate bond matures in two years, provides a coupon of 6% paid annually, and has a yield of 8% per annum with continuous compounding. The yield on a similar (but risk-free) bond is 5% per annum with continuous compounding. Assume that the expected recovery rate in the event of a default is 40%, that default can only occur at 0.5 years and 1.5 years, the recovery amount is paid immediately at default, and that the bond's principal amount is \$100.
    a. What is the value of the risk-free bond today, $B(0, 2)$? (Calculate the number.)
    b. What is the value of the corporate bond today, $\tilde{B}(0, 2)$? (Calculate the number.)
    c. Assuming the default probability is the same in the first and second years, what is the default probability?
    d. Find the asset swap spread s. (assume semiannual payments).

13. Alpha Corporation's corporate bond matures in two years, has a coupon of 6% paid annually, and an 8% per annum yield with continuous compounding. Assume that the risk-free zero curve is flat at 6% per annum with continuous compounding, that defaults can occur only at times 0.5 years and 1.5 years, that the recovery rate is 40%, and the bond principal is \$100.
    a. What is the value of the risk-free bond today, $B(0, 2)$? (Calculate the number.)
    b. What is the value of the corporate bond today, $\tilde{B}(0, 2)$? (Calculate the number.)
    c. If the default intensity is the same in the first and second years, what is the default intensity?

14. Consider a five-year CDS. Assume the recovery rate is 40%, the default intensity is 1%, the risk-free rate is 5%, the premium is paid quarterly, the default leg and accrual leg are paid immediately after default, and that default occurs in the middle of the period. Find the CDS spread. You may use a spreadsheet.

15. Consider a five-year CDS. Assume the recovery rate is 40%, the default intensity is 1%, the risk-free rate is 5%, the premium is paid quarterly, and the default leg and accrual leg are paid immediately after default. Find the CDS spread using simulation.
16. Suppose that the risk-free zero curve is flat at 5% per annum with continuous compounding. Consider a three-year CDS. Assume that default can only occur halfway through each year with default intensity $\lambda = 1\%$, the recovery rate is 40%, premium payments are made annually at the end of each year and the default leg and accrual leg are paid immediately after default.
    a. Find the CDS spread s.
    b. Assume that a CDS has a notional principal of \$1 million. If default occurs at 2.5 years, find the premium leg, accrual leg, and default leg.
    (You may assume $s = 0.6\%$. In fact, it is calculated by a.)
    What cash flows will the protection buyer give to the protection seller and when?
    What cash flows will the protection seller give to the protection buyer and when?

17. Assume that Alpha Corporation has a default intensity (hazard rate) of 3%. A two-year plain vanilla CDS is structured such that the CDS spread s is paid annually. Assume that the risk-free zero curve is flat at 6% per annum with continuous compounding, that defaults can occur only at times 0.5 years and 1.5 years, that the recovery is 40%, and the accrual interest is paid when default occurs.
    a. What is the CDS spread, s?
    b. Assume instead that in the market the spread is 200 basis points for the CDS contract, the notional principal of the CDS is \$1 million, any defaults occur in 1.5 years, and the recovery rate is 40%. List the cash flows and their timing for the seller of the CDS.

18. A CDO squared contains 40 CDO mezzanine tranches. Each CDO contains 1,000 mortgages with \$1 principal for each mortgage. Assume the recovery rate is 60%. There is an equity tranche, mezzanine tranche, and senior tranche within each CDO structure. The attachment for the equity tranche is 0% and the detachment is 5%. The attachment for the mezzanine tranche is 5% and the detachment is 25%. The attachment for the senior tranche is 25% and the detachment is 100%. Assume that 600 mortgages default inside 20 of the CDOs, 300 mortgages default inside another 10 CDOs, and 100 mortgages default inside the remaining 10 CDOs. Assume

that the CDO squared has three tranches: the attachment for the equity tranche is 0% and the detachment is 5%. The attachment for the mezzanine tranche is 5% and the detachment is 25%. The attachment for the senior tranche is 25% and the detachment is 100%.

a. What is the total principal amount in the CDO squared?
b. How many dollars of principal of the CDO squared remain after all these defaults occur?
c. Does the senior tranche of the CDO squared become affected? If so, by how much?

19. A CDO contains 100 corporate bonds with $1 principal each. Assume the recovery rate for each bond is 20% in the event of default. There is an equity tranche, mezzanine tranche, and senior tranche within each CDO structure. The attachment for the equity tranche is 0% and the detachment is 5%. The attachment for the mezzanine tranche is 5% and the detachment is 25%. The attachment for the senior tranche is 25% and the detachment is 100%. Assume that the CDO has a five-year maturity, and that any defaults occur at the middle of a year. The default and the accrual interest are paid at default. The premium is paid at the end of the year. Assume that the equity tranche spread is 30%, the mezzanine tranche spread is 3%, and the senior tranche spread is 1%. Further assume that there are three defaults each year. Calculate the mezzanine tranche premium leg, default leg, and accrual leg in the third year.

End-of-Chapter Problems (Challenging Problems)

This chapter discussed the fundamental pricing principal of financial derivatives: Expected payoff of A to B = Expected payoff of B to A (with expected payoffs expressed as risk-neutral present values). This principal can be used to value a swap.

A ***swap*** is a derivative contract through which two parties exchange a sequence of cash flows. Therefore, a swap is a portfolio of forward contracts. Section 5.7 discussed different types of swaps. We defer the valuation of various swaps here: interest rate swaps, foreign currency swaps, equity swaps, forward swaps, and swaptions. The formulas have been given elsewhere (e.g., the CFA curriculum) without details. However, the proofs are not shown in most finance books and financial mathematics books. So, we will derive the proofs here.

Topic 1 Valuing a plain vanilla interest rate swap

A ***plain vanilla interest rate swap*** is when one party agrees to pay a stream of floating interest rate payments (e.g., LIBOR rates) in exchange for receiving a fixed-rate stream of payments from the other party. The size of the payments is scaled by the notional principal of the swap and is valid for the time period specified in the swap agreement. Chapter 7 developed the price formula. The difference here is to include the day convention, although the formula is the same in terms of the mathemtics.

At the swap's initiation, the fixed rate is specified in the contract. In an informationally efficient market, the fixed rate of a plain vanilla interest rate swap is the rate that initially equates the ***value of the interest rate swap*** to zero for both sides of the agreement. After the initiation of the swap, similar to forward contracts, the agreement will take on a positive value to one side and a negative value to the other side as interest rates change.

1. A one-year quarterly-paid plain vanilla interest rate swap is initiated today. Assume that annualized 90 days through 360 days LIBOR spot rates are 1%, 1.5%, 2%, and 2.5%, respectively. Find the annualized fixed rate that makes the value of the interest rate swap equal to 0 at initiation.

   a. This swap is equivalent to the two parties swapping two bonds with equal values: Party A offers the cash flows of a one-year quarterly fixed coupon bond to Party B, and B offers a one-year quarterly floating-rate bond to A. The coupon rate of the fixed coupon bond is the same as the fixed rate for this swap. The principal amounts and of each bond are equal and occur on the same date so no principal cash flows are exchanged; they merely serve to scale the size of the fixed and floating interest rate payments. In fact, the interest rate payments are netted so that the only payment made each period is from the party receiving a smaller rate than they are paying.
   Assume that both bonds have a principal amount of $1. The present value of the fixed coupon, C, and the principal ($1) can be used to find the value of this fixed coupon bond by discounting each cash flow at the market interest rate corresponding to its longevity:

$$V_{\text{fixed}} = \frac{C}{1 + R_{90} \times \frac{90}{360}} + \frac{C}{1 + R_{180} \times \frac{180}{360}} + \frac{C}{1 + R_{270} \times \frac{270}{360}} + \frac{C + 1}{1 + R_{360} \times \frac{360}{360}}$$

   b. Explain why the value of the floating rate bond is always equal to its face value on the date of its coupon reset (i.e., $V_{\text{float}} = 1$).

c. Based on the fundamental principal that the expected and discounted payoff of A to B = V_{fixed}, and the expected and discounted payoff of B to A = V_{float}, are equal:
Inserting the value of the floating rate note ($1) into the left side of the equation from a:

$$\frac{C}{1 + R_{90} \times \frac{90}{360}} + \frac{C}{1 + R_{180} \times \frac{180}{360}} + \frac{C}{1 + R_{270} \times \frac{270}{360}} + \frac{C + 1}{1 + R_{360} \times \frac{360}{360}} = 1.$$

Show that the annualized fixed rate is

$$r_{\text{fixed}} = c \times 4 = \frac{1 - \frac{1}{1 + R_{360} \times \frac{360}{360}}}{\frac{1}{1 + R_{90} \times \frac{90}{360}} + \frac{1}{1 + R_{180} \times \frac{180}{360}} + \frac{1}{1 + R_{270} \times \frac{270}{360}} + \frac{1}{1 + R_{360} \times \frac{360}{360}}} \times 4$$

Note that the quarterly payment is notional principal × $\frac{r_{\text{fixed}}}{4}$, where Z_i is the present value of $1 to be received on the ith payment date, which is calculated based on LIBOR rates. Defining Z_i as the present value of $1 to be received on the ith payment date (which is calculated based on LIBOR rates), a formula for the fixed interest rate per period for an N-period plain vanilla interest rate swap is given by:

$$r_N = \frac{1 - Z_N}{Z_1 + Z_2 + \ldots + Z_N}.$$

2. As time passes, the ***value of the plain vanilla interest rate swap*** is not necessarily zero. On floating-rate reset dates the value of the swap is the difference between the value of the fixed-rate side and the principal value of the floating rate note (e.g., $1). However, between reset dates, the swap's value must include any changes in the value of the both sides of the swap.
Consider a one-year quarterly interest rate swap with a fixed rate of 2.47% at initiation on January 1. At the time of the swap's initiation, the 90-day LIBOR rate was 1%. Then, 30 days later, on January 31, the 60-, 150-, 240-, and 330-day LIBOR rates (annualized) are 1.5%, 2%, 2.5%, and 3%, respectively. Assume that the notional principal amount is $30,000,000. Find the value of the swap to the fixed-rate payer for January 31.
   a. Calculate the quarterly payment for the fixed-rate payer. Then, using the LIBOR rates at January 31, find the value of the

payment for the fixed-payer side of the swap V_{fixed} as for January 31.

b. Calculate the first payment from the floating rate payer which happens 90 days from initiation (i.e., April 1). Recall that at the time of each payment, the present value of the floating rate payer's future payment stream always returns to par (its face value). Using the January 31 LIBOR rates provided above, find the value of the floating-rate payer's payment, V_{float}, as of January 31.

c. Find the value of the swap to the fixed-rate payer $V = V_{\text{float}} - V_{\text{fixed}}$ as of January 31.

Topic 2 Valuing a Currency Swap

There are four kinds of currency swaps between two specified currencies (although generically there are only two types, each viewed from their two sides):

- Pay fixed rate in one currency and receive fixed rate in the other currency.
- Pay fixed rate in one currency and receive floating rate in the other currency.
- Pay floating rate in one currency and receive fixed rate in the other currency.
- Pay floating rate in one currency and receive floating rate in the other currency.

Consider first the example of a swap that pays a fixed rate in one currency and receives a fixed rate in another currency as our example. Note that currency swap fixed-rates calculations in one currency serve the same role in modeling as the fixed rates in the plain vanilla interest rate swap previously discussed. The fixed currency rate, r_N, has the formula: $r_N = \frac{1 - Z_N}{Z_1 + Z_2 + \ldots + Z_N}$, where Z_i is the present value of $1 to be received on the ith payment date calculated from LIBOR rates in the corresponding currency.

The cash flows of the currency swap are as follows.

At initiation of the currency swap, the two parties exchange notional principals based on today's exchange rate. During the life of the swap, Party A pays the corresponding fixed rate for one currency based on its LIBOR rates for its currency (just as in the case of a plain vanilla interest rate swap) and receives the corresponding fixed rate in the other currency based on its LIBOR rates for its currency (just as in the case of a plain vanilla interest rate swap). At the end of the swap agreement, the two parties exchange the original principal amounts.

3. A one-year quarterly paid currency swap is initiated today between U.S. dollars and British pounds with a notional principal amount of $30 million. Assume that the annualized 90-day, 180-day, 270-day, and 360-day LIBOR spot rates for U.S. dollars are 1%, 1.5%, 2%, and 2.5%, respectively, and that the annualized 90-days through 360-days LIBOR spot rates for British pounds are 2%, 3%, 4%, and 5%, respectively. Find the annualized fixed rate for both sides that sets the value of the currency swap equal to 0 at the swap's initiation using the following steps.
   a. Find the fixed rate on a one-year quarterly plain vanilla interest rate swap in U.S. dollars.
   b. Assume that the exchange rate today is 0.71 British pounds per U.S. dollar. What is the notional British pound principal amount?
   c. Find the fixed rate on a one-year quarterly plain vanilla interest rate swap in British pounds.

4. As time passes, the net value of the currency swap will change from zero. Assume that a one-year quarterly currency swap has a fixed rate of 2.47% in U.S. dollars and that the fixed rate is 4.88% in British pounds at initiation. Now, 300 days later, the current 60-day LIBOR rate (annualized) for U.S. dollars is 1.5% and for British pounds is 2.5%. At the last settlement date, the 90-day LIBOR rate in U.S. dollars was 1.2% and in British pounds was 2.2%. Assume that the notional principal amount is $30,000,000, that the exchange rate is 0.70 British pounds per U.S. dollar now, and that the exchange rate was 0.71 British pounds per dollar at initiation. Find the value of a currency swap for each of the four swaps:
   a. Receive U.S. dollars fixed and pay British pounds fixed.
   b. Receive U.S. dollars floating and pay British pounds fixed.
   c. Receive U.S. dollars fixed and pay British pounds floating.
   d. Receive U.S. dollars floating and pay British pounds floating.

Topic 3 Valuing an Equity Swap

The ***equity swap*** is an exchange of equity returns and LIBOR rates. For example, Party A will pay (to Party B) a stream of fixed rates (or for the case of an equity swap based on floating rates pay LIBOR) on the contract's notional value and will receive from Party B a stream of percentage increases in a specified equity index applied to the same notional value. The swap contract calls for the equity swap to be in effect for a pre-specified period of time. The fixed rate per period, r_N, has the same formula as in the case of a plain vanilla interest rate swap: $r_N = \frac{1 - Z_N}{Z_1 + Z_2 + \ldots + Z_N}$, where Z_i is the present value of $1 to be received on the ith payment date calculated from LIBOR rates.

In the market for equity swaps, competition tends to drive the fixed rate to be based entirely on interest rates prevailing at the initiation of the contract and the value of the swap contract to be near zero to both sides. In practice, sometimes equity swap rates offer a small premium above the LIBOR rates such as three basis points to attract a counterparty.

5. Assume that the annualized 90-day through 360-day spot rates are 1%, 1.5%, 2%, and 2.5%, respectively. Find the fixed percentage rate for an equity swap per period assuming that it is fairly priced.
6. A fairly priced one-year $30 million principal value equity swap has a fixed quarterly rate 0.006175. The counterparty pays the quarterly return on a pre-specified equity index for the one-year period. The index's spot price is 2800 at initiation. After 30 days, the index stands at 2850 and the current 60-, 150-, 240-, 330-day LIBOR rates (annualized) are 1.5%, 2%, 2.5%, and 3%, respectively. Find the new value of the swap to the fixed-rate payer.
   a. Calculate the quarterly payment for the fixed-rate payer of a plain vanilla interest rate swap using $1 as a face (notional) value. Then, using the current LIBOR rates, find the value of the payments of the fixed-payer side of the swap, V_{fixed}, as of today.
   b. Find the value of the equity payments today, V_{equity}, if $1 is invested at initiation.
   c. The value of the swap to the fixed-rate payer is $V = (V_{\text{equity}} - V_{\text{fixed}}) \times$ notional principal.

Topic 4 Forward Swaps

A ***forward interest rate swap*** is a commitment to enter into an interest rate swap at some point in the future.

7. Assume that a forward swap has a maturity of one year. At that time, the counterparties will enter into a one-year quarterly paid interest rate swap with a fixed rate. Derive the formula for the fixed rate.

Topic 5 Swaptions

There are two kinds of ***swaptions:***

- A ***payer swaption*** is the center to enter into a pre-specified interest rate swap at some date in the future as the fixed-rate payer at a rate specified in the swaption.
- A ***receiver swaption*** is the center to enter into a pre-specified interest rate swap at some date in the future as the fixed-rate receiver (i.e., the floating-rate payer) at a rate specified in the swaption.

8. Explain why a payer swaption is equivalent to a put option on a coupon bond. What is the maturity of the bond? What is the maturity of the put option? What is the put option's strike price?
9. Explain why a receiver swaption is equivalent to a call option on a coupon bond. What is the maturity of the bond? What is the maturity of the put option? What is the call option's strike price?

Notes

1 See https://www.businessnewsdaily.com/10353-cdo-financial-derivatives-economic-crisis.html (accessed May 2020).
2 See https://www.ft.com/content/9c33cea0–6ceb-11e9–80c7–60ee53e6681d (accessed May 2020).
3 David X. Li, "On Default Correlation: A Copula Function Approach", The Journal of Fixed Income, March 2000.
4 The above CDS quotes are real quotes from an example in O'Kane's book.

Index

For Product Safety Concerns and Information please contact our EU representative GPSR@taylorandfrancis.com
Taylor & Francis Verlag GmbH, Kaufingerstraße 24, 80331 München, Germany

www.ingramcontent.com/pod-product-compliance
Ingram Content Group UK Ltd.
Pitfield, Milton Keynes, MK11 3LW, UK
UKHW022212190926
475702UK00003B/314
* 9 7 8 0 3 6 7 4 1 0 3 9 1 *